Sustainable Development Goals Series

The **Sustainable Development Goals Series** is Springer Nature's inaugural cross-imprint book series that addresses and supports the United Nations' seventeen Sustainable Development Goals. The series fosters comprehensive research focused on these global targets and endeavours to address some of society's greatest grand challenges. The SDGs are inherently multidisciplinary, and they bring people working across different fields together and working towards a common goal. In this spirit, the Sustainable Development Goals series is the first at Springer Nature to publish books under both the Springer and Palgrave Macmillan imprints, bringing the strengths of our imprints together.

The Sustainable Development Goals Series is organized into eighteen subseries: one subseries based around each of the seventeen respective Sustainable Development Goals, and an eighteenth subseries, "Connecting the Goals", which serves as a home for volumes addressing multiple goals or studying the SDGs as a whole. Each subseries is guided by an expert Subseries Advisor with years or decades of experience studying and addressing core components of their respective Goal.

The SDG Series has a remit as broad as the SDGs themselves, and contributions are welcome from scientists, academics, policymakers, and researchers working in fields related to any of the seventeen goals. If you are interested in contributing a monograph or curated volume to the series, please contact the Publishers: Zachary Romano [Springer; zachary.romano@springer.com] and Rachael Ballard [Palgrave Macmillan; rachael.ballard@palgrave.com].

Eduardo G. Pereira · Alberto J. Fossa ·
Thomas L. Muinzer
Editors

Carbon Capture Utilization and Storage

Law, Policy and Standardization Perspectives

palgrave
macmillan

Editors
Eduardo G. Pereira
Universidade de São Paulo
São Paulo, Brazil

Alberto J. Fossa
Universidade de São Paulo
São Paulo, Brazil

Thomas L. Muinzer
University of Aberdeen
Aberdeen, UK

ISSN 2523-3084 ISSN 2523-3092 (electronic)
Sustainable Development Goals Series
ISBN 978-3-031-81271-2 ISBN 978-3-031-81272-9 (eBook)
https://doi.org/10.1007/978-3-031-81272-9

Cover illustration: Henglein and Steets Alamy Stock Photo

This Palgrave Macmillan imprint is published by the registered company Springer Nature Switzerland AG
The registered company address is: Gewerbestrasse 11, 6330 Cham, Switzerland

Acknowledgements

We gratefully acknowledge the support of the RCGI, Research Centre for Greenhouse Gas Innovation (23.1.8493.1.9), hosted by the University of São Paulo (USP) and sponsored by FAPESP, São Paulo Research Foundation (2020/15230-5) and Shell Brasil, as well as the strategic importance of the support given by ANP (Brazil's National Oil, Natural Gas and Biofuels Agency) through the R&DI levy regulation.

Contents

Notes on Contributors

About the Editors

Eduardo G. Pereira is a worldwide recognised scholar specialising in Natural Resources and Energy Law. He is a founding partner at the International Energy Law Training and Research Company as well as at the International Energy Law Advisory Group. He has been active in the natural resources and energy industry for more than 15 years and is an international expert on oil, gas and energy contracts and regulations. His experience in this area—both academic and practical—is extensive. He has practical experience in over 50 jurisdictions covering America, Europe, Africa and Asia. Dr. Pereira concluded his doctoral thesis on oil and gas joint ventures at the University of Aberdeen (Scotland). He conducted postdoctoral research at the Oxford Institute for Energy Studies (University of Oxford, UK), the Scandinavian Institute of Maritime Law (University of Oslo, Norway), the Institute of Energy and Environment (University of São Paulo, Brazil) and more recently at the Canadian Institute of Resources Law (University of Calgary, Canada). He possesses positions as a full-time, part-time, honorary, adjunct and/or visiting scholar in several leading academic institutions around the world. He is also a series editor for the Global South Series at the Journal of Sustainable Development Law and Policy and an associate editor of OGEL. He is also the author and editor of several leading oil and gas textbooks. Further information about his profile can be found at www.eduardogpereira.com.

Alberto J. Fossa is an electrical engineer and a Master in Energy and Doctor of Science from the University of São Paulo (USP). Fossa is a business consultant in Strategic Management and Management of the Technology and researcher of the Energy and Environment Institute of the University of São Paulo (IEE-USP) in the field of Quality Infrastructure and Energy Management. Fossa is the head of Brazilian delegation on ISO TC 301 Energy Management & Energy Savings, chair of Brazilian Technical Committee ABNT CB116 of Energy Management and Savings and an Energy expert from United Nations Industrial Development Organization (UNIDO).

Thomas L. Muinzer FRSA is a Reader in Energy Transition Law at the University of Aberdeen, Scotland. His teaching, research and consultancy work focuses most particularly on climate and energy decarbonisation law and policy.

Contributors

Owen L. Anderson is a distinguished Oil & Gas scholar and co-academic director of the Kay Bailey Hutchison Energy Center at the University of Texas at Austin and the Eugene Kuntz chair in Oil, Gas & Natural Resources emeritus and George Lynn Cross research professor emeritus at the University of Oklahoma. Portions of this chapter are adapted from an article published in Summer 2023, *Carbon Sequestration: A Fresh Look at an Essential Tool in the War on Climate Change*, Oil, Gas & Energy Law Intelligence, OGEL 3 (2023)—Carbon Capture and Storage | OGEL Journal (2023).

Rune Tjomsås Andersen specialises in tax law and company law and has broad experience in these fields in both the public and private sectors. He has in-depth expertise on petroleum and energy taxes, property tax, general corporate taxation and international tax law. He has worked in the Norwegian Ministry of Finance's Tax Law Department and has extensive experience of law-making and committee work.

Elena I. Athwal Harvard Law School, LL.M | Research Fellow, UNESCO Chair of Environmental Law & Sustainable Development | Research Associate, UN Working Group on Business & Human Rights | Vice President, The Diwan.

Nathaniel Babajide is a senior fellow of the Higher Education Academy (SFHEA) with a specialisation in Energy Economics, Transition and Sustainability. His research interests include energy economics, management and policy, energy access improvement, energy innovation, climate change responses and socio-economic development for developing countries. He is currently investigating how low-carbon energy systems could enhance sustainable development in developing countries.

Guilherme Baccari University of São Paulo (USP), Research Centre for Greenhouse Gas Innovation (RCGI – USP), Institute of Psychology (IP – USP).

Soibi O. Bob-Manuel holds an LL.B. from Kwame Nkrumah University, Kumasi, Ghana. She completed an LL.M in Oil and Gas Law (with distinction) from Robert Gordon University, Aberdeen. She was called to the Nigerian Bar in 2012 and practised law in Nigeria. Her practice areas included corporate commercial, oil and gas law, intellectual property, energy and natural resources, dispute resolution and regulatory compliance. Soibi is completing a PhD in Law at the University of Aberdeen, United Kingdom. Her PhD focuses on the legal application of clean energy technology and the intersection of climate change law, intellectual property rights and public goods theory.

Juliana Costa has a law degree from the Federal University of Rio de Janeiro and is a lawyer in the Oil, Gas & Biofuels and Infrastructure & Projects practices at Veirano Advogados. With more than 5 years of experience in the sector, Juliana has in-depth knowledge of the national regulation of oil, natural gas and renewable solutions related to the industry. She stands out for her ability to advise companies on drafting and negotiating contracts and providing legal advice on complex regulatory issues.

Carter Czaikowski (JD) works for the law firm Rose LLP in Calgary Alberta, Canada and has a special interest in construction law including energy infrastructure issues.

Lauren Davies advises clients on the development and financing of cross-border, complex projects in the energy, natural resource and infrastructure sectors, acting for clients in the international and US markets. In particular Lauren has extensive experience in project finance, advising project companies, sponsors, host governments and lenders in connection with the development and financing of projects in the oil and gas, petrochemical, energy transition, mining and metals, LNG, port and infrastructure sectors, continuing to advise throughout the entirety of a transaction lifecycle and across the full spectrum of her clients' business needs.

Alexandre de Barros Gallo is a researcher at the Research Centre for Greenhouse Gas Innovation (RCGI), based at the University of São Paulo. He is currently engaged in the areas of energy management, GHG mitigation and associated standardisation mechanisms. He holds a double degree in Engineering from the Polytechnic School of the University of São Paulo and the Ecole Nationale Supérieure des Mines de Paris, a Master's degree in Energy Science and a PhD in Energy Analysis and Planning from the Institute of Energy and Environment of the University of São Paulo, including a postgrad research period at Imperial College London.

Hannah Hylton Edwards is an advocate attorney-at-law qualified to practise law in both the Republic of Trinidad and Tobago and the Republic of Guyana. She is also an undergraduate and postgraduate lecturer and tutor with the University of the West Indies St. Augustine Campus. Hannah has a Master of Laws degree in Public Law (with distinction) from the University of the West Indies St. Augustine Campus. She also holds a Legal Education Certificate from the Hugh Wooding Law School and graduated as a Dean's List Honouree from the UWI obtaining her Bachelor of Laws with Upper Second Class Honours.

Saelese Haynes is an attorney-at-law of 13+ years' call, researcher and a national scholar of Trinidad and Tobago. She has been called to both the Bar of Trinidad of Tobago and to the Bar of England and Wales. She is the head of Chambers at Titan Chambers in Trinidad which she founded in 2018 in an effort to streamline and further explore dynamic and developing areas of the law including Eco Technology and Intellectual Property. She has contributed to several works of scholarship in the field of Oil and Gas including *The Palgrave Handbook of Natural Gas and Global Energy Transitions* (©2022, Palgrave Macmillan) and the *Journal* *of Energy & Natural Resources Law*. Saelese is passionate about projects that change the legal landscape in Trinidad and Tobago with a focus on sustained and sustainable development.

Randy Hendrika is the managing associate and a founding member of Latief & Co Law Firm. He has an extensive practical experience in advising and representing clients from various industries, including oil and gas, banking and finance, financial technology, retail and e-commerce and foreign direct investments. Several notable experiences in the oil and gas sector among others include the preparation and negotiation of Production Sharing Contracts (PSC), Farm-in/out Agreement, Facility Sharing Agreement, and in the banking and finance sector, preparation and negotiation of multi-million dollar facility agreements and its security documents.

Allan Ingelson is a professor in the Faculty of Law at the University of Calgary. He is the editor of the Canadian Energy Law Service (Carswell). He holds memberships in the Canadian Bar Association, the Association of International Petroleum Negotiators, the Academy of Legal Studies in Business and the Law Society of Alberta. He has lectured on energy and environmental regulation in Lima, Peru; Houston, Texas; Delhi, India; Beijing, China; Sao Paulo, Brazil; Colombia, South America and Doha, Qatar.

Felipe Villasuso Lago is a senior compliance officer at AIQ Artificial Intelligence (a joint venture between Abu Dhabi National Oil Company "ADNOC" and Presight AI), where he is responsible to ensure compliance with Ethics and Compliance policies, assessment of new technologies and compliance with AI and Energy regulations. Before joining AIQ, Felipe was the governance lead at ADNOC Logistics & Services PLC in Abu Dhabi and senior manager at PwC Legal in Dubai and Qatar. Felipe is a PhD candidate at the London South Bank University.

Anton Latief is the founding partner of Latief & Co Law Firm (LCO) in Jakarta, Indonesia. Prior to establishing LCO, he has successfully established the Energy Practice Group of one of Indonesia's prominent law firms and had the opportunity to lead the Indonesia Legal Department of a French oil supermajor after being assigned in their Gas and Power division in Paris, France.

With more than 23 years of experience, he has been involved in the development of major Indonesia's and international upstream oil and gas projects such as the Yamal LNG (Russia), Shah Deniz—Phase 2, the Mahakam PSC and Abadi LNG Project.

Cylon Liaw is Consultant and Researcher in the fields of Energy Policies, Standardization, and Public Policies, with experience in the Government of the State of Rio de Janeiro and at the Research Center for Innovation in Greenhouse Gases (RCGI), a project funded by Shell and Fapesp, supported by USP. He holds a Master's in Science (focus on Energy) from IEE/USP (2018), with an exchange in Oil Management and Economics at IFP School. He has a Bachelor's in Administration from FEA—USP (2011), with an exchange in International Relations at the University of Coimbra and in Administration at Laurea University of Applied Sciences. He is currently a doctoral candidate in the Graduate Program in Energy at IEE-USP, including a postgrad research period at Erasmus University Rotterdam in 2023.

Oliver Lowson is a senior legal counsel actively engaged on hydrogen/ammonia, carbon capture, SAF and new energies greenfield projects in the UAE and internationally. Oliver Lowson is a Project Development lawyer with experience advising sponsors, lenders, contractors, IOCs/NOCs and governments on the procurement, construction and operation of major projects in the energy, infrastructure, industrial and real estate sectors in the UK, MENA, USA, and CIS.

Karen Louise Mascarenhas is a deputy coordinator of the project "Social Perception and Scientific Diplomacy in Technological Transitions Towards a low-carbon Society" and director of Human Resources and Institutional Communication, for the Research Centre for Greenhouse Gas Innovation (RCGI) (www.usp.br/rcgi), based at the Polytechnic School of the University of São Paulo (USP), founded by FAPESP and SHELL. She holds a PhD from the Institute of Psychology of the University of São Paulo, with part of her research at Imperial College London and the Sustainable Gas Institute. She has a Master's degree in Social Psychology, a degree in Psychology and a postgraduate degree in Administration from the Getúlio Varga Foundation.

Onyemaechi Titilayo Ogiri is a lecturer in the Faculty of Law at Anchor University, Lagos, Nigeria, and a doctoral researcher at the Faculty of Law, University of Ibadan, Nigeria. She is a barrister and solicitor of the

Supreme Court of Nigeria. She has an LLB from the Enugu State University. She also attended the Nigerian Law School. Onyemaechi holds an LLM from University of Ibadan. Email: onyemaechitbn@gmail.com

Opeyemi Omotuyi is a law lecturer at the Federal University Oye Ekiti, Nigeria. Having practised in the academia for over eight years, she has built a niche for herself in the area of sustainable development. Her research expertise borders on the deployment of law as an instrument of social change particularly, the realisation of the global sustainable development agenda. Some of her research outputs in the area of climate change, energy transition, oil and gas decommissioning, among others, have been published as scholarly articles in various reputable journals. She has also authored various book chapters on environmental protection, EHS management and renewable energy, among others.

Rune Omdahl works with major Norwegian and international clients on energy-related projects and transactions. With specialist expertise in the areas of petroleum law, contract law and product sales agreements, transport law, company law and international negotiations, he is a sought-after advisor for industrial projects at home and abroad. He also works on financial issues relating to the petroleum sectors and matters relating to financial default.

Peter Kayode Oniemola is a senior lecturer at the Faculty of Law, University of Ibadan, Nigeria. He is the Energy Law programme coordinator at the Centre for Energy, Petroleum Economics and Law (CPEEL), University of Ibadan. He is a graduate of Bachelor of Laws (LLB) of University of Ilorin, Nigeria. He holds a Master of Laws (LL.M) Degree of the University of Ibadan. He also holds a PhD in Law from the University of Aberdeen. Peter was appointed as McArthur Foundation scholar in the Faculty of Law, University of Ibadan, from March 2008 to March 2009.

Tara Righetti is the occidental chair in Energy and Environmental Policy at the School of Energy Resources and the Haub School of Environment and Natural Resources and a professor in the College of Law. She is the co-director of the Nuclear Energy Research Center. Professor Righetti served as a fulbright research scholar for the 2021-2022 academic year at the Université de Lille in Lille, France. She previously directed the Professional Land Management concentration from January 2016 to August 2020. Prior to joining the University of Wyoming in the fall of 2014, she served in legal and executive roles in a privately owned upstream oil and gas company.

Jonathan Roberts Senior Associate at Herbert Smith Freehills advising on energy projects driving economic development and the energy transition.

Haline Rocha Being a geoscientist and researcher with MSc and PhD in Energy Studies, Haline Rocha has explored how geosciences can support the energy transition, such as through subsurface carbon storage, pal eoclimate interpretations and natural hydrogen research. Prior to joining S&P Global Commodity Insights as a principal research analyst, Haline worked as a postdoctoral researcher analysing potential reservoirs and caprocks to CO_2 geological storage in the Paraná Basin, Brazil—focused on unconventional reservoirs (e.g. shale and coal beds from Irati and Rio Bonito Formations) through geochemistry, petrography and porous media analysis, also working as an independent consultant assessing potential decarbonisation routes and technologies to the hard-to-abate sectors in Brazil.

Cristina F. A. Rodrigues is an associate professor at University Fernando Pessoa, Porto, Portugal, and University Agostinho Neto, Luanda, Angola. She is a consultant geologist in coal, oil and natural gas exploration. Supervised and co-supervised 6 PhD theses. Over her 25-year career, her main fields of expertise are organic petrology and geochemistry of conventional (oil and natural gas) and non-conventional reservoirs (coal and coalbed methane), and basin analysis and modelling. She has organised 4 international congresses and meetings and coordinated 7 research projects in the domains of expertise. She has been the correspondent member of the Academy of Sciences of Lisbon since 2020 and is also a member of the International Committee for Coal and Organic Petrology-ICCP since 1999.

Martha M. Roggenkamp is an emeritus professor of Energy Law at the University of Groningen. She established the Groningen Centre of Energy Law and Sustainability and has been developing a Master programme on energy law and leading the executive postdoctoral joint degree on energy law by the North Sea Energy Law Partnership. Currently she is a distinguished honorary professor of energy law at the

University of Pay, France, and is chairing the Dutch Energy Law Association. She has published widely on energy law issues (e.g. CCS, offshore wind and energy law in Europe) since the early 1990s.

Edmilson Moutinho dos Santos Graduated in Economics (1990) and Electrical Engineering (1988) from the University of São Paulo, Edmilson Moutinho has his Master's degrees in Energy Management and Policy (University of Pennsylvania; 1993) and in Energy Systems Planning (State University of Campinas; 1992) and a PhD in Energy Economics from the French Petroleum Institute and Université de Bourgogne (1997). He is currently an associate professor at the Institute of Energy and Environment at the University of São Paulo (IEE-USP).

Tina Soliman Hunter BA (Hons); G Dip A (LIS); M App Sc (Dist); JD (Hons); PhD (Bergen). Professor of Energy and Natural Resources Law, and Co-Director, *Transforming Energy & Markets Research Centre*, Macquarie University.

M. J. Lemos de Sousa is a professor at University Fernando Pessoa, Porto, Portugal. Previously he has been a professor, head of Department and Research, Department of Geology, Faculty of Sciences, University of Porto. He has supervised 16 PhD thesis. Over his 50-year career, his fields of expertise are organic petrology and geochemistry (coal, oil and natural gas), and Carboniferous stratigraphy and palaeobotany. He has participated in 8 research projects and has organised 10 international congresses and meetings. He has been a member of the Academy of Sciences of Lisbon since 2000 and vice-president of the Sciences Class 2016–2021. He has been a full member of the International Committee for Coal and Organic Petrology-ICCP (1971–2004), honorary member (2005) and ICCP president (1995–1999). Reinhard Thiessen Medal Award (2001).

Ellen Swarbrick is a trainee solicitor with experience in the development and financing of projects in the energy, natural resource and infrastructure sectors, as well as cross-border M&A in the energy sector and general corporate work. Her experience includes advising on transactions involving traditional energy sources as well as energy transition matters involving renewable fuels and low-carbon hydrogen.

Colombo Tasinari graduated in Geology (1975), obtained his Master's (1981) and PhD (1988) and had been a full professor (1996) from the University of São Paulo (USP). He is a professor at the Institute of Energy and Environment (IEE) at the USP. He coordinates USP's High Resolution Geochronology Laboratory (GeoLab-SHRIMP) and the CNPQ Research Group on the Geological Storage of CO_2 at IEE-USP. He was a director of USP's Institute of Geosciences (2008–2011) and Institute of Energy and Environment (2015–2019). He is a member of three Sciences Academy and has around 200 scientific publications, with more than 5,000 citations and a Web of Sciences H-Factor index of 33.

Nathalia Weber University of São Paulo (USP), Research Centre for Greenhouse Gas Innovation (RCGI – USP), CCS Brasil.

Alistair Wishart advises on high-value, complex and innovative construction and development projects in traditional and energy transition matters in a variety of jurisdictions including Uzbekistan, Turkey, Azerbaijan, the U.S. and the Middle East. Alistair's practice encompasses both traditional energy and petrochemicals projects as well as energy transition and renewable energy projects. He has a wealth of experience in drafting and negotiating engineering, procurement and construction (EPC) contracts, as well as guiding clients through the ongoing issues that typically arise during the development and operational stages of large-scale energy and infrastructure projects.

Lan WU is a researcher who earned her doctoral degree from the Faculty of Law at The Chinese University of Hong Kong, specialising in energy law and climate law. Her doctoral research focuses on the legal challenges of integrating renewable energy-sourced electricity into power grids, particularly in the context of electricity market reforms in China. She served as the energy and climate fellow at the Strategic Partnership and Policy team of the United Nations Development Programme in China. Currently, she is a lecturer at the Faculty of Law of Ocean University of China.

Constantinos Yiallourides is an associate professor of international law at Macquarie University, Sydney, specialising in law of the sea, offshore energy and environmental protection. He is the deputy director of the University's Centre for Environmental Law and the research leader in law of the sea at the British Institute of International & Comparative Law, London. He is also an honorary lecturer at the University of Aberdeen (Scotland, UK) and was previously a visiting researcher at the University of Tokyo, Waseda University, Kobe University (Japan) and the University of Hamburg (Germany). Constantinos is a practising consultant in international law, advising states and other entities on a broad set of issues of public international law.

List of Figures

Chapter 18

Chapter 21

Chapter 23

List of Tables

Chapter 3

Chapter 6

Chapter 9

Chapter 10

Chapter 18

Chapter 21

CHAPTER 1

Introduction

Thomas L. Muinzer, Eduardo G. Pereira, and Alberto J. Fossa

1 The Focus of This Book: Carbon Capture Utilization and Storage

The world is at a period in history in which the international consensus of nations understands that it is necessary to move toward sustainable living and economies. This understanding is driven by a range of factors, including international law and agreements, notably the Paris Agreement of 2015. A *sustainable development* agenda has emerged internationally, with initial roots in the Brundtland Report of 1987,[1] and

[1] World Commission on Environment and Development, *Our Common Future* (UN, 1987); commonly known as the Brundtland Report.

T. L. Muinzer (✉)
University of Aberdeen, Aberdeen, UK
e-mail: thomas.muinzer@abdn.ac.uk

E. G. Pereira · A. J. Fossa
Universidade de São Paulo, São Paulo, Brazil
e-mail: afossa@newencreative.com.br

E. G. Pereira et al. (eds.), *Carbon Capture Utilization and Storage*, Sustainable Development Goals Series,
https://doi.org/10.1007/978-3-031-81272-9_1

the United Nations' *Sustainable Development Goals* have become prominent features of international and national governance.[2] Such drivers have impacted energy law, policy and governance. Among their many effects, they have created momentum for the deployment of zero-carbon and negative-carbon technologies to capture carbon dioxide by 2050.

In 2003 the United States commenced a Carbon Sequestration Leadership Forum, and encouraged Annex I and non-Annex I signatory countries to the United Nations Framework Convention on Climate Change to become members. This Forum has developed and grown, and its origins emphasize the link between climate governance and carbon capture, utilization, and storage (CCUS) technology deployment.[3] It seems likely that fossil fuels will continue playing a substantial role in the global energy mix in the short term, and it is apparent that a rapid deployment of CCUS can assist in the transition to Net-Zero emissions significantly. This will require the ongoing research and development of necessary technologies, and an ability to bring those technologies to market in a scalable way around the world. Operating in concert with a pronounced shift toward the decarbonization of fuel supply, and with increased energy efficiency measures, CCUS affords a major technology option that can assist in the attainment of Net-Zero emission reductions on a scale required by the Paris Agreement.

According to the United Nations, CCUS will contribute about one-sixth of required carbon emission reductions by 2050, and 14% of cumulative emissions reductions between 2015 and 2050.[4] This is a substantial contribution, and depending on political will, the condition of technological development, and the status of deployment barriers—notably the condition of relevant markets—it is not inconceivable that this could be a conservative estimate. It should be emphasized that CCUS presents itself as an attractive option to governments and entrepreneurial companies around the world not only because it contributes to the direct reduction of carbon emissions from source points; it also provides a

[2] Thomas L Muinzer and Colin Mackie, "The Sustainable Development Goals", pp. 15–25, in Eduardo G Pereira, Thomas L Muinzer and Patrick R Baker (eds), *Energy Law and the Sustainable Development Goals* (Routledge, 2024).

[3] See further the website of the Carbon Sequestration Leadership Forum: https://www.cslforum.org/.

[4] United Nations, *Carbon Capture and Storage: A Key to Climate Change Mitigation* (UN, 2015), 3.

platform for low-carbon hydrogen production, and it can assist in the removal of carbons from hard-to-abate emissions sectors, including in the steel, cement and other industries that pose particular decarbonization challenges.[5]

CCUS typically involves capturing greenhouse gases directly from the point of emission, particularly at industrial facilities that exploit fossil or biomass fuels. It can also involve capture from the atmosphere. Captured carbons can be used directly on-site for industrial purposes, compressed and transported for use in relevant industrial applications, or injected into deep aquifers which trap the carbon for permanent storage. Other potential uses of captured carbons include usage as feedstock to produce synthetic fuels, chemicals, and building materials. It can also be sold to oil companies for enhanced oil recovery, although this activity is generally unhelpful in relation to the Net-Zero transition, as it enables greater extraction of oil from mature oilfields and thus tends toward the negation of the emissions savings accrued by the CCUS element of the operation chain.[6] Aside, then, from its contribution to global Net-Zero emissions goals, another particular merit of CCUS is the use of the captured carbons for industrial purposes, which may constitute a potential revenue stream for CCUS facilities.

International support has been and will continue to be, a significant determinant in the ongoing development of CCUS. To date, some degree of joined-up thinking by significant international actors, the stimulation of international research and development, and transnational financing have all played a role in galvanizing the expansion of CCUS. The Intergovernmental Panel on Climate Change, the G8, and the International Energy Agency have all played supporting roles, as has the aforementioned Carbon Sequestration Leadership Forum, and the Global Carbon Capture and Storage Institute.[7] The International Energy Agency has

[5] On hard-to-abate emissions sectors, see further IRENA, *Decarbonising Hard-To-Abate Sectors with Renewables: Perspectives for the G7* (IRENA, 2024).

[6] See further Christophe McGlade, "Can CO2-EOR Really Provide Carbon-Negative Oil?" (International Energy Agency, 2019).

[7] See further Jürgen-Friedrich Hake and Olga Shenk, "International Cooperation in Support of CCUS", pp. 311–327 in Wilhelm Kuckshinrichs and Jürgen-Friedrich Hake (eds), *Carbon Capture, Storage and Use: Technical, Economic, Environmental and Societal Perspectives* (Springer, 2015).

emphasized and summarized the potential utility of CCUS in clean energy transitions in the following terms:

> CCUS can be retrofitted to existing power and industrial plants, allowing for their continued operation. It can tackle emissions in hard-to-abate sectors, particularly heavy industries like cement, steel or chemicals. CCUS is an enabler of least-cost low-carbon hydrogen production, which can support the decarbonisation of other parts of the energy system, such as industry, trucks and ships. Finally, CCUS can remove CO_2 from the air to balance emissions that are unavoidable or technically difficult to abate.[8]

As a consequence of international support and governance innovation, the Carbon Capture and Storage ('CCS') element of CCUS has been incentivized strategically under international climate law via the Clean Development Mechanism, which was first introduced under the Kyoto Protocol of 1997.[9] International development of incentives for CCS demonstration and the pooling and dissemination of knowledge from research and development activities have become a feature of international CCUS governance.[10]

In relation to the future of CCUS, scientists have commented optimistically as follows, with reference to the low-carbon energy transition:

> Carbon capture, utilization, and storage (CCUS) is a vital technology for the large-scale industrial reduction of carbon dioxide emissions and is considered a foundational technology for achieving carbon neutrality in the next 30 to 50 years... We have every reason to believe that, with the collective effort of the vast scientific elite, the green low-carbon Earth that humanity hopes for will arrive sooner rather than later.[11]

It is clear, then, that the CCUS sector has the potential to enjoy a strong future in the immediate term if its development continues to be

[8] International Energy Agency, "Carbon Capture, Utilisation and Storage", online briefing, available here: https://www.iea.org/energy-system/carbon-capture-utilisation-and-storage.

[9] Kyoto Protocol, Article 12.

[10] Hake and Shenk, *supra.*, n.7.

[11] Weifeng Lv, Tiyao Zhou, Yongbin Wu and Xiaoqing Lu, *Advances in Carbon Capture, Utilization and Storage Technologies (CCUS)*, (MDPI, 2024), ix.

carefully supported and expanded. It is also, however, necessary to recognize that scholars and commentators have pointed out certain challenges confronting CCUS. This is emphasized by observations from R.V. Shahi, Former Secretary of the Ministry of Power in the Government of India:

> Critics say large-scale deployment is unproven and decades away from being commercialized. They say that it is risky and expensive, and that a better option is renewable energy. Some environmental groups point out that CCS technology leaves behind dangerous waste material that has to be stored, just like nuclear power stations.[12]

Furthermore, while CCUS embodies a useful means of decarbonizing energy in principle, energy consumption and efficiency can pose challenges, particularly in relation to CCS. Again, Shahi has emphasized as follows:

> Another limitation of CCS is its energy penalty. The technology is expected to use between 10% and 40% of the energy produced by a power station. Wide-scale adoption of CCS may erase efficiency gains in… power plants… and increase resource consumption[.][13]

This is an important consideration. It is also possible, however, that where a particular power station is employing CCS technology, this could potentially reduce its emissions enormously, and so any extra expense in terms of efficiency loss may be more than offset by operational gains arising from CO_2 reductions.

Environmental safety and the environmental impact of CCUS are also relevant issues. Life Cycle Assessment has been applied by researchers in order to investigate differing environmental consequences associated with CCS deployment, and this work has highlighted that although CCS provides an environmentally useful means of decreasing CO_2 emissions, it can increase other negative environmental impacts significantly at the same time.[14] A further relevant matter concerns *public acceptance*, where

[12] R V Shahi, "Carbon Capture and Storage: Can or Should We Be So Indifferent?", pp. 3–11, in *Carbon Capture, Storage and Utilization: A Possible Climate Change Solution for Energy Industry*, eds Malti Goel, M Sudhakar and R V Shahi (Routledge, 2018), 6.

[13] Shahi, ibid., 6.

[14] Andrea Schreiber, Petra Zapp and Josefine Marx, "Environmental Aspects of CCS", pp. 101–126 in Wilhelm Kuckshinrichs and Jürgen-Friedrich Hake (eds), *Carbon Capture,*

it is widely understood that public acceptance is an important component of any augmented CCS roll-out, and, by implication, an important component of any broader CCUS roll-out engaging expanded utilization capacities.[15] Thus, as well as the prominent headline considerations of technological scope and capability, market scalability, and associated matters, there are a range of other significant issues that must be taken into account and factored into CCUS strategy and regulation in going forward.

In addition to the variety of opportunities and challenges influencing the deployment and expansion of CCUS systems, it is also necessary to recognize the component parts of the CCUS technology chain. Lv and others summarize that "CCUS technology primarily encompasses four areas: carbon capture, carbon transportation, carbon utilization, and carbon storage."[16] Experts have also stressed that "[a]ccording to the different climate change roadmaps (IEA, IPCC), Carbon Capture Storage (CCS) will play a key role in... climate change mitigation policy"[17]; however, its "development raises a trade-off between the deployment of large-scale projects (learning by replication), and the preservation of a large portfolio of competing technologies (learning by diversity),

Storage and Use: Technical, Economic, Environmental and Societal Perspectives (Springer, 2015).

[15] See further Diana Schumann, "Public Acceptance", pp. 221–251, in Wilhelm Kuckshinrichs and Jürgen-Friedrich Hake (eds), *Carbon Capture, Storage and Use: Technical, Economic, Environmental and Societal Perspectives* (Springer, 2015). Xu et al. have summarized significant elements of the interpretation of public acceptance in relation to CCUS as follows: "Whether CCUS can be deployed on a large scale... depends not only on technological advancement and energy consumption but also on public acceptance, which is regarded as an informal form of authorization for energy technology. Extensive studies have demonstrated that public acceptance is primarily determined by perceived risks (PR) and perceived benefits (PB). Specifically, PR is a predictor of public protest, and PB is more closely related to public acceptance. PR and PB have received increasing academic attention within this context, but so far, no consensus has been reached on how to measure them." Yinghua Xu, Bingsheng Liu, Yuan Chen, and Shijian Lu. "Public perceived risks and benefits of carbon capture, utilization, and storage (CCUS): Scale development and validation." *Journal of Environmental Management* 347 (2023): 119109.

[16] Weifeng Lv et. al., *supra* n.11, ix.

[17] X Galiègue, Audrey Laude and N Béfort, "Mapping CCUS Technological Trajectories and Business Models: The Case of CO_2-Dissolved", pp. 27–46, in Karine Ballerat-Busserolles, Ying Wu and John J Carroll, *Cutting-Edge Technology for Carbon Capture Utilization and Storage* (Wiley, 2018), 27.

on each of its steps (capture, transport, storage)."[18] In terms of large-scale projects and their impact on broader markets, it has been the case unfortunately that as such projects have emerged, they have tended to employ enhanced oil recovery, with a negative consequence being that the development of broader business models has been constrained:

> Although EOR[19] has provided a first feasible business model for CCS, CCS has still to prove its economic viability on a large variety of carbon emitters (power plant, industrial and bioenergy sources). A competing business model for CCS is to find other carbon uses and energy sources, better adapted to medium and small carbon sources.[20]

These observations catch the spirit of current market limitations and tensions between both *large-scale* CCS project forms in their own right, and between *large/medium/small* deployment opportunities more generally.

This book has a particular interest in law and regulation. In relation to these issues and CCUS, it has been the case that relatively sophisticated legal and regulatory regimes have started to emerge within individual countries (such as Germany), regional spheres (such as the EU), and internationally, with the result that multi-level governance regimes are beginning to crystallize and develop.[21] Further, as noted above, component parts of the CCUS technology chain can be distinguished, and this in turn can have a bearing on the legal/regulatory regime or associated principles at issue in a particular instance, and how they are conceptualized. For example, in interpreting and assessing CCS law, effective legal regimes of both a national and international character are required to adequately engage and cover: long-term liability for storage; regulation of transport; the treatment of stored carbon under emissions trading regimes; issues of property ownership; public engagement.[22] Thus, specific areas of the

[18] X Galiègue et. al., ibid., 27.

[19] i.e., 'Enhanced Oil Recovery'.

[20] X Galiègue et. al., *supra*., n.17, 27.

[21] See further Dirk Uwer and Daniel Zimmer, *Carbon Capture and Storage: The legal Landscape of Climate Change Mitigation Technology* (Globe Law and Business Ltd, 2020).

[22] Ian Havercroft, Richard Macrory and Richard Stewart (eds), *Carbon Capture and Storage: Emerging Legal and Regulatory Issues* (Bloomsbury, 2018).

CCUS chain tend to necessitate bespoke legal requirements. This is highlighted further, for example, by the work of El-Kady and others, which finds significant shortcomings in frameworks governing the specific issue of CO_2 pipeline transportation.[23] Attention to nuances between CCS and CCU (i.e., Carbon Capture and Utilization) issues, and the consequent nuances in regulatory regimes that can be necessary to each, may also need to be balanced carefully, so that one is not prioritized at the expense of the other, with the result that one of the two regulatory strands may be less usefully developed.[24]

In terms of the *utilization* component of CCUS in particular, which has emerged more recently than CCS as a prevalent area of interest and practice, it is useful to be clear on what this entails, not least because law and regulation are required to provide a framework that is responsive to CCU needs, some of which lie beyond the exclusive parameters of narrowly drawn CCS regime framing. This utilization element has been summarized as follows:

> Carbon utilization is the process of using captured CO_2 as a resource to make value-added products, and it is also an important aspect of climate mitigation. Generally, there are three categories carbon utilization technologies can be divided into: chemical technologies, biological technologies, and mineralization processes.[25]

Chemical technologies typically produce products or are otherwise utilized in the chemical, polymer, fuel and beverage and food industries,

[23] Ahmed Hamdy El-Kady, Md Tanjin Amin, Faisal Khan, and Mahmoud M. El-Halwagi. "Analysis of CO2 Pipeline Regulations from a Safety Perspective for Offshore Carbon Capture, Utilization, and Storage (CCUS)." *Journal of Cleaner Production* (2024): 140734.

[24] See, e.g., the discussion of EU CCS and CCU approaches in Kim Talus and Reza Maddahi, "Carbon Capture and Utilization Under EU Law: Impermanent Storage of CO_2 in Products and Pre-combustion Carbon Capture." *The Journal of World Energy Law & Business* (2024): jwae009.

[25] Reza Mahmoudi Kouhi, Mohammad Milad Jebrailvand Moghaddam, Faramarz Doulati Ardejani, Aida Mirheydari, Soroush Maghsoudy, Fereshte Gholizadeh, and Behrooz Ghobadipour, "Carbon Utilization Technologies & Methods", pp. 1–50, in Ali Ahmadian, Ali Elkamel and Ali Almansoori (eds), *Carbon Capture, Utilization, and Storage Technologies: Towards More Sustainable Cities* (Springer, 2024), 1.

and biological technologies typically produce products or are otherwise utilized in the bio-alcohol, bio-fuel, and bio-plastic sectors.[26] The third category, carbon mineralization, involves converting CO_2 into solid matter, including carbonate and silicate matter. One innovative use of carbon materialization at the present time involves making enhanced concrete blocks by injecting captured CO_2 into the concrete prior to it hardening. Riedl and others explain that "[t]his reaction creates calcium carbonate, which not only strengthens the concrete allowing for less cement to be used, but also permanently sequesters the CO_2."[27] Given the form and character of utilization, then, it is clear that law and regulation require regime-framing suitable to its needs, and that aspects of this framing will differ from narrowly drawn CCS regime design applicable to, e.g., the injection of carbon into deep aquifers for permanent storage.

2 The Content, Scope, and Significance of This Book

In exploring CCUS in detail, this book places its emphasis on issues of law, policy, and standardization. It is particularly concerned to address and explore the means by which governments, companies, and wider society are endeavoring to use and/or can use, CCUS initiatives to advance the low-carbon energy transition and augment sustainable development. Relatedly, the work also places some emphasis on outlining and engaging with general opportunities and constraints operating on CCUS deployment.

The preceding section has discussed CCUS in the context of international governance. The Paris Agreement of 2015 has gradually intensified international decarbonization efforts, with Article 2 requiring international signatories to pursue efforts to keep global warming to 1.5 degrees Celsius above pre-industrial levels of warming. The international community, operating under the terms of the United Nations Framework Convention on Climate Change, is focused on implementing this sea-change in decarbonization intensity. This in turn has placed an increasing

[26] Kouhi et. al., ibid., 1–2.

[27] Danielle Riedl, Zachary Byrum, Serena Li, Hélène Pilorgé, Peter Psarras and Katie Lebling, "5 Things to Know About Carbon Mineralization", World Resources Institute, 22 June 2023 (unpaginated online article): https://www.wri.org/insights/carbon-mineralization-carbon-removal.

emphasis on CCUS research, development, and deployment, and the relevant literature in the area has struggled to keep pace with these cutting-edge changes. Cumulatively, this text advances research in the area by providing a leading-edge statement on CCUS that assists in bringing scholarship and understanding up to date against the backdrop of this fast-changing landscape.

Part I provides an introduction to CCUS and fundamental issues and concepts of importance. Particular attention is accorded to the issues of climate change, the low-carbon energy transition, sustainable development, and the connection and interaction between CCUS and Net-Zero concerns. Technological developments over recent years, the condition of CCUS technology at the present time, and the outlook for CCUS technology into the future are also considered, and the types of products that can capture carbon and the significance of such utilization are addressed. The issue of standardization and associated matters is also addressed, where the importance of the construction of a normative framework that promotes the elimination of technical barriers and favors the penetration and knowledge of CCUS solutions by stakeholders comes to the fore. Legal and policy analysis also addresses CCUS regulatory barriers and opportunities, particularly those involving the development of public policies and laws to encourage the adoption of CCUS technologies. Relatedly, barriers and opportunities in the sphere of financing and economics are considered.

Part II examines specific CCUS initiatives and experiences in selected jurisdictions. It provides case studies that illustrate approaches to CCUS initiatives and identify opportunities and constraints acting on host governments and energy companies seeking to advance the energy transition via CCUS. In affording evaluation of best practices across a variety of jurisdictions, a given chapter in this section can be: referred to in isolation to provide insight in relation to a jurisdiction of particular interest; referred to with reference to another chapter(s) to provide comparative insight across two or more important jurisdictions; read as part of a read-through of the full section, to provide a comprehensive body of knowledge across a broad geographic range. The book concludes with reflections, and with recommendations relevant to the ways in which host governments and energy industries might use CCUS initiatives to advance economic development and sustainability goals in going forward.

This body of work makes innovative contributions by presenting a consistent history of the CCUS technologies under development, by interrogating the connection between CCUS and the low-carbon energy transition, and by incorporating legal, regulatory, and normative issues into its considerations. The chapters delving into the approaches to CCUS adopted within specific jurisdictions serve to provide concrete examples of CCUS opportunities and challenges in practice, that can inform law, policy, and standardization in the sector in going forward.

Researchers have found that different levels of research and development across the world, including in relation to component elements within broader CCUS project chains, mean that the current overall state of CCUS technology and market development is complex and difficult to read.[28] Thus it has been advanced that that comprehensive global databases of current research in its different phases from concept to commercial use are necessary.[29] This can improve information-share within the research and development community, enhance joined-up thinking, and provide greater information for investors to base decisions on in going forward.[30] While it is no doubt correct that such global databases would be helpful, this book assists in partially mitigating such problems, by pooling knowledge and insight across a range of themes, concepts, and jurisdictions within one reference text.

It also appears to be the case that domestic and regional CCUS legal and policy regimes may benefit from a strong degree of harmonization. For example, analysis of legal and regulatory approaches in Europe, China, and the Middle East has identified similarities and divergences,[31] and it would be important as CCUS becomes increasingly deployed in a globalized context that legal-regulatory divergences do not act as substantial barriers to market. Again, this book assists in pooling knowledge in the area so that similarities and divergences can be more clearly viewed and interpreted across a range of jurisdictions, via both the case study chapters in Part II and the work's general concern to address the issue of standardization.

[28] Bartosz Dziejarski, Renata Krzyżyńska, and Klas Andersson, "Current Status of Carbon Capture, Utilization, and Storage Technologies in the Global Economy: A Survey of Technical Assessment." *Fuel* 342 (2023): 127776.

[29] Dziejarski et al., ibid.

[30] See further Dziejarski et al., ibid.

[31] Hao Zhang, "Regulations for Carbon Capture, Utilization and Storage: Comparative Analysis of Development in Europe, China and the Middle East." *Resources, Conservation and Recycling* 173 (2021): 105722.

In the narrower context of the legal profession and lawyerly work itself, it is notable that law in the area can be complex and challenging. A good example of this is provided by the "legal and commercial primer" on CCUS published by Lee and others in the *Texas Journal of Oil, Gas and Energy Law*.[32] This 35-page analysis details aspects of current and amended CCUS law in the United States, and demonstrates the highly technical and evolving nature of this area of legal practice.[33] By focusing in significant part on law and policy considerations, this book helps to elucidate core aspects of this under-studied and challenging field. More broadly, extant research has suggested that CCUS technology itself is in a relatively good and well-developed stage, and that constraints on its scalability to market arise chiefly in the economic and social spheres, serving to partially inhibit progress.[34] This includes in relation to law and policy, where "effective and comprehensive CCUS law and regulation" and a "predictable and enduring policy environment" appear to require more extensive, directed and sophisticated development.[35] Again, in pooling relevant knowledge and insight, this book provides a useful resource for governments, energy industry actors, researchers and others who seek to progress these economic and social spheres in a direction that can better enable CCUS.

[32] Austin Lee, James McAnelly, Elizabeth McGinley, and Jarrod Gamble. "The Way Forward: A Legal and Commercial Primer on Carbon Capture, Utilization, and Sequestration." *Texas Journal of Oil, Gas, and Energy Law* 16 (2021): 43.

[33] See generally Lee et. al., ibid.

[34] Hope McLaughlin, Anna A. Littlefield, Maia Menefee, Austin Kinzer, Tobias Hull, Benjamin K. Sovacool, Morgan D. Bazilian, Jinsoo Kim, and Steven Griffiths. "Carbon Capture Utilization and Storage in Review: Sociotechnical Implications for a Carbon Reliant World." *Renewable and Sustainable Energy Reviews* 177 (2023): 113215.

[35] McLaughlin et. al., ibid., 27.

CHAPTER 2

Climate Change, Just Energy Transition and Sustainable Development

Thomas L. Muinzer and Constantinos Yiallourides

1 Introduction

This chapter begins by introducing the problem of anthropogenic climate change, and addresses the role of international climate law and policy in tackling this problem. It subsequently considers the concept of the *just energy transition*, and the significance of the principle of *sustainable development* and the associated international Sustainable Development Goals. The analysis then focuses on Carbon Capture, Utilisation and Storage, examining its form and function in the context of the low-carbon transition. Its deployment to date and its future potential are considered. It is determined that it may have a useful and significant role to play in

T. L. Muinzer (✉)
University of Aberdeen, Aberdeen, UK
e-mail: thomas.muinzer@abdn.ac.uk

C. Yiallourides
School of Law, Macquarie University, Sydney, Australia

E. G. Pereira et al. (eds.), *Carbon Capture Utilization and Storage*, Sustainable Development Goals Series,
https://doi.org/10.1007/978-3-031-81272-9_2

ongoing efforts, provided that it is not employed as a vehicle to disincentivise or reduce the deployment of other much-needed renewable and sustainable low-carbon technologies.

2 Climate Change

Anthropogenic climate change is one of the great problems facing humanity.[1] In order to tackle the phenomenon and manage the impacts arising from it, the substantial mitigation of greenhouse gas releases into the atmosphere, such as carbon dioxide (CO_2) and methane (CH_4), and extensive impact adaptation measures, must be implemented over the current decade and the decades to follow.[2] Climate law and policy endeavours to engage with and redress the problem of anthropogenic climate change, including through the establishment of an extensive international framework.[3] The first major international-level instrument to engage the problem substantially was the United Nations Framework Convention on Climate Change (UNFCCC), which was adopted by the United Nations (UN) in 1992. This manifested in one legal framework an international effort to cooperate in combatting anthropogenic global warming. In drawing common international intentions and standards together, the UNFCCC was an important step in international cooperation. The UNFCCC was also subject to weaknesses, however. These included an absence of robust action plans and concrete greenhouse gas emissions reduction targets.[4]

[1] See, e.g., IPCC, *Climate Change 2022: Impacts, Adaptation and Vulnerability. Contribution of Working Group II to the Sixth Assessment Report of the Intergovernmental Panel on Climate Change*, eds. H-O Pörtner, DC Roberts, M Tignor, ES Poloczanska, K Mintenbeck, A Alegría, M Craig, S Langsdorf, S Löschke, V Möller, A Okem, and B Rama (Cambridge, UK and New York, NY, USA: Cambridge University Press, 2022), 3056.

[2] Chris Turne, Anne-Gaelle Ausseil, and Linda Broadhurst, 'Urgent Need for an Integrated Policy Framework for Biodiversity Loss and Climate Change' (2020) 4 Nature Ecology & Evolution 8, 996–996.

[3] Daniel Bodansky, Jutta Brunnée, and Lavanya Rajamani, *International Climate Change Law* (Oxford University Press, 2017).

[4] See further J von Stein, 'The International Law and Politics of Climate Change: Ratification of the United Nations Framework Convention and the Kyoto Protocol' (2008) 52 (2) Journal of Conflict Resolution 243.

The Kyoto Protocol was agreed in 1997 in order to strengthen the international climate regime, coming into force in February 2005.[5] Here, developed nations listed in Annex I to the UNFCCC committed to binding greenhouse gas reduction targets of "at least 5 per cent below 1990 levels in the commitment period 2008 to 2012."[6] This notable progress was impeded, though, by the non-ratification of the Protocol by certain nations. While the European Union did ratify the agreement, notable absentees included the USA, China and India, namely the other most abundant global greenhouse gas emitters. This undermined the progress that could be made under the terms of the Protocol. In 2002 both China and India did join the agreement, however they did not receive binding reduction targets. The USA continued to avoid ratification.

The Kyoto Protocol's capacity to drive change was also weakened by the level of its headline 5% reduction target, which was set much too low to reduce the overall harmful trajectory of global anthropogenic climate change sufficiently.[7] Nor did the Protocol map out the detail of how its signatories were to sustain their progress after the Protocol's initial 2008–2012 period elapsed. The negotiation of the post-2012 compliance phase was drawn out over subsequent years in a faltering and unproductive manner. A 'Bali Road Map' was agreed at the 2007 UN Climate Change Conference in Bali, Indonesia, which provided a practical path that was intended to guide the UN towards further binding targets[8]; however, it was hoped that the targets could be cohered at the UN's Copenhagen Conference 2009, but sufficient agreement could not be arrived at.[9] A Copenhagen Accord was approved, which acknowledged a common desire to keep rising global temperatures below 2 °C above pre-industrial

[5] Kyoto Protocol to the United Nations Framework Convention on Climate Change.

[6] Ibid., Article 3(1).

[7] Thomas L Muinzer, 'What Do We Mean When We Talk About National 'Climate Change Acts' and How Important Are They in the Context of International Climate Law?' in Thomas L Muinzer (ed.), *National Climate Change Acts* (Hart: UK, 2021), 11–42, 29.

[8] The Bali Roadmap draws together a range of decisions and plans, including the Bali Action Plan, Decision 1/CP.13.

[9] The 15th Session of the Conference of the Parties (COP 15) to the United Nations Framework Convention on Climate Change.

levels,[10] however the international community was unable to agree on legally binding reduction targets, and action plans were not established outlining a clear path forward. Subsequent conferences at Cancun and Durban did not resolve this impasse.[11]

Significant progress was finally made in December 2012 in Doha, Qatar, when the 'Doha Amendment to the Kyoto Protocol' was agreed by the Kyoto signatories.[12] Here, the Kyoto parties established a second Kyoto commitment period, to run from 1 January 2013 to 31 December 2020. An 'overall commitment' to emissions cuts for the industrialised nations was set at 18% below 1990 baseline levels for this timeframe.[13] The 2013–2020 compliance period was initially formulated as a loose arrangement that the signatories planned to firm up over time, and so it was not yet on a par with the more consolidated 2008–2012 compliance period, which is one reason why a comparably vast timespan features in the second period (i.e. covering 2013–2020).

In Paris in 2015 there was an intention amongst the international community to further develop legally binding commitments, which the European Union and many other actors had hoped would follow in the spirit of Kyoto to include substantive top-down emissions reduction targets that could extend the 2008–2012 Kyoto approach. Further deadlock due to a lack of forthcoming diplomatic agreement resulted in a failure amongst the international community to agree on collective mitigation target action, and the Paris Agreement was cohered and agreed in a lighter form.[14] While the agreement did not rise to the challenge of extending Kyoto, it did obligate signatories to prevent dangerous and irreversible levels of climate change. This was construed as an obligation to hold rising temperatures to 2 °C maximum warming above pre-industrial levels, with an intention to "pursue efforts" to hold it beneath 1.5 degrees.[15] The Paris Agreement has since superseded the

[10] The Copenhagen Accord of 18 December 2009, Decision-/CP.15 (United Nations, 2009), Article 1.

[11] The 16th and 17th Sessions of the Conference of the Parties to the United Nations Framework Convention on Climate Change, which took place in Mexico and South Africa, respectively.

[12] Kyoto Protocol (Amendment) Decision 1/CMP.8.

[13] This amended Article 3 of the Kyoto Protocol.

[14] Adoption of the Paris Agreement, FCCC/CP/2015/L.9/Rev.1.

[15] Paris Agreement, Article 2(1)(a).

Kyoto Protocol agreement. The frequently reluctant USA did sign the Paris Agreement, and it remains a signatory despite an intention of former President Donald Trump to disengage from it. Japan, New Zealand and the Russian Federation are also signatories, each of whom is a nation that did have binding target commitments under the first Kyoto Protocol period but that had damagingly gone on to refuse to adopt equivalent commitments under the second 2013–2020 period.[16] While the Paris Agreement has significant content and merits,[17] the divergence between this 2015 treaty and the 2013–2020 Kyoto Protocol period served to split the International Climate Law Regime into two parallel tracks over the years of 2015–2020, with the Paris Agreement ultimately becoming the dominant track and replacing Kyoto once 2020 had concluded.[18] This resulted in a weaker top-down international climate law regime that paired out top-down emissions reduction targets in favour of giving states a freer hand to set their own reduction levels internally on a more relaxed and pragmatic basis.[19]

3 The Just Energy Transition

Set against the backdrop of the anthropogenic climate change challenge and the legal-political frameworks that have been devised to redress it, work engaging the notion of the "*just energy transition*" (JET) has endeavoured to interpret, assess and structure the low-carbon transition of energy systems with reference to moral action and just outcomes.[20] Wang and Lo describe the emergence of the JET as "a discourse that

[16] Muinzer, *supra.*, n.7, 29–30.

[17] Daniel Klein, et al., eds., *The Paris Agreement on Climate Change: Analysis and Commentary* (Oxford University Press, 2017).

[18] See further 'Splitting of the ICL Regime into Two Parallel Tracks', 34–38, in Muinzer, *supra.*, n.7.

[19] See 'The ICL Regime Now and Going Forward', 38–39, in Muinzer, ibid.

[20] J Herberg, E Drewing, JL Reinermann, J Radtke, M LaBelle, A Stojilovska and K Gürtler, 'Energy Spaces: Bridging Scales and Standpoints of Just Energy Transitions' 25(2) Journal of Environmental Policy & Planning, 135–141 (introduction to a special issue on the JET, according particular analytical attention to transregional and spatial aspects of energy transitions).

centers social justice in energy transition to respond to the dual challenges of climate change and sustainability".[21] García-García et al. have also reviewed the JET as a term and concept, advancing the following definition:

> A 'just energy transition towards a low carbon economy' can be defined as a long-term technological and socio-economic process of structural shift that affects the generation, distribution, storage and use of energy and causes rearrangements at micro (innovation), meso (social networks, rules and technical elements) and macro (exogenous environment) levels, while also ensuring that the desired socioeconomic functions can be accomplished through decarbonised and renewable means of energy production and consumption, safeguarding social justice, equity and welfare.[22]

The International Institute for Sustainable Development has also provided a normative definitional outline, as follows:

> A just energy transition is a negotiated vision and process centred on dialogue, supported by a set of guiding principles, to shift practices in energy production and consumption. It aims to minimize negative impacts on workers and communities with stakes in high-carbon sectors that will wind down, and to maximize positive opportunities for new decent jobs in the low-carbon growth sectors of the future. It strives to ensure that the costs and benefits of the transition are equitably shared.[23]

In spite of such extant definitions, research has clarified that conflicting approaches to interpretation and an associated general absence of a common technical understanding of the JET that can be replicated across the social scientific literature persists.[24] This hinders research progress in

[21] X Wang and K Lo, 'Just Transition: A Conceptual Review' (2021) 82 Energy Research & Social Science 102291, 5.

[22] Pablo Garcia-Garcia, Óscar Carpintero and Luis Buendía, 'Just Energy Transitions to Low Carbon Economies: A Review of the Concept and its Effects on Labour and Income' (2020) 70 Energy Research & Social Science 101664, 5.

[23] IISD, *Real People, Real Change: Strategies for Just Energy Transitions* (IISD, 2018), 2.

[24] Thomas L Muinzer, 'Challenges in Research Approaches to the 'Just Energy Transition' in Legal Studies and Other Branches of the Social Sciences' (2023) 16(1) The Journal of World Energy Law & Business 8–17.

the area.[25] For one thing, it prevents social scientists from utilising the concept in a precise and mutually intelligible way in order to provide or otherwise directly inform principles that conclusions can be deduced from, and which might yield in turn empirically testable propositions.[26] Additional work, including in particular from Huang and Liu, has found that the low-carbon energy transition itself primarily embodies a response to headline problems posed by climate change and sustainable development.[27] The authors have found that a substantial body of discourse on the JET has both crystallised and organised itself with reference to climate change and sustainable development drivers, and they note that framing in this sphere has a tendency to locate social justice at the centre of energy transitions.[28]

4 Sustainable Development and the UN Sustainable Development Goals

The common understanding of "*sustainable development*" has its international origin in the Brundtland report, produced by the Brundtland Commission in 1987.[29] Here, sustainable development is defined as "development that meets the needs of the present without compromising the ability of future generations to meet their own needs".[30] The UN has since established 17 SDGs to assist in driving the concept of sustainable development in practice. These SDGs are accompanied by

[25] Thomas L Muinzer, Eduardo G Pereira and Patrick Baker, 'Climate Change and a Just Energy Transition' in Eduardo G Pereira, Thomas L Muinzer and Patrick Baker (eds), *Energy Law and the Sustainable Development Goals: Host Government Instruments for Sustainability in Oil and Gas Operations* (Taylor & Francis: UK forthcoming).

[26] See generally Muinzer, *supra*., n. 24.

[27] P Huang and Y Liu, 'Toward Just Energy Transitions in Authoritarian Regimes: Indirect Participation and Adaptive Governance' (2021) 64 Journal of Environmental Planning and Management 1, 1.

[28] Huang and Liu, ibid., 1–2.

[29] Report of the World Commission on Environment and Development, *Our Common Future* (1987) (Brundtland Report).

[30] *Our Common Future*, ibid., Chapter 2, paragraph 1.

169 targets. The SDGs establish abstract goals that function as aspirational benchmarks.[31] As they emerged, the goals and targets were 'grounded in international law and made consistent with existing commitments expressed in various international agreements and other soft law instruments'.[32]

The UN's General Assembly adopted the 17 SDGs in 2015, with an emphasis on the achievement of a more sustainable world by the target date of 2030.[33] As time has advanced, the perspective applied to the SDG intentions is being stretched increasingly beyond 2030, as the world continues to look and plan ahead.[34] Immediately prior to the current SDG era, the UN General Assembly had agreed to a Millennium Declaration in 2000.[35] This established the Millennium Development Goals (MDGs), which were in operation from 2000 to 2015, and were informed by the principle of sustainable development; however, the MDGs did not place sustainable development as prominently to the fore as the subsequent SDGs. The MDGs were also restricted primarily to developing countries, whereas the SDGs span nations in general. In addition to including a goal to "ensure environmental sustainability", the MDGs engaged in the following matters: eradication of extreme poverty and hunger; achievement of universal primary education; promotion of gender equality and empowerment of women; reduction of child mortality; improvement of maternal health; combatting of HIV/AIDS, malaria and

[31] Thomas L Muinzer and Colin Mackie, 'The Sustainable Development Goals' in Eduardo G Pereira, Thomas L Muinzer and Patrick Baker (eds), *Energy Law and the Sustainable Development Goals: Host Government Instruments for Sustainability in Oil and Gas Operations* (Taylor & Francis: UK forthcoming).

[32] Rakhyun E Kim, 'The Nexus Between International Law and the Sustainable Development Goals' (2016) Review of European, Comparative & International Environmental Law 15, 15.

[33] A Fleming, RM Wise, H Hansen and L Sams, 'The Sustainable Development Goals: A Case Study' (2017) 86 Marine Policy 94–103, 94.

[34] Detlef P van Vuuren, Caroline Zimm, Sebastian Busch, Elmar Kriegler, Julia Leininger, Dirk Messner, Nebojsa Nakicenovic et al., 'Defining a Sustainable Development Target Space for 2030 and 2050' (2022) 5(2) One Earth 142–156; IIASA and TWI2050, *The World in 2050 Initiative: Transformations to Achieve the Sustainable Development Goals* (IIASA, Austria, 2018).

[35] UN General Assembly, *United Nations Millennium Declaration* (UNGA Resolution A/RES/55/2, 18 September 2000).

other diseases; development of a global partnership for development.[36] Fukuda-Parr notes that the MDGs "commanded more authority than previous goals because they were cast as a comprehensive set of priorities",[37] and they formed a crucial foundation out of which the more far-reaching SDGs developed, with the latter foregrounding the principle of sustainable development more prominently.

The SDGs were established in UN Resolution A/RES/70/1 of 2015, entitled *Transforming our World*.[38] Each Goal is fleshed out by a range of targets and intentions.[39] In 2017, another UN Resolution developed the framework further, endeavouring to fine-tune and augment the articulation of sustainable development by building in a system of indicators.[40] Here, it was thought that the inclusion of indicators would allow progress towards the SDG targets and intentions to be measured more effectively.[41] The 17 SDGs establish aspirational benchmarks designed to assist in the realisation of sustainable development in practice across the world. Swain has emphasised that the goals are "non-binding", insofar as "signatories committed to 'respecting national policies and priorities' with 'each Government setting its own national targets'".[42] Swain summarises that "shared responsibility for SDGs outcomes is collective, extending to all the leaders, UN agencies, multilateral and bilateral aid agencies,

[36] *United Nations Millennium* Declaration, ibid.

[37] Sakiko Fukuda-Parr, 'Sustainable Development Goals' in Thomas G Weiss and Sam Daws (eds), *The Oxford Handbook on the United Nations* (2 edn, online, OUP 2018) 1, 1–2.

[38] United Nations General Assembly, *Transforming Our World: The 2030 Agenda for Sustainable Development* (UNGA Resolution A/RES/70/1, 21 October 2015).

[39] See further Goal 1–17 and sub-paragraphs at United Nations, A/RES/70/1, ibid., 14–27.

[40] United Nations, A/RES/71/313 *Resolution Adopted by the General Assembly on 6 July 2017, Work of the Statistical Commission Pertaining to the 2030 Agenda for Sustainable Development* (UN, 2017).

[41] Ibid., Annex ("Global indicator framework for the Sustainable Development Goals and targets of the 2030 Agenda for Sustainable Development"). For further discussion of the form and contribution of indicators, see Tomáš Hák, Svatava Janoušková, and Bedřich Moldan, 'Sustainable Development Goals: A Need for Relevant Indicators' (2016) 60 Ecological Indicators 565–573.

[42] Ranjula Bali Swain, 'A Critical Analysis of the Sustainable Development Goals' in Walter Leal Filho (ed.), *Handbook of Sustainability Science and Research* (Cham: Springer, 2018), 341–355, 343.

and numerous other private sector, nongovernmental and civil society actors".[43] Thus, although they are significant, the SDGs are not themselves legally binding and enforceable within states unless a particular state chooses to implement them directly in national law, and they are also spread somewhat thinly insofar as they have been described as "global in nature and universally applicable".[44] In other words, they are not applied sharply and specifically to (e.g.) national governments.

While the principle of sustainable development places a pronounced degree of emphasis on environmental protection in the context of socio-economic development,[45] research has suggested that the environmental dimension of sustainability is not as strongly foregrounded within the SDG framework as it needs to be in order to reflect this characteristic appropriately.[46] Further, researchers have addressed the manner in which the SDGs interact and intersect, identifying "tensions at the heart of the Goals" that can mean that "progress towards some risks jeopardising, at least without the imposition of appropriate measures and integrated planning, progress on others".[47] Relatedly, it should be noted that while the goals can be viewed in isolation, they are also fundamentally co-dependent. Nilsson et al. find that "[i]mplicit in the SDG logic is that the goals depend on each other", noting that if countries "ignore the overlaps and simply start trying to tick off targets one by one, they risk perverse outcomes".[48] As an example, Nilsson et al. point out that "using coal to improve energy access (goal 7) in Asian nations, say, would accelerate climate change and acidify the oceans (undermining goals 13 and 14), as well as exacerbating other problems such as damage to health

[43] Swain, ibid., 343.

[44] Resolution A/RES/70/1, *supra.*, n.38, 13.

[45] A conceptual discussion of the principle is provided in Carlos Alberto Ruggerio, 'Sustainability and Sustainable Development: A Review of Principles and Definitions' (2021) 786 Science of the Total Environment 147481.

[46] Mathis Wackernagel, Laurel Hanscom, and David Lin, 'Making the Sustainable Development Goals Consistent with Sustainability' (2017) 5 Frontiers in Energy Research 18, 1.

[47] Duncan French and Louis J Kotzé, eds., *Sustainable Development Goals: Law, Theory and Implementation* (Edward Elgar Publishing, 2018), 2.

[48] Måns Nilsson, Dave Griggs, and Martin Visbeck, 'Policy: Map the Interactions Between Sustainable Development Goals' (2016) 534.7607 Nature News 320, 320.

from air pollution (disrupting goal 3)".[49] Muinzer and Mackie have also considered the linguistic construction of the goals, and have pointed out tensions arising in the realm of meaning that can pose complications relating to their practical operationalisation.[50]

In terms of Carbon Capture, Utilisation and Storage (CCUS), it is notable that certain SDGs have particular relevance to oil and gas governance. Goal 7, concerning "Affordable and Clean Energy", endeavours to ensure that governments and industry are acting to improve and expand access to affordable, reliable and sustainable energy.[51] Goal 9, engaging "Industry, Innovation and Infrastructure", also places a burden on governments and industry to establish suitably resilient energy infrastructure designed to ensure that energy deployment occurs in harmony with the sustainable needs of the planet and society ("sustainable industrialization"), and in a way that fosters sustainably sensitive innovation.[52] Given the need to transition rapidly away from fossil fuel emissions, this potentially affords a role for CCUS where fossil fuel emissions are occurring; in such cases, the implication is that emissions should be captured and sequestered where possible. Goal 13 also necessitates appropriate "Climate Action" to be taken,[53] Goal 3 requires the promotion of healthy lives and well-being,[54] and Goal 15 requires the protection and restoration of ecosystems, sustainable forest management, the reversal of land degradation, and the cessation of biodiversity loss.[55] Thus the exacerbation of anthropogenic climate change via harmful fossil fuel emissions must be mitigated against on these grounds.

[49] Nilsson et al., ibid., 320–321.

[50] Muinzer and Mackie, *supra.*, n.31.

[51] Resolution A/RES/70/1, *supra.*, n.38, "Goal 7".

[52] Resolution A/RES/70/1, ibid., "Goal 9".

[53] Ibid., Goal 13, "Climate Action".

[54] Ibid., Goal 3, "Good Health and Well-Being".

[55] Ibid., Goal 15, "Life on Land". Goal 14 applies equivalent protections to oceans, seas and marine resources.

5 Decarbonising Energy amid Rising Demand and Security Concerns

Anthropogenic emissions of CO_2 are the largest driver of climate change.[56] The main source of emitted CO_2 globally is the combustion of fossil fuels, such as coal, crude oil, and natural gas, for energy production.[57] Global demand for energy is forecast to increase by 66% by 2030, with fossil fuels accounting for a staggering 80% of the energy mix in 2022.[58] Scientific scenarios suggest that, even with substantial increases in energy derived from other sources, fossil fuels will remain a major energy source in the twenty-first century.[59] The consequences—global temperature rise of 1.4–5.8 °C and increasing extreme weather events—could prove disastrous for future generations.[60] As concerns about human-induced climate change intensify, the need to decarbonise energy systems and 'avoid a climate disaster' becomes increasingly salient.[61]

Ensuring large CO_2 emission reductions requires a fundamental shift in the way we generate, transport, and use energy.[5] Energy-hungry nations must meet their commitments under the Paris Agreement to reduce substantially CO_2 emissions and keep global temperature rise

[56] IPCC Report: 'Climate Change 2022: Mitigation of Climate Change' (IPCC, 2022) (Summary for Policymakers), www.ipcc.ch/report/ar6/wg3/; IPCC, 'Special Report: Global Warming of 1.5 °C' (IPCC 2018) Ch.2, www.ipcc.ch/sr15/chapter/chapter-2/.

[57] A Shahbazi and BR Nasa, 'Carbon Capture and Storage (CCS) and its Impacts on Climate Change and Global Warming' (2016) 7 Journal of Petroleum & Environmental Biotechnology 291.

[58] BP Energy Outlook 2023, www.bp.com/content/dam/bp/business-sites/en/global/corporate/pdfs/energy-economics/energy-outlook/bp-energy-outlook-2023.pdf; IEA, 'CO2 Emissions in 2022', www.iea.org/reports/co2-emissions-in-2022.

[59] Anissa Nurdiawati and Frauke Urban, 'Towards Deep Decarbonisation of Energy-Intensive Industries: A Review of Current Status, Technologies and Policies' (2021) 14 Energies 2408; Stephen F. Lincoln, 'Fossil Fuels in the 21st Century' (2005) 34(8) Ambio 621–627; Robert H Socolow and Stephen Wilson Pacala, 'A Plan to Keep Carbon in Check' (2006) 295(3) Scientific American 50.

[60] IPCCS Report: 'Climate Change 2022: Mitigation of Climate Change' (Summary for Policymakers), www.ipcc.ch/report/ar6/wg3/.

[61] UNEP, Emissions Gap Report 2022, www.unep.org/resources/emissions-gap-report-2022.

below 2 °C.[62] At the same time, nations must safeguard their energy security, namely the 'uninterrupted availability of energy sources at an affordable price'.[63] Recent events have highlighted the urgent need for new pathways to low-carbon energy transition. The COVID-19 pandemic and the war in Ukraine, in particular, have disrupted energy markets, leading to price instability, energy shortages, and heightened energy security concerns worldwide.[64] Losing just a fraction of the world's fossil fuels amidst the war in Ukraine caused significant economic and social disruptions. The scale of these disruptions has also highlighted the need for a just and orderly transition away from fossil fuels to avoid future energy shortages and higher prices.[65] Renewable power is fast increasing, but it remains insufficient to meet significant energy demands, particularly in developing economies.[66] Thus, these events have reminded us that any successful and enduring low-carbon energy transition must address both energy security and sustainability objectives. CCUS has a potential capacity to expedite low-carbon energy transitions, while addressing energy supply and sustainability concerns. This issue is considered in detail in the sections to follow.

[62] Daniel Bodansky, 'The Paris Climate Change Agreement: A New Hope?' (2016) 110(2) American Journal of International Law 288–319; see further '2. Climate Change', above.

[63] IEA, 'Energy Security'. www.iea.org/topics/energy-security.

[64] IEA, World Energy Outlook 2022. https://www.iea.org/reports/world-energy-outlook-2022/executive-summary; Jinfang Tian, 'Global Low-Carbon Energy Transition in the Post-COVID-19 Era' (2022) 307 Applied Energy 1.

[65] BP Energy Outlook 2023; Klaus Hubacek et al., 'Russia–Ukraine War Has Nearly Doubled Household Energy Costs Worldwide—New Study' (*The Conversation*, 17 February 2023), https://theconversation.com/russia-ukraine-war-has-nearly-doubled-household-energy-costs-worldwide-new-study-200104.

[66] Ofélia Araújo and José Luiz de Medeiros, 'How Is the Transition Away from Fossil Fuels Doing, and How Will the Low-Carbon Future Unfold?' (2021) 23 Clean Technologies and Environmental Policy 1385–1388; Philip Ringrose, *How to Store CO2 Underground: Insights from Early-Mover CCS Projects* (Springer, 2020), 6.

6 CCUS: A Bridge to a Low-Carbon Future?

CCUS has emerged as a potentially crucial tool which could be deployed at scale to reduce the impact of CO_2 emissions on global temperatures.[67] The potential role of CCUS in low-carbon energy transitions has been acknowledged by the International Energy Agency (IEA), the International Renewable Energy Agency (IRENA), and the Intergovernmental Panel on Climate Change (IPCC).[68] In a 2020 Report, the IEA estimated that achieving net-zero by 2050 would require 50% more CCUS deployment, alongside other serious decarbonisation targets.[69] The IEA also estimated in its Sustainable Development Scenario, where global CO_2 emissions from the energy sector fall to zero on a net basis by 2070, that CCUS would account for 15% of the cumulative reduction in emissions.[70] CCUS appears to offer one technically feasible, and policy-desirable, pathway to reduce the carbon footprint of fossil fuel-based energy production while also ensuring sustainability.[71]

Carbon capture and storage (CCS) involves capturing CO_2 emissions at their source (including power generation or industrial facilities that use either fossil fuels or biomass for fuel) and securely storing them deep underground. The IPCC's Special Report on CCS in 2005 served as a catalyst for CCS to enter the low-carbon energy transition arena. Most international prospective scenarios have since proposed CCS as a crucial tool in low-carbon energy transition.[72]

[67] Tiina Koljonena et al., 'The Role of CCS and Renewables in Tackling Climate Change' (2009) 1 Energy Propedia 4323–4330; Muriel Cozier, 'The UN COP21 Climate Change Conference and the role of CCS' (Society of Chemical Industry, 2015).

[68] IPCC, 'Special Report on Carbon Dioxide Capture and Storage' (Cambridge University Press, 2005); IEA, 'Carbon Capture and Storage: The Solution for Deep Emissions Reductions' (EIA Report, 2015); IEA and IRENA, 'Perspectives for the Energy Transition: Investment Needs for Low-Carbon Energy System' (2017).

[69] IEA 2020, Ibid., 14.

[70] IEA, 'Special Report on Carbon Capture Utilisation and Storage: CCUS in Clean Energy' (EIA, 2020), 44–54, www.iea.org/reports/ccus-in-clean-energy-transitions.

[71] Global CCS Institute, 'Global Status of CCS: Targeting Climate Change' (2019 Global CCS Report); Iasonas Ioannou et al., 'Trade-offs Between Sustainable Development Goals in Carbon Capture and Utilisation' (2023) 16 Energy & Environmental Science 113, 122; IEA, 'Special Report on Carbon Capture Utilisation and Storage' (2020), 13–29.

[72] For an overview, see Ringrose (2020), 126–128.

In recent years, the 'usage' component of CCUS has also entered the value chain. This would involve using part of the captured CO_2 as a raw material for enhanced oil recovery (EOR) or to manufacture new products, before permanently storing the rest of the CO_2 in subsurface geological locations. This is why the technology is now often referred to as 'CCUS' (the term used in this chapter).[73] According to the IEA, CCUS is a 'group of technologies' which reduces emissions in key sectors and removes CO_2 to balance emissions that are 'technically difficult or prohibitively expensive to eliminate'.[74] CCUS may be more economically attractive than CCS, given that it 'offers the potential for new valuable products'.[75] Indeed, CCUS aims at monetising CO_2 to reduce the net abatement cost or even result in profits.[76] For example, CO_2 EOR is a well-established technology, developed by the petroleum industry, which monetises CO_2 by enhancing the extraction of residual hydrocarbons from mature wells with declining oil and gas production.[77] CO_2 EOR's subsurface geological injection was first undertaken in 1972 in the Permian Basin in West Texas.[78] At present, at least 56 oil fields in the Permian basin in West Texas use this method. Together, they produce about 200,000 barrels of oil per day and account for 85% of CO_2 EOR-produced oil in the United States.[79] CO_2 EOR is extensively utilised across the world's maturing oil basins to increase reservoir production.[80]

This 'EOR' perspective to CCUS comes amid a global context where fossil fuel use is being increasingly challenged and where renewable power

[73] Isabelle Czernichowski-Lauriol et al., 'The Gradual Integration of CCUS into National and Regional Strategies for Climate Change Mitigation, Energy Transition, Ecological Transition, Research and Innovation: an overview for France' (GHGT-15 Conference, 15th–18th March 2021 Abu Dhabi, UAE).

[74] IEA, 'Special Report on Carbon Capture Utilisation and Storage', 13–15.

[75] Ibid.

[76] Czernichowski-Lauriol et al. (2021), 31.

[77] US Chamber of Commerce, 'CO2 Enhanced Oil Recovery' (Institute for 21st Century Energy, 2013), 4–5; Van Bergen et al., 'Worldwide Selection of Early Opportunities for CO2-EOR and CO2-ECBM' in J Gale and Y Kaya (eds), *Proceedings of the 6th International Conference on Greenhouse Gas Control Technologies* (GHGT-6) (2003), 639–644.

[78] US Chamber of Commerce (2013), 4–5.

[79] K Michael et al., 'CO2 Storage in Saline Aquifers II—Experience from Existing Storage Operations' (2009) 1 Energy Propedia 1973–1980, 1973.

[80] Ibid.

is rapidly accelerating. On the one hand, renewable energy proponents often fear that discussions on mitigating the impact of continued fossil fuel use through CCUS could divert attention or slow down the growth of renewable energy and energy efficiency measures.[81] To those groups, 'clean-up' options, like CCUS, are only opportunities for stakeholders with vested interests in fossil fuels to seek to prolong their involvement in the industry for as long as possible, while extracting maximum value from fossil fuel investments.[82] This remains a serious concern, and such practices must be guarded against. As Elliott writes, 'the best way to store carbon is to leave it in the ground, and to look elsewhere for energy'.[83]

On the other hand, CCUS proponents consider that the transition from conventional to renewable energy is not fast enough and that CO_2 technologies are more viable in the short term than renewable energy, with the latter sometimes depicted as being 'still far-off'.[84] Today, fossil fuels remain the dominant supply of energy production globally. In spite of an urgent need to decarbonise, it is possible that they will maintain this status for some time. In this context, if CO_2 emissions reduction is urgent, then considering CCUS is also urgent as a 'transitional mitigation option…bridging the transition from today's fossil-intensive energy system to a post-fossil system with sizable contributions from renewables'.[85]

7 International Oil Companies Pivot to CCUS

International oil companies (IOCs), in particular, have shown a keen interest in CCUS. Shareholders and the public are adding increasing pressure on IOCs to reduce their carbon footprint. As a result, IOCs' oil and

[81] C Armeni, 'Carbon Capture and Storage in the Sub-Seabed and Sustainable Development: Please Mind the Gap' (2013) 27 Ocean Yearbook 1, 15.

[82] Centre for International Environmental Law, 'Confronting the Myth of Carbon-Free Fossil Fuels: Why Carbon Capture Is Not a Climate Solution' (2021), https://www.ciel.org/issue/carbon-capture-and-storage/; Damien Gayle, 'Carbon Capture Is Not a Solution to Net Zero Emissions Plans, Report Says' (*Guardian*, 1 September 2020), www.theguardian.com/environment/2022/sep/01/carbon-capture-is-not-a-solution-to-net-zero-emissions-plans-report-says.

[83] Elliott (2020), 39–49.

[84] Ibid.; see also Fang Wang et al. (2021), 2.

[85] IPCC, 'Special Report on Carbon Dioxide Capture and Storage' (2005), 357.

gas investments could become unsustainable in the future due to falling carbon prices, potentially leading to stranded assets. To mitigate these concerns, oil companies have proactively adopted strategies to adapt to the energy transition and ensure business viability in an evolving energy sector. For instance, the Norwegian national oil company Statoil (now known as 'Equinor') was a pioneer in injecting CO_2 in the sub-seabed for oceanic storage.[86] Since 1996, the company has been injecting CO_2 into a geologic formation beneath the North Sea.[87] This was motivated by a Norwegian tax on CO_2 emissions, which encouraged oil companies to store rather than release the extracted CO_2 into the atmosphere.[88]

The Sleipner project's success in injecting and storing about one million tons (Mt) of CO_2 annually has significantly contributed to the understanding of geologic CO_2 storage in sub-seabed locations and is used as a case-study worldwide.[89] Currently, approximately 35 commercial CCUS projects capture nearly 45 Mt of CO_2 annually worldwide. This is equivalent to the carbon offset achieved by planting 2 billion trees each year, as calculated by Imperial College London.[90] Between 1996 and 2020, nearly 200 Mt of CO_2 has been successfully stored.[91] While this is a significant achievement, the IEA's net-zero scenario calls for the storage of over 1.6 billion tonnes (Gt) of CO_2 per year by 2030. This is

[86] H Kongsjorden et al., 'Saline Aquifer Storage of Carbon Dioxide in the Sleipner Project' (1997) 17(5/6) Waste Management 303–308, 303–304.

[87] Ibid.

[88] T Torp and K Brown, 'CO2 Underground Storage Costs as Experienced at Sleipner and Weyburn' (2005) Proceedings of the 7th International Conference on Greenhouse Gas Control Technologies (GHGT-7), September 5–9, 2004, Vancouver, Canada, Volume I, 531–540.

[89] T Torp and J Gale, 'Demonstrating Storage of CO2 in Geological Reservoirs: The Sleipner and SACS Projects' (2004) 29 Energy 1361–1369.

[90] Cited in Geoscience Australia, 'Submission to the Inquiry into the 2009 and 2013 Amendments to the 1996 Protocol to the Convention on the Prevention of Marine Pollution by Dumping of Wastes and Other Matter, 1972 (London Protocol)' (10 March 2023), 7–8.

[91] K Zhang et al., 'Extension of CO2 Storage Life in the Sleipner CCS Project by Reservoir Pressure Management' (2022) 108 Journal of Natural Gas Science and Engineering 1.

comparable to the carbon offset achieved from planting 80 billion trees annually.[92]

8 The Multifaceted Role of CCUS in Achieving Net-Zero Targets

As mentioned earlier, achieving deep CO_2 emission reductions requires a significant transformation in how we generate, transport, and consume energy. It has also been discussed that the concept of the JET indicates that this transformative process must be undertaken with reference to equitable and just societal outcomes, albeit that the JET has yet to be delineated by analysts in a suitably coherent, consolidated form.[93] This change in how we generate, transport, and consume energy is possible only through the deployment of a range of technologies. Amongst these, CCUS aids in directly cutting emissions in vital energy and industrial sectors. CCUS also plays a crucial role in removing CO_2 to offset emissions which are difficult to eliminate, making it an essential component of achieving net-zero targets.[94] The IEA notes that 'a faster transition requires more CCUS' and highlights four essential roles of CCUS in the low-carbon energy transition: (a) addressing emissions from established energy infrastructure; (b) serving as a solution for sectors with emissions that are challenging to mitigate; (c) acting as a foundation for low-carbon hydrogen production; and (d) enabling the removal of carbon from the atmosphere.[95]

First, CCUS can retrofit current power and industrial plants which, if left unchecked, could emit 600 Gt of CO_2 in the next 50 years—equivalent to 17 years of current annual emissions.[96] The IEA notes that 'CCUS is the only alternative to retiring existing power and industrial

[92] Samantha McCulloch, 'Carbon Capture in 2021: Off and Running or Another False Start?' (EIA, 2021), www.iea.org/commentaries/carbon-capture-in-2021-off-and-running-or-another-false-start; Climate Neutral Group Insights, 'What Exactly Is 1 Tonne of CO2?' (2022), www.climateneutralgroup.com/en/news/what-exactly-is-1-tonne-of-co2-v2/.

[93] '3. The Just Energy Transition', above.

[94] Fang Wang et al., 'Technologies and Perspectives for Achieving Carbon Neutrality' (2021) 2(4) Innovation 2.

[95] IEA, 'Special Report on Carbon Capture Utilisation and Storage' Ch.2.

[96] Ibid., 13–17, 151.

plants early or repurposing them to operate at lower rates of capacity utilisation or with alternative fuels'.[97] CCUS would allow existing plants to operate with reduced emissions. This would contribute to energy security by supporting diverse power generation and integrating variable renewables. Retrofitting existing plants with CCUS can also help to preserve employment and economic prosperity in regions reliant on emissions-intensive industry, while avoiding the economic and social disruption of early retirements.[98]

Second, CCUS could provide a critical mitigation tool for 'hard-to-abate' emissions.[99] Heavy industries, like cement, iron, and steel, are responsible for nearly 20% of today's global CO_2 emissions. In these sectors, it has been suggested that reaching net-zero emissions without CCUS is 'virtually impossible'.[100] For instance, cement production releases considerable CO_2 emissions, not necessarily related to fossil fuel usage, which amount to over 4% of all energy sector emissions.[101] Given the lack of alternative cement production methods, capturing and storing these emissions becomes paramount.[102] Likewise, in the chemical industry, CCUS is often seen as the most effective option for reducing emissions from fertiliser (ammonia) and methanol production.[103]

Third, CCUS can help expedite the growth of low-carbon hydrogen production to meet demands in transport, industry, and buildings.[104] CCUS could aid the production of clean hydrogen from coal or natural gas, today's predominant energy sources. Thus, CCUS would offer a technically feasible route to introduce hydrogen into new markets.

[97] Ibid., 21.

[98] Ibid.

[99] IEA, 'The Challenge of Reaching Zero Emissions in Heavy Industry' (IEA, 2020), www.iea.org/reports/achieving-net-zero-heavy-industry-sectors-in-g7-members/executive-summary.

[100] Ibid., 13, 44, 63–67.

[101] Ibid., 23; see also IPCC, 'Special Report on Carbon Dioxide Capture and Storage' (2005), 77–83, 113, 161.

[102] Fiona Harvey, 'Cement Makers Across World Pledge Large Cut in Emissions by 2030' (*Guardian*, 12 October 2021), www.theguardian.com/business/2021/oct/12/cement-makers-across-world-pledge-large-cut-in-emissions-by-2030-co2-net-zero-2050.

[103] IEA, 'Special Report on Carbon Capture Utilisation and Storage', 23, 65.

[104] Ibid., 71–76.

Presently, CCUS-produced hydrogen is approximately half the cost of hydrogen production via renewable-powered electrolysis.[105]

Fourth, CCUS can facilitate carbon removal, direct or indirect, from the atmosphere. Various methods exist for atmospheric carbon removal, from nature-based solutions like afforestation to technological interventions anchored in CCUS. These technologies include 'negative carbon' bioenergy with carbon capture and storage (BECCS) and direct air capture (DAC).[106] The IEA notes that, while nature-based and technology-based approaches can complement one another, technology-based approaches often present advantages.[107] For instance, they are, unlike trees, resilient against fires that could release CO_2 into the atmosphere. Yet, views differ on the need for and viability of BECCS and DAC.[108] As a result, their carbon removal potentials, costs, and environmental impacts are largely undefined.[109]

9 Conclusion

The discussion in this chapter has shown that the development and implementation of CCUS technologies cannot be extricated from the pressing issue of anthropogenic climate change, which has resulted in a significant body of international political-legal architecture that is compelling a transition away from fossil fuel reliance in the energy sector. This momentum away from fossil fuels is reinforced by the set of ideas embodied by the JET, an interpretive concept that presupposes that the low-carbon transition of energy systems must be structured and implemented with reference to moral action and just outcomes. Set against this backdrop, it has been seen that the UN has established 17 SDGs, and that certain SDGs have particular relevance for oil and gas governance, contributing to the necessity for extensive and ongoing fossil fuel mitigation. Any

[105] Ibid., 23; see also IEA, 'Global Hydrogen Review 2021' (2021), 7, 20.

[106] IEA, 'Special Report on Carbon Capture Utilisation and Storage', 77–85.

[107] Ibid.

[108] Jocelyn Timperley, 'Climate Change: Can We Really Take CO2 Back Out the Air?' (*BBC*, 22 March 2023), www.bbc.com/future/article/20230321-ipcc-the-rising-need-for-negative-emissions.

[109] IPCC, 'Special Report: Global Warming of 1.5 °C' (2018) Ch.2; Elliott (2020), 28–31.

detailed assessment of the role and future of CCUS must take account of these issues, and be informed by them.

In the low-carbon energy transition, CCUS technologies are a crucial element within a portfolio of critical technological options aimed at reducing atmospheric CO_2 levels. Thus, CCUS can complement other decarbonisation measures.[110] There is no escaping the fact that if nations continue to rely on carboniferous fossil fuels, they are contributing to the very issue they are seeking to reverse: anthropogenic climate change. Nations must navigate the shift away from fossil fuels and CCUS cannot be used as an incentive to expand fossil fuel usage. While CCUS deployment may be appropriate to cut emissions in vital energy and industrial sectors, it does not relieve States of their obligations in the Paris Agreement to stabilise greenhouse gas emissions by reducing dependence on fossil fuels, thereby diminishing the need for CCUS in the first place. In all, CCUS offers one of the many critical technological solutions to put the global energy system on track for net-zero emissions. However, CCUS should not lead to a reduction of efforts needed to support energy-saving policies, renewable energy, and other safe and sustainable low-carbon technologies.

[110] Hon Chung Lau, 'The Role of Carbon Capture and Storage in the Energy Transition' (2021) 35 Energy Fuels 7364–7386.

CHAPTER 3

Developments and Evolution of CCUS Technologies: A Review

Cristina F. A. Rodrigues, Haline Rocha, Colombo Tasinari, and M. J. Lemos de Sousa

1 Introduction

The energy transition has undeniably been the main topic at international meetings in recent years due to its acknowledged role in mitigating climate change. Climate change, specifically global warming, is currently

C. F. A. Rodrigues (✉) · M. J. L. de Sousa
Fundação/Universidade Fernando Pessoa, Praça 9 de Abril, 349, 4249-004 Porto, Portugal
e-mail: cfrodrig@gmail.com

H. Rocha · C. Tasinari
IEE-USP, Institute of Energy and Environment, University of São Paulo, Av. Prof. Luciano Gualberto 1289, Sao Paulo, SP 05508-080, Brazil

C. F. A. Rodrigues · M. J. L. de Sousa
MARE-Centro de Ciência do Mar e Ambiente, URI Coimbra, Coimbra, Portugal

Academia das Ciências de Lisboa, Lisboa, Portugal

E. G. Pereira et al. (eds.), *Carbon Capture Utilization and Storage*, Sustainable Development Goals Series,
https://doi.org/10.1007/978-3-031-81272-9_3

considered to be closely related to CO_2 emissions, which are derived mainly from fossil fuel energy sources. Shifting energy production away from fossil fuels and towards carbon-neutral sources is the basic concept of the current energy transition scenario. Nevertheless, without adopting policies to achieve an equilibrium between the energy and climate demands, it is relentless to put together a worldwide strategy for the energy transition.

Eliminating fossil fuels from the global energy mix has been the focus of the so-called Green Corporations for the past few years. However, it is widely acknowledged that fossil fuels (oil, natural gas, and coal) will continue to be the primary energy sources over the next 50 years or longer.[1,2] This scenario will call for the development of advanced low-carbon fossil fuel energy technologies in addition to the standard strategy of avoiding and/or reducing CO_2 emissions. The First World Climate Conference, held in 1979, was the primary initiative officially developed to discuss international climate challenges. Over time, several measures, mainly political in nature, were discussed, such as the Kyoto Protocol in 1997.[3] However, it was only in 2015 (COP21), with the Paris Agreement, that a consensual position was achieved. The long-term goal of keeping the rising average temperature well below 2 °C, and ideally no more than 1.5 °C, over pre-industrial levels inspired the development of the Paris Climate Agreement, another global political action plan developed to put the world on track to prevent dangerous climate changes. To achieve this main goal, the following must be accomplished: (1) global emissions must peak as soon as possible, recognizing that this will take longer for developing countries, and (2) carbon neutrality must be achieved in the second half of this century.

In the COP28 held in 2023 in Dubai, "negotiators from nearly 200 countries recognized, for the first time, the need to transition away from

[1] Rodrigues, Cristina, Henrique Pinheiro, and Manuel Lemos de Sousa. 2022. Clean Energy Transition Challenge: The Contributions of Geology. In: Transitioning to Affordable and Clean Energy. Edited by Edwin C. Constable. Transitioning to sustainability series 7. Basel: MDPI, Switzerland, pp. 47–94.

[2] IEA. 2023. Annual Energy Outlook AEO2023. U.S. Energy Information Administration. 46p.

[3] UN. 1997. Kyoto Protocol to the United Nations Framework Convention on Climate Change. (Document FCCC/CP/1997/L.7/Add.1). Kyoto (Japan), 1–10 December 1997, third session. 24p.

fossil fuels", meaning that "we didn't turn the page on the fossil fuel era" but instead "is the beginning of the end".[4]

The concept of carbon neutrality is currently quite popular throughout the world due to the knowledge that both the global population growth and the need for energy supply are predicted to expand over the coming decades. Although the terms carbon neutral and carbon free are often confused, they refer to distinct aspects of climate action. The former relates to companies, processes, and products that can become carbon neutral when they calculate their CO_2 emissions and compensate for what they have produced via carbon offsetting initiatives. On the other hand, carbon-free products, services, and companies are those that do not generate any CO_2 emissions during the entire supply chain, including all the raw materials, logistics, and packaging. The carbon neutrally new approach implied the establishment of the carbon offset concept, together with carbon offset credit, carbon offset project, and carbon offset programme. Carbon offset refers to reductions or removals of CO_2 emissions or other greenhouse gases (GHGs) used to compensate for emissions made elsewhere. The carbon offset credit consists of a transferrable instrument certified by governments or independent certification bodies to represent an emission reduction of one metric tonne of CO_2 or an equivalent amount of other GHGs. A carbon offset project entails verified environmental conservation or protection activities that reduce, avoid, or remove CO_2 or GHG emissions from the atmosphere. Carbon offset programmes can range from international or governmental to independent non-governmental organizations (NGOs), which regulate carbon offsets under the Kyoto Protocol.

Globally, scientists predict that up to 10 $GtCO_2$ will need to be removed annually from the atmosphere by 2050, with an increased removal capacity of up to 20 $GtCO_2$ per year by 2100.[5,6] In this new

[4] UN. 2023. United Nations Framework Convention on Climate Change: Conference of the Parties Serving as the Meeting of the Parties to the Paris Agreement. (Document FCCC/PA/CMA/2023/L.17). United Arab Emirates, first global stocktake. 21p.

[5] National Academies of Sciences, Engineering, and Medicine. 2019. Negative Emissions Technologies and Reliable Sequestration: A Research Agenda. Washington, DC: The National Academies Press. 510p. https://doi.org/10.17226/25259

[6] Lebling, Katie, Eliza Northrop, Colin McCormick, and Elizabeth Bridgwater. 2022. Towards Responsible and Informed Ocean-Based Carbon Dioxide Removal: Research and Governance Priorities. Report. Washington: World Resources Institute. 100p.

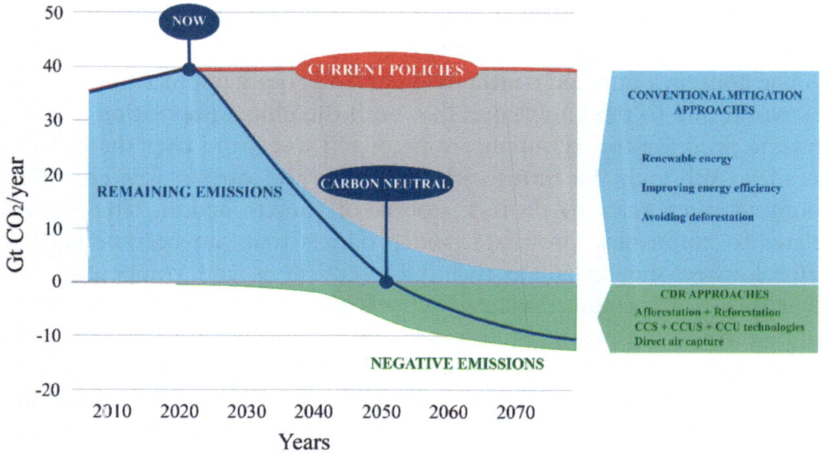

Fig. 1 CO_2 emissions scenarios (Adapted from Lebling et al. [2022][7])

context, it is clear that to reach carbon neutrality in the second half of this century, we would need more than just reducing CO_2 emissions, we will also need to deploy technologies to remove CO_2 from the atmosphere.

The conventional mitigation initiatives used in recent years to reduce emissions, such as improving energy efficiency, increasing renewable energy production, and avoiding deforestation, are of great significance, yet will not be sufficient to meet the climate goals established in the Paris Agreement (Fig. 1). As a result, carbon dioxide removal (defined as carbon removal or even CDR) technologies have emerged as the most realistic approach in all climate mitigation scenarios. Therefore, expanding the available options and capacity for carbon removal[8] is the new climate focus, which must include both natural approaches (such as trees/plants restoration and agricultural soil management) and more technological and engineered approaches (like carbon mineralization, underground reservoir sequestration, ocean storage, concrete products, and direct air capture). Although all CO_2 removal options need to be applied, reservoir geological sequestration is currently the solution that can contribute most to the goal of carbon neutrality (Table 1).

[7] See footnote 6.

[8] Ibid.

Table 1 CO_2 removal options

CDR option	*Carbon dioxide removal potential ($GtCO_2e$/year)*	
	Minimum	*Maximum*
Afforestation and Reforestation	0.01	14
Soil Carbon Sequestration	0.1	13
Biochar	0.03	1
Accelerated Weathering	0.001	18
Direct Air Capture	0.0004	16
Terrestrial BECCS	0.04	32
Aquatic BECCS	1.2	53
Ocean Fertilization	1	11
EOR "Cumulative Storage CO_2e"	0.05	370
Ocean Storage	0.01	0.9

Adapted from Martin et al. [2017][9]

In 2007, the European Commission Report had already pointed out that achieving the goal of Limiting Global Climate Change to 2 °C: the way ahead for 2020 and beyond[10] by reducing CO_2 emissions would be impossible without CCS. Finally, in 2015, with the Paris Agreement, CCUS technologies, and not only CCS, were identified as one of the potential solutions for achieving carbon neutrality.

2 History of CCS/CCUS/CUU Technologies

CCS technologies are often presented as a new technology to reduce CO_2 emissions. However, CCS technology is a powerful tool that applies a well-established technology in a new and innovative way. This technology existed long before what is now reflected. Since the 1920s, CO_2 capture

[9] Martin, Derek, Katelyn Johnson, Andrew Stolberg, Xilin Zhang, and Carissa DeYoung. 2017. Carbon Dioxide Removal Options: A Literature Review Identifying Carbon Removal Potentials and Costs. A project submitted in partial fulfillment of the requirements for the degree of Master of Science (Natural Resources and Environment) at the University of Michigan. 169p.

[10] EC. 2007. Communication from the Commission to the Council, the European Parliament, the European Economic and Social Committee and the Committee of the Regions. Commission of the European Communities, Brussels, 10.1.2007, COM(2007) 2 final. 13p.

has been employed to separate CO_2 from saleable methane (CH_4) gas in natural gas reservoirs. Scientists have known for decades that CO_2 could trap heat and warm the planet. Nonetheless, Callendar[11] was the first to link human activities to global warming, leading to the proposal of the first CO_2 capture plant in 1938.[12,13,14] The use of supercritical CO_2 as a solvent has long been acknowledged. Given that supercritical CO_2 is a solvent and is miscible with oil, injecting CO_2 into an oil reservoir (at high pressures and temperatures) reduces the interfacial tension between the two fluids, increases the viscosity of the CO_2, reduces capillary forces, and, finally, reduces the oil viscosity, allowing for an easier displacement of residual oil that would not otherwise be recovered.[15,16] The CO_2 injection used to improve oil recovery is now known as enhanced oil recovery (CO_2-EOR). In 1952, Whorton, Brownscombe, and Dyes of the Atlantic Refining Company granted the first patent for CO_2-EOR technology.[17]

The first large-scale project (Scurry Area Canyon Reef Operating Committee—SACROC) to inject CO_2 into underground geological structures was implemented in 1972 by Chevron in the oilfield (Permian Basin) in Scurry County (West Texas, USA). The injected CO_2 used in

[11] Callendar, G.S. 1938. The Artificial Production of Carbon Dioxide and Its Influence on Temperature. Quarterly Journal of the Royal Meteorological Society 64(275): 223–240.

[12] Hawkins, Ed, and Phil. D. Jones. 2013. On Increasing Global Temperatures: 75 years After Callendar. Quarterly Journal of the Royal Meteorological Society 139(677): 1961–1963. https://doi.org/10.1002/qj.2178.

[13] Anderson, Thomas R., Ed Hawkins, and Philip D. Jones. 2016. CO2, the Greenhouse Effect and Global Warming: From the Pioneering Work of Arrhenius and Callendar to Today's Earth System Models. Endeavour 40(3): 178–187.

[14] Anderson, Steven T. 2016. Cost Implications of Uncertainty in CO2 Storage Resource Estimates: A Review. Natural Resources Research 26(2): 137–159. https://doi.org/10.1007/s11053-016-9310-7.

[15] Parker, Michael E., James P. Meyer, and Stephanie R. Meadows. 2009. Carbon Dioxide Enhanced Oil Recovery Injection Operations Technologies (Poster Presentation). Energy Procedia 1(1): 3141–3148. https://doi.org/10.1016/j.egypro.2009.02.096.

[16] Ricky, Emanuel, Grant Charles Mwakipunda, Edwin E. Nyakilla, Naswibu A. Kasimu, Chao Wang, and Xingguang Xu. 2023. A Comprehensive Review on CO2 Thickeners for CO2 Mobility Control in Enhanced Oil Recovery: Recent Advances and Future Outlook. Journal of Industrial and Engineering Chemistry 126(696): 69–91. https://doi.org/10.1016/j.jiec.2023.06.018.

[17] Ibid.

the SACROC project was captured in several natural gas processing plants in the Val Verde area near the oilfield.[18,19,20,21,22]

However, Marchetti[23] first proposed the concept of geological storage of anthropogenic CO_2 emissions, as a greenhouse gas mitigation solution, in 1977, applying the entire chain CO_2 capture, transport, and storage strategy. For around 20 years, few studies have been done, but CCS technologies have gained legitimacy as a result of individual efforts by several researchers.[24,25,26,27,28] In 1996, the Sleipner Project[29] was developed

[18] Langston, M.V., S.F. Hoadley, and D.N. Young. 1988. Definitive CO2 Flooding Response in the SACROC Unit. SPE Enhanced Oil Recovery Symposium 6: 27–34. https://doi.org/10.2118/17321-ms.

[19] Denney, Dennis. 2007. Smart-Well Technology at the SACROC CO2 EOR Project. Journal of Petroleum Technology 59(08): 47–50. https://doi.org/10.2118/0807-0047-JPT.

[20] White, Don. 2009. Monitoring CO2 Storage During EOR at the Weyburn-Midale Field. The Leading Edge 28(7): 838–842. https://doi.org/10.1190/1.3167786.

[21] GCCSI. 2016. Global Status of CCS Special Report: Introducing to Industrial Carbon Capture and Storage. Melbourne, Australia: Global CCS Institute. 16p.

[22] Ma, Jinfeng, Lin Li, Haofan Wanga, Yi Du, Junjie Ma, Xiaoli Zhang, and Zhenliang Wang. 2022. Carbon Capture and Storage: History and the Road Ahead. Engineering 14(7): 33–43.

[23] Marchetti, Cesare. 1977. On Geoengineering and the CO2 Problem. Climatic Change 1(1): 59–68.

[24] Baes Jr., C.F., S.E. Beall, D.W. Lee, and G. Marland. 1980. The Collection, Disposal, and Storage of Carbon Dioxide. In: Interactions of Energy and Climate: Proceedings of an International Workshop Held in Münster, Germany, March 3–6, 1980. Edited by W. Bach, J. Pankrath and J. Williams. Springer Netherlands, pp. 495–519.

[25] Gunter, William D., Ernest H. Perkins, and Tom J. McCann. 1993. Aquifer Disposal of CO2-Rich Gases: Reaction Design for Added Capacity. Energy Conversion and Management 34(9–11): 941–948.

[26] Hendriks, C.A., and Kornelis Blok. 1993. Underground Storage of Carbon Dioxide. Energy Conversion and Management 34(9–11): 949–957.

[27] Bachu, Stefan, W.D. Gunter, and E.H. Perkins. 1994. Aquifer Disposal of CO2: Hydrodynamic and Mineral Trapping. Energy Conversion and Management 35(4): 269–279.

[28] Harrison, Wendy J., Richard F. Wendlandt, and E. Dendy Sloan. 1995. Geochemical Interactions Resulting from Carbon Dioxide Disposal on the Seafloor. Applied Geochemistry 10(4): 461–475.

[29] Celia, Michael A. 2017. Geological Storage of Captured Carbon Dioxide as a Large-Scale Carbon Mitigation Option. Water Resources Research 53(5): 3527–3533. https://doi.org/10.1002/2017WR020841.

as the first international scientific and large-commercial demonstration of CCS. Equinor, under the Norwegian CO_2 emissions tax policy,[30,31] was responsible for this project. During this time, Equinor extracted natural gas from a geological formation under the North Sea, which had a high CO_2 content. The CO_2 was separated during a natural gas purification process and subsequently injected into deep saline aquifers for storage, contributing to the reduction of CO_2 emissions.[32,33,34,35]

CCS technologies were considered an alternative in 1997 to achieve the United Nations Framework Convention on Climate Change's goals. The first climate change agreement, the Kyoto Protocol,[36] stated that reducing CO_2 emissions would require the implementation of several measures, including fuel switching, energy efficiency improvements, and the swapping of fossil fuels with renewable energies, but this would be insufficient. Back in 1997, the Kyoto Protocol stated that additional measures, such as CO_2 sequestration technologies, would be required, and that the best scenario would be a combination of all the described measures. The Kyoto Protocol motivated the use of CO_2 sequestration technologies in general terms, but it was unclear about CO_2 geological storage. Furthermore, the Kyoto Protocol did not address an essential issue regarding the allocation of credits when CO_2 sequestration takes

[30] Larsen, Bodil Merethe, and Runa Nesbakken. 1997. Norwegian Emissions of CO2 1987–1994: A Study of Some Effects of the CO2 Tax. Environmental and Resource Economics 9(3): 275–290.

[31] Christiansen, Atle Christer. 2001. Climate Policy and Dynamic Efficiency Gains: A Case Study on Norwegian CO2-Taxes and Technological Innovation in the Petroleum Sector. Climate Policy 1(4): 499–515. https://doi.org/10.3763/cpol.2001.0150.

[32] Kongsjorden, Helge, Olav Kårstad, and Tore A. Torp. 1998. Saline Aquifer Storage of Carbon Dioxide in the Sleipner Project. Waste management 17(5–6): 303–308.

[33] Gale, John, Niels Peter Christensen, Annette Cutler, and Tore A. Torp. 2001. Demonstrating the Potential for Geological Storage of CO2: The Sleipner and GESTCO Projects. Environmental Geosciences 8(3): 160–165.

[34] Torp, A., and John Gale. 2003. Demonstrating Storage of CO2 in Geological Reservoirs: The Sleipner and SACS Projects. In: Greenhouse Gas Control Technologies-6th International Conference. Edited by John Gale and Y. Kaya. Japan: Pergamon, pp. 311–316.

[35] Solomon, Semere. 2007. Carbon Dioxide Storage: Geological Security and Environmental Issues—Case Study on the Sleipner Gas Field in Norway. Bellona report. 126p.

[36] See footnote 3.

place in a country other than the one where it was generated.[37] Despite the Kyoto Protocol's unclear position, governments in major oil operator countries (Canada, USA, and Norway) invested significantly in research and development of CO_2 capture and storage technology during the 1990s and early 2000s.

During the first decade of 2000, several industrial-scale projects have been developed, as have several important smaller pilot-scale field experiments.[38] In that scenario and taking into consideration that the United Nations Framework Convention on Climate Change formally acknowledged the need for CO_2 geological storage to achieve the carbon neutrally by 2050 goal, the European Parliament and the European Council proposed a European Union Directive on Geological Storage of CO_2, the so-called Directive 2009/31/EC unveiled on 23 April 2009.[39] This directive was created to define a regulatory framework for geological sequestration/storage of CO_2, including the conditions for delivering storage permits and promoting the climate change mitigation goal. It was also pointed out that "this technology should not be serve as an incentive to increase the share of fossil fuel power plants".[40] CO_2 geological storage formations are typically chosen to have pressure and temperature conditions that allow the CO_2 to exist in so-called supercritical or liquid form, such as deep saline aquifers, mature, and depleted oil and gas reservoirs, and unmineable coal seams (Fig. 2).

[37] Torvanger, Asbjorn, Kristin Rypdal, and Steffen Kallbekken. 2005. Geological CO2 Storage as a Climate Change Mitigation Option. Mitigation and Adaptation Strategies for Global Change 10(4): 693–715.

[38] GCCSI. 2011. The Global Status of CCS: 2011. 220p.

[39] EC. 2009. Directive 2009/31/EC of the European Parliament and of the Council of 23 April 2009 on the geological storage of carbon dioxide. European Commission, Brussels. Available at: http://eur-lex.europa.eu/LexUriServ/LexUriServ.do?uri=OJ:L:2009:140:0114:0135:EN:PDF. Accessed 10 December 2012.

[40] Ibid., page 14.

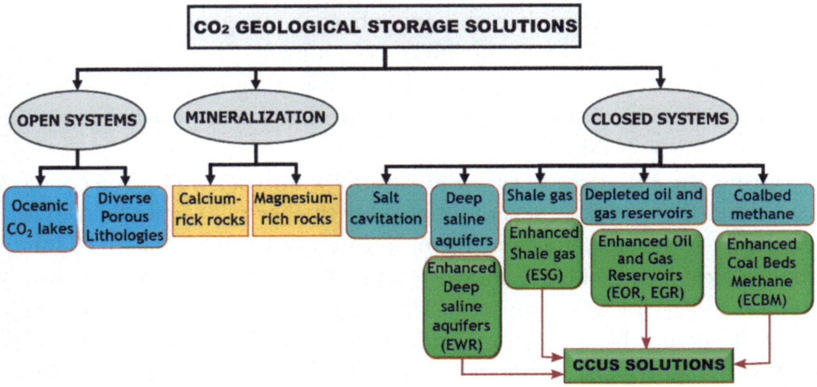

Fig. 2 CO_2 geological storage (Adapted from Lemos de Sousa and Rodrigues [2008][41])

CCS technologies were initially used in the pure sequestration approach and were mainly applied to deep saline aquifers. Later, in this CCS approach, it was expanded to mature and depleted oil and gas reservoirs. Nevertheless, several significant obstacles stand in the way of the expansion of CCS projects in deep saline aquifers. Some large-scale integrated projects have been put on hold or cancelled over the years for economic and policy reasons.[42] It is clear that those shortcomings need to be addressed before CCS projects can be scaled up and turned into an effective climate solution. Furthermore, it is undeniable that a policy support will provide industry confidence to continue investing in CCS technologies, ensuring ongoing innovation and, ultimately, helping to increase the technique's efficiency and reduce costs. Recognizing that

[41] Lemos de Sousa, Manuel João, and Cristina Fernanda Alves Rodrigues. 2008. "CO2 Abatement: State of Art". CIAGEB—global change, energy, environment and bioengineering RDID&D Unit. Porto, Fernando Pessoa University (Personal communication).

[42] See footnote 37.

solving these challenges will take time, but the urgency of CO_2 reduction is now a reality, alternative CO_2 abatement options[43,44] have been assessed (Fig. 3).

The concept of CCS technologies has evolved over the last few years, mainly driven by the need to find alternative ways of reducing CO_2 to help meet the 2050 carbon neutrality target. First of all, the concept of carbon capture encompasses a set of technological processes that trap or capture CO_2 from an emitting source. Currently, CO_2 can be captured from several industries, heat and power plants, and waste energy facilities, or it can be separated from biogas plants, or it can be captured directly from the air. In addition to the well-known CCS concept, two other concepts have emerged: Carbon Capture, Utilization, and Storage (CCUS), and Carbon Capture and Utilization (CCU). CCS technologies, the most widespread and technically demonstrated approach, include CO_2 capture, injection, and storage in an underground geological structure with the distinct goal of permanently storing CO_2. CCUS comprises methods and technologies for capturing CO_2 from an emitting source, recycling CO_2 for use, and evaluating safe and long-term storage options. This technical approach emerged to make it an economically viable option to compensate for the high costs of those technologies. However, enhanced oil recovery (EOR) is the only activity at scale that involves both utilization and storage. CCU technologies have recently gained a lot of attention in both the public and private sectors for capturing CO_2 and converting it, as a renewable carbon feedstock, into a potentially commercially viable product rather than permanently storing it. In fact, CCU technologies are those that capture and use CO_2; the long-term effects of storage are not the goal of this approach. In CCU, CO_2 can be used as a renewable product in biological or chemical conversion processes to manufacture or synthesize new products, such as chemicals, fuels, types of cement, and plastics, as well as in industrial processes as a solvent,

[43] EASAC (European Academies Sciences Advisory Council). 2013. Carbon Capture and Storage in Europe. German National of Sciences Leopoldina, Halle (Saale) (EASAC Policy Report 20).

[44] CSLF (Carbon Sequestration Leadership Forum). 2011. What Is Carbon Utilization? CCS in Focus: Carbon Capture and Storage. 7p.

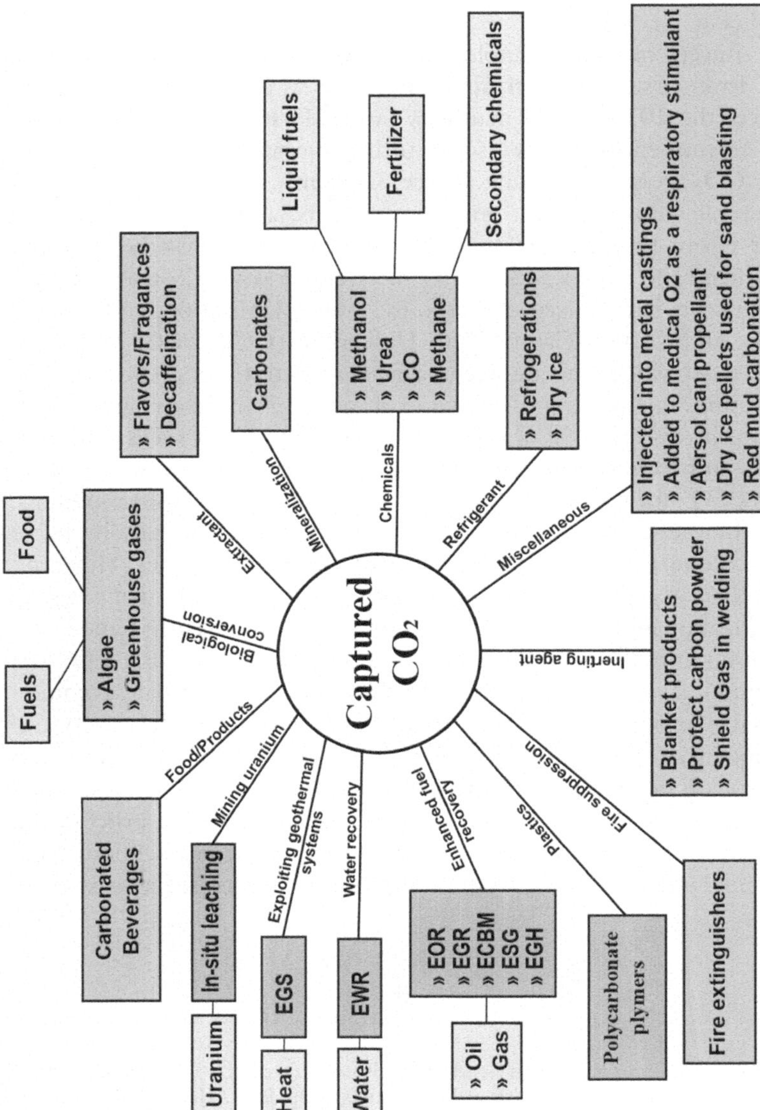

Fig. 3 Alternative CO_2 abatement options (Adapted from CSLF [2011][45])

[45] Ibid.

thereby reducing emissions to the atmosphere.[46,47] Thus, given their diverse applications, CUU technologies must be monitored and assessed individually to identify their different contributions to climate change mitigation. CCU does not permanently store CO_2. Instead, the CO_2 is temporarily kept out of the atmosphere and it will be re-emitted at the end of the product's life cycle, which can be a few days or years.[48,49,50]

Finally, it is critical to draw attention to how the three concepts outlined above are now being applied. Some authors[51] use the three concepts interchangeably as if they were synonymous, while another group[52] combines CCU and CCS, establishing that CCUS technologies have two options: (i) the CO_2 captured can be used in a range of applications, such as the production of plastics and chemicals, or (ii) it can be injected into geological formations for permanent and safe storage. In this chapter, CCUS technologies are used in the original accession, which states that these technologies consist of a set of processes involving CO_2 capture, compression (when necessary), transportation, utilization

[46] Kätelhön, Arne, Raoul Meys, Sarah Deutz, Sangwon Suh, and André Bardow. 2019. Climate Change Mitigation Potential of Carbon Capture and Utilization in the Chemical Industry. Proceedings of the National Academy of Sciences 116(23): 11187–11194. https://doi.org/10.1073/pnas.1821029116.

[47] Peres, Christiano B., Pedro M.R. Resende, Leonel J.R. Nunes, and Leandro C. de Morais. 2022. Advances in Carbon Capture and Use (CCU) Technologies: A Comprehensive Review and CO2 Mitigation Potential Analysis. Clean Technologies 4(4): 1193–1207. https://doi.org/10.3390/cleantechnol4040073.

[48] Abanades, J. Carlos, Eduward S. Rubin, Marco Mazzotti, and Howard J. Herzog. 2017. On the Climate Change Mitigation Potential of CO2 Conversion to Fuels. Energy & Environmental Science 10(12): 2491–2499. https://doi.org/10.1039/c7ee02819a.

[49] De Kleijne, Kiane, Steef V. Hanssen, Lester van Dinteren, Mark A. Huijbregts, Rosalie van Zelm, Heleen de Coninck. 2022. Limits to Paris Compatibility of CO2 Capture and Utilization. One Earth 5(2): 168–185.

[50] Thielges, Sonja, Barbara Olfe-Kräutlein, Alexander Rees, Joschka Jahn, Volker Sick, and Rainer Quitzow. 2022. Committed to Implementing CCU? A Comparison of the Policy Mix in the US and the EU. Frontiers in Climate 4: 943387. https://doi.org/10.3389/fclim.2022.943387.

[51] Hong, Wan Yun. 2022. A Techno-Economic Review on Carbon Capture, Utilisation and Storage Systems for Achieving a Net-Zero CO2 Emissions Future. Carbon Capture Science & Technology 3: 100044. https://doi.org/10.1016/j.ccst.2022.100044.

[52] Jeffery, Louise, Niklas Höhne, Mia Moisio, Thomas Day, and Benjamin Lawless. 2020. Options for Supporting Carbon Dioxide Removal. Discussion Paper. Germany: New Climate Institute. 27p.

to make the project economically viable, and, finally, permanent storage. As a result, geological formations are the only possibilities that meet all these criteria.

3 CCUS Technologies as a CO_2 Removal Option

Climate change mitigation's main goal is to reduce energy sector emissions by 80–100% between now and 2050, which will require the widespread deployment of low-carbon technologies.[53] To effectively address climate change, a diverse approach involving strategies and technologies will be needed. Furthermore, as the world's population continues to grow, the demand for energy supply is likely to increase significantly over the coming decades. As a result, carbon removal technologies will most likely be necessary due to the technological and economic challenges associated with reducing emissions in certain energy sectors. In this scenario, CO_2 removal technologies (Fig. 2) play a critical role in the carbon-neutral process, both in the short and long term. In the short term, CO_2 removal technology can compensate for residual emissions that cannot be reduced since abatement technologies are not yet available or are not cost-effective at scale. Given the long lifetime of CO_2 in the atmosphere, CO_2 removals will be required in the long term to lower the high concentration of CO_2 in the atmosphere caused by previous emissions.[54]

As previously stated, CCUS technologies are worldwide acknowledged as the most efficient carbon removal options, both technically and in terms of CO_2 abated.[55] The CCUS process chain encompasses four main steps: selecting the CO_2 extraction/capture method, identifying

[53] See footnote 5.

[54] IPCC. 2023. Sections. In: Climate Change 2023: Synthesis Report. Contribution of Working Groups I, II and III to the Sixth Assessment Report of the Intergovernmental Panel on Climate Change. Edited by The Core Writing Team, Hoesung Lee and José Romero. Geneva, Switzerland, pp. 35–115. https://doi.org/10.59327/IPCC/AR6-9789291691647.

[55] IPCC. 2018. Global Warming of 1.5 °C. An IPCC Special Report on the impacts of global warming of 1.5 °C above pre-industrial levels and related global greenhouse gas emission pathways, in the context of strengthening the global response to the threat of climate change, sustainable development, and efforts to eradicate poverty. Edited by Working Group I Technical Support Unit. Special Report. Cambridge, United Kingdom: Cambridge University Press. 616p. https://doi.org/10.1017/9781009157940.

the CO_2 transportation method, establishing the CO_2 utilization process, and finally selecting the CO_2 geological storage reservoir. Other intermediate processes are required to effectively complete the CCUS process, such as identifying the CO_2 source, deciding (if necessary) the CO_2 cleaning/purification technique, and establishing (when needed) the CO_2 compressing procedure.[56]

The purpose of this chapter is to stress the various geological storage alternatives rather than to detail capture technologies and transportation options throughout the CCUS technology chain. However, some general features will be addressed.

3.1 CO_2 Capture Technologies

The CO_2 capture process is highly complex and traditionally quite expensive, and it is closely related to CO_2 properties, such as solubility, density, reactivity, heat capacity, and pressure sensitivity. CO_2 is highly soluble in water and other solvents, but at reservoir conditions exceeding the critical temperature and pressure, supercritical CO_2 solubility in water is extremely low.[57] CO_2 is denser than air and other gases, and while it is stable, it can react with specific chemicals. For example, the interaction of water-saturated supercritical CO_2 with small sulphur dioxide (SO_2) and oxygen (O_2) contents resulted in mineral corrosion and/ or mineral precipitation. CO_2 has a low heat capacity, which increases as pressure increases while decreasing mobility at constant temperatures. CO_2 is extremely sensitive to pressure fluctuations, which may affect its solubility and density.[58,59]

[56] Davoodi, Shadfar, Mohammed Al-Shargabi, David A. Wood, Valeriy S. Rukavishnikov, and Konstantin M. Minaev. 2023. Review of Technological Progress in Carbon Dioxide Capture, Storage, and Utilization. Gas Science and Engineering 117: 205070. https://doi.org/10.1016/j.jgsce.2023.205070.

[57] Borhani, Tohid N., and Meihong Wang. 2019. Role of Solvents in CO2 Capture Processes: The Review of Selection and Design Methods. Renewable and Sustainable Energy Reviews 114: 109299. https://doi.org/10.1016/j.rser.2019.109299.

[58] Zhang, Yingying, Xiaohua Lu, and Xiaoyan Ji. 2019. Carbon Dioxide Capture. Deep Eutectic Solvents 15: 297–319. https://doi.org/10.1002/9783527818488.ch15.

[59] Osman, Ahmed I., Mahmound Hefny, M.I.A. Abdel Maksoud, Ahmed M. Elgarahy, and David W. Rooney. 2021. Recent Advances in Carbon Capture Storage and Utilisation Technologies: A Review. Environmental Chemistry Letters 19(1): 797–849. https://doi.org/10.1007/s10311-020-01133-3.

CO_2 capture technologies have been extensively explored and developed in recent years, and different options are now technically feasible, including industrial separation, post-combustion, pre-combustion, oxy-fuel combustion, chemical looping combustion, and direct air capture.[60] Nevertheless, for the power industry, available CCS/CCUS technologies include post-combustion capture, pre-combustion capture, and oxy-fuel combustion capture.

The post-combustion carbon capture involves the separation and capture of CO_2 from a flue gas (mixture of nitrogen (N_2), CO_2, water (H_2O), O_2, sulphur oxide (SO_x), nitrogen oxide (NO_x), and heavy metals among others) produced from the combustion of fossil fuels (coal, natural gas, or oil) with normal air. This technology's key advantage is that it does not require any changes to the combustion system. The post-combustion CO_2 capture efficiency is assumed to be 90% vol, but the energy consumption is high because the partial pressure of CO_2 in typical flue gases is low (Table 2).[61,62]

Table 2 CO_2 removal efficiency and energy consumption of CO_2 capture of different CO_2 capture technologies

CO_2 capture technology	*CO_2 removal efficiency (% vol.)*	*Energy consumption for CO_2 capture (GJ/t CO_2)*
Industrial Separation	90	5.00
Post-Combustion	90	4.14
Pre-Combustion	90	3.35
Oxy-Fuel Combustion	>90	4.05
Chemical Looping Combustion (CLC)	96 to 99	0.95
Direct Air Capture (DAC)	85 to 93	5.25

Adapted from Hong [2022][63]

[60] See footnote 50.

[61] Chao, Cong, Yimin Deng, Raf Dewil, Jan Baeyens, and Xianfeng Fan. 2021. Post-combustion Carbon Capture. Renewable and Sustainable Energy Reviews 138: 110490. https://doi.org/10.1016/j.rser.2020.110490.

[62] Zanco, Stefano E., José-Francisco Pérez-Calvo, António Gasós, Beatrice Cordiano, Viola Becattini, and Marco Mazzotti. 2021. Postcombustion CO2 Capture: A Comparative Techno-Economic Assessment of Three Technologies Using a Solvent, an Adsorbent, and a Membrane. ACS Engineering Au 1: 50–72. https://doi.org/10.1021/acsengineeringau.1c00002.

[63] See footnote 50.

The pre-combustion carbon capture technique focuses mainly on extracting CO_2 from fossil fuels before the combustion processes generate energy.[64] In this context, CO_2 is a by-product that is being removed. The pre-combustion process is essentially the same for coal, oil, and natural gas. However, when coal or oil are used, additional stages of gas purification are required to remove impurities, such as ash and sulphur compounds.[65,66,67] The CO_2 concentration captured by the pre-combustion process is quite high (90% vol), and the separation method is simple to implement, resulting in low energy consumption (Table 2).

The oxy-fuel combustion carbon capture involves nearly pure oxygen (95%–99%), extracted from the air at very high temperatures (about 1,300 °C to 1,400 °C), which is used as an oxidant additive to the combustion system to produce flue gas. The flue gases produced from the combustion process are mainly represented by CO_2 and H_2O vapour with only trace amounts of other gases (such as SO_2 and NO_x). This technology can capture more than 90% vol of CO_2, and the energy consumption associated with the CO_2 separation is relatively low, but the additional energy consumption related to the oxygen generation system must be reflected in the overall process (Table 2).[68,69]

[64] GCCSI. 2012. CO2 Capture Technologies. Post Combustion Capture (PCC). January 2012. 16p.

[65] Theo, Wai Lip, Jeng Shiun Lim, Haslenda Hashim, Azizul Azri Mustaffa, and Wai Shin Ho. 2016. Review of Pre-combustion Capture and Ionic Liquid in Carbon Capture and Storage. Applied Energy 183(C): 1633–1663. https://doi.org/10.1016/j.apenergy.2016.09.103.

[66] Valiani, Saba, Nassim Tahouni, and M. Hassan Panjeshahi. 2017. Optimization of Pre-combustion Capture for Thermal Power Plants Using Pinch Analysis. Energy 119(C): 950–960. https://doi.org/10.1016/j.energy.2016.11.046.

[67] Chen, Zexiong. 2022. A Review of Pre-combustion Carbon Capture Technology. In: Proceedings of the 2022 7th International Conference on Social Sciences and Economic Development (ICSSED 2022). Edited by Z.P. Xu and J.D. Bi. Advances in Economics, Business and Management Research: Atlantis Press, vol. 215, pp. 524–528. https://doi.org/10.2991/aebmr.k.220405.086.

[68] Nemitallah, Medhat A., Mohamed A. Habib, Hassan M. Badr, Syed A. Said, Aqil Jamal, Rached Ben-Mansour, Esmai M.A. Mokheimer, and K. Mezghani. 2017. Oxy-Fuel Combustion Technology: Current Status, Applications, and Trends. International Journal of Energy Research 41(12): 1670–1708. https://doi.org/10.1002/er.3722.

[69] Yadav, Sujeet, and S.S. Mondal. 2022. A Review on the Progress and Prospects of Oxy-Fuel Carbon Capture and Sequestration (CCS) Technology. Fuel 308(4): 122057.

3.2 CO_2 Transport Technologies

The CO_2 transportation is a crucial part of the CCUS supply chain, connecting CO_2 sources/capture locations and storage sites. Pipelines, ships, trucks, or rails are the currently available methods of transportation (Table 3), depending on the location of the storage sites, the distance between the CO_2 sources and the storage sites, the amount of CO_2 to be transported, and the composition of the captured and purified CO_2 gas.[70,71,72,73] CO_2 can be transported in different forms, including gaseous phase, liquid phase, dense phase, supercritical phase, and solid phase. Additionally, the first four CO_2 methods are more suited to large-scale transportation. However, gaseous and liquid transports are typically employed for short distances, while dense-phase and supercritical transports are suitable for long distances.[74]

Pipeline transport is the most common and mature method of CO_2 transport in large-scale CCUS projects. Until now, it has been reported that CO_2 transport by pipelines could be done over long distances on land (>2,500 km), as well as in the sea (up to 2.2 km deep). CO_2 transport pipelines require pressures between 86 and 152 bar to avoid two-phase flow regimes and CO_2 density increase, and temperatures between 13 °C and 43 °C to prevent damage to the external pipe coating and to avoid

[70] Wang, Haixia, Jusheng Chen, and Qingling Li. 2019. A Review of Pipeline Transportation Technology of Carbon Dioxide. IOP Conference Series: Earth and Environmental Science 310(3): 032033. https://doi.org/10.1088/1755-1315/310/3/032033.

[71] Nazeri, Mahmoud, Hooman Haghighi, Craig Mckay, Dale Erickson, and Suling Zhai. 2021. Impact of CO2 Specifications on Design and Operation Challenges of CO2 Transport and Storage Systems in CCUS. Paper Presented at the SPE Offshore Europe Conference & Exhibition, Virtual, September 2021. Paper Number: SPE-205472-MS. https://doi.org/10.2118/205472-MS.

[72] Simonsen, Kenneth René, Dennis Severin Hansen, and Simon Pedersen. 2024. Challenges in CO2 Transportation: Trends and Perspectives. Renewable and Sustainable Energy Reviews 191(C): 114149. https://doi.org/10.1016/j.rser.2023.114149.

[73] Zhou, Yi, Fei Xie, Dan Wang, Yuxin Wang, and Ming Wu. 2024. Carbon Capture, Utilization and Storage (CCUS) Pipeline Steel Corrosion Failure Analysis: A Review. Engineering Failure Analysis 155: 107745.

[74] Lu, Hongfang, Xin Ma, Kun Huang, Lingdi Fu, and Mohammadamin Azimi. 2020. Carbon Dioxide Transport Via Pipelines: A Systematic Review. Journal of Cleaner Production 266: 121994. https://doi.org/10.1016/j.jclepro.2020.121994.

Table 3 CO_2 transportation alternatives

Transportation method	*Operating conditions (Pressure and Temperature)*	*Phase*	*Capacity*
Pipeline	86 bar and 152 bar 12 °C and 43 °C	Vapour, dense phase	~100 Mt CO_2/year 6500 km of pipeline transport in operation
Ship	17 bar −40 °C	Liquid	>70 Mt CO_2/year
Truck	20 bar −20 °C	Liquid	>1 Mt CO_2/year
Railway	20 bar −20 °C	Liquid	>3 Mt CO_2/year

Adapted from Baroudi et al. [2021],[75] Kearns et al. [2021],[76] Hong [2022],[77] McKaskle et al. [2022][78]

effect on the metal used to construct the pipelines at very low temperatures.[79,80] Transporting CO_2 over its critical pressure (73.8 bar) and above its critical temperature (31.1 °C) results in a supercritical phase, while transporting CO_2 above the critical pressure but below the critical temperature results in a dense phase.[81]

CO_2 ship transport provides safe, reliable, and flexible CO_2 transportation well-suited for long distances and low medium volumes. Ship transport is now considered an alternative option for large-scale transport for large onshore distances or overseas, where CO_2 pipelines are unavailable or are not technically or economically feasible. In a ship carrier, CO_2

[75] See footnote 78.

[76] See footnote 75.

[77] See footnote 50.

[78] See footnote 82.

[79] Onyebuchi, Victor E., Athanasios Kolios, Dawid P. Hanak, Chechet Biliyok, and Vasilije Manovic. 2018. A Systematic Review of Key Challenges of CO2 Transport via Pipelines. Renewable and Sustainable Energy Reviews 81(2): 2563–2583. https://doi.org/10.1016/j.rser.2017.06.064.

[80] Kearns, David, Harry Liu, and Chris Consoli. 2021. Technology Readiness and Costs of CCS. Global CCS Institute, 50p.

[81] Ibid.

is transported as a liquid at low temperatures to retain its high density and pressure rates while minimizing CO_2 volume. Thus, the CO_2 is transported at pressures around 17 bar and temperatures of—40 °C.[82] One factor that can overestimate the conventional costs of ship transport compared to pipeline transport is the liquefaction facility always required to reduce CO_2 volume.[83]

CO_2 truck and railway transports are flexible and reliable methods for small-scale CO_2 transportation, from 4 tonne to 30 hundred ton, over short distances, usually limited to 332 km for trucks and 1,609 km for railways.[84,85] Truck and railway tankers transport CO_2 at pressures and temperatures of 20 bar and −20 °C, respectively.[86] Both trucks and railway tankers are insulated to keep the temperature stable and prevent evaporative losses.[87] Nonetheless, it should be noted that CO_2 truck and railway transports are not been considered in CCUS projects due to key constraints, such as route options, dangerous substance transport, and potential disturbance to local populations.[88]

3.3 CO_2 Utilization and Storage Technologies

As previously stated, this document addresses the CO_2 storage topic from the standpoint of CO_2 being stored permanently. Thus, the present work's main focus consists of underground geological storage options.

[82] See footnote 50.

[83] Baroudi, Hisham Al, Adeola Awoyomi, Kumar Patchigolla, Kranthi Jonnalagadda, and E.J. Anthony. 2021. A Review of Large-Scale CO2 Shipping and Marine Emissions Management for Carbon Capture, Utilisation and Storage. Applied Energy 287(46): 116510.

[84] See footnote 50.

[85] Stolaroff, Joshuah K., Simon H. Pang, Wenqin Li, Whitney G. Kirkendall, Hannah M. Goldstein, Roger D. Aines, and Sarah E. Baker. 2021. Transport Cost for Carbon Removal Projects with Biomass and CO2 Storage. Frontiers in Energy Research 9: 639943. https://doi.org/10.3389/fenrg.2021.639943.

[86] See footnote 50.

[87] McKaskle, Ray, Carrie Beitler, Katherine Dombrowski, and Kevin Fisher. 2022. The Engineer's Guide to CO2 Transportation Options. Proceedings of the 16th Greenhouse Gas Control Technologies Conference (GHGT-16) 23–24 October 2022. 15p. Available at SSRN: https://ssrn.com/abstract=4278858 or http://dx.doi.org/10.2139/ssrn.4278858.

[88] See footnote 50.

Geological sequestration of CO_2 can be defined as the systematic storage of significant amounts of CO_2 captured from fossil fuels combustion or processing or directly from the air in stable, deep, porous, and "highly" permeable geological formations for long-term storage, preventing its release into the atmosphere. The most interesting geological formations are depleted hydrocarbon fields (oil and natural gas), deep saline aquifers, unmineable coal seams, basaltic/igneous formations, organic-rich shales, and salt caverns. However, to clearly understand a CO_2 underground geological reservoir, it is crucial to understand the CO_2 underground behaviour.[89,90,91] CO_2 in underground geological storage exhibits several properties that cannot be overlooked, including density, solubility, viscosity, corrosion, and pressure and temperature sensitivity. According to Bachu,[92] CO_2's high density allows it to respond to gravitational forces, causing it to sink and spread more quickly than other gases. CO_2 can dissolve in water and react with specific minerals found in many types of porous rock reservoirs, affecting their long-term stability. CO_2 has a lower viscosity than water, which might influence fluid flow behaviour in underground reservoirs, and it can sometimes form a distinct phase from water.[93] The effect of CO_2 corrosion must be assessed using two different approaches. The first is the acidic nature of CO_2, which can cause corrosion in some materials used in storage-site infrastructure, such

[89] Zheng, Huanda, Yi Zhong, Zhiping Mao, and Laijiu Zheng. 2018. CO2 Utilization for the Waterless Dyeing: Characterization and Properties of Disperse Red 167 in Supercritical Fluid. Journal of CO2 Utilization 24: 266–273. https://doi.org/10.1016/j.jcou.2018.01.014.

[90] Ali, Muhammad, Nilesh Kumas Jha, Nilanjan Pal, Alireza Keshavarz, Hussein Hoteit, and Mohammad Sarmadivaleh. 2022. Recent Advances in Carbon Dioxide Geological Storage, Experimental Procedures, Influencing Parameters, and Future Outlook. Earth-Science Reviews 225(14): 103895. https://doi.org/10.1016/j.earscirev.2021.103895.

[91] Liu, Guizhen, Bofeng Cai, Qi Li, Xian Zhang, and Tao Ouyang. 2022. China's Pathways of CO2 Capture, Utilization and Storage Under Carbon Neutrality Vision 2060. Carbon Management 13(1): 435–449. https://doi.org/10.1080/17583004.2022.2117648.

[92] Bachu, Stefan. 2008. CO2 Storage in Geological Media: Role, Means, Status and Berries to Deployment. Progress in Energy and Combustion Science 34(2): 254–273.

[93] Mojtaba, Seyyedi, Rostami Behzad, Nazari Moghaddam Rasoul, and Rezai Mohammad. 2014. Experimental Study of Density-Driven Convection Effects on CO2 Dissolution Rate in Formation Water for Geological Storage. Journal of Natural Gas Science and Engineering 21(11): 600–607. https://doi.org/10.1016/j.jngse.2014.09.020.

as pipelines and wellbore casing.[94] The second approach focuses on the impact of CO_2 brine-rock interactions on the rock reservoir. It can cause mineral grain corrosion through mineral dissolution or precipitation in the pore space of rock reservoirs. In both previous situations, the rock's physical properties will change.[95] Additionally, it is known that CO_2 changes its state from solid to liquid, gas, or supercritical, depending on the temperature and pressure. Thus, CO_2 remains gaseous below 0.5 MPa and changes to liquid above this pressure. CO_2 coexists in three states at 56.56 °C and 0.5 MPa (triple point). CO_2 reaches its critical point at 31.10 °C and 7.38 MPa. Above this temperature and pressure, it changes into a dense liquid state, acting as a supercritical state.[96]

Nevertheless, CO_2 geological storage can encompass significant investment due to infrastructure requirements, such as pipelines, compressors, pumps, and other facilities, especially if developed on a large scale involving sites and a sizable infrastructure. So, it is very likely that there will be high costs associated with the implementation of this technology on a large scale. Thus, CCS technologies, which initially focused only on the permanent storage of CO_2, have undergone a conceptual change to CCUS technologies, which implies introducing the "utilization" component before storing the CO_2, so that costs are minimized. The extraction of hydrocarbons (oil and natural gas), water, or minerals takes place simultaneously with the storage of CO_2 (Table 4), which can significantly lower the cost threshold for CO_2 geological storage from a climate policy perspective.

The CO_2-EOR and CO_2-EGR technologies have been developed to enhance oil and gas recovery from depleted conventional reservoirs. They are technically mature due to the knowledge accumulated over several decades in oil and natural gas production. Coalbed methane and shale gas have been recognized as excellent unconventional gas reservoirs since the

[94] Kalam, Shams, Teslim Olayiwola, Mohammed M. Al-Rubaii, Benedict I. Amaechi, Mohammad S. Jamal, and Abeeb A. Awotunde. 2021. Carbon Dioxide Sequestration in Underground Formations: Review of Experimental, Modeling, and Field Studies. Journal of Petroleum Exploration and Production Technology 11(2): 303–325. https://doi.org/10.1007/s13202-020-01028-7/978-3-319-27019-7_7.

[95] Wdowin, Magdalena, and Wojciech Franus. 2016. Determination of CO2-Brine-Rock Interactions for Carbon Dioxide Sequestration Using SEM–EDS Methods. Geologic Carbon Sequestration 119–133. https://doi.org/10.1007/978-3-319-27019-7_7.

[96] See footnote 89.

Table 4 CO_2 geological storage and utilization technologies

Name of technology	*Geological type*	*Use type*
CO_2-EOR	Depleted oil reservoirs	Enhanced oil recovery
CO_2-ECBM	Deep unmineable coal seams	Enhanced coalbed methane recovery
CO_2-EGR	Depleted natural gas reservoirs	Enhanced natural gas recovery
CO_2-ESG	Organic-rich shales	Enhanced shale gas recovery
CO_2-EGS	Geothermal systems	Exploiting geothermal resources
CO_2-EWR	Deep saline aquifers	Water recovery
$CO_2 + O_2$ In situ uranium leaching	Uranium ore deposits	Mining uranium

Adapted from Davoodi et al. [2023][97]

1990s. However, due to their typically low permeabilities, closely related to their high organic content and subsequent adsorption mechanisms, CO_2-ECBM and CO_2-ESG technologies have only been considered as economic options for CCUS technologies since the turn of the century, with the implementation of multi-lateral horizontal wells. The remaining four options (Table 4) are not as well-established technically or economically, but they are already starting to gain ground in the CCUS field as more studies and tools continue to emerge.

3.3.1 CO_2-EOR

CO_2-EOR has become an important option for the productive utilization of CO_2 emissions. CO_2 is used to enhance oil recovery in mature fields, extending their life cycle. This process involves injecting CO_2 to pressurize oil reservoirs to facilitate additional oil recovery. Depending on reservoir characteristics, primary production can result in recovering up to 20% of the original oil in place. Secondary recovery, also known as water flooding, has the potential to recover an additional 15% to 20% of the original oil remaining in place. The water used in the secondary stage is typically the water produced from the reservoir, which is recycled back into the reservoir. As a result, after the primary and secondary

[97] See footnote 55.

recovery stages, a significant amount (approximately 65%) of the original oil in place remains in the reservoir, because the oil has been bypassed by the water, which does not mix with the oil.[98] CO_2-EOR is a type of tertiary oil recovery that could potentially recover a further 10% to 20% of the original oil in place.[99,100,101] Typically, the EOR response is around 0.6, which means that for every tonne of CO_2 injected, 0.6 tonnes of additional oil are recovered.[102]

The CO_2-EOR process changes the physical properties of the oil to enable higher production by improving volumetric sweep and displacement efficiency. This process is mainly influenced by the CO_2 interaction with the oil, which is determined by its miscibility properties. In EOR operations, the CO_2 can be used in both miscible and immiscible states, depending on reservoir pressure, temperature, and oil properties. The ability of CO_2 to become miscible in oil and form a homogeneous liquid, allowing a single-phase flow to develop, is determined by the minimum miscibility pressure (MMP). However, there are three types of miscibility between the CO_2 and the reservoir oil: (1) first contact process, (2) vaporization gas-drive process or high-pressure gas drive, and (3) condensation

[98] Melzer, L. Stephen. 2012. Carbon Dioxide Enhanced Oil Recovery (CO2 EOR): Factors Involved in Adding Carbone Capture, Utilization and Storage (CCUS) to Enhanced Oil Recovery. CO2 Consultant and Annual CO2 Flooding Conference Director. Midland. 17p.

[99] Gozalpour, F., S.R. Ren, and B. Tohidi. 2005. CO2 Eor and Storage in Oil Reservoir. Oil & Gas Science and Technology 60(3): 537–546. https://doi.org/10.2516/ogst:2005036.

[100] Manrique, E.J., C.P. Thomas, R. Ravikiran, M. Izadi Kamouei, M. Lantz, J. Romero, and V. Alvarado. 2010. EOR: Current Status and Opportunities. SPE Improved Oil Recovery Symposium. Tulsa: Oklahoma, USA. Pper Number: SPE-130113-MS. https://doi.org/10.2118/130113-ms.

[101] Elmabrouk, Saber Kh., Husen E. Bader, and Walid Mohamed Mahmud. 2017. An Overview of Power Plant CCS and CO2-EOR Projects. Proceedings of the International Conference on Industrial Engineering and Operations Management. Rabat: Marocco. 12p.

[102] See footnote 36.

gas-drive process or enriched gas drive.[103,104,105] At and above the MMP, the dynamic miscibility between the CO_2 and the oil is achieved with time as displacements take place through single or multiple contacts. The intermediate and higher molecular weight hydrocarbons in oil are vaporized into the CO_2 (vaporization gas-drive process) and part of the injected CO_2 is dissolved in the oil (condensation gas-drive process). Below the MMP, CO_2 is immiscible in oil, but the CO_2 will still dissolve in the oil, causing a swelling effect on the oil, reducing its viscosity and interfacial tension. In general, the miscible state of CO_2 in oil will result in higher oil recovery than the immiscible state.[106]

The main goal of screening criteria for CO_2-EOR is to identify a suitable storage site and develop a profitable allocation plan for CCUS projects. The study by Shaw and Bachu[107] was one of the first consistent efforts to define CO_2-EOR screening criteria in oil reservoirs. Initially, screening criteria were developed for CO_2 miscible floods, which were generally considered more promising, offering higher incremental oil production compared to CO_2 immiscible floods. However, it was important not to underestimate the potential of CO_2 immiscible floods, so different criteria were established for both CO_2 miscible and CO_2 immiscible floods (Table 5). Screening parameters are mainly related to reservoir rock and fluid properties. The following parameters were proposed by

[103] GCCSI. 2013. Technical Aspects of CO2 EOR and Associated Carbon Storage. 14p.

[104] Verma, Mahendra K. 2015. Fundamentals of Carbon Dioxide-Enhanced Oil Recovery (CO2-EOR)—A Supporting Document of the Assessment Methodology for Hydrocarbon Recovery Using CO2-EOR Associated with Carbon Sequestration. Open-File Rep. 2015-1071. United States Geological Survey, Reston, Virginia. 19p.

[105] El-hoshoudy, Abdelaziz Nasr, and Saad Desouky. 2018. CO2 Miscible Flooding for Enhanced Oil Recovery. In: Carbon Capture, Utilization and Sequestration. Edited by Ramesh K. Agarwal. IntechOpen, Chapter 5, pp. 79–93. https://doi.org/10.5772/intechopen.79082.

[106] Ajayi, Temitope, Jorge Salgado Gomes, and Achinta Bera. 2019. A Review of CO2 Storage in Geological Formations Emphasizing Modelling, Monitoring and Capacity Estimation Approaches. Petroleum Science 16(5–6): 1028–1063.

[107] Shaw, J., and S. Bachu. 2002. Screening, Evaluation, and Ranking of Oil Reservoirs Suitable for CO2-Flood EOR and Carbon Dioxide Sequestration. Journal of Canadian Petroleum Technology 41(09). https://doi.org/10.2118/02-09-05.

Taber et al.[108]: oil gravity, oil viscosity, oil composition, oil saturation, formation type, net thickness, average permeability, depth, and temperature. Additionally, the reservoir pressure at the beginning of the CO_2 flood was recommended by Rivas et al.,[109] and porosity was included by Al Adasani and Bai.[110] Subsequent studies[111,112] suggested combining traditional screening criteria with statistical analysis and machine learning techniques to assist engineers in the selection of the most suitable EOR methods based on parameter similarities. More recently, Rosiani et al.[113] proposed an interdependency between screening parameters, indicating that they influence and are related to each other.

Once the oil reservoirs have been screened for the CO_2-EOR candidates, the next step is to develop a design for the optimal recovery efficiency of the flooding process. A critical task involves selecting CO_2 flooding and injection methods, which depend on reservoir geology, fluid and rock properties, timing relative to water flooding, and well configuration.[114] Several injection methods have been proposed, including continuous CO_2 injection, water-alternating-gas (WAG) injection, carbonated water injection (CWI), CO_2 foam injection, and cyclic (Huff n' Puff)

[108] Taber, J.J., F.D., Martin, and R.S. Seright. 1997. EOR Screening Criteria Revisited—Part 1: Introduction to Screening Criteria and Enhanced Recovery Field Projects. SPE Reservoir Engineering 12(03): 189–198. https://doi.org/10.2118/35385-pa.

[109] Rivas, O., S. Embid, and F. Bolivar. 1994. Ranking Reservoirs for Carbon Dioxide Flooding Processes. SPE Advanced Technology Series 2(01): 95–103. https://doi.org/10.2118/23641-pa.

[110] Al Adasani, Ahmad, and Baojun Bai. 2011. Analysis of EOR Projects and Updated Screening Criteria. Journal of Petroleum Science and Engineering 79(1–2): 10–24. https://doi.org/10.1016/j.petrol.2011.07.005.

[111] Anguita, Davide, Alessandro Ghio, N. Greco, Luca Oneto, and Sandro Ridella. 2010. Model Selection for Support Vector Machines: Advantages and Disadvantages of the Machine Learning Theory. The 2010 International Joint Conference on Neural Networks (IJCNN). 8p. https://doi.org/10.1109/ijcnn.2010.5596450.

[112] Leon, Florin, Sabina-Adriana Floria, and Costin Bădica. 2017. Evaluating the Effect of Voting Methods on Ensemble-Based Classification. 2017. IEEE International Conference on Innovations in Intelligent Systems and Applications (INISTA). Gdynia: Poland. 6p. https://doi.org/10.1109/inista.2017.8001122.

[113] Rosiani, Diyah, Asep Kurnia Permadi, Hasian Parlindungan Siregar, Agus Yodi Gunawan, and Tutuka Ariadji. 2022. A New CO2-EOR Methods Screening Model Based on Interdependency Parameters. Applied Sciences 12(8): 3937. https://doi.org/10.3390/app12083937.

[114] See footnote 103.

Table 5 Screening criteria for CO_2 miscible flooding and immiscible flooding in CO_2-EOR projects

Criteria	*CO_2 miscible flooding*	*CO_2 immiscible flooding*
Depth (m)	488–4074	350–2590
Temperature (°C)	27–127	28–92
Original pressure	>MMP	<MMP
Permeability (mD)	>5	>0.1
Oil gravity (°API)	22–45	11–35
Viscosity (cP)	<10	0.6–592
Fraction of oil remaining (%)	>20	30–70
Porosity (%)	3–37	>30
Net thickness (m)	<40	10–20

Adapted from Bachu [2016],[115] Sun et al. [2017],[116] Wang et al. [2023][117]

[115] Bachu, Stefan. 2016. Identification of Oil Reservoirs Suitable for CO2-EOR and CO2 Storage (CCUS) Using Reserves Databases, with Application to Alberta, Canada. International Journal Greenhouse Gas Control 44: 152–165.

[116] Sun, Lili, Hongen Dou, Zhiping Li, Yongle Hu, and Xining Hao. 2017. Assessment of CO2 Storage Potential and Carbon Capture, Utilization and Storage Prospect in China. Journal of the Energy Institute 91(6): 970–977. https://doi.org/10.1016/j.joei.2017.08.002.

[117] Wang, Peng-Tao, Xi Wu, Gangke Ge, Xiaoyan Wang, Mao Xu, Feiyin Wang, Yang Zhang, Haifeng Wang, and Yan Zheng. 2023. Evaluation of CO2 Enhanced Oil Recovery and CO2 Storage Potential in Oil Reservoirs of Petroliferous Sedimentary Basin, China. Science and Technology for Energy Transition 78(3): 1–12. https://doi.org/10.2516/stet/2022022.

CO_2 injection.[118,119,120,121,122,123,124] Each injection process has its own reservoir management, operational, and economic characteristics. The continuous CO_2 injection method involves the continuous injection of a predetermined volume of CO_2 without any additional fluids. To maximize gravity separation, CO_2 injection is sometimes followed by a lighter gas, such as N_2. Compared to other CO_2 injection methods, this process has a simpler injection system and a higher level of injectivity. It is generally used to promote gravity drainage of reservoirs containing medium to light oil, as well as reservoirs that are strongly water-wet or are sensitive to water flooding. This process enables higher oil recovery in the early stages of CO_2 flooding.[125,126] The continuous CO_2 injection process can allow for rapid penetration of CO_2 into production wells through highly permeable formations, promote gravity separation, and significantly reduce sweep efficiency.[127] In the WAG injection process, a predetermined volume of CO_2 is injected in cycles alternating with equal volumes of water. The WAG process promotes a higher sweep

[118] Zhou, Dengen, Meisonh Yan, and Wm. Marc Calvin. 2012. Optimization of a Mature CO2 Flood—From Continuous Injection to WAG. SPE Improved Oil Recovery Symposium. Tulsa: Oklahoma, USA. https://doi.org/10.2118/154181-ms.

[119] Talebian, Seyedeh Hosna, Rahim Masoudi, Isa Mohd Tan, and Pacelli Lidio Zitha. 2014. Foam Assisted CO2-EOR: A Review of Concept, Challenges, and Future Prospects. Journal of Petroleum Science and Engineering 120: 202–215. https://doi.org/10.1016/j.petrol.2014.05.013.

[120] Afzali, Shokufe, Nima Rezaei, N., and Sohrab Zendehboudi. 2018. A Comprehensive Review on Enhanced Oil Recovery by Water Alternating Gas (WAG) Injection. Fuel 227(05): 218–246. https://doi.org/10.1016/j.fuel.2018.04.015.

[121] See footnote 104.

[122] Esene, Cleverson, Nima Rezaei, Amer Aborig, and Sohrab Zendehboudi. 2019. Comprehensive Review of Carbonated Water Injection for Enhanced Oil Recovery. Fuel 237: 1086–1107. https://doi.org/10.1016/j.fuel.2018.08.106.

[123] Al-Shargabi, Mohammed, Shadfar Davoodi, David A. Wood, Valeriy S. Rukavishnikov, and Konstantin M. Minaev. 2022. Carbon Dioxide Applications for Enhanced Oil Recovery Assisted by Nanoparticles: Recent Developments. ACS Omega 7(12): 9984–9994.

[124] Tan, Yongsheng, Qi Li, Liang Xu, Abdul Ghaffar, Xiang Zhou, Pengchun Li. 2022. A Critical Review of Carbon Dioxide Enhanced Oil Recovery in Carbonate Reservoirs. Fuel 328: 125256.

[125] See footnote 117.

[126] See footnote 122.

[127] Ibid.

efficiency during the CO_2 injection stage. The low mobility ratio slows down the gas breakthrough time and reduces the amount of CO_2 used during injection, thereby reducing the cost of purchasing CO_2. This process is suitable for most low permeability reservoirs or where there are contrasting permeabilities between different layers.[128] Strategic planning, including the definition of amounts of injected water and CO_2, injection periods, and well controls, should be undertaken to optimize a WAG injection project. Disadvantages have been identified, such as in heterogeneous reservoirs with high oil viscosity, in which the injection volumes of water and CO_2 should be low.[129] CWI is the simplest CO_2 injection process in EOR, considering that CO_2 is more soluble in water compared to other common gases.[130] It involves the injection of water saturated with CO_2 up to a maximum of 3 wt% to 5 wt%. This process shows a higher oil recovery than the WAG process, mainly due to CO_2 mass transfer from the carbonated water to the oil phase, leading to a higher mobilization of the oil phase and a better areal and microscopic sweep efficiency. The dissolved CO_2 in the oil phase increases the oil mobility by swelling the oil and reducing both oil viscosity and interfacial tension, leading to lower residual oil saturation. One of the disadvantages of the CWI process is the low amount of CO_2 used, which is less than 6 to 7 times when compared to other CO_2 injection processes.

The CO_2 foam injection process is a dispersion of a relatively large volume of gas bubbles (CO_2) as a non-wetting fluid in a small volume of a continuous wetting phase. The continuous wetting phase is water containing a surfactant (foaming agent) at a specific concentration above the critical micelle concentration.[131] Foams are unstable and break quickly without a foaming agent. A CO_2 foam is therefore essentially a mixture of gaseous CO_2, water, and a foaming agent. Foam quality is closely related to bubble size. CO_2 foam quality decreases as bubble

[128] See footnote 103.

[129] See footnote 119.

[130] See footnote 121.

[131] Majeed, Talha, Muhammad Shahzad Kamal, Xianmin Zhou, and Theis Solling. 2021. A Review on Foam Stabilizers for Enhanced Oil Recovery. Energy Fuels 35(7): 5594–5612. https://doi.org/10.1021/acs.energyfuels.1c00035.

size increases and foams become more unstable.[132,133] In oil reservoirs, CO_2 foam is generated by consecutive injections of a surfactant solution and gaseous CO_2.[134] In addition, CO_2 foams can also be generated by injecting the surfactant dissolved in supercritical CO_2 into the reservoir, without the need for a liquid injection. CO_2 foam has been introduced to reduce formation damage in water-sensitive formations and to enable rapid cleanup.[135] Nevertheless, the main application of CO_2 foam in the EOR process is to reduce the mobility of gaseous CO_2, which leads to an improvement in the sweep efficiency and a delay in the break-through.[136] Some limitations can be pointed out for the CO_2 foam injection process, such as the fact that this injection method is often used in highly heterogeneous reservoirs characterized by highly permeable and porous layers and oil-depleted layers with high water saturation.[137] The CO_2 cyclic injection process, also known as "huff and puff", involves using a single well for both injecting and producing. This means that well-to-well connectivity is not required.[138] This process consists of three stages: CO_2 injection stage (huff process), soaking stage, and oil production stage (puff process).[139,140] During the CO_2 injection stage, CO_2 is injected into the reservoir at a high rate, displacing some of the mobile oil

[132] Sheng, James J. 2013. Chapter 11—Foams and Their Applications in Enhancing Oil Recovery. Enhanced Oil Recovery Field Case Studies. pp. 251–280. https://doi.org/10.1016/b978-0-12-386545-8.00011-7.

[133] Ibrahim, Ahmed Farid, and Hisham A. Nasr-El-Din. 2020. CO2 Foam for Enhanced Oil Recovery Applications. In: Foams—Emerging Technologies. Edited by Huijin Xu. Chen Yang and Dengwei Jing. IntechOpen. https://doi.org/10.5772/intechopen.89301.

[134] See footnote 118.

[135] See footnote 132.

[136] See footnote 119.

[137] See footnote 122.

[138] Gamadi, Talal D., James J. Sheng, and Mohamed Y. Soliman. 2013. An Experimental Study of Cyclic Gas Injection to Improve Shale Oil Recovery. SPE Annual Technical Conference and Exhibition. New Orleans: Louisiana, USA. https://doi.org/10.2118/166334-ms.

[139] Sheng, James J. 2017. Optimization of Huff-n-Puff Gas Injection in Shale Oil Reservoirs. Petroleum 3(4): 431–437. https://doi.org/10.1016/j.petlm.2017.03.004.

[140] Shabib-Asl, Abdolmohsen, Shengnan (Nancy) Chen, and Sixu Zheng. 2022. Performance of CO2 Foam Huff and Puff in Tight Oil Reservoirs. Frontiers in Energy Research 10: 826469. https://doi.org/10.3389/fenrg.2022.826469.

and water into a further location near the well, which reduces water saturation and increases oil permeability. This stage is short compared to the other two, and at the end, the reservoir pressure is higher than the initial pressure. In the soaking stage, the well is shut down to allow for CO_2 dissolution and diffusion, improving the oil's physical properties. A longer soaking stage duration requires a higher CO_2 volume to achieve higher oil recovery. In the production stage, developed after an optimum soaking time, the injection well will be converted into a production well. Therefore, any free CO_2 that was not dissolved into the oil will be produced as a gas phase, along with the swelled oil (containing dissolved CO_2) and some oil located further away from the well, which is produced due to the pressure drop's driving force.[141,142] The cyclic injection process is mainly used in low-injectivity reservoirs, such as thigh sandstones and shales. This injection process's main goal is to promote production mechanisms such as oil viscosity reduction, oil swelling, reservoir re-pressurization, and solution gas drive.[143] There are several factors and operational parameters that can influence this injection process, including injection pressure, number of cycles, soaking time, injection fluid composition, oil viscosity, CO_2 solubility in oil, and differential production pressures.[144,145] This CO_2 injection is an effective, low-risk, and low-cost process since it does not require the construction of new special injection wells. However, some setbacks can occur, such as possible corrosion of equipment and the need to separate and capture recycled CO_2 from oil.[146]

[141] Ibid.

[142] Ding, MingChen, Yefei Wang, Dexin Liu, Xia Wang, Hailong Zhao, and Wuhua Chen. 2021. Enhancing Tight Oil Recovery Using CO2 Huff and Puff Injection: An Experimental Study of the Influencing Factors. Journal of Natural Gas Science and Engineering 90(8): 103931. https://doi.org/10.1016/j.jngse.2021.103931.

[143] See footnote 138.

[144] See footnote 141.

[145] Song, C., C.R. Clarkson, H. Hamdi, and Amin Ghanizadeh. 2021. Comparison of CO2 and Lean Gas Cyclic Injection ('Huff-n-Puff') in Artificially-Fractured Shale Core Samples. Unconventional Resources Technology Conference. Houston: Texas, USA. Paper number: URTEC-2021–5309-MS. https://doi.org/10.15530/urtec-2021-5309.

[146] See footnote 139.

3.3.2 CO_2-ECBM

CO_2-ECBM is a promising technology to reduce anthropogenic CO_2 emissions. CO_2-ECBM has similarities to CO_2 tertiary flooding, one of the most successful EOR methods, but the CH_4 displacement and CO_2 storage mechanisms are quite different from those in CO_2-EOR reservoirs.[147] This technology involves injecting CO_2 into coal seams, especially unmineable coal seams, primarily to displace natural gas, which is mainly CH_4.[148,149] The main perceived benefits of CO_2-ECBM are to store CO_2, reduce coalbed methane production time, improve coalbed methane recovery, and increase natural gas reserves.[150,151] Coalbed methane, also known as coal seam gas, is characterized as an unconventional reservoir, as mentioned before, due to the specific nature of gas storage and fluid circulation. A coal seam is a porous medium reservoir with a unique microstructure that allows the storage of a volume of gas much higher than its pore volume capacity.[152] Gases are stored in coal seams through different mechanisms: (1) physical adsorption on pore internal surface areas in a condensed form similar to a liquid state, (2) absorption within the molecular structure, and (3) as free gas in pores and the network of natural fractures (cleat system).[153] The former corresponds to over 95% to 98% of the total gas stored, and the absorbed gas

[147] Jikich, Sinisha A., Grant S. Bromhal, William Neal Sams, F. Burcu Gorucu, Turgay Ertekin, and Duane H. Smith. 2004. Economics for Enhanced Coalbed Methane (ECBM) and CO2 Sequestration with Horizontal Wells. Proceedings of SPE Eastern Regional Meeting. Charleston: West Virginia. Paper Number: SPE-91391-MS. https://doi.org/10.2523/91391-ms.

[148] Shi, J.Q., and Durucan, S. 2005. CO2 Storage in Deep Unminable Coal Seams. Oil & Gas Science and Technology 60(3): 547–558. https://doi.org/10.2516/ogst:2005037.

[149] IEAGHG. 2007. Status Report on CO2-ECBM Research Activities. Report Number 2007/TR1. February 2007. 5p.

[150] See footnote 146.

[151] Connell, Luke D., Zhejun Pan, Michael Camilleri, Meng Shangzhi, David Down, John Carras, Zhang Wenzhong, Fu Xiaokang, Guo Benguang, Cameron Briggs, and Nicholas Lupton. 2014. Description of a CO2 Enhanced Coal Bed Methane Field Trial Using a Multi-Lateral Horizontal Well. International Journal of Greenhouse Gas Control 26: 204–219. https://doi.org/10.1016/j.ijggc.2014.04.022.

[152] Rodrigues, Cristina Fernanda Alves. 2002. The Application of Isotherm Studies to Evaluate the Coalbed Methane Potential of the Waterberg Coalfield, South Africa. Faculdade de Ciências do Porto, 298p. (Tese de doutoramento).

[153] Ibid.

and the free gas account for the remaining 2% to 5%.[154] Fluid movement in coal is described by Darcy's Law as gases flow through the cleat system, and by diffusion mechanisms as gases flow from internal surface areas of the organic microporous structure. Traditionally, CH_4 release from coal seams is closely related to the overall pressure reduction, which is usually achieved by dewatering the coal seam through pumping.[155] The pressure reduction induced by water removal is followed by gas diffusion through the coal matrix, which is controlled by the gas concentration differential, and eventually, the free gas flows through the cleat system in response to the pressure gradient.[156] This dewatering process and subsequent pressure reduction allows for less than 50% recovery of the CH_4 in place. However, in the CO_2-ECBM process, the CO_2 is used to displace the adsorbed CH_4 molecules and to increase the production of CH_4 without the usual reduction in the reservoir pressure.[157] During the injection of CO_2 into a CH_4-bearing coal seam, the CO_2 molecules compete with the CH_4 molecules for the same adsorption space. CH_4 molecules will detach from the pore's internal surface areas as CO_2 is injected due to a decrease in the CH_4 partial pressure in the free gas phase. The displaced CH_4 is then free in the cleat system and can flow to a production well. As CO_2 is known to have a higher affinity for the microporous structure of organic matter than CH_4, the injected CO_2 will then be adsorbed

[154] Rodrigues, Cristina Fernanda, and Manuel João Lemos de Sousa. 2002. The Measurement of Coal Porosity with Different Gases. In: Proceedings of the Symposium on Geology of Coal-Bearing Strata, Papers from the 31st International Geological Congress, Rio de Janeiro, Brazil, August 6–7, 2000. Edited by W. Kalkreuth. Amsterdam: International Journal Coal Geology 48(3–4): 245–251.

[155] Godec, Michael, George Koperna, and John Gale. 2014. CO2-ECBM: A Review of its Status and Global Potential. Energy Procedia 63: 5858–5869. https://doi.org/10.1016/j.egypro.2014.11.619.

[156] Rodrigues, Cristina Fernanda Alves, Maria Alzira Pimenta Dinis, and Manuel João Lemos de Sousa. 2015. Review of European Energy Policies Regarding the Recent "Carbon Capture, Utilization and Storage" Technologies Scenario and the Role of Coal Seams. Berlin, Heidelberg. Environmental Earth Sciences 74(3): 2553–2561. https://doi.org/10.1007/s12665-015-4275-0.

[157] Robertson, Eric P. 2010. Enhanced Coal Bed Methane Recovery and CO2 Sequestration in the Powder River Basin. Idaho Falls: Idaho National Laboratory. 25p. (INL/EXT-10-18941).

by coal.[158,159,160] The concept of CO_2-ECBM was first proposed by Puri and Yee,[161] and since then several projects have been developed but were generally considered unsuccessful. This failure was mainly related to economic issues due to high uncertainties in reservoir property estimates. These reservoir properties can be summarized as follows: permeability anisotropy (vertical and horizontal), reservoir heterogeneity (rock type), CO_2-water relative permeability, stress dependence of cleat permeability and porosity, and coal matrix swelling and shrinking effects.[162] It is well known that coal permeability is typically relatively low, which is closely related to the reservoir properties mentioned above. This low coal matrix permeability is very likely to decrease with CO_2 injection due to coal swelling caused by CO_2 adsorption. The use of vertical wells for CO_2 injection, which is severely limited because most coal seams are very thin, resulting in small contact areas with the reservoir, is one of the operational reasons for the lack of success mentioned above. Therefore, a key challenge to the success of CO_2-ECBM is the development of techniques to manage the coal swelling caused by CO_2 injection.[163] In recent decades, a new drilling technology has been used. This advanced technology consists of multi-lateral horizontal wells drilled to follow the reservoir for CO_2 injection, maximizing the contact area with the coal seam reservoir and thus minimizing the effects of any permeability reduction due to coal swelling that might occur during CO_2 injection. Ultimately, Jikich et al.[164] suggested an increase of up to 3 to 5 times in drainage areas and up to 10 times in productivity indices in horizontal wells compared to vertical wells, as well as a higher sweep efficiency.

[158] See footnote 151.

[159] See footnote 147.

[160] Rodrigues, Cristina Fernanda, Maria Alzira Pimenta Dinis, and Manuel João Lemos de Sousa. 2016. Coal Gas Adsorption/Desorption Isotherms Versus Diffusion Process. Olney, Bucks MK, UK. International Journal of Global Warning 9(4): 467–485. https://doi.org/10.1504/IJGW.2016.076332.

[161] Puri, R., and D. Yee. 1990. Enhanced Coalbed Methane Recovery. SPE Annual Technical Conference and Exhibition. https://doi.org/10.2118/20732-ms.

[162] See footnote 146.

[163] See footnote 150.

[164] See footnote 146.

Reservoir screening criteria are essential for selecting the most suitable coal seams for CO_2 storage and potential ECBM.[165,166,167,168,169,170] General geological factors such as reservoir homogeneity, background structure, permeability, depth (including temperature and pressure), reservoir geometry, and in situ gas saturation composition must be considered. The coal seam reservoir should be laterally continuous and vertically isolated from surrounding layers, i.e. the coal seam should be as homogeneous as possible, as well as it is necessary to avoid faulted and folded areas. The permeability of coal seams is typically low, but a minimum of 1 mD to 5 mD should be required. The depth of the reservoir should be adequate since it will affect pressure and temperature. For example, shallow coal seams tend to have low pressures and consequently low gas content. However, gas content increases as temperature decreases. In deep coal seams, permeability decreases but can be improved by CO_2 injection. In addition, CO_2 becomes a supercritical fluid above its critical point, i.e. temperatures of 31.1 °C and pressures of 7.4 MPa, and the operational CO_2-ECBM project becomes extremely expensive. Therefore, the most suitable depth for the CO_2-ECBM process is between 300 and 1500 m.[171,172] Thick (concentrated) coal seams are generally more suitable than thin (stratigraphically dispersed) coal seams. In terms of gas saturation conditions, CH_4-saturated coal seams are the most suitable, since undersaturated coal reservoirs delay CH_4 production. However, the injection of CO_2 can reduce the delay in CH_4 production by increasing

[165] IEAGHG. 1999. Enhanced Coal Bed Methane Recovery with CO2 Sequestration. Report Number PH3/3. August 1999. 139p.

[166] See footnote 151.

[167] See footnote 147.

[168] See footnote 154.

[169] Corum, Margo D., Kevin B. Jones, and Peter D. Warwick. 2013. CO2 Sequestration Potential of Unmineable Coal—State of Knowledge. Energy Procedia 37: 5134–5140. https://doi.org/10.1016/j.egypro.2013.06.428.

[170] Talapatra, Akash. 2020. A Study on the Carbon Dioxide Injection into Coal Seam Aiming at Enhancing Coal Bed Methane (ECBM) Recovery. Journal of Petroleum Exploration and Production Technology 10(5): 1965–1981. https://doi.org/10.1007/s13202-020-00847-y.

[171] See footnote 164.

[172] Tomić, Lola, Vesna Karovic Maricic, Dušan Danilović, and Miroslav Crnogorac. 2018. Criteria for CO2 Storage in Geological Formations. Podzemni radovi 2018(32). 61–74. https://doi.org/10.5937/PodRad1832061T.

the reservoir gas saturation.[173,174] In addition, several other parameters specific to coal characteristics can influence both CO_2 storage and ECBM recovery. These parameters include moisture content, mineral matter content, maceral composition, and coal rank. Moisture content can significantly affect adsorption capacity, adsorption phase density, and mixture adsorption behaviour.[175] Both moisture content and mineral matter content compete with gas for the same space in the organic microporous structure, therefore gas storage decreases as they increase. In general, according to the International Union of Pure and Applied Chemistry,[176] macerals are characterized by micropores (<2 nm), mesopores (>2 and <50 nm) and macropores (>50 nm). As mentioned above, adsorption is the dominant gas storage mechanism in coal seams, which means that storage takes place mainly in the internal pore surface areas. Therefore, the maceral group, which is mainly characterized by micropores (vitrinite group), must be highly represented in the coal seam. Coal rank also affects its storage behaviour, since the aromatization of the organic structure increases with the rank increase, at least up to 2% Rr (vitrinite random reflectance). The aromatization process allows the development of a microporous structure, thus increasing the internal pore surface area. As a result, coals with ranks ranging between 0.8%Rr and 1.5%Rr are the best option for CO_2-ECBM processes.[177,178]

Unconventional reservoirs, such as coalbed methane and shale gas, have additional advantages. Due to their high organic matter content, they can also act as source rocks, and the geological parameters, trap, and seal, do not play a relevant role in the reservoir storage process. In organic-rich reservoirs, CO_2 entrapment is mainly achieved by the matrix of organic matter adsorption behaviour, ensuring the safe and permanent storage of CO_2.

A critical step that must be taken into account in a CO_2-ECBM project is the composition of the injected gas. Pure CO_2 should be the best

[173] See footnote 164.

[174] See footnote 151.

[175] See footnote 147.

[176] International Union of Pure and Applied Chemistry. 1999. Pure and Applied Chemistry 66(8): 1739–1758.

[177] See footnote 164.

[178] See footnote 152.

option due to its higher affinity to organic microporous structure than other gases.[179,180] Additionally, it is also important to use CO_2 removal technologies to achieve carbon neutrality by 2050. However, due to CO_2 supply and cost constraints, mixtures of CO_2 and N_2 are currently being used. Moreover, it is important to understand that the injection of pure CO_2 displaces CH_4 through a relatively sharp front, due to the higher adsorption affinity of the former than the latter, whereas the injection of pure N_2 results in a smoother displacement front, meaning that the N_2 moves faster than the CH_4, leading to a premature breakthrough of N_2.[181,182,183] The N_2 injected into a coal seam tends to remain in the coal system, leading to a decrease in the CH_4 partial pressure.[184] Therefore, if gas mixtures (CO_2 + N_2) are used in the injection process, the CH_4 produced will be contaminated with N_2 until the breakthrough of the CO_2 takes place (Table 6).[185,186]

3.3.3 CO_2-EWR

The traditional method of CO_2 storage in deep saline aquifers has been used since the turn of the century. This technology, like the ones

[179] See footnote 152.

[180] Gunter, William D., Mathew J. Mavor, and John R. Robinson. 2005. CO2 Storage and Enhanced Methane Production: Field Testing at Fenn-Big Valley, Alberta, Canada, with Application. Greenhouse Gas Control Technologies 7 I: 413–421. https://doi.org/10.1016/b978-008044704-9/50042-2.

[181] Pini, Ronny, Dorian Marx, Luigi Burlini, Giuseppe Storti, and Marco Mazzotti. 2011. Coal Characterization for ECBM Recovery: Gas Sorption Under Dry and Humid Conditions, and Its Effect on Displacement Dynamics. Energy Procedia 4: 2157–2161. https://doi.org/10.1016/j.egypro.2011.02.101.

[182] See footnote 1.

[183] Zhou, Fengde, Furgan Hussain, and Yildiray Cinar. 2013. Injecting Pure N2 and CO2 to Coal for Enhanced Coalbed Methane: Experimental Observations and Numerical Simulation. International Journal of Coal Geology 116–117: 53–62. https://doi.org/10.1016/j.coal.2013.06.004.

[184] Schepers, Karine, Anne Oudinot, and Nino Ripepi. 2010. Enhanced Gas Recovery and CO2 Storage in Coal Bed Methane Reservoirs: Optimized Injected Gas Composition for Mature Basins of Various Coal Rank. SPE International Conference on CO2 Capture, Storage, and Utilization. New Orleans: Louisiana, USA. Paper Number: SPE-139723-MS. https://doi.org/10.2118/139723-ms.

[185] See footnote 181.

[186] See footnote 183.

Table 6 Screening criteria for CO_2-ECBM

Criteria	*Suitable condition*
Thickness (m)	>1
Depth (m)	300–1500
Temperature (°C)	35–100
Pressure (MPa)	>7.5
Matrix porosity (%)	>0.5
Permeability (mD)	>1mD
Ash content (%)	<25
Methane content (m^3/ton)	2.5–50
Vitrinite content	High
Coal rank (%Rr)	0.6–1.5

Adapted from Rodrigues [2002],[187] Yu et al. [2017],[188] Pratama et al. [2017],[189] Sun et al. [2017],[190] Tomić et al. [2018][191]

mentioned above, is seen as a potential technique to reduce carbon emissions through the removal of CO_2, thereby enabling the mitigation of global climate change. Deep saline aquifers are reservoirs, such as those found in oil and gas fields. However, in contrast to depleted oil and gas reservoirs, where there is already an extensive and detailed database of site characteristics and site operations produced through the first and second recovery processes, CO_2 storage in deep saline aquifers is constrained by data unavailability, no experience with long-term commercial-scale CO_2

[187] See footnote 152.

[188] Yu, Han, Qingzhe Jiang, Zhaozheng Song, Qian Ma, Bo Yuan, and Huanxi Xiong. 2017. The Economic and CO2 Reduction Benefits of a Coal-to-Olefins Plant Using a CO2-ECBM Process and Fuel Substitution. Royal Society of Chemistry Advances 7(79): 49975–49984. https://doi.org/10.1039/c7ra07994j.

[189] Pratama, Edo, Mohd suhaili Ismail, and Syahrir Ridha. 2017. Screening Criteria of Optimum Carbon Dioxide Injection for Enhanced Coalbed Methane Recovery and Prediction of Carbon Dioxide Storage Capacity: A Case Study in South Sumatera Basin, Indonesia. Chemical Engineering Transactions 56: 997–1002. https://doi.org/10.3303/CET1756167.

[190] See footnote 115.

[191] See footnote 172.

storage projects, and ultimately high cost.[192,193,194,195,196] In addition, CO_2 storage in deep saline aquifers can show serious problems related to continuous large-scale injection of CO_2, leading to limitations in CO_2 injection and the safety of CO_2 storage projects. These problems are mainly related to reservoir pressure build-up phenomena, which increase the risk of hydraulic caprock fracturing and the reactivation of existing faults and can induce seismicity. All of these problems can lead to CO_2 leakage.[197,198] As a result, it can be very difficult for investors, and especially decision-makers, to determine realistic capacity, risk, and associated costs. Therefore, a new geoengineering approach to CO_2 geological utilization and storage, called CO_2-EWR, has been developed as an alternative to traditional CO_2 storage in deep saline aquifers. The CO_2-EWR process is similar to CO_2-EOR, where the CO_2 injection is used as a displacement fluid to enhance water recovery, and CO_2 is simultaneously stored underground.[199] Furthermore, CO_2-EWR operational procedures are similar to CO_2-EOR, but with much larger well spacing, well injectivity, and flow rate of a single well. There are significant advantages to CO_2-EWR technologies over traditional deep saline aquifer storage technologies. The production of saline water can be used to control and

[192] See footnote 13.

[193] See footnote 14.

[194] Liu, Danqing, Ramesh Agarwal, and Yilian Li. 2016. Numerical Simulation and Optimization of CO2-Enhanced Water Recovery by Employing a Genetic Algorithm. Journal of Cleaner Production 133: 994–1007.

[195] Yang, Lin, Haodong Lv, Dalin Jiang, Jingli Fan, Xian Zhang, Weijun He, Jinsheng Zhou, and Wenjing Wu. 2020. Whether CCS Technologies Will Exacerbate the Water Crisis in China? —A Full Life-Cycle Analysis. Renewable and Sustainable Energy Reviews 134: 110374. https://doi.org/10.1016/j.rser.2020.110374.

[196] Wei, Ning, Xiaochun Li, Zhunsheng Jiao, Philip H. Stauffer, Shengnan Liu, Kevin Ellett, and Richard H. Middleton. 2022. A Hierarchical Framework for CO2 Storage Capacity in Deep Saline Aquifer Formations. Frontiers in Earth Science 9(1315): 1–21. https://doi.org/10.3389/feart.2021.777323.

[197] Ibid.

[198] Yang, Zhijie, Tianfu Xu, Fugang Wang, Yujie Diao, Xufeng Li, Xin Ma, and Hailong Tian. 2019. A Study on the CO2-Enhanced Water Recovery Efficiency and Reservoir Pressure Control Strategies. Geofluids 2019(6053756): 1–17. https://doi.org/10.1155/2019/6053756.

[199] See footnote 196.

mitigate the excessive build-up of reservoir pressure through a reasonable engineering design of the production wells to achieve the safety and stability of a large-scale CO_2 geological storage project.[200,201] In addition, the produced deep saline water can be used in several processes that can generate substantial profits and thus could help to offset the initially indicated costs of active pressure management for CO_2 storage projects, while effectively alleviating water scarcity in arid areas.[202,203,204] Therefore, the deep saline water, after treatment, can be used for a variety of applications, such as cooling water for thermal power plants, irrigation for agriculture, and possibly drinking water, depending on the amount of total dissolved solids, organic products, and other substances that need to be removed.[205,206] In addition, Veil et al.[207] suggested that it could also be used in geothermal energy production, which will be discussed later in this chapter, and for the extraction of valuable minerals such as lithium and potassium. Selecting the most suitable reservoirs for CO_2-EWR projects is also a complex process, involving general geological conditions, as well as reservoir characteristics, including the caprock (Table 7).

[200] Li, Qi, Ya-Ni Wei, Guizhen Liu, and Hui Shi. H. 2014. CO2-EWR: A Cleaner solution for Coal Chemical Industry in China. Journal of Cleaner Production 103(8): 330–337. https://doi.org/10.1016/j.jclepro.2014.09.073.

[201] See footnote 198.

[202] Kobos, Peter H., Malynda A. Cappelle, Jim L. Krumhansl, Thomas A. Dewers, Andrea McNemar, and David J. Borns. 2011. Combining Power Plant Water Needs and Carbon Dioxide Storage Using Saline Formations: Implications for Carbon Dioxide and Water Management Policies. International Journal of Greenhouse Gas Control 5(4): 899–910. https://doi.org/10.1016/j.ijggc.2011.03.015.

[203] Veil, John A., Christopher B. Harto, and Andrea T. McNemar. 2011. Management of Water Extracted From Carbon Sequestration Projects: Parallels to Produced Water Management. SPE Americas E&P Health, Safety, Security, and Environmental Conference. Houston: Texas, USA. Paper Number: SPE-140994-MS. https://doi.org/10.2118/140994-ms.

[204] See footnote 200.

[205] See footnote 202.

[206] See footnote 203.

[207] See foonote 203.

Table 7 Screening criteria of CO_2-EWR

Criteria	*Suitable condition*
Reservoir type	Sandstones and carbonates
Reservoir thickness (m)	>20
Depth (m)	800–2500
Temperature (°C)	>35
Pressure (MPa)	>7.5
Porosity (%)	>10
Permeability (mD)	>20
Caprock thickness (m)	>20
Caprock integrity	Rather no fault
Salinity (g/L)	>3

Adapted from IEAGHG [2009] and[208] Sun et al. [2017][209]

3.3.4 CO_2-EGS

Geothermal energy is a renewable and reliable source of energy that makes use of the natural heat of the earth to heat buildings or to generate electricity. Geothermal energy has several advantages, including high stability, which means it is not seasonal, high utilization, safety, and low operating costs compared to other renewable energy sources, such as solar and wind. In addition, due to its seasonal independence, geothermal energy can be used for baseload supply without additional energy storage.[210] There are several different sources of geothermal energy, such as hydrothermal, magmatic, and hot dry rocks (bedrock that mainly consists of intact granite or other crystalline basement rocks with high temperatures of around 150 °C and 600 °C).[211,212] Among the geothermal energy

[208] IEAGHG. 2009. CCS Characterisation Criteria. Technical Study, Report Number 2009/10. July 2009. 112p.

[209] See footnote 115.

[210] Sowizdzal, Anna, Pawel Gladysz, Trond Andresen, Maciej Miecznik, Bjorn S. Frengstad, Marcin Liszka, Anna Chmielowska, Marcel Gawron, Sigurd W. Levseth, Leszek Pajak, Lars A. Stenvik, and Barbara Tomaszewska. 2021. CO2-Enhanced Geothermal Systes for Climate Neutral Energy Supply. Trondheim Conference on CO2 Capture, Transport and Storage. Trondheim: Norway, pp. 277–283.

[211] Wu, Yu, and Pan Li. 2020. The Potential of Coupled Carbon Storage and Geothermal Extraction in a CO2-Enhanced Geothermal System: A Review. Geothermal Energy 8(1): 19.

[212] Tagliaferri, Mauro, Pawel Gładysz, Pietro Ungar, Magdalena Strojny, Lorenzo Talluri, Daniele Fiaschi, Giampaolo Manfrida, Trond Andresen, Anna Sowizdzat. 2022.

source options, the hydrothermal geothermal energy source is consistently mature due to more than 100 years of experience, and more recently the hot dry rock geothermal energy source.[213,214] Despite this, the exploitation of this energy source is still very low when taking into account the huge resources available worldwide. Geothermal energy has been largely confined to conventional hydrothermal systems, which are characterized by natural fracture networks that allow fluid circulation, and consequently, geothermal energy can be extracted by drilling into these hot fluids.[215] Undeniably, most geothermal sources occur in rocks that lack fracture networks, which induce permeability and fluid circulation.[216] The idea of Enhanced Geothermal Systems (EGS) is mainly based on the concept of hot dry rock geothermal sources, which are characterized by an enormous geothermal potential. Enhanced geothermal systems have been developed by injecting H_2O into deep wells to create permeability through hydraulic fracturing (stimulation), and extracting geothermal energy by circulating water through a system of injecting and producing wells.[217,218] The use of water as a heat transmission fluid has been shown to have serious drawbacks: (1) the injection of water into hot rock (characterized by high temperatures) can cause strong dissolution and precipitation effects, which could change fracture permeability, making the operational process unstable and complex, (2) the stimulated reservoir volume is very small, (3) the costs associated with drilling deep wells and injecting cold fluids at high pressure are high, (4) water is a sparse and valuable commodity, which is significantly lost (10% to

Techno-Economic Assessment of the Supercritical Carbon Dioxide Enhanced Geothermal Systems. Sustainability 14(24): 16580. https://doi.org/10.3390/su142416580.

213 Ibid.

214 Pajak, Leszek, Anna Sowizdzał, Pawel Gładysz, Barbara Tomaszewska, Maciej Miecznik, Trond Andresen, Bjorn S. Frengstad, and Anna Chmielowska. 2021. Multi-Criteria Studies and Assessment Supporting the Selection of Locations and Technologies Used in CO2-EGS Systems. Energies 14(21): 7683. Doi: 10.3390/ en14227683.

215 Pruess, Karsten. 2006. Enhanced Geothermal Systems (EGS) Using CO2 as Working Fluid—A Novel Approach for Generating Renewable Energy with Simultaneous Sequestration of Carbon. Geothermics 35(4): 351–367.

216 Pruess, Karsten, and Nicolas Spycher. 2010. Enhanced Geothermal Systems (EGS) with CO2 as Heat Transmission Fluid—A Scheme for Combining Recovery of Renewable Energy with Geologic Storage of CO2. Lawrence Berkeley National Laboratory. 6p.

217 Ibid.

218 See footnote 203.

20%) during the circulation system.[219,220] Given the enormous disadvantages listed above, combined with the need to reduce/remove CO_2 emissions from the atmosphere, Brown[221] proposed a new EGS technology. This new technology, known as CO_2-EGS, uses CO_2 as a heat transmission fluid instead of the more traditional H_2O. Thus, CO_2-EGS technology can offset the high costs typically associated with CCS projects by combining geothermal energy production with geological storage of CO_2. CO_2 is a promising alternative heat transmission fluid due to its thermodynamic and hydrodynamic properties, which allow the introduction of several improvements, such as (1) large buoyancy force due to high density differences between the cold CO_2 in the injection well and the hot CO_2 in the production well, leading to a reduction in the power consumption of the fluid circulation system, (2) low viscosity, which can result in high flow velocities for a given pressure gradient, and (3) low reactivity, as CO_2 is less effective as a solvent for minerals, eliminating or at least reducing dissolution and precipitation problems typical of water systems.[222,223,224]

3.3.5 CO_2-EGH

Natural gas hydrates are ice-like crystalline compounds that are characterized by a network of water in which individual guest molecules are entrapped. Gas hydrates are formed by the van der Waals-forced entrapment of guest gas molecules in hydrogen-bonded water molecules,

[219] See footnote 215.

[220] See footnote 211.

[221] Brown, Donald W. 2000. A Hot Dry Rock Geothermal Energy Concept Utilizing Supercritical CO2 Instead of Water. In: Proceedings of the Twenty-Fifth Workshop on Geothermal Reservoir Engineering. Stanford University: California, pp. 233–238.

[222] Ibid.

[223] Randolph, Jimmy, Benjamin M. Adams, Thomas H. Kuehn, and Martin O. Saar. 2012. Wellbore Heat Transfer in CO2-Based Geothermal Systems. Transactions - Geothermal Resources Council 36: 549–554.

[224] Lei, Hongwu. 2022. Performance Comparison of H2O and CO2 as the Working Fluid in Coupled Wellbore/Reservoir Systems for Geothermal Heat Extraction. Frontiers in Earth Science 10: 819778.

which can stabilize the entire structure.[225,226] Essential conditions for hydrate formation include the presence of suitably sized and shaped guest molecules, the appropriate amount of water, low temperatures (a few degrees above 0 °C), and high pressures (a few MPa). The formation of gas hydrates is a crystallization process involving nucleation and crystal growth, followed by a large accumulation phase, which is mainly a physical rather than a chemical process.[227] The most common guest molecules for gas hydrate formation are ethane, methane, butane, propane, nitrogen, and carbon dioxide. However, the conditions required for CO_2 hydrate formation are less stringent than for other gases, e.g. at a temperature of 0 °C, the pressure required for CO_2 hydrate formation is 3.8 MPa, for N_2 hydrate formation 15 Ma, and 10 MPa for O_2 hydrate formation.[228] Gas hydrates have been noted for their substantial gas storage capacity. They can reach about 170 volumes of gas per volume of hydrate at standard room temperature and pressure, which provides a higher density to the gas storage. Furthermore, gas hydrates are incredibly abundant worldwide in the form of clathrates, for example in the deep sea, continental margins, and permafrost regions.[229,230] The amount of CH_4 stored in natural

[225] Torré, Jean-Philippe, Christophe Dicharry, Marvin Ricaurte, Delphine Daniel-David, and Daniel Broseta. 2011. CO2 Capture by Hydrate Formation in Quiescent Conditions: In Search of Efficient Kinetic Additives. Energy Procedia 4(4): 621–628. https://doi.org/10.1016/j.egypro.2011.01.097.

[226] Gambelli, Alberto Maria, Xhino Rushani, Daniela Pezzolla, Federico Rossi, and Giovanni Gigliotti. 2023. Production of CO2 Hydrates in Aqueous Mixtures Having (NH4)2SO4 at Different Concentrations; Definition of Consequences on the Process Evolution, Quantification of CO2 Captured and Validation of Hydrates Production as Technique for Ammonium Removal from Waste Water. Sustainability 15(12): 9841. https://doi.org/10.3390/su15129841.

[227] Srivastava, Shubhangi, Ann Mary Kollemparembil, Viktoria Zettel, Timo Claßen, Bernhard Gatternig, Antonio Delgado, and Bernd Hitzmann. 2022. Experimental Investigation of CO2 Uptake in CO2 Hydrates Formation with Amino Acids as Kinetic Promoters and Its Dissociation at High Temperature. Scientific Reports 12(1): 8359. https://doi.org/10.1038/s41598-022-12538-1.

[228] Ndlovu, Phakamile, Saeideh Babaee, Paramespri Naidoo, and Kuveneshan Moodley. 2024. Kinetic Studies of Gas Hydrates for CO2 Capture in the Presence of Nanoparticles. Industrial Engineering Chemistry Research 63(9): 3867–3879. https://doi.org/10.1021/acs.iecr.3c04061.

[229] Du, Hongxing, Yiqun Zhang, Bo Zhang, Shouceng Tian, Gensheng Li, and Panpan Zhang. 2023. Study of CO2 Injection to Enhance Gas Hydrate Production in Multilateral Wells. Energy 283(6964): 129078. https://doi.org/10.1016/j.energy.2023.129078.

[230] See footnote 226.

gas hydrates has been estimated to be significantly greater than that of all proven fossil fuels. Thus, gas hydrates have the potential to become an important energy source due to their high energy density and lower CO_2 emissions compared to coal and oil, which could simultaneously help optimize the global energy mix and contribute to mitigating climate change.[231,232] There are several conventional mining methods for hydrate extraction, including depressurization, thermal stimulation, and chemical inhibitor injection.[233,234,235,236] All of these production techniques are dissociation-based methods, i.e. there is a risk that CH_4 will be released into the atmosphere due to its rapid decomposition, thereby accelerating the greenhouse effect rather than mitigating it.[237,238] Therefore, to eliminate these potential CH_4 releases into the atmosphere, and considering that CO_2 storage in solid form is less susceptible to disturbances and more secure,[239,240] a non-destructive method of CH_4 hydrate recovery using

[231] See footnote 225.

[232] Pandey, Jyoti Shanker, Charilaos Karantonidis, Adam Paul Karcz, and Nicolas von Solms. 2020. Enhanced CH4-CO2 Hydrate Swapping in the Presence of Low Dosage Methanol. Energies 13(20): 5238. https://doi.org/10.3390/en13205238.

[233] Ni, Liu, and Zhao Jingde. 2011. Characteristics of CO2 Hydrate Production with Mechanical Enhancement. 2011. International Conference on Computer Distributed Control and Intelligent Environmental Monitoring (CDCIEM). pp. 1445–1448. https://doi.org/10.1109/cdciem.2011.176.

[234] Ma, Zhiwei W., Peng Zhang, Huashan S. Bao, and Shuai Deng. 2016. Review of Fundamental Properties of CO2 Hydrates and CO2 Capture and Separation Using Hydration Method. Renewable and Sustainable Energy Reviews 53(6): 1273–1302. https://doi.org/10.1016/j.rser.2015.09.076.

[235] Almenningen, Stian, Josef Flatlandsmo, Martin A. Fernø, and Geir Ersland. 2017. Multiscale Laboratory Verification of Depressurization for Production of Sedimentary Methane Hydrates. SPE Journal 22(01): 138–147. https://doi.org/10.2118/180015-pa.

[236] See footnote 228.

[237] Lee, Yohan, Young-ju Seo, Taewoong Ahn, Jaehyoung Lee, Joo Yong Lee, Se-Joon Kim, and Yongwon Seo. 2017. CH 4—Flue Gas Replacement Occurring in sH Hydrates and Its Significance for CH4 Recovery and CO2 Sequestration. Chemical Engineering Journal 308: 50–58. https://doi.org/10.1016/j.cej.2016.09.031.

[238] See footnote 232.

[239] Zhang, Yongtao, Fulin Chen, Yan He, and Fei Wang. 2022. Enhanced CO2 Hydrate Formation via Biopromoter Coupled with Initial Stirring Activation. Fuel 330: 125713. https://doi.org/10.1016/j.fuel.2022.125713.

[240] See footnote 228.

CO_2 injection has been proposed.[241] This new technology, the so-called replacement technology, consists of the replacement of molecules of CH_4 with an equal number of molecules of CO_2. The main reasons for this exchange process are: (1) the larger size of CO_2 molecules than CH_4 molecules allows an easier capture process in the former, (2) the CO_2 hydrate is more stable than the CH_4 hydrate under favourable hydrate formation conditions, (3) the enthalpy of hydrate formation is higher in CO_2 (−54.49 kJ/mol) than in CH_4 (−57.98 kJ/mol). Thus, when the CO_2 is injected in the proximity of natural CH_4 hydrates, the exchange process takes place spontaneously.[242,243] This CO_2-enhanced gas hydrate technology is considered to be one of the most promising approaches to achieving carbon neutrality, both for energy production and CO_2 removal. Although several projects have been developed,[244,245,246] there is still a significant gap between the current production potential and the threshold required for commercial development of gas hydrates. The development of CO_2-enhanced gas hydrate technology is still limited, mainly due to the random induction period and the slow formation rate of hydrates. Therefore, various methods, including mechanical and chemical approaches, are being investigated to improve the formation rate of hydrates.[247,248]

3.3.6 $CO_2 + O_2$ In situ Uranium Leaching

Nuclear energy was highlighted as one of the cleanest, if not the cleanest, and most efficient energy sources. Nuclear energy is classified as a low-carbon or environmentally friendly energy source, meaning that

[241] See footnote 237.

[242] See footnote 232.

[243] See footnote 226.

[244] Li, Jin-fa, Jian-liang Ye, Xu-wen Qin, Hai-jun Qiu, Neng-you Wu, Hai-long Lu, Wen-wei Xie, Jing-an Lu, Fei Peng, Zhen-qiang Xu, Cheng Lu, Zeng-gui Kuang, Jian-gong Wei, Qian-yong Liang, Hong-feng Lu, and Bei-bei Kou. 2018. The First Offshore Natural Gas Hydrate Production Test in South China Sea. China Geology 1(1): 5–16.

[245] See footnote 232.

[246] Sinehbaghizadeh, Saeid, Agus Saptoro, and Amir H. Mohammadi. 2022. CO2 Hydrate Properties and Applications: A State of the Art. Progress in Energy and Combustion Science 93: 101026. https://doi.org/10.1016/j.pecs.2022.101026.

[247] See footnote 227.

[248] See footnote 239.

there is no production of CO_2 emissions or is minimal throughout its entire supply chain (from production to utilization). Nowadays, it is used in several applications behind power generation, including organometallic synthesis, catalysis, medicine, radiometric dating, and industrial X-rays.[249,250] Additionally, its high energy density allows to produce sustainable and stable electricity supply.[251] Uranium is the raw material for nuclear energy and is widely distributed throughout the world in relatively low concentrations. According to the International Atomic Energy Agency (IAEA) uranium deposits can be classified as primary uranium minerals, secondary uranium minerals, and formation deposits, based on their geological setting, genesis of mineralization, and ultimately, economic significance from high to low uranium fractions.[252] Primary uranium minerals are represented by uraninite or pitchblende (fraction of 88,1wt%), coffinite (fraction of 70.4 wt%), brannerite (fraction of 59.8 wt%); secondary uranium minerals are autunite, carnotite, torbernite, and uranophane, with uranium fractions below 50 wt%; deposits in formations, such as sandstone, quartzite, phosphorite, lignite, and black shale are also frequent with uranium fractions below 1 w%.[253] However, the most economically viable and widely exploited is the sandstone-hosted uranium deposit. It accounts for up to 60% of global uranium production.[254,255] Nevertheless, uranium is a toxic heavy metal and radioactive element with toxicity six orders of magnitude greater

[249] Peluzo, Bárbara Maria Teixeira Costa, and Elfi Kraka. 2022. Uranium: The Nuclear Fuel Cycle and Beyond. International Journal of Molecular Sciences 23(9): 4655. https://doi.org/10.3390/ijms23094655.

[250] Fuxin, Zheng, Yanguo Teng, Yuanzheng Zhai, Jingdan Hu, Junfeng Dou, and Rui Zuo. 2023. Geo-Environmental Models of In-Situ Leaching Sandstone-Type Uranium Deposits in North China: A Review and Perspective. Water 15(6): 1244. https://doi.org/10.3390/w15061244.

[251] Li, Guihe, and Jia Yao. 2024. A Review of In Situ Leaching (ISL) for Uranium Mining. Mining 4(1): 120–148. https://doi.org/10.3390/mining4010009.

[252] Degueldre, Claude. 2017. Uranium as a Renewable for Nuclear Energy. Progress in Nuclear Energy 94: 174–186. https://doi.org/10.1016/j.pnucene.2016.03.031.

[253] Ibid.

[254] Qiu, Wenjie, Yun Yang, Jian Song, Weimin Que, Zhengbang Liu, Haicheng Weng, Jianfeng Wu, and Jichum Wu. 2023. What Chemical Reaction Dominates the CO2 and O2 In-Situ Uranium Leaching? Insights from a Three-Dimensional Multicomponent Reactive Transport Model at the Field Scale. Applied Geochemistry 148: 105522.

[255] See footnote 251.

than its radioactivity, and as a result, the use of conventional open pit or underground mining can lead to uranium leaching into groundwater, producing a serious hazard to the surrounding ecological environment of the mining area.[256] In this context, in situ leaching (ISL), also known as in situ recovery (ISR), emerged as a cleaner, greener, and cheaper way of producing uranium. It was tried on an experimental basis in the early 1960s and the first commercial project became operational in 1974.[257] SL involves selecting suitable uranium deposits (high permeability to the solution used); drilling well-positioned (taking into account the shape of the deposit) wells, including injection and production wells; injecting a solution (or lixiviant) capable of dissolving the uranium; pumping the uranium-bearing solution (pregnant solution); and finally, treating the produced solution to separate the uranium and the leaching solution, which is then recycled and re-injected into the ore deposit.[258,259,260,261] This ISL method offers several advantages over the traditional methods mentioned above, namely: depth is not a critical parameter; works directly within natural ore deposits, avoiding extensive surface disturbance and excavation; short construction time and minimal infrastructure costs; safe and simple mining process; minimal environmental impact due to avoidance of waste rock, radioactive dust, and emissions; increases the industrial value of low-grade sandstone-hosted uranium deposits.[262] The ISL method uses different leaching solutions such as acid leaching (pH <

[256] Tewari, Rajesh, Nele Horemans, Robin Nauts, Jean Wannijn, May Van Hees, and Hildegarde Vandenhove. 2015. Uranium Exposure Induces Nitric Oxide and Hydrogen Peroxide Generation in Arabidopsis Thaliana. Environmental and Experimental Botany 120: 55–64. https://doi.org/10.1016/j.envexpbot.2015.08.

[257] International Atomic Energy Agency (IAEA). 2009. World Distribution of Uranium Deposits (UDEPO) with Uranium Deposit Classification. IAEA-TECDOC-1629. Vienna: Austria, 128p. ISBN: 978-92-0-110509-7.

[258] Xu, Lechang, Naizhong Liu, and Guofu Zhang, 2011. Waste Water Treatment of CO2 + O2 in-situ Leaching Uranium. The New Uranium Mining Boom, pp. 31–37. https://doi.org/10.1007/978-3-642-22122-4_4.

[259] See footnote 252.

[260] Asghar, Fiaz, Zhanxue Sun, Gongxin Chen, Yipeng Zhou, Guangrong Li, Haiyan Liu, and Kai Zhao. 2020. Geochemical Characteristics and Uranium Neutral Leaching through a CO2 + O2 System—An Example from Uranium Ore of the ELZPA Ore Deposit in Pakistan. Metals 10(12): 1616. https://doi.org/10.3390/met10121616.

[261] See footnote 251.

[262] See footnote 251.

2), alkaline leaching (pH > 9), neutral leaching (pH: 6.8–8.2, including $CO_2 + O_2$ leaching and weak acid leaching), and bioleaching (solutions containing specific types of microorganisms).[263] The neutral $CO_2 + O_2$ in situ leaching is currently considered to be one of the most effective methods for the exploitation of uranium ore deposits compared to the other techniques, which can be applied to low permeability, low grade, high carbonate, and high salinity uranium ore deposits and have a potentially lower environmental footprint.[264,265,266,267] In addition, the ISL process can also play a proactive role as a CO_2 removal option, rather than simply being a technique used to avoid CO_2 emissions.

4 Conclusions

In the new climate-energy scenario, which will have to be developed to achieve carbon neutrality by 2025, it is essential to ensure the development of an efficient strategy that seeks to address both components. It is recognized that secure, sustainable, and economically viable energy is needed to meet industrial and domestic needs, but it must be produced alongside efforts to protect the environment. In addition, it is also true that the current world's energy mix is heavily reliant on fossil fuels, approximately 80%. The need to move towards a carbon-neutral system requires implementing an efficient and sustainable energy transition strategy. However, the effective development of competitive low-carbon energy, namely renewables, will take time since some critical features still need to be improved. In this context, the world's energy players are making enormous technical and economic efforts to achieve this transition as soon as possible without compromising either the environment or the energy supply. So, it is crucial to undertake ongoing

[263] See footnote 251.

[264] See footnote 260.

[265] Yang, Yun, Wenjie Qiu, Zhengbang Liu, Jian Song, Jianfeng Wu, Zhi Dou, Jinguo Wang, and Jichum Wu. 2022. Quantifying the Impact of Mineralogical Heterogeneity on Reactive Transport Modeling of CO2 + O2 in-situ Leaching of Uranium. Acta Geochimica 41(12): 50–63.

[266] See footnote 254.

[267] Zhang, Haitao, Tong Zhang, and Yupeng He. 2023. Reactive Transport Model of Uranium by CO2 + O2 in situ Leaching. Environmental Science and Pollution Research 30(24): 65976–65989.

initiatives that will help not only to avoid CO_2 emissions but also to seek alternatives to reduce and, above all, to remove CO_2 emissions. This new strategic scenario led to the development of the concept of carbon dioxide removals. CCS technologies have been officially recognized by the Paris Agreement as a powerful tool that cannot be overlooked to effectively achieve carbon neutrality by 2050. After several years of investment in CCS technologies, it became clear that the high costs associated with infrastructures would discourage potential investors. Facing the unavoidable need to reduce CO_2 emissions, the unpopular CCUS technologies and even the CCU technologies gained their own space within the scope of CO_2 abatement technologies. CCUS technologies are currently considered one of the best solutions to achieve carbon neutrality, as they can both produce a valuable resource and remove CO_2 emissions from the atmosphere, thereby lowering the cost threshold for CO_2 geological storage infrastructures. Additionally, CCUS technologies can be applied not only in the traditional CO_2-EOR projects, which are used to improve oil and gas recovery from depleted fields, it can also be used in CO_2-ECBM and CO_2-ESR, allowing to enhance natural gas (mainly CH_4) recovery from coal seams and shale gas, respectively. CO_2-EWR technologies, also considered a CO_2 removal option, are applied in deep saline aquifers. However, instead of the traditional injection of CO_2 and subsequent dissolution in water, CO_2-EWR technologies focus on water production while injecting CO_2. The water that is produced can be treated and used for agriculture, industry, and even for domestic use. Less popular, but also highly efficient, are the CO_2-EGS, CO_2-EGH, and $CO_2 + O_2$ in situ uranium leaching technologies. CO_2-EGS technologies consist of producing natural heat from the earth to heat buildings and generate electricity while storing CO_2. CO_2-EGH technologies consist of the CH_4 recovery by injecting CO_2, which replaces the CH_4 in the CH_4 hydrate. Finally, $CO_2 + O_2$ in situ uranium leaching technologies consist of injecting a solution of $CO_2 + O_2$ to induce dissolution of the uranium ore body to recover uranium from the solution that is pumped back to the surface. CCUS technology can act as a bridge for developing more environment-friendly energy technologies and as a mechanism for reducing the CO_2 emissions of fossil fuel power plants and other CO_2 emitting sources.

Acknowledgements The authors are indebted to both Fundação Fernando Pessoa and MARE-Centro de Ciência do Mar e Ambiente/URI Coimbra /

Portugal) for permitting them to include the present work in their research programmes. Financed by Fundação para a Ciência e Tecnologia (Portugal) is also gratefully acknowledged. Thanks are also due to the RCGI—Research Centre for Gas Innovation, hosted by the University of São Paulo (USP) and sponsored by FAPESP—São Paulo Research Foundation (2014/50279-4 and 2020/15230-5) and Shell Brasil, and the strategic importance of the support given by ANP (Brazil's National Oil, Natural Gas and Biofuels Agency) through the R&D regulation. We also gratefully acknowledge the institutional support of the Institute of Energy and Environment (IEE) and the University of São Paulo (USP), Brazil.

CHAPTER 4

Financial Considerations for CCUS Developments

Lauren Davies, Alistair Wishart, and Ellen Swarbrick

1 Net-Zero and the Financing Gap

There is increasing pressure on economies across the world to accelerate the transition to a less carbon-intensive economy. This is reflected in the number of governments setting net-zero emissions at the top of their policy agenda: the EU has set a legally binding target on its member states to achieve net-zero greenhouse gas emissions by 2050, and, as at the time of writing, a further 28 countries have adopted legally binding net-zero targets,[1] with many more countries and regions worldwide committing to net-zero or low-emission targets.

Carbon capture, utilisation and storage (**CCUS**) is a key part of the transition for many industries, and particularly those industries which are not able to electrify or switch to less carbon-intensive fuels, for instance,

[1] 'Net Zero Scorecard', https://eciu.net/netzerotracker, accessed 16 September 2024.

L. Davies (✉) · A. Wishart · E. Swarbrick
Vinson & Elkins LLP, London, UK
e-mail: ldavies@velaw.com

E. G. Pereira et al. (eds.), *Carbon Capture Utilization and Storage*, Sustainable Development Goals Series,
https://doi.org/10.1007/978-3-031-81272-9_4

thermal power generation, oil and gas production and certain heavy industries such as steel, cement and chemical production. CCUS is also integral to the production of less carbon-intensive fuels, being required for the production of 'blue' hydrogen. Blue hydrogen can in turn improve demand in certain sectors, facilitating the uptake of the even lower carbon 'green' hydrogen, which may share the same infrastructure.

The scale of development and investment required in CCUS projects to allow industry to meet net-zero goals is immense. The International Energy Agency estimates that the amount of CO_2 captured per year will need to increase from 2023 levels of around 50 million tonnes (**Mt**) CO_2 per year[2] to over a gigatonne of CO_2 per in order to reach its net-zero emissions by 2050 scenario.[3] The cost of this development is expected to be vast, with McKinsey & Company anticipating that annual global investment in CCUS technology required to achieve net-zero targets will be between US$120 billion and US$150 billion per year by 2035.[4] This investment will need to be supported and financed by a range of investors, from governments and other public bodies through to developers and commercial banks. Taking into account CCUS development planned at the end of 2023, the IEA estimates that there will still be a capture capacity shortfall of around 565 Mt of CO_2 per year and a storage capacity shortfall of around 385 Mt of CO_2 per year in 2030.[5]

In order to maintain the pace of development of CCUS projects to meet net-zero objectives by 2030 and beyond, billions of dollars will need to be invested. However, attracting the scale of investment required to meet net-zero objectives is complicated through the economic challenges of the current state of CCUS technology and projects. Overcoming the challenges of CCUS will give an easier path to returns for investors and financiers, unlocking the investment required to bridge the

[2] 'Carbon Capture, Utilisation and Storage' (www.iea.org, 11 July 2023), https://www.iea.org/energy-system/carbon-capture-utilisation-and-storage, accessed 16 September 2024.

[3] Ibid.

[4] 'The World Needs to Capture, Use and Store Gigatons of CO_2: Where Are How?' (www.mckinsey.com, 5 April 2023), https://www.mckinsey.com/industries/oil-and-gas/our-insights/the-world-needs-to-capture-use-and-store-gigatons-of-co2-where-and-how, accessed 16 September 2024.

[5] See n 2.

CCUS financing gap and to allow the development of CCUS at the scale required.

2 The Economics of CCUS Projects—Costs

While CCUS is widely accepted as being integral for countries to reach their net-zero targets, it is still an incredibly immature market. The range of project types and locations creates sizeable uncertainty when it comes to costs, which can in turn lead to difficulties obtaining financing. For global capture capacity to increase more than seven times over 10 years from 2023 levels, over US$150 billion in capital expenditure is required.[6]

The capital requirements of CCUS projects will depend on the type of project being developed, and the types of financing required for those projects are therefore likely to vary from project to project. A CCUS enhancement to an existing power or industrial project is likely to have a lower capital expenditure than a greenfield project and is particularly popular for industrial processes in hard-to-abate sectors, such as hydrogen and ammonia production. Compared to replacing existing hydrogen and ammonia production plants with an entirely different technology of electrolysers, along with the adjustments in infrastructure required, retrofitting with CCUS offers a cost-effective way to reduce emissions while the market matures and demand for low-carbon products increases. In fact, the IEA has estimated that retrofitting CCUS technology to existing power and industrial plants may mitigate 8 gigatonnes of CO_2 emissions in 2050. CCUS can therefore be an important stepping stone to the production of renewable fuels. In such cases, the existing revenue streams from the underlying project may allow for debt financing to be used to support the development of the CCUS project.

Conversely, the capital costs are likely to be substantial for a 'full value chain' CCUS project which entails developing the CCUS project itself, along with the associated transportation network and storage facility. Historically, oil and gas companies with large balance sheets have been able to finance these projects using their own funds or with corporate financing. In terms of the CCUS project itself, economics will be affected by the varying concentrations of the CO_2 that need to be captured in that particular project. A higher concentration of CO_2 emissions results

[6] Wood Mackenzie, 'What's Shaping CCUS Project Costs?', https://www.woodmac.com/news/opinion/ccus-project-costs/.

in a lower energy requirement to separate the CO_2 particles out. In other words, the most carbon-intensive processes will see better project economics from including CCUS technology. Emissions streams from ethanol production or natural gas processing may offer better returns for a CCUS project compared to cement production, for example.[7]

For the required infrastructure, transportation and storage costs are variable and will depend heavily on location and infrastructure type. Captured CO_2 may be transported via HGVs and ships for smaller quantities, but most commonly, larger quantities are transported by pipeline. Extensive pipeline networks already exist around the world; that being said, the estimated CO_2 transportation infrastructure needed to support longer term CCUS deployment over the next 30–40 year is about 100 times what currently exists.[8] The geological storage of CO_2 not otherwise utilised involves costs related to injection and monitoring, although storage costs can be negative if the CO_2 is injected into oilfields in enhanced oil recovery.

Capital expenditure on transport and storage infrastructure in particular can be reduced through the development of shared hubs or CCUS 'clusters'. Standalone CCUS projects are economically viable only for the most significant CO_2 emitters. Such projects would include refineries, for which the sheer scale and cost of CO_2 justifies the development of dedicated CCUS and associated transportation infrastructure. For smaller emitters, the cost of developing these additional facilities far exceeds the cost of emitting CO_2. One way to overcome this is through the development of industrial clusters. Industrial clusters are premised on the economic principle of economies of scale, whereby many smaller CO_2 emitters will be physically located in close proximity to each other, increasing demand for, and reducing per unit cost of, transportation and storage infrastructure. Clusters are widely accepted as being integral to reducing capital costs of CCUS projects, and many governments and private companies have supported cluster development for CCUS and associated production processes.[9] This will allow different financing

[7] Baylin-Stern & Bergout (2021), Globe and Mail Editorial Board, https://www.iea.org/commentaries/is-carbon-capture-too-expensive, accessed 16 September 2024.

[8] IEA, referenced by Global CCS Institute (2018), https://www.globalccsinstitute.com/wp-content/uploads/2018/12/Global-CCS-Institute-Fact-Sheet_Transporting-CO2-1.pdf, accessed 16 September 2024.

[9] Northern Lights Cluster in Norway (partnership between our partner Equinor, Shell and Total); PORTHOS, Netherlands' Port of Rotterdam CCUS Backbone Initiative; Zero Carbon Humber; Net Zero Teesside UK.

models to be developed for separate parts of the value chain. For example, a transportation network may be financed using similar models to an oil and gas pipeline, whereas a carbon capture project can be financed according to the revenue source. Additionally, lower capital costs for carbon capture projects using shared infrastructure may allow for smaller or more specialised companies to develop standalone carbon capture projects.

In the medium-to-longer term, as has been seen in the development of other renewable energy sources such as solar and wind electricity generation, advances in technology, economies of sale and the development of efficiencies due to an improvement in technical know-how are likely to all result in a fall in the cost of carbon capture projects, and a reduction in the diminishing of returns on investment over time. Nevertheless, costs of CCUS are currently a major barrier to the wider adoption of the technology. These will need to be offset by revenue streams to make projects economically viable in the near term.

3 The Economics of CCUS Projects—Revenues

Critically, the principal driver of any financing of a CCUS project is the economic story. The large capital requirements of a CCUS project will mean that, for all but the largest developers with an extensive balance sheet, an element of financing will likely be required to pay for the up-front costs of developing a CCUS project. A key principle of any financing, whether it is debt or equity financing, is whether the asset being financed will be able to give a return on investment, or be able to repay the debt. This, in turn, will be driven by the source of revenue for the CCUS project.

Currently, the main use, and value, of captured CO_2 is in enhanced oil recovery, where the captured CO_2 is injected into the reservoir to produce more oil extraction than would conventionally be achieved. The value of the captured CO_2 from the CCUS project feeding enhanced oil recovery is derived from the value of the additional oil that is extracted, rather than the value of the CO_2 itself. Direct sales of CO_2 for other uses are typically smaller scale, with carbon pricing being insufficient to support large-scale carbon capture projects. Prior to 2020, the vast majority of CCUS projects were developed to utilise the captured CO_2

for enhanced oil recovery, with the majority of these located near hydrocarbon reservoirs in the US, Canada and Norway.[10] These projects have been economically successful not only for the additional value associated with enhanced recovery of oil, but also as a result of lower capital costs. These projects are developed in close proximity to the oil reservoirs and existing CO_2 infrastructure clustered around the reservoirs allows for economies of scale.

More recent technological and policy developments have resulted in new potential direct or indirect sources of value for CO_2 being captured from CCUS projects. The CO_2 captured in CCUS projects can be used or sold as feedstock for a range of new synthetic fuels, such as renewable methanol which reduces the overall carbon intensity of these fuels. CCUS facilities are also one of the key means of decarbonising certain hard-to-abate sectors, cost-effectively reducing the carbon intensity of thermal power generation, natural gas processing and industrial processes, such as cement, iron, steel and ammonia production. This technology can be used in the development of new projects, for instance the Net Zero Teeside Power project in the United Kingdom (**UK**), which is scheduled to become the first commercial-scale gas-fired power station with CCUS in the world, but can also be retrofitted to existing power and industrial facilities.

The revenue stream for a carbon capture project, whether this is an enhancement to an existing project, or integrated into a new-build project, will likely be the sales of the offtake from the project. For a thermal power generation project, the revenue stream will likely be the sale of electricity or water, whereas for an industrial project, it will be the sale of the goods produced by the project, such as refined metals, cement or chemicals. In such industrial markets, it remains to be seen whether these products will be able to command a higher price to offset the costs of building and operating the carbon capture project (a so-called green premium). There are indications that this 'green premium' is possible and is currently being developed in the European steel market. However, if the offtake is sold at the current market price of non-carbon-abated products, the return on investment will naturally fall well below expected levels due to the additional costs of the CCUS technology. While we will

[10] 'CCUS in Clean Energy Transitions: A New Era for CCUS (www.iea.org, 11 July 2023), https://www.iea.org/reports/ccus-in-clean-energy-transitions/a-new-era-for-ccus, accessed 16 September 2024.

discuss government support elsewhere in this chapter, increased government incentives, such as carbon pricing and trading schemes, contracts for difference and other subsidy regimes are being used or are in development to compensate investors for these lower returns for carbon-abated projects.[11]

Recent policy developments and government support have led to the potential for transportation networks and storage networks to be developed separately from the underlying carbon capture projects. These are large, and often very capital-intensive projects, with naturally monopolistic qualities, meaning that it is often inefficient for a single company to develop an extensive, dedicated transportation and storage network for a single carbon capture project. By developing carbon hubs or clusters, a shared transportation and storage network can be instituted, using existing gas network models. Both transportation and storage networks may be multi-user, with the networks charging a fee for the transportation of CO_2, and ultimately its storage. These networks are ideal prospective recipients of government support and structuring; for example, the UK has introduced a transport and storage regulatory investment business model. This is a 'user-pays' model based on the existing regulated asset base model used in other regulated utility models in the UK, such as gas or water. The user fee will be regulated to allow the developer to recover its costs of developing and operating the network, as well as generating an allowed return on investment to incentivise development. Additionally, the UK's proposed model will include government support for insufficient revenue during the development phase of the CCUS hub, as well as limited government support for certain project risks not capable of being insured against, in each case to provide additional revenue certainty for developers and financiers.[12]

[11] For example, the UK is to implement a new import carbon pricing mechanism by 2027. See 'New UK Levy to Level Carbon Pricing' (*Gov.UK*), https://www.gov.uk/government/news/new-uk-levy-to-level-carbon-pricing#:~:text=The%20UK%20is%20to%20implement,to%20support%20the%20decarbonisation%20drive.&text=Goods%20imported%20into%20the%20UK,those%20produced%20in%20the%20UK, accessed 16 September 2024.

[12] For more detail please see HM Government, 'CCUS Net Zero Investment Roadmap, Capturing Carbon and a Global Opportunity' (April 2023), https://assets.publishing.service.gov.uk/media/64a29b7d06179b00131ae94e/ccus-investment-roadmap.pdf, accessed 16 September 2024.

4 Government Support

For standalone CCUS projects, the absence of a market and price for the sequestration of carbon means it is difficult to form a revenue stream to support investment in and financing of these projects. As a result, the success of these projects will, at least in the short-to-medium term, rely on government support. In certain jurisdictions, there is a carbon tax or a requirement for entities to surrender emissions credits for each tonne of CO_2 they produce. For instance, within Europe, the EU Emissions Trading Scheme has led to an actively traded secondary market for emissions credits. To date, such trading schemes have not been sufficient to create a standalone revenue stream for CCUS projects, but have allowed investors to reduce the cost of abating CO_2 from other processes.

Government support can take many forms, but ultimately seeks to compensate investors for the higher cost of a carbon-abated project, or to provide a revenue stream for a standalone carbon sequestration project. Grants are the most straightforward form of government support, and many governments are developing grant platforms to support new CCUS projects, for instance the EU's Innovation Fund. Although, government grants will often only comprise a small part of the capital required.

Subsidy schemes may also be used by governments to compensate investors for the lower returns on or for the unprofitable element of a CCUS project. These may take the form of direct subsidies, as is the case with the Dutch government's sustainable energy transition subsidy scheme (SDE++), which has been used to provide €2 billion of subsidies for the development of the Port of Rotterdam Transport Hub and Offshore Storage project.

Subsidies may also be less direct, through contracts for difference or government revenue support. In the UK, the proposed model for supporting the generation of carbon-abated electricity using CCUS technology is similar to the contracts for difference model used for the development of renewable electricity. Under this model, the government will pay both an availability fee and a generation fee to the generator, reflecting the costs of the project, and an agreed price for the generated electricity, allowing a guaranteed minimum rate of return for investors. Similarly, for CCUS projects in industrial carbon-abated facilities, the UK government is proposing a model which provides a grant for a proportion of the total capital costs of the project. In addition, it is proposing an operating expenditure payment, whereby developers are paid a sum, which is the difference between the 'strike price' (a price reflecting the cost of investing in the CCUS technology) and the 'reference price' (a measure of the average market price), with an agreed return on investment.[13]

Tax credits or reductions may be used to compensate or incentivise developers, such as in the US, where section 45Q of the Internal Revenue Code creates an emission reduction tax credit based on the CO_2 being utilised or stored. This allows for CCUS projects to be developed on a corporate basis to utilise tax credits within the group or to utilise tax investor or tax equity financing using the emission reduction tax credit as the principal source of return on investment for the CCUS project.

While government support is essential to the development of CCUS projects, and particularly to the development of the required capital-intensive transport and storage networks, it creates an inherent risk for investors. Governments and policies change over time, and there is therefore a risk of discontinuation of government support to all CCUS projects reliant on it, particularly those which are not able to provide a standalone revenue stream. Government support will need to be long-lasting for investors to recoup their investment and generate a return to incentivise investors to develop new CCUS projects. Similarly, the legal and fiscal regime, and particularly the respect for the rule of law may be critical in the development of government-supported CCUS projects. Investors will be more willing to support projects where government support is likely to remain on the original contracted terms throughout the life of the project, despite intervening governmental change. Nevertheless, as has been seen in government support for renewable electricity and other similar regimes, it is likely that the terms of government support may be renegotiated in the future as carbon pricing develops, the risks associated with CCUS projects are more fully understood and mitigated, and the cost of CCUS projects falls.

[13] UK Dept. for Business, Energy & Industrial Strategy 'Carbon Capture, Usage and Storage: Industrial Carbon Capture Business Models Summary' (Dec. 2022), https://assets.publishing.service.gov.uk/media/639c59bbd3bf7f7f95e05734/industrial-carbon-capture-business-model-summary-december-2022.pdf, accessed 16 September 2024.

5 Sources of Finance for CCUS Projects

The availability of financing for a CCUS project will depend on a variety of factors, particularly the revenue model of that project, the commercial feasibility of the project, the availability of government support, the technical and geographical challenges of the project and the risks associated with that particular project. These will change from project to project, and region to region, and so will be assessed on an individual project basis.

As stated above, the majority of CCUS projects developed to date have been enhanced oil recovery projects, largely developed by oil companies with a strong operating history, positive cashflows and a substantial balance sheet. In addition, the revenue streams from enhanced oil recovery CCUS projects have been intrinsically linked with the oil production projects. These projects have historically been financed on the balance sheet of the developer, whether through equity or corporate financing, although many historic CCUS projects have relied on some form of government support, such as direct grants from the US Department of Energy in the case of the Petro Nova project, and carbon credits in the case of the Gorgon LNG project.

Potential sources of finance for a CCUS project are as follows:

- **Equity**: Equity is the developer or an investor providing their own funds for the development of a project. The motivations of a developer or an investor may vary, with most investors looking for a return on investment on the equity spent, while others wish to develop a CCUS project to decarbonise their general portfolio or support one of their other operations. Equity investment can come from shareholders, strategic investors, venture capital or private equity, with each investor having different motivations and seeking different returns on their investment.
- **Debt Finance**: Commercial banks are likely to have a large part to play in the financing and development of CCUS projects. Developers may be able to raise debt on their balance sheets, if they have existing revenues from other sources, to be used to develop CCUS projects. Developers could also use construction or project finance to develop a CCUS project off-balance sheet. Other more specialised products, such as export finance, may also be utilised by developers for certain projects. The environmental impact of a

CCUS project would also make such projects and developers ideal candidates for sustainability-linked or green financing, where the debt proceeds will need to be utilised to fund sustainable projects. Commercial banks have increasingly focused on financing the energy transition in support of climate change goals, and so there is substantial liquidity from the banking sector to support the financing of CCUS projects. Although, there are also challenges in deploying this finance to support CCUS projects, as is discussed later on in this Chapter. Historically, few CCUS projects have been financed on their own merits, with projects such as Petro Nova being financed by the Japan Bank for International Cooperation and Mizuho to support the development of the Japanese technology used in the project.

- **Capital Markets**: A developer could raise financing through the capital markets either by selling its shares on a public market, or by issuing bonds. This provides access to substantial capital pools on attractive terms. Share issuances represent a long-term pool of capital without a repayment requirement, making it suitable for earlystage non-revenue generating phases of development, though the sale of shares reduces the ownership of the company for the existing shareholders. Debt issuances in the form of bonds or loan notes represent debt capital on substantially less restrictive terms than an equivalent project financing or even a corporate financing. Debt issuances will require repayment, and so will often need a revenue stream to ensure the repayment of the debt, leading to the debt capital markets often being unavailable for construction phase projects.

  Typically, these instruments are not usually used for early-stage projects or for newer technologies, as investors are generally unwilling to take the technology risks or the construction risks inherent in these companies, and will generally not value a non-revenue generating company highly, making this a very risky prospect for investors. Equally, bondholders will expect a high return on their investment as a result of the risk that they are taking, meaning that interest rates may have to be substantial to ensure a successful debt issuance, which may cause financial issues for a pre-revenue company. Furthermore, there are often substantial disclosure and other rules for companies which have instruments listed on a public exchange, which may be unsuitable for early-stage or smaller developers.

For larger or more established developers, particularly those with multiple diversified revenue streams, the capital markets may be an attractive way of raising capital. Capital markets give access to large and diverse pools of capital, and allow these companies to issue long-term debt or equity in order to raise large sums. CCUS developers may be able to issue project bonds for the development of a specific CCUS project or issue green or sustainable bonds to raise funds to support their CCUS development programme, though these come with additional conditions as to the use of the funds, and may be adversely impacted if the CCUS projects that they fund do not perform as expected.

- **Government Support**: Government support can take many forms, from research and development grants, to revenue contracts for development, loan guarantees or direct equity investment in the project. Government support for CCUS projects also extends to less direct funding methods, such as CO_2 pricing mechanics and emissions trading schemes creating a value for the CO_2 priced and stored.

Government policy can also be a form of indirect support for CCUS projects. For instance, the development of carbon clusters and revenue models for transportation and storage systems allows the development of a communal transportation network, which, in turn, allows developers to implement non-value chain CCUS projects far more easily. This has made its way into policy in several states, for example in the UK, where the £1 billion Carbon Capture and Storage Infrastructure Fund has been established to support the development of transportation and storage facilities and industrial carbon capture projects in the UK, focusing on several carbon hubs across the UK.[14]

One major drawback to government support in the development of CCUS projects is the consistency and availability of government

[14] UK Dept. for Energy Security and Net Zero and UK Dept. for Business, Energy & Industrial Strategy 'The Carbon Capture and Storage Infrastructure Fund: An Update on Its Design (May 2021) (accessible webpage)' (*uk.gov*, 7 May 2021 (updated 16 Dec. 2022)), https://www.gov.uk/government/publications/design-of-the-carbon-capture-and-storage-ccs-infrastructure-fund/the-carbon-capture-and-storage-infrastructure-fund-an-update-on-its-design-accessible-webpage#:~:text=The%20Carbon%20Capture%20and%20Storage%20Infrastructure%20Fund%20(CIF)%20was%20first,Spending%20Review%20in%20November%202020, accessed 16 September 2024.

support during the life of the project. The reliance on government support for a revenue stream for a CCUS project creates the risk that the project will not be economically viable if the government support were to cease or be withdrawn. It is likely that in the nearer-to-mid-term, CCUS projects will need to rely on government support while the CO_2 market develops. Governments with a clear policy focus on supporting CCUS projects, and achieving their carbon emissions goals, as well as having a history of consistent government support for other industries, are likely to attract more investment from private developers relying on that government support. Initial government support to de-risk initial investment, eventually leading to a thriving industry, is a model that has been used before around the world to develop solar and wind technology, so there is substantial precedent for the success of such a model in the longer term.

Some of the most common forms of government support that could be used to support the CCUS industry are as follows:

- **Public-Private Partnerships**: Public-private partnerships are generally long-term arrangements between a private developer and a public entity for the development of public infrastructure. These partnerships can take many forms, but range from joint venture agreements between a private developer and a public entity, to a contractual arrangement where the private developer finances and develops the project in exchange for a contractual revenue stream from the operation of the project of the public entity. Public-private partnerships generally share the risks of development, allowing the private developer to de-risk CCUS projects during the early stage of the technological development. Where the government provides a revenue stream for the operation of the project, this may provide the revenue model for the project to be financed.

  Forms of public-private partnerships have been used for the development of CCUS projects in recent years, for instance, the funding of the Petra Nova project in Texas was composed of direct grants from the US Department of Energy. The funding was used to develop a post-combustion carbon capture facility at a coal-fired plant to demonstrate the use of carbon capture technology at scale and de-risk construction and technological issues for future CCUS projects through lessons learned at the plant.

Government support and public-private partnerships have many overlapping similarities and goals, particularly the risk-sharing between government and private developers and financial support. Many of the government support mechanisms described in this section may be a form of public-private partnership.

- **Grants and Subsidies**: Grants are direct payments from the government to a company, often used to support research and development of new technologies within a country or region. Grant funding may cover specific aspects of a project, such as research or the installation of a specific technology. By providing some of the capital cost via grant, the government is, in effect, de-risking the project for the investor by lowering the overall capital expense for the developer, allowing them to develop a CCUS project more easily or to generate a return on the investment. Grants may cover a small part of the overall cost of the project, for instance the US Department of Energy awarded a grant of US$ 167 million to the Petro Nova project as part of the Clean Coal Power Initiative to prove and accelerate the development of commercial large-scale carbon capture and storage technology. Similarly, the UK government has offered grant funding of up to £20 million to support the development, demonstration and de-risking of novel CCUS technologies in the UK, focused on research and early-stage development and demonstration technology.[15] This allows developers to trial and develop early-stage technology where there may not yet be a clear revenue case for the CCUS project, as the developer may still be testing the technology's effectiveness or refining its construction process. Grant funding allows developers to de-risk the technology and construction process through demonstration and trial projects. More recent government policy, focused on achieving the net-zero targets, has led to more comprehensive grant schemes. The EU's Innovation Fund, which is funded through the EU's emissions trading scheme, supports the

[15] UK Dept. for Energy Security and Net Zero and UK Dept. for Business, Energy & Industrial Strategy 'Carbon Capture, Usage and Storage (CCUS) Innovation 2.0 Competition: Call 2 (Closed to Applications)' (*uk.gov*, 31 May 2022 (updated 28 June 2023)), https://www.gov.uk/government/publications/carbon-capture-usage-and-storage-ccus-innovation-20-competition-call-2, accessed 16 September 2024.

development of projects over a 10-year period, and may provide up to 60% (for regular grants) or up to 100% (for competitive bidding) of relevant costs, which usually covers capital and operational costs minus revenues for the first 10 years of operation, though 60% of the grant funding is linked to verified emissions avoidance of the project during its lifetime.[16]

Subsidies or other indirect government support may impact the revenue story of a CCUS project, particularly those that either reduce the up-front capital costs for the investors, or provide certainty of income for the project. The UK's proposed dispatchable power agreement business model provides an example of this, where a low-carbon power plant containing a carbon capture facility may acquire a contract for difference with the UK government. Once the plant is operational, the project will be paid an availability payment for low-carbon electricity generation capacity, and a variable payment for the costs of the electricity generated.[17] For industrial carbon capture facilities, the UK government has proposed a hybrid business model where the government may fund a capital grant of up to 50% of the total capital costs, with a revenue support contract with carbon storage payments on capital expenditure, operating costs and an expected rate of return, and a carbon capture strike price based on a fixed trajectory reference price.[18]

- **Carbon Tax and Emissions Trading Schemes**: Carbon taxation or carbon pricing schemes both create incentives for companies to develop CCUS projects to decarbonise their operations, as well as create an economic incentive for such CCUS developers. The impact of carbon taxation and pricing schemes on the development of CCUS projects has already been seen in Norway, where Equinor (formerly Statoil) developed the Sleipner carbon capture project principally to gain relief from the Norwegian carbon tax.

[16] 'Climate Action: What Is the Innovation Fund?' (*climate.ec.europa.eu*), https://climate.ec.europa.eu/eu-action/eu-funding-climate-action/innovation-fund/what-innovation-fund_en, accessed 16 September 2024.

[17] UK Dept. for Business, Energy & Industrial Strategy 'Carbon Capture, Usage and Storage: Dispatchable Power Agreement Business Model Summary' (Nov. 2022), https://assets.publishing.service.gov.uk/media/6373993e8fa8f559604a0b8b/ccus-dispatchable-power-agreement-business-model-summary.pdf, accessed 16 September 2024.

[18] See n 13.

- **Taxation**: Government fiscal policy may also play an important role in both incentivising the development of CCUS projects and supporting their financing. In the US, Section 45Q of the Revenue Code provides a monetary basis for CO_2 captured, with tax credits for the 2024 calendar year of US$27.75 per metric tonne of qualified CO_2 which is disposed of in secure geological storage, and US$13.88 per metric tonne of qualified CO_2 which is used as a tertiary injectant in a qualified enhanced oil or natural gas recovery project and disposed of in secure geological storage. This gives an equity revenue stream for the CCUS project allowing for a higher rate of return for investors and the potential for financing.
- **Government Guarantees**: Governments may also provide support for CCUS projects through guarantees of commercial loans, or through direct lending to the developer, in order to achieve certain policy goals, most usually net-zero emissions goals or the promotion of national exports. CCUS financing may benefit from loans or guarantees issued by export credit agencies where the CCUS project benefits national exports, or where national exports are used in the development of that CCUS project. Other CCUS projects may be able to benefit from government infrastructure or climate-linked guarantees, such as the UK's sovereign infrastructure guarantee, which gives sovereign backing to the guarantees issued by the UK Infrastructure Bank. The guarantee from a government of the debt of a CCUS project or a developer would greatly increase its attractiveness to financiers, and the involvement in a quasi-governmental organisation in the financing would be additional comfort to the developers and the financiers of the government's commitment to supporting the project. These products are not always available to developers and often come with restrictions on the financing, particularly in the case of export credit agencies.

The proposed make-up of a finance group also plays an important role in the availability of finance. Development banks and export credit agencies are an integral part of the international project finance market; many projects will not have come online without their involvement. At a high-level, development banks are established to promote international and regional economic co-operation. They are focused on emerging markets and aim to attract private capital to projects that are important from a socio-economic perspective. Development banks offer a range of products, in the absence of which many commercial banks may not be able (or willing) to lend to a project.

Export credit agencies, like development banks, also offer a wealth of products and services to support the development and financing of energy and infrastructure projects across the world and often fill a funding gap that private-sector lenders alone cannot fulfil. Development banks and/or export credit agencies are likely to play a significant role in the development of CCUS projects in developing jurisdictions, for example by providing commercial and political support to mitigate certain political or non-commercial risks that commercial banks are unable and/or unwilling to bear. Government policies in support of net-zero targets have resulted in development banks and export credit agencies being encouraged to support the development of low-carbon projects within their region or to support the development and export of low-carbon technologies and businesses within their home nation. With this political focus, it is expected that development banks and export credit agencies will be integral in the development of CCUS projects in developing or less stable regions.

6 Challenges to Raising Capital for CCUS Projects

Historically, most CCUS projects have been financed using a combination of equity financing and government support; very few CCUS projects have used external sources of financing specifically for the project. The corporate financing approach may be an option for oil majors and developers with a strong operating history supported by positive cashflows, but may not be available for all developers. For example, many developers simply do not have the balance sheet, historic financial statements or evidence of past performance required by lenders to support the substantial volume of capital necessary to develop a CCUS project, and so will need to rely on the revenues generated by the CCUS project to access capital. Project finance offers an attractive alternative and has been used to finance the construction and development of large-scale complex industrial and energy facilities across the world. For a project financing, a project's shareholders (its **sponsors**) incorporate a special purpose vehicle (an **SPV**), the sole business of which is to develop and operate a project and to borrow funds from project financiers.[19] This SPV has no operating history and, until it becomes operational, no revenues or cashflow. Project financiers therefore rely solely on operational cashflows for the repayment of their debt, subject to any agreed sponsor support during the completion phase. Because there is limited (or no) recourse to the sponsors or their assets, project finance lenders and their advisers conduct extensive due diligence to ensure that the project's projected cashflows are based on reasonable assumptions. This diligence process helps to identify potential project-related risks. Appropriate risk mitigation measures are then considered and implemented to the extent required for the bankability of the project, through (among other means) the development of a robust project contractual structure.

The project finance model has long been employed as a means of raising significant volumes of funds for capital-intensive energy and infrastructure projects across the world. In many circumstances, a project would not be viable without the ability to raise external debt through a project finance structure and/or without the involvement of key project finance debt and credit providers. While project finance has played a vital role in the exploitation of traditional oil and gas assets for some time, as well as in the development of solar, wind and other renewable projects, there is a compelling need for this financing structure to be deployed at a larger scale in the development of projects required to accelerate the transition from fossil fuels to energy sources with lower carbon emissions. The inherent benefits of the non-recourse (or limited recourse, as applicable) project finance model over the more traditional corporate financing, along with the well-established use of carbon capture technologies in industry to date and, more recently, the successful implementation of a number of large-scale CCUS projects, should encourage the rapid growth of the carbon capture project finance market.

[19] Many carbon capture projects located in the United States are structured and developed to incorporate some element of tax equity in a manner that seeks to comply with the requirements of the relevant section(s) of the Internal Revenue Code, in order to benefit from emission reduction tax credit(s). Key structuring considerations in connection with such available tax credits go beyond the scope of this Chapter.

7 Revenue

A project's ability to generate a long-term revenue stream, at a level sufficient to service debt and meet project operating costs until final maturity of the debt, is a potential barrier to accessing the substantial volumes of external debt often associated with project finance transactions. A project will not attract financing unless the financial model, based on reasonable assumptions and supported by relevant evidence, projects a consistent revenue stream for the term of the debt at a level that meets certain debt service coverage ratios. Moreover, in order to be viable from the standpoint of the sponsors, a project will also need to demonstrate that it will meet their desired return on investment and projected revenues are a key part of this.

In a more 'traditional' limited-recourse project financing, sponsors are required to demonstrate the availability of an offtake market for the term of the debt, plus a buffer. Financiers require comfort that the assumptions used in the financial model to project revenues are appropriate, in particular in terms of sales price and volume (i.e., availability of purchasers in the market to take product). Project lenders have been reluctant to take volume risk and customarily require the SPV to enter into long-term agreement(s) that guarantee the offtake of a substantial portion of the production, unless there is a proven long-term liquid market for the product (demonstrated by availability of short-term and spot contracts and supported by market diligence). Financiers are more willing to take price risk, provided that prices are linked to market indices or otherwise based on market principles and provided that the sales arrangements provide for take-or-pay damages to ensure a continued project revenue stream. As with the 'supply-or-pay' damages regime, parties will seek to set the take-or-pay damages at a level that ensures the continued payment of debt service and operating costs. Financiers also require comfort as to the identity of those offtakers, and any offtaker guarantor, in terms of creditworthiness and experience.

The CCUS market will take a different approach to revenue stream, which will be heavily influenced by the structure and scope of the project under development. The revenue case for CCUS projects is a barrier to accessing capital—equity requires a return on investment, whereas lenders will require repayment of their debt. Without a clear source of revenue, it is likely that CCUS projects, particularly for sequestration, will struggle to attain financing in the shorter to medium term until this barrier is overcome. CCUS projects are often capital-intensive, meaning that there will be a requirement for large amounts of capital to be deployed, which will only occur if there is a clear case for a return on investment for both equity and lenders. Financiers to CCUS projects will assess the tariff agreed between the CCUS developer and the emitter(s), as well as the pricing associated with any transfer of environmental credits from the project to the emitter or any third party (and, moreover, any ability to transfer credits to a financier may also open up certain financing products and structures). Financiers will need comfort that the tariff, along with any price paid for relevant environmental credits, is sufficient to service their debt and pay operating costs for the tenor of the debt. Financiers require transparency on the projected operating costs and will require comfort that the assumptions on which these are based, for example any monitoring, research and development costs and any land lease payments associated with the sequestration of carbon, as well as costs relating to any transportation, are reasonable.

For CCUS projects with a CO_2 market or a co-operating project, for instance where the CO_2 is used as a feedstock or used in enhanced oil recovery, financing is likely to be more accessible. As government policy develops, it is likely that sequestered CO_2 will eventually have a value, whether this is through tax equity, an emissions trading scheme, or other government support. Investors and financiers are likely to be more favourable to a market-based CO_2 valuation, such as a carbon or emissions trading market, than a direct government taxation on CO_2 or government-backed purchase of CO_2, as policy may change over time and revenue streams may be withdrawn. Governments with a history of consistent support for other industries are more likely to be attractive to investors and financiers, due to a perceived regulatory stability for any government-based revenue stream for the project.

Being a relatively nascent industry, a project's revenue stream may also be closely aligned with the availability of government incentives and protections. Transparency on carbon taxes and availability of environmental credits, subsidies and government funding is vital. This is particularly the case for certain CCUS projects, where the availability of these governmental credits is critical for the generation of revenue and where otherwise cashflows can be unstable and unpredictable. Diversification of revenue streams may improve a project's bankability. If a CCUS project contemplates the sale of some volume of carbon, for example to nearby facilities for use in chemical or fuel processes or in advanced oil recovery techniques, then financiers will require comfort as to the availability of a long-term purchaser (and the terms of such purchase arrangement) that is able to take that carbon. If that portion of revenue is key for the project economics, financiers will also analyse the potential implications on the CCUS project and its projected revenue stream if that offtaker is not able to take the carbon, due to shutdowns, bankruptcy or any other reason.

For CCUS transportation and storage projects, if these operate on a merchant basis, the revenue source is clearer, where users of the transportation and storage network are charged a fee for CO_2 transported and/or stored. In the early stage of CCUS development, where the user base may not be sufficient to support the massive capital expenditure incurred for such transportation and storage networks, it is likely that government support will also be required to incentivise investors and financiers to develop these much-needed networks.

Where a CCUS project is an enhancement project to an existing or new-build industrial or power generation project, the CCUS equipment is often seen as an additional capital expense to be funded and may lower an investor's rate of return on the project if financed through the offtake revenue of that project.

It remains to be seen whether charging a higher price for a more sustainable product (a 'green premium') will be commercially viable, and no such green premium yet exists for any CCUS abated product. There is a comparison in the demand for renewable energy private power purchase agreements for businesses seeking to use only renewable energy, as well as a nascent market for green steel, decarbonised using hydrogen or recycling, driven by a demand for automobile companies looking to decarbonise their supply chain. It is entirely possible that an increasing focus on the carbon intensity of a business's supply chain will result in a demand for premium decarbonised products in the future, but that does not address the initial issue of deployment of CCUS projects at scale now to allow the world to meet its net-zero targets.

8 Project Risks

While the revenues generated by a CCUS project are a key determining factor in whether an investor or financier is willing to invest or lend into a CCUS project, they are not the sole considerations. Investors and lenders will consider the full scope of project risks when contemplating an investment or loan to determine whether the project is likely to be constructed and successfully operated so that it produces the expected revenue stream to generate the return or repay the debt.

The development of projects necessary for the transition away from fossil fuels requires collaboration and risk-sharing amongst a number of stakeholders. While the project finance model is well-suited for such an approach, it is necessary for market participants to also take a holistic view of the development of a broader CCUS industry rather than evaluating opportunities on a project-by-project basis. That being said, each project is unique and bankability or investment considerations will ultimately take into account a number of factors, such as the contractual and commercial structure, the project's geographical location, project development and operational timelines, project costs, sources of funding and available government support, in addition to considerations as to the identity, expertise and creditworthiness of the sponsors and key project parties (among other things).

The financing structure(s) available for carbon capture projects will also be influenced by the scope of the proposed project. For example, projects developed as standalone CCUS projects involving all or any elements of sequestration, compression and transportation, usage and/or storage attract different bankability considerations and strategies compared to carbon capture units that form part of an integrated facility for which broader financing is sought. Where the process forms part of an integrated project, the risk profile will also change according to the applicable carbon capture process applied, technology used, method and location of carbon utilisation or sequestration and the capital costs of the broader project under development. Bankability considerations will also differ according to whether the relevant carbon capture application is introduced to an existing project or whether it is implemented as part of a new development. Carbon capture projects implemented within existing clusters or hubs may benefit from lower development costs, through the use of shared infrastructure, but if the carbon capture technology is not embedded within a broader project, then the contractual structure between the emitter and the technology provider will be more closely scrutinised. As is the case in more traditional project-financed transactions, a developer must work closely with their financial and legal advisers in the early phases of a project to develop a bankable corporate and commercial structure.

Below we explore some of the key bankability themes that prevail throughout large-scale, limited- (or non-) recourse project finance transactions. We also consider the potential application of these themes to the carbon capture finance market.

9 Completion Risk

Financiers and their consultants conduct detailed diligence on a project's proposed construction strategy, focusing on completion risks and the availability of, and ability to implement, relevant risk mitigants. Once a project has achieved completion and is operating at the required capacity and generating revenues at the required level, financiers become more flexible with the risks they are willing to assume. Until this milestone, there are a number of risks that might arise during the construction phase of a project including, but not limited to, the risk of project cost overruns, late completion, inadequate performance or failure to meet required specifications, contractor insolvency and performance risk, risks associated with the use of process technology and any co-completion and interface risk. Project lenders require comfort that these risks have been mitigated, whether through the allocation of as much risk as possible to other project parties or, where available, through additional sponsor support. Broadly speaking, if a project is not completed on time or within budget, or does not meet the minimum performance levels and design capacity requirements, then there could be adverse consequences on its ability to repay its lenders or generate a return on investment. These are key risks that financiers are not willing to take without a suitable risk mitigation package.

This is a particularly acute concern in CCUS projects, due to the emergence of new technology, the lack of deployment at scale, and the risk taken on other projects in the value chain for the operation of a CCUS project. While we will explore these points in detail later on in the Chapter, at present there is a considerable level of risk in both the cost and timing implications of CCUS projects, as well as a concern as to the performance of the technology. As with any newly developed technology, developers will learn efficiencies and pitfalls during the construction phase of initial projects, though lenders and investors are hesitant to invest into projects with substantial or unknown cost overrun or delay risk. A particular risk for CCUS is the performance of the technology, as an underperforming CCUS project may totally disqualify itself from any revenue stream (e.g. the UK's dispatchable power agreement business model envisages a minimum 80% CO_2 capture rate for the effectiveness of the dispatchable power agreement charges) and may cause substantial reputational damage to lenders and investors. The nature of a CCUS project is such that, unlike an underperforming power or industrial project, lenders and investors are unlikely to be willing to compromise on the performance levels of the project.

Financiers and their consultants assess the likelihood of project cost overruns during the construction phase, which requires detailed analysis on the anticipated project costs and components thereof, alongside comfort as to available (and committed) funding from shareholder equity and external debt. A project's financial model and project (construction) budget will be scrutinised from this perspective. Financiers also assess the risk of project delays and the likelihood of any delayed completion, focusing on the potential consequences on debt service and loan repayments resulting from the same. Further analysis focuses on commissioning, performance testing and operational risks, as the financiers require comfort that, once completed, the project will meet, and operate at, the necessary performance levels required to generate the projected revenue (whether that be from carbon sales, e.g. from enhanced oil recoveries, or relevant tariffs or incentives). If applicable to a project, the risk of delay in completing a unit or relevant asset on the ability for another to complete and/or operate will be assessed (i.e., co-completion risk), as will interface risk management and the risk of misalignment between any onshore and offshore scopes. Financiers will look to the reputation, creditworthiness and past performance of the contractor(s) engaged to undertake the construction of the relevant carbon capture facilities, along with the track record of the project's sponsors and any project management consultant in successfully overseeing, implementing and operating projects in the same or similar industries and of a similar scale before.

In the nearer term, many of these risks may be more difficult to quantify, due to a lack of a track record of successful CCUS projects using the most recent technology. A particular risk for sequestration projects is both the potential injection rate for the geology of the project, as well as the risk of leakage of CO_2, and as at present, the consequences of such issues have not yet been fully assessed or quantified.

Lenders are generally very reluctant to take on such unquantifiable risks, and so will look to ensure that they are repaid, even in a downside scenario. They will seek to rely on substantial amounts of equity in the project company, which may come from investors or may even be co-contributed government grants or other funds, as well as guarantees of their investment, either from the sponsors of the project, or potentially, in the nearer term, from the government. Governments appear to be willing to support this approach to achieve their net-zero policy goals, for example with the UK's sovereign infrastructure guarantee providing UK government backing for guarantees issued by the UK Infrastructure Bank. It is likely that sponsor support and even government support will be crucial to de-risk CCUS projects for financiers in the nearer term. Guarantees may also be provided by export credit agencies or other multilateral agencies, such as the Multilateral Investment Guarantee Agency, though such guarantees may only be available for specific projects, and may only cover certain political or non-commercial risks.

Sponsors may be asked to take some level of completion risk. Sponsor completion support takes a number of forms including the provision of completion guarantees, whereby the sponsor(s) guarantee(s) the repayment of the entirety of the outstanding debt if the loans are accelerated due to the occurrence of an event of default, including if completion of the project does not occur by a longstop date. Another form of sponsor support may be a commitment to fund additional equity in the borrower on the occurrence of certain trigger events occurring prior to the completion of the project (such as project cost overruns, delays and underperformance). The debt service undertaking falls in between the completion guarantee and the equity contribution commitment and requires the sponsors to service the outstanding debt in accordance with the original repayment schedule following the occurrence of an event of default (rather than accelerating the repayment of all outstanding loans in full, as is the case with a completion guarantee, but subject to some exceptions). Project lenders will also assess the credit of any sponsor providing such completion support. If the sponsor does not have an investment grade credit rating or is otherwise deemed risky from a credit perspective, project lenders may require letters of credit and/or parent company guarantees (from creditworthy entities) to backstop any of these completion support arrangements.

The technical 'Completion' of the CCUS project for the purposes of the financiers will be the date on which a number of conditions have been satisfied, and these conditions will be heavily negotiated before execution of the finance documents. The conditions typically include a performance-based test, requiring the borrower to demonstrate that it can operate the project at a certain level of performance for a certain period of time, thereby ensuring that the project will generate revenue at the required level to repay the lenders and pay operating costs. This will be particularly important where government support mechanics contain conditions to access government funding, particularly around the capture or injection rate of CO_2, as these will be fundamental performance criteria for the success of the project, and the ability for the lenders to be repaid. Other conditions include demonstration of coverage ratios, insurances, permits and material project documents in place, provisional (or substantial) completion under the construction contract, funding of any reserve bank accounts, compliance with applicable laws and other pre-agreed conditions. If co-completion risk for other related CCUS projects elsewhere in the chain is applicable, for instance transport and storage infrastructure for any captured carbon, financiers may include a condition requiring that all infrastructure necessary for the operation of the project as a whole has been completed, including tests to demonstrate adequate operations as between the various interfaces.

The requirement for, and form of, sponsor support, or even government support, is heavily influenced by the specifics of the project under consideration, as well as the nature of the completion risks identified and the availability of other completion risk mitigants. The nature of sponsor support required for the bankability of a project is closely linked to a project's contracting strategy and structure. For smaller or more novel technologies, a CCUS developer may not have the means to provide such completion support, and therefore financiers will need to be more flexible when assessing the completion risk mitigant package. Again, the requirement for, and level of, sponsor completion support will be heavily influenced by the project under consideration and its overarching risk profile.

As has been seen from the development of other new technologies, particularly renewable energy and other complex industrial facilities, as sponsors and developers gain experience in constructing and operating CCUS projects, it is likely that these risks will become known and quantifiable. While it may be the case that certain key risks, for instance the consequence of CO_2 leakage, remain a covered risk, it is likely that the amount of government and sponsor support required for CCUS projects will diminish as lenders and developers are able to fully understand and quantify the risks inherent to a CCUS project.

10 Completion Phase Contracting Approach

The engineering, procurement and construction approach for a project, along with the terms of the contract(s), will be scrutinised by the financiers. These construction arrangements represent the most significant portion of project costs and also carry the most completion-related risks. At a high-level, the use of a fixed-price turnkey engineering, procurement and construction (**EPC**) contract, implemented by a reputable and creditworthy contractor (or consortium of contractors) with a single-point of responsibility and tight controls over changes to the scope and timetable, has a higher probability of raising external finance than other contracting approaches, such as separate engineering, procurement and construction contracts or pricing terms with lower cost certainty. A number of factors influence whether it is feasible for a project to execute a robust lump sum, turnkey EPC contract, including the project scope and the number of interfaces.

The contracting strategy can be driven by a number of considerations, including whether there is a pool of available contractors willing and able to take on the relevant parts of the project on a lump sum basis under one single EPC contract, as well as the different specialist skills required for different elements of a large project. If the CCUS project is part of a large, complex industrial project involving the development of a carbon capture unit alongside a number of standalone units and infrastructure, for instance a co-located or associated processing project, then multiple contractors under more than one construction contract are likely to be required. If the CCUS project also involves building the infrastructure for the transportation and sequestration of captured carbon, different contractors may be engaged for the different parts of the scope, particularly if the different elements of the project involve very different specialities. In this case, financiers will analyse co-completion and interface risks in particular. Different EPC structures can be bankable, provided that the lenders get comfortable with the level of interface risk between the different EPC contracts and the ability of the project company to manage those interfaces.

11 Key Risks in CCUS Projects

While carbon capture technologies have been in use for some time, the development of CCUS projects is still relatively novel, and therefore financiers are likely to require substantial diligence before they commit to lend to a project. This will be true for lenders providing corporate loans, as well as financiers funding through a project finance model (and also for equity investors), and therefore many of the themes discussed in this Chapter will be applicable for different types of financing structure. As is evident from some of the key risk considerations outlined below, the successful implementation of a suitable finance model in the development of CCUS projects requires dedication and expertise from all industry participants.

In order to address the numerous workstreams and specialisms involved in the implementation and financing of energy and infrastructure projects, many of the key traditional and renewable energy project finance 'players' (including banks, export credit agencies and development banks, among others) possess the necessary skillset to identify, analyse and mitigate key project risks and employ dedicated specialist internal departments to assist with their diligence processes. The use of independent experts to conduct due diligence on behalf of the financier group (e.g. from a legal, technical, market, ESG and insurance perspective, as applicable), and the involvement of the project sponsors' advisors in the early development and structuring phases of a project, and throughout the documentation phase, helps to ensure that risks are properly identified and allocated (and mitigants documented) in a bankable manner. This existing depth of expertise in the market, together with the necessary understanding of the flexibilities and structures required to develop and successfully execute an appropriate risk-sharing structure in project-financed transactions, will prove invaluable in the development of projects in the energy transition space, particularly those that present new challenges and those requiring tailor-made financing solutions.

This section examines a non-exhaustive list of the key risks that will be analysed by financiers before committing financing to a CCUS project. Financiers will also analyse risks associated with the operation and maintenance of the project, financial and commercial terms (such as the risks associated with changes to interest rates, exchange rates and commodity prices), tax changes, events of force majeure and the availability of insurance coverage (e.g. to cover leakage risks), among other themes, and each project will be assessed on its own risk profile.

12 Technology Risk

While falling under the more general 'completion risk' umbrella, process technology-related risks are often analysed on a standalone basis. Underperformance of the process technology in any manner could, by itself, cause a project to fail. Financiers will therefore pay close attention to a project's chosen technology.

Lenders will analyse the track record of the technology provider, including its success rate for implementation of the carbon capture technology in precedent projects. Project financiers require comfort that the technology is proven and has been successfully implemented on other projects of a similar scale and, where applicable, in similar markets. The proven track record of CCUS technology generally, in particular through its consistent and successful long-term usage in industry and more recently through its association with power generation, is a significant benefit for the development and financing of CCUS projects compared to a number of other types of project in the energy transition space (including those for which technologies are being tested at scale but remain unproven at this stage). Moreover, some financiers may be prepared to take some level of technology risk, such as scale-up risk, on projects in the energy transition space, particularly where there is an internal, political and/or ESG-related reason to support such projects.

However, the picture for technology risks with CCUS projects is nuanced because carbon can be captured using many different processes. Some processes involve well-established technologies that are tried and tested at industrial scale, whereas other capture methods are nascent. For example, chemical absorption and physical separation methods are more established, mature technologies, while chemical looping and direct separation methods are still only emerging from pilot stage demonstration projects. In practice, the capture method chosen for a particular project will depend on a number of factors, including the composition and nature of the gas stream from which the carbon will be captured. As carbon capture is considered for a wider range of applications and industrial processes, project financiers will continue to require due diligence on the technology risks involved for each project.

Financiers and investors will be more comfortable with the technical risks associated with carbon capture technologies that have been repeatedly implemented at scale. However, as with other fast-moving areas of technological development, lenders to CCUS projects will also need to be satisfied that the project is using the right technology, despite the rapid change in the industry and the possibility that a better, more efficient, cheaper technology is within reach.

13 Performance Risk

A key risk to the financing of CCUS projects is the performance risk. Particularly for new technologies, and those that have not yet been deployed at scale, financiers will be concerned about the impact of the project not performing as expected, both from a financial and a reputational standpoint. Due to a likely reliance on some form of government support, or the generation of carbon credits, underperformance of a CCUS project may cause a fundamental issue for the ongoing revenue stream of a CCUS project. This may not be comparable with other projects, where underperformance leads to less revenue generated, and a potential re-sizing of the debt case. Where revenues are directly linked to a performance level, as with many projects funded on policy grounds, failure to achieve adequate performance may result in no revenue being generated at all. Additionally, as financiers are seeking to invest and finance CCUS projects on a policy as well as economic basis and may have repackaged some CCUS loans into green or sustainable instruments, the failure of a CCUS project to perform may also cause reputational issues for the financiers. It is likely that the technology and performance risk will be one of the risk areas that financiers will focus on most eagerly.

14 Geotechnical Risk

One particular type of risk that must be evaluated and properly mitigated in any project is geotechnical risk, or risks associated with the ground conditions underneath the project site. For a typical industrial project, this risk includes the possibility of contamination of the soil at the project site, the risk of the discovery of underground obstructions or cavities, the suitability of the site for construction of the particular project facility, the depth and characteristics of bedrock under the surface of the site and other such considerations. Unforeseen ground conditions are common causes of delay and cost overruns on construction projects. Typically, the risk can be mitigated by conducting geotechnical surveys that can be can identify potential issues before construction starts. If an EPC contractor has sufficient information about site conditions, then often the EPC contract can allocate the risk of any unforeseen site conditions to the EPC contractor. Project lenders and their technical advisors also take comfort from site investigations and geotechnical survey reports.

CCUS projects that include sequestration of CO_2 have significant inherent geotechnical risks that are more difficult and more expensive to assess. The method of sequestration or storage of carbon used by the CCUS project can lead to risks that some financiers may consider unpriceable. Some methods of carbon storage are more effective and more well understood than others. For example, injecting captured CO_2 into used oil wells or mined coal mines, known as geological sequestration, has a higher expected retention rate and longer expected storage times compared to direct injection of CO_2 into the ocean, known as oceanic sequestration. At present, CCUS storage facilities are often treated similarly to oil and gas deposits, with countries, such as the UK, having their offshore decommissioning obligations apply to carbon storage projects in the same way as oil and gas. While policy has not yet developed in this area beyond non-accessibility of government support, there may be policy developments in the future to address the leakage of carbon from these carbon storage projects. Such decommissioning risks and potential leakage liability are likely to be substantial, so developers and financiers should also focus on storage, leakage and decommissioning liability and ensure that they are sufficiently protected against this. As it is likely that much of this policy will be derived from existing oil and gas policy, which is a mature industry, there will be substantial precedent on how developers and lenders have assigned these risks in the oil and gas sector to draw on to develop mitigation and risk allocation for CCUS projects.

Geotechnical risks relating to the operational stage of a CCUS project can also be significant and to a greater degree than for a normal industrial project. Injecting CO_2 into a geological formation increases the pressure in that formation, which can lead to seismic events, such as earthquakes, that are triggered by the additional pressure. Injection of CO_2 into pre-existing faults in the subsurface or the creation of new fractures in subsurface rock can also cause seismic events. Financiers and their technical advisors will scrutinise the project due diligence relating to the sequestration site. These inherent risks will continue to become better understood as the industry develops, but the need for ongoing monitoring of sequestration sites is likely to continue to affect the cost of CCUS projects into the future.

15 Supply Risk

Financiers require comfort that the key feedstock or other material will be supplied to the project from the date on which the project requires such feedstock and at all times thereafter, at least until repayment of the loans (plus a buffer). To address this risk and to mitigate the risk of a lack of or insufficient supply, lenders customarily require the borrower to enter into long-term supply agreements with reputable and creditworthy suppliers. Financiers will also need comfort that these suppliers have access to the relevant natural resource or other feedstock at the required volumes for the term of the supply arrangements. Financiers will often diligence the availability of reserves and the ability to extract or otherwise supply these volumes. In most project finance transactions, suppliers must commit to supply minimum volumes or, if they do not supply at the required volume or specification, pay damages (which might be set at a level that can sustain the project's debt service and operating expenses for a certain period). The availability of alternative supply will also be considered, although the uncertainty of such arrangements, particularly supply and pricing certainty, means that financiers are less likely to rely on alternative arrangements as a mitigant even for freely traded products with an available market value, such as crude oil.

Unique bankability risks arise in the CCUS market. If the carbon capture unit relies on carbon emissions from an existing facility, then financiers will consider the likelihood of supply interruptions from that facility, particularly where there is no scope to mitigate non-supply risk through seeking input from a substitute supplier. For example, lenders will assess how long the facility has been operating, its maintenance and shutdown schedules, the identity and performance of that emitter, as well as whether the anticipated life is sufficient to meet the CCUS supply requirements for the term of the proposed project financing. When carbon capture is integrated within the design of a new facility, financiers will assess, among other things, the potential risk of shutdowns at that facility (that could be caused, e.g. by issues with raw feedstock supply or relevant product offtake arrangements and associated storage capacity), as well as more general risks connected with the development of the facility as a whole. Similarly, if the project is a transportation or storage network, financiers may assess the end users, though in less detail, to understand their ability to provide CO_2 for transportation and storage. Financiers will assess the implications of non-supply and shutdowns on a project's ability to benefit from governmental credits or subsidies (i.e., its revenue stream and cashflow available for debt service).

16 Political and Legal Risk

Financiers require comfort that political and legal risk is mitigated. For projects located in emerging markets, considerations might be focused on the likelihood of discriminatory action, expropriation, war or civil disturbance, terrorism, sanctions breaches, non-issuance, non-availability or revocation of key permits, licences and governmental consents, change in law and tax changes, among other things. The mitigants required to address such risks will differ from project-to-project, but in most cases, it is imperative for the relevant government to demonstrate support for the project. This remains true for the carbon capture market. Government policies and incentives must be deployed to attract private capital, with the aim of encouraging and strengthening the role of private-sector investment as the industry matures. As witnessed in the development of energy projects in the early renewables markets, governments must play a significant role in the de-risking of private-sector investment through the provision of tax credits, incentives, feed-in-tariffs, contracts for difference, grants and other benefits.

Financiers also require comfort that all of the necessary permits and consents are in place. For example, carbon capture projects will need to demonstrate that permits required for the transportation and sequestration of carbon are available and will not be revoked.

17 Environmental, Social and Governance (ESG) Risk; Reputational Risk

ESG considerations play an important role in project finance transactions, with project financiers having imposed strict environmental and social requirements on projects for many years. For example, many projects are required to demonstrate that they meet standards set by the International Finance Corporation (**IFC**) and the International Bank for Reconstruction and Development (the **World Bank Group**), which differ according to the relevant classification under which the project falls. Moreover, a substantial number of commercial banks and other financial institutions operating in emerging and developed project finance markets have adopted the Equator Principles. The main aim of the Equator Principles is to ensure that the projects financed by these institutions are developed in a socially responsible manner and reflect sound environmental management practices. Application of the Equator Principles to a project should also benefit the banks reputationally by encouraging a more transparent environmental and social risk assessment process.

Private-sector investors, such as banks and financial institutions, remain under pressure from shareholders, stakeholders and external groups to deploy capital towards projects that meet environmental, social and governance targets (and away from investments in fossil fuels). As reflected throughout this Book, CCUS technologies play a key role in the global drive to net-zero through their ability to significantly reduce CO_2 emissions, and carbon capture projects should remain a compelling prospect for financiers looking to invest in projects that satisfy internal and external sustainability and ESG requirements. On the other hand, CCUS projects are not always viewed favourably by the public, with some groups arguing that they are a short-term decarbonisation solution and a means of enabling the continued development and operation of carbon-intensive activities. Risks associated with the permanent sequestration of captured carbon will be at the forefront of any ESG analysis, as well as ongoing monitoring (alongside research and development) to ensure no release from the storage site. If the project contemplates an element of carbon sales, then ESG considerations relating to the transportation and sale arrangements will also be key. A number of factors and external pressures will influence the perceived ESG-related risks on a particular project, and project financiers will approach ESG risk based on an assessment of the unique profile of each project.

18 The Future of CCUS

Collaboration and risk-sharing among industry stakeholders are vital elements in the successful development and financing of energy and infrastructure projects. Throughout the decades, the project finance model has been pivotal for parties in assessing risks and allocating these among the most appropriate stakeholder(s) in a manner that both attracts financing and also delivers necessary shareholder returns. Such collaboration must continue so that project developers have access to the necessary external funding and investment required to support a successful carbon capture project finance market. Government policies are required to incentivise private investments that are required to pursue net-zero goals. In order to launch some of these carbon capture projects, project developers, financiers and other key stakeholders have to be prepared to depart from some of the more traditional project financing concepts and to maintain flexibility when assessing the unique risk profile of each project.

In the nearer term, government support is likely to be essential in developing CCUS projects at the pace required to meet net-zero carbon emissions goals. Government support will de-risk carbon projects and provide revenue streams to encourage private investment, as well as supporting the development of newer technologies and larger scale projects. Government policy to support the development of carbon transportation and storage hubs or networks will likely support the scale-up of carbon capture projects, as this allows developers to split the value chain and to develop smaller or more specialised projects while transferring the risks and capital requirements of operation or storage. In the longer term, it will be vital to establish a value for CO_2 captured and stored in order to create a revenue story able to attract investment and finance even if direct government support is wound down. It is likely that carbon credits and carbon taxation policies will create a market for carbon, either directly or indirectly, which will result in a value being placed on the storage of carbon.

Similar policies have been successful in supplementing the costs of renewable energy development, particularly for wind and solar energy, and other renewable forms of electricity generation. These policies reduced the initial costs for development and supported initial revenue streams, allowing renewable energy to compete with conventional thermal energy generation while the costs of renewable technology fell. As cost reductions due to advanced technologies, increased efficiencies and economies of scale allowed renewable energy to compete directly with conventional thermal generation sources, the government support was scaled back. This model could be replicated with CCUS, with government support allowing CCUS projects to be developed at scale, with the development in technology and lessons learned on these initial projects contributing to cost reductions for subsequent CCUS projects. There is also a growing push for corporations to decarbonise their production chains, and so there may be an increased demand for decarbonised products in many currently carbon-intensive industries, particularly for thermal power generation, metal refining and production, cement production and certain chemical processes. On the demand side, customers willing to pay a premium for a low-carbon product may also lead to a financial incentive for developers to create CCUS-enhanced projects to capitalise on this demand. Again, government support for transportation and storage networks and hubs will dramatically reduce the costs of creating decarbonised products and allow them to compete more effectively with non-decarbonised products, as the manufacturer will not need to develop capital-intensive private transport and storage facilities for the captured CO_2.

Access to capital may also be supplemented in states that are willing to contribute more from their public balance sheets. The Middle East is already home to several existing carbon capture facilities, together storing around 3.7 Mt CO_2 per year, and has a geographical advantage for CCUS projects, being a key oil and gas producing region, so possessing the geology suitable for the underground storage of CO_2. The Middle East also has a long history of supporting strategically important projects, and, with COP28 being hosted in Dubai in 2023, many countries in the region have committed to net-zero goals. As has been seen with hydrogen in Saudi Arabia, the Middle East has potential to be a first-mover and market leader in CCUS if it is willing to deploy its substantial public finances and state-owned companies to the development of CCUS.

Until now, the vast majority of CCUS development has taken place in the United States, with additional development in Canada and Europe. Outside of this, few countries have developed CCUS projects to date, although Australia, Brazil, China, Saudi Arabia and the United Arab Emirates, each have at least one CCUS project in development. A commonality with many of these projects is that they have been developed in oil producing countries largely to support enhanced oil recovery. There are many other countries with the geography and technical capacity developed from a mature oil industry which would be suitable for the development of additional CCUS projects, both to support enhanced oil recovery and to sequester carbon in depleted oil and gas reserves underground. However, in regions with more stretched public finances or which do not have a historic track record of political or regulatory stability, it may be difficult for government support to be deployed to support CCUS projects at scale. In order to develop CCUS projects in these regions at the pace required to meet net-zero goals, external support may be required, both from national governments, multilateral institutions and sponsors. It is likely that the initial CCUS projects developed in these regions will continue to be driven by external economics, such as enhanced oil recovery or other feedstock projects, in the near term, while the global carbon economy develops.

With a clear focus from developers, financiers, investors and governments on delivering on net-zero targets, there is a large pool of capital available for the development and financing of CCUS projects if some of these barriers can be overcome. Risk-sharing and the co-development of CCUS projects between these stakeholders will be vital, as will clear and structured government policies and support programmes. As the number of CCUS projects under development and in operation grows, the data gathered from the construction and performance of these projects will guide many of these stakeholders in understanding and quantifying the risks associated with CCUS projects and may de-risk future developments or allow for mitigants to be developed or imported from other similar industries.

CHAPTER 5

The Connection and Role Between CCUS and Net Zero

Peter Kayode Oniemola, Onyemaechi Titilayo Ogiri, and Soibi O. Bob-Manuel

1 Introduction

The concept of Net Zero, crucial in the context of global climate action and has been viewed to entail state where the amount of greenhouse gases emitted into the atmosphere is balanced by an equivalent amount that is removed.[1] The achievement of Net Zero is a key goal in the fight against climate change, as it is essential for stabilising global temperatures

[1] Shelly Welton, 'Neutralizing the Atmosphere' (2022) 132 The Yale Law Journal 171–249, 171.

P. K. Oniemola (✉)
Faculty of Law, University of Ibadan, Ibadan, Nigeria
e-mail: petermola@yahoo.com

O. T. Ogiri
Faculty of Law, Anchor University, Lagos, Nigeria

S. O. Bob-Manuel
University of Birmingham, Birmingham, UK

E. G. Pereira et al. (eds.), *Carbon Capture Utilization and Storage*, Sustainable Development Goals Series,
https://doi.org/10.1007/978-3-031-81272-9_5

owing to the release of greenhouse gas emissions from anthropogenic activities.[2] The actualisation of balance is attainable through a combination of reducing existing emissions and by implementation of strategies to remove carbon dioxide from the atmosphere. Carbon Capture, Utilisation, and Storage (CCUS) emerge as an essential technology that can be employed, offering pathways for capturing and storing carbon dioxide emissions from industrial and energy-related sources and potentially utilising the captured carbon in ways that are beneficial.[3]

The Paris Agreement, a landmark in international climate framework under the UNFCCC that replaced the Kyoto Protocol. It has a significant role in shaping the trajectories towards global Net Zero.[4] The Paris Agreement was adopted by 196 nations at the United Nations Climate Change Conference (COP21) in 2015.[5] The Paris Agreement envisions limiting global warming to well below 2 degrees Celsius above pre-industrial levels, with a further goal of reducing the increase to 1.5 degrees Celsius.[6] The significance of the Paris Agreement lies in its establishment of a global framework for climate action, calling for significant reductions in greenhouse gas emissions and encouraging the transition to a low-carbon in the various sectors of the economy.[7]

[2] Thomas Hale, Stephen M Smith, Richard Black, Kate Cullen, Byron Fay, John Lang and Saba Mahmood, 'Assessing the Rapidly-Emerging Landscape of Net Zero Targets' (2022) 22(1) Climate Policy 18–29, 18. https://doi.org/10.1080/14693062.2021.2013155.

[3] Alexandre Gallo, Eduardo G. Pereira, Alberto Fossa, Hannah Hylton-Edwards, Thomas Muinzer, Edmilson Moutinho dos Santo, Clara Dybwad and Cylon Liaw, 'Towards the International Standardization of Carbon Dioxide Capture, Transportation, Utilization and Storage (CCUS) Technologies: Current Challenges and Future Directions' (2021) 12(2) The Journal of Sustainable Development, Law and Policy 157–189, 157. https://doi.org/10.4314/jsdlp.v12i2.2.

[4] Vila Johansson, 'Just Transition as an Evolving Concept in International Climate Law' (2023) 35(2) Journal of Environmental Law 229–249.

[5] United Nations Climate Change, 'The Paris Agreement', https://unfccc.int/process-and-meetings/the-paris-agreement, accessed 21 November 2023.

[6] Article 2(1)(a) Paris Agreement 2015.

[7] Peter Kayode Oniemola, 'A Proposal for Transnational Litigation against Climate Change Violations in Africa' (2021) 38(2) Wisconsin International Law Journal 304–305.

It is the contention of this chapter that CCUS is intricately linked to the objectives of the Paris Agreement.[8] The technology is viewed as essential for meeting the ambitious targets set out in the provisions of Article 2 of the Paris Agreement, especially if considerations are given to the current reliance on fossil fuels and the emissions intensity of various industrial processes. CCUS can play a variety of roles in this regard. It can directly reduce emissions from heavy industries and power generation, sectors where alternatives are either costly or currently insufficient in meeting the needs of the society.[9] Looking at it in the context of energy transition, CCUS provides a means to decarbonise existing infrastructure, thereby offering a platform of transiting smoothly to renewable sources without essentially carrying out a complete overhaul of current systems. Furthermore, the integration of CCUS in achieving Net Zero is not only limited to emission reduction; but also encompasses the utilisation aspect.[10] This involves the conversion of captured carbon into valuable products, which creates an avenue for a circular carbon economy, a novel approach to managing emissions.[11] Utilisation therefore provides economic incentives for adopting CCUS technologies and also helps in reducing the carbon footprint by replacing materials and fuels that would have otherwise emitted additional greenhouse gases.

The role of CCUS in the context of the Paris Agreement and the broader Net Zero goal is worthy of examination. This is because it

[8] Peter Freudenstein, Louis Uzor and Chirtoph Beuttler, 'The Pertinent Role of Carbon Dioxide Removal and Direct Air Capture in Achieivng the Goals of the Paris Agreement' in Oxford Institute for Energy Studies *Carbon Capture, Utilization and Storage (CCUS): Barriers, Enabling Frameworks and Prospects for Climate Change Mitigation* (2022) 1–71, 63.

[9] Eduardo G. Pereira, Alberto Fossa, Carlos Eduardo Pellegrino Cerri, Opeyemi Omotuyi, Hannah Hylton-Edwards, Edmilson Moutinho dos Santos, Alexandre Gallo, Cylon Liaw and Bruna Toguedani, 'Advancing Sustainability Through Carbon Capture, Utilization and Storage (CCUS) Technologies: The Hydrogel Case Study' (2023) 14(1) The Journal of Sustainable Development, Law and Policy 218–250, 218.

[10] SC Dike and NK Akani, 'Promoting Carbon, the Capture, Storage and Utilization (CCSU) in Nigeria: Lessons from the UK' (2020) 6(2) Journal of International Law and Jurisprudence 184–192.

[11] Eduardo G. Pereira, Alberto Fossa, Carlos Eduardo Pellegrino Cerri, Opeyemi Omotuyi, Hannah Hylton-Edwards, Edmilson Moutinho dos Santos, Alexandre Gallo, Cylon Liaw and Bruna Toguedani, 'Advancing Sustainability Through Carbon Capture, Utilization and Storage (CCUS) Technologies: The Hydrogel Case Study' (2023) 14(1) The Journal of Sustainable Development, Law and Policy 218–250, 220.

represents a bridge between the current carbon intensive state of global industries and the desired carbon-neutral future. By providing a practical approach for emission reductions from the most challenging sectors, and offering an opportunity for turning carbon emissions into valuable products, CCUS can be a critical component in strategies aimed at combating climate change and achieving the ambitious goals set out in the Paris Agreement but the attendant costs involved, uncertainties and the need to build sufficient capacity requires global attention in the climate change regime and Net Zero campaign. This chapter essentially examines the move towards Net Zero, global transition to Net Zero and CCUS, the potential role of the Paris Agreement on CCUS, the future of CCUS in Net zero and draws a conclusion on connection and role between CCUS and Net Zero.

2 Energy Transition and the Move Towards Net Zero

We are in the era of energy transition and Net Zero push, which represent a global shift from fossil fuel-based energy systems to cleaner, more sustainable sources, in response to the urgent need to mitigate climate change.[12] This transition is driven by the goal of achieving Net Zero emissions[13] as it is imperative that the amount of greenhouse gases emitted into the atmosphere is balanced by an equivalent amount removed. Clean technologies are central to this shift, encompass a range of innovations, from renewable energy sources such as solar and wind to energy efficiency solutions and electrification.[14] Many countries, such as China, have embraced the use of renewable energy as a major source of energy, going as far as implementing a legal framework to establish the renewable

[12] Donald Zillman, Lee Godden, LeRoy Paddock and Martha Roggenkamp (eds), *Innovation in Energy Law and Technology: Dynamic Solutions for Energy Transitions* (Oxford, 2018; online edn, Oxford Academic, 24 May 2018), https://doi.org/10.1093/oso/9780198822080.001.0001, accessed 21 November 2023.

[13] Albert C Lin, 'Making Net Zero Matter' (2022) 79(2) Washington and Lee Law Review 679–767.

[14] Emadul Islam, Asheer Shah and Tariq A Karim, 'Role of Renewable Energy Policy in Ensuring Net-Zero Carbon Emissions and Energy Sustainability: A Bangladesh Perspective' in HH Wu, WY Liu and MC Huang (eds) *Moving Toward Net-Zero Carbon Society* (Springer, 2023) https://doi.org/10.1007/978-3-031-24545-9_4.

energy system.[15] CCUS stands out among these technologies because of its exceptional potential and the difficulties it brings in its implementation. CCUS solutions reduce emissions from industries where direct reduction is more difficult,[16] while renewables such as solar and wind become more cost-effective and broadly adopted. These come from manufacturing and power facilities that rely on fossil fuels. Deployed technology will need to trap carbon dioxide emissions at the source, put them to use in other ways, or store them underground.

The fact that CCUS is behind other clean technologies in terms of development[17] is one of its biggest obstacles. Its wide-scale deployment faces several challenges such as high costs, technological complexity, and need for significant infrastructure development.[18] There are also challenges related to public perception, regulatory frameworks, and ensuring safe and permanent storage of captured carbon.[19] Despite these challenges, the potential of CCUS in contributing to Net Zero goals has been viewed as very significant.[20] It is one of the few technologies with the capabilities of decarbonising heavy industries and can also be used for direct air capture,[21] a critical tool for offsetting unavoidable

[15] Peter Kayode Oniemola, 'Legal Response to Support Renewable Energy in China' (2014) 32 Journal of Energy and Natural Resources Law 179–202.

[16] Sampo Kouria, Eemeli Tsuparia, Janne Kärkia, Sebastian Teira, Risto Sormunenb, Timo Arponenc and Mari Tuomaala, 'The Potential for CCUS in Selected Industrial Sectors – Summary of Concept Evaluations in Finland' (2017) 114 Energy Procedia 6418–6431 (13th International Conference on Greenhouse Gas Control Technologies, GHGT-13, 14–18 November 2016, Lausanne, Switzerland).

[17] International Energy Agency, Energy Technology Perspectives 2020: Special Report on Carbon Capture Utilisation and Storage—CCUS in Clean Energy Transitions 28.

[18] Hao Zhang, 'Regulations for Carbon Capture, Utilization and Storage: Comparative Analysis of Development in Europe, China and the Middle East' (2021) 173 Resources, Conservation & Recycling, The Chinese University of Hong Kong Faculty of Law Research Paper No. 2021-38.

[19] Tara Righetti and Matteo Fermeglia, 'Call for Papers: OGEL Special Issue on "Carbon Capture and Storage"' (OGEL, 16 November 2022), https://www.ogel.org/news.asp?key=734, accessed 22 November 2023.

[20] Deniz Yagmur Kayhan, 'Importance of Carbon Capture, Utlization and Storage (CCUS) to Achieve Net-Zero Emissions with Actions in Industry' (2023) (International Conference on Turkiye's Contributions to the UN Decade of Ocean Science for Sustainable Development, 26 September 2023).

[21] Neil Segel, 'Direct Air Capture Facilities and Production of Carbon-Neutral Hydrocarbons' (2022) 52 Texas Environmental Law Journal 83–110.

anthropogenic emissions.[22] Therefore, much effort is being directed at the development, which includes research into more efficient and cost-effective capture methods, exploring new utilisation pathways, and framing policies and incentives to support its deployment.[23]

While energy transition and push towards Net Zero are being propelled by a variety of clean technologies, CCUS is singled out because of its unique capabilities and the substantial work needed for its development and widespread implementation within the framework of international law on climate change and opportunities offered for collaboration through Net Zero. Addressing its challenges and harnessing its potential will be crucial for achieving global climate goals that have been a major consideration in international climate change negotiations.

3 Global Transition to Net Zero and CCUS

Achieving Net Zero has gained global momentum, with countries setting ambitious targets, commonly aiming for 2050.[24] The strategies to reach Net Zero are diverse and have been a function of the unique economic, geographical, and political contexts of each nation. In the global movement for Net Zero, CCUS is emerging as a critical technology for deployment. The significance of the technology is underscored in the context of industries that are challenging to decarbonise, such as steel, cement, chemical production, coal and oil and gas. These industries contribute significantly to CO_2 emissions worldwide, and options for cutting these emissions are quite limited. CCUS technology provides a

[22] Peter Kayode Oniemola, 'Considering Carbon Capture and Storage for the Nigerian Petroleum Industry? Is a Complementary Liability Regime Required?' (2023) 21(3) Oil, Gas and Energy Law Journal 3.

[23] Kelsey Simpkins, 'New Method could Lead to Cheaper, More Efficient Ways to Capture Carbon' (University of Colorado Boulder, 16 March 2022), https://www.colorado.edu/today/2022/03/16/new-method-could-lead-cheaper-more-efficient-ways-capture-carbon, accessed 21 November 2023.

[24] Over 100 countries have either set a net-zero target or are considering such targets. Heleen L van Soest, Michel GJ den Elzen and Detlef P van Vuuren, 'Net Zero Emission Targets for Major Emitting Countries Consistent with the Paris Agreement' (2021) 12 Nature Communications 1–9, https://doi.org/10.1038/s41467-021-22294-x, accessed 21 November 2023.

viable solution by capturing the carbon dioxide produced, either using it in other processes or by underground storage.[25]

Beyond industrial applications, CCUS is very useful for CO_2 removal directly from the atmosphere using technologies such as Direct Air Capture (DAC).[26] Direct Air Capture is essentially used in offsetting emissions that are difficult or impossible to eliminate by other means. This aspect of CCUS is becoming increasingly recognised in domestic mitigation strategies as countries incorporate it not only as a tool for emission reduction but also as a major component in their climate policies.[27] Integrating CCUS into national climate strategies often involves ranges of policy support, financial incentives, and investment in research.[28] Countries such as the United States,[29] Canada,[30] Norway,[31] United Kingdom, and Australia[32] are leading the way, with policies and regulatory frameworks that encourage both the development and the deployment of CCUS technologies to advance their mitigation commitments. These

[25] IEA, 'Carbon Capture, Utilisation and Storage', https://www.iea.org/energy-system/carbon-capture-utilisation-and-storage, accessed 22 November 2023.

[26] Dovlev Pinhas, 'Making Products out of Thin Air: Accelerating Direct Air Capture Technologies' (2023) 38 Journal of Environmental Law and Litigation 207–232.

[27] United States Government Accountability Office, 'Decarbonisation: Status, Challenges, and Policy Options for Carbon Capture, Utilization, and Storage' (Report to Congressional Addressees, September 2022).

[28] Tracy Hester, 'Legal Pathways to Negative Emissions Technologies and Direct Air Capture of Greenhouse Gases' (2018) 48 Environmental Law Reporter 10413–10432.

[29] Rudianto Rimbono and Raldi H Koestoer, 'Briefed on the Development and Progress of CCUS Projects in China, USA, Western Europe, Russia, Norway, India, and Indonesia: A Journal Review' (2023) 1(1) Journal of Character and Environment 1–11. https://doi.org/10.61511/aes.v1i1.2023.42.

[30] Romany M Webb and Michael B Gerrard, 'The Legal Framework for Offshore Carbon Capture and Storage in Canada' (2021) Sabin Center for Climate Change Law, Columbia Law School 1–28.

[31] Jo-Kristian S. Roettereng, 'The Foreign Policy of Carbon Sinks: Carbon Capture and Storage as Foreign Policy in Norway' (2014) 63 Energy Procedia 6927–6944.

[32] Jocelyn Taylor, Aleksandra Kalinowski and Andrew Feitz, 'CCUS in Australia' (16th International Conference on Greenhouse Gas Control Technologies, GHGT-16, 23–27 October 2022, Lyon, France) 1–12.

include frameworks on carbon pricing mechanisms, government investments, grants, and subsidies aimed at making CCUS more economically viable.[33] For example, Canada enacted the Carbon Capture and Storage Statutes Amendment Act 2010,[34] which provides for active government participation in CCUS. Canada also enacted the Carbon Capture and Storage Funding Act which provides for a funding of $2 million for the construction and operation of carbon capture and storage projects. The United States enacted the Carbon Dioxide Transportation Infrastructure Finance and Innovation Act (CIFIA). This Act provides funding for 'large-capacity, common-carrier carbon dioxide (CO_2) transport projects'.[35] There are policies even at the State level in the United States. The California Air Resources Board (CARB) introduced CCUS incentives into the Low-Carbon Fuel Standard (LCFS) Regulation. This regulation made it possible for facilities that captured CO_2 to generate tradable credits.[36]

The economic and industrial implications of the use of CCUS are also significant. For countries that have heavy carbon intensive industries, CCUS offers an avenue for the reduction of emissions while maintaining industrial output and competitiveness. The technological development and deployment are expected to create and bring about the creation of jobs and economic development opportunities, particularly in areas heavily reliant on fossil fuel industries. However, to fully realise potential of CCUS in the global transition to Net Zero, it will involve overcoming series of challenges. The development of the necessary infrastructure for transporting and storing captured CO_2 is quite challenging. For example, considerations have to be given to the use of existing pipeline networks and identifying suitable geological storage sites. There is also the negative public scepticism and regulatory barriers that can operate

[33] IEA, 'Carbon Capture, utilisation and Storage', https://www.iea.org/energy-system/carbon-capture-utilisation-and-storage, accessed 22 November 2023.

[34] Section 2(1) of the Carbon Capture and Storage Statutes Amendment Act 2010 amended the Mines and Minerals Act.

[35] Loan Programs Office, 'Carbon Dioxide Transportation Infrastructure', https://www.energy.gov/lpo/carbon-dioxide-transportation-infrastructure, accessed 22 November 2023.

[36] Clara Galeazzi, Grace Law and John P Holdren, 'Carbon Capture, Utilization, and Storage: Carbon Dioxide Transport Costs and Network-Infrastructure Considerations for a Net-Zero United States' (Paper, Belfer Center for Science and International Affairs, Harvard Kennedy School, 20 July 2023) 1–45, 16.

as impediments to the deployment of CCUS projects. To address such concerns, it becomes imperative that there is transparent communication and measures are put in place for robust regulatory frameworks to ensure environmental safety and gain public trust.

As the world moves towards Net Zero, the role of CCUS is becoming increasingly central focus with more attention being paid. The potential of CCUS to address emissions from hard-to-abate sectors and directly remove CO_2 from the atmosphere positions it as a key technology in the global climate framework. Therefore, the success will hinge on technological advancements, international collaboration, strategic policy frameworks, and a balanced approach to addressing economic, environmental, and societal needs which may be sought after through the instrumentalities of the current climate change regime.

4 The Influence of the Paris Agreement on CCUS

The Paris Agreement was adopted in 2015, making it the cornerstone in the global effort to combat climate change taking over from the Kyoto Protocol. This section examines the potential role of the Paris Agreement in the promotion of CCUS to meet net zero commitments. It examines relevant provisions of the Paris Agreement on mitigation, climate financing, technology transfer, and capacity building.

4.1 General Mitigation Obligations

The Paris Agreement was adopted in 2015, making it the cornerstone in the global effort to combat climate change taking over from the Kyoto Protocol. The goal of the Paris Agreement is to limit global warming to well below 2 degrees Celsius.[37] This ambitious objective is also backed by the Nationally Determined Contributions (NDCs), which set essentially roadmaps on how countries plan to reduce emissions and mitigate climate impacts. In these evolving strategies, Carbon Capture, Utilisation, and Storage (CCUS) is gaining prominence as a key technology. For example,

[37] Article 2(1) Paris Agreement.

The Kingdom of Saudi Arabia,[38] The United Kingdom,[39] and the United States[40] all have included CCUS in their NDCs.

CCUS's role in the context of the Paris Agreement has far reaching implications. Primarily, it offers a practical solution for reducing carbon emissions from industrial sources, which are among the most challenging sectors to decarbonise the economy. Notable of these are Industries such as cement, steel, and chemicals, vital for economies, which are significant contributors to global CO_2 emissions. The effect of the CCUS will be to enable industries to capture and either use or store the emitted CO_2, thereby reducing their carbon footprint without compromise of industrial output. This aspect is critical for countries that are heavily industrialised or those that are developing and expanding their industrial sectors that are largely carbon intensive. The focus of the Paris Agreement is on ambitious emission reductions and should push countries to explore beyond traditional emission reduction methods. The CCUS therefore emerges as a mechanism for capturing emissions but also for removing CO_2 directly from the atmosphere. As identified earlier, technologies like Direct Air Capture (DAC), though still in their early stages of development, hold immense potential in the long-term as strategy of achieving Net Zero emissions, which is also a goal connected to the aspirations of the Paris Agreement.[41]

The increasing inclusion of CCUS in NDCs by countries such as the Kingdom of Saudi Arabia, the UK, the United States, and many others is a testament to its relevance. Nations are not only looking at CCUS from a technological perspective but also considering the policy,

[38] Update First Nationally Determined Contribution of Kingdom of Saudi Arabia 2021, https://unfccc.int/sites/default/files/resource/202203111154---KSA%20NDC%202021.pdf, accessed 22 November 2023.

[39] United Kingdom of Great Britain and Northern Ireland's Nationally Determined Contribution 2022, https://assets.publishing.service.gov.uk/government/uploads/system/uploads/attachment_data/file/1109429/uk-nationally-determined-contribution.pdf, accessed 22 November 2023.

[40] The United States of America Nationally Determined Contribution: Reducing Greenhouse Gases in the United States: A 2030 Emissions Target, https://unfccc.int/sites/default/files/NDC/2022-06/United%20States%20NDC%20April%2021%202021%20Final.pdf, accessed 23 November 2023.

[41] International Energy Agency, 'Direct Air Capture: A Key Technology for Net Zero', https://iea.blob.core.windows.net/assets/78633715-15c0-44e1-81df-41123c556d57/DirectAirCapture_Akeytechnologyfornetzero.pdf, accessed 23 November 2023.

legal, and financial frameworks needed to support the deployment in meeting the objectives of mitigations. Therefore, developing incentives for private sector investment in CCUS regulations for the safe and effective storage of captured carbon, and fostering international collaboration for technology transfer and research will need to be prioritised.

Examining the broader context of the Paris Agreement, CCUS is more than a mere technological solution to climate change. It represents a collaboration envisioned in the UNFCCC and should be such that there is cooperation between governments, industries, and communities. It can be seen as a bridge between the current state of industrial emissions and future of a low-carbon world. As countries continue to refine and implement their NDCs, the role of CCUS is expected to expand overtime and aligning with the global commitment to a sustainable and climate-resilient future that should be reckoned with. This alignment not only highlights the versatility and necessity of CCUS in current international climate strategies but also underscores its potential to evolve and adapt within the dynamic landscape of global climate policies going by the obligations.

4.2 Climate Financing Obligations

Climate finance emerges as a crucial obligation under the Paris Agreement.[42] Nations have to put up with the challenges and costs associated with transitioning to a low-carbon economy. The costs associated with mitigation are enormous.[43] This financing is important for both adaptation and mitigation strategies, which implication include advanced technologies such as CCUS. The Paris Agreement recognise the significant of financial resources and urges developed countries support developing nations in climate actions.[44] This has as it bases the principle of common but differentiated responsibilities and respective capabilities which is well articulated in the Paris Agreement.[45] The expectations that financing

[42] Patricia Galvao Ferreira, 'Climate Finance and Transparency in the Paris Agreement: Key Current and Emerging Legal Issues' (2018) CIGI Paper 195, 1–11.

[43] Signe Krogstrup and William Oman, 'Macroeconomic and Financial Policies for Climate Change Mitigation: A Review of the Literature' (2019) IMF Working Paper WP/19/185, 1–58, 11, https://www.imf.org/-/media/Files/Publications/WP/2019/wpiea2019185-print-pdf.ashx, accessed 31 December 2024.

[44] Article 9 Paris Agreement 2015.

[45] Article 2(2) Paris Agreement 2015.

under the Paris Agreement will come from a variety of sources, such as public, private, bilateral, and multilateral, and well as alternative sources.[46] Kenya, for example, established a Climate Fund for mobilising support from a variety of sources.[47] In 2014, it developed about US$ 3.2 billion worth of projects and programmes pooling resources from public, private, international, and domestic sources.[48]

The effective implementation on CCUS will be reliant on climate finance. The cost of procurement of CCUS technologies is high and can constitute a barrier. It was found that carbon storage cost $2011 per tonne of CO_2 in the United States.[49] The costs only cover the initial outlay for technology development and deployment of ongoing expenses related to operation, maintenance, and the infrastructure needed for transportation and storage of captured carbon. For many countries, particularly the developing ones, these costs can be prohibitively high, creating a barrier to adopting CCUS technologies.[50] The high cost of CCUS is ascribed to numerous variables such as the changing nature of the technology itself which is yet to benefit from the economies of scale that could bring down costs. Furthermore, major investment in adapting and upgrading facilities is required to integrate CCUS into the existing industrial processes. Research and development also require financial backing.

[46] Articles 6(4)(b) and 11(4) Paris Agreement.

[47] UNEP, 'Climate Finance Law: Legal Readiness for Climate Finance' (Report and Findings of Workshop, London, 9–11 March 2018), 12.

[48] Vincent Mutie Nzau, 'Climate Change Financing in Kenya' (2014) International Institute for Environment and Development, http://www.jstor.org/stable/resrep01573, accessed 23 November 2023.

[49] Derek Vikara, Chung Yan Shih, ShangMin Lin, Allison Guinan, Timothy Grant, David Morgan and Donald Remson, 'U.S. DOE's Economic Approaches and resources for Evaluating the Cost of Implementing Carbon Capture, Utlization, and Storage (CCUS)' (2017) 5(4) Journal of Sustainable Energy Engineering 307–340, 324.

[50] In Indonesia, CCUS was included into the NDC but the high cost is a major challenge that could hinder its implementation. To ensure that the high cost does not disrupt the CCUS plan, carbon trading has been considered as a mechanism to offset the huge cost. Adanya T Amelinda and Subiakto Soekarno, 'Financial Feasibility Study of Carbon Capture, Utilization, and Storage Project in West Java, Indonesia' (2023) 8(3) European Journal of Business & Management Research 215–220. https://doi.org/10.24018/ejbmr.2023.8.3.1969.

When it comes to underdeveloped countries, where financial constraints make widespread implementation of CCUS technologies difficult, climate finance emerges as a crucial method for doing so. Grants, low-interest loans, and financial incentives for private investment are only a few examples of available funding mechanisms. To guarantee that nations not only have access to CCUS technologies but also have the necessary expertise and legal frameworks supporting its deployment, this initiative will focus on capacity building, technological transfer, and support for policy formulation.

The development of CCUS technologies and the realisation of their promise are inextricably intertwined with the Paris Agreement's climate finance needs. Climate financing is crucial for ensuring that all nations, regardless of wealth, can participate in and reap the rewards of the global transition to a low-carbon future through the CCUS. This approach aligns with the overarching goals of the Paris Agreement, facilitating a collaborative and equitable path towards global climate change mitigation and Net Zero.

4.3 Technology Transfer Obligations

It is acknowledged that a global response to climate change requires a collaborative effort in sharing knowledge, expertise, and technology. The Paris Agreement places a strong emphasis on technology transfer, particularly for sustainable and climate-friendly technologies.[51] The provision of the agreement is crucial for promoting CCUS using it as a technology essential for reducing global greenhouse gas emissions. Technology transfer within the framework of the Paris Agreement is beyond the deployment of hardware from developed to developing countries. It also entails sharing the know-how, skills, and methodologies necessary for effective implementation and utilisation of advanced technologies such as the CCUS. This transfer is particularly important for CCUS because the technology is a complex one and still much evolving. There is a requirement of possessing sophisticated knowledge in the field of engineering, geology, and chemistry, among other disciplines, to be effectively implemented.

[51] Article 10 Paris Agreement.

For countries with limited resources or expertise in CCUS, technology transfer becomes the game changer to advance its deployment. It can enable these countries to leapfrog to advanced carbon capture methods without having to invest heavily in the initial research and development phase that could be prohibitively expensive. This is crucial because many of these countries are faced with both the challenge of developing their economies while putting in place measures for reducing emissions. Moreover, technology transfer in the context of CCUS will not only give room for mitigation but also about economic development. This can bring about new industries and job opportunities, particularly in areas such as engineering, plant operation, and environmental monitoring. By implications, this can result in developing countries benefiting in diversification of their economies and increase their technological capacities.[52]

The Paris Agreement can foster this transfer through various mechanisms such as financial support, capacity building initiatives, and cooperative research and development efforts.[53] It is expected that developed countries should take the lead and provide support to developing countries to help them adopt and implement CCUS technologies. This support can take the forms of funding, joint research projects, training and capacity building programs, and favourable trade terms for CCUS-related equipment and technology.[54] Through environmental collaboration and support, technology transfer obligations in Paris Agreement can significantly boost the global deployment of CCUS. This not only helps meet the climate goals but also promotes sustainable economic development, making it a win-win for both developed and developing nations. Consequently, technology transfer under the Paris Agreement can be a

[52] John H Barton, 'Mitigating Climate Change Through Technology Transfer: Addressing the Needs of Developing Countries' (2008) Energy, Environment and Development Programme: Programme Paper 08/02, https://www.chathamhouse.org/sites/default/files/public/Research/Energy,%20Environment%20and%20Development/1008barton.pdf.

[53] Article 6(8) Paris Agreement.

[54] Anna Belianska, Nadja Bohme, Kailhao Cai, Yoro Diallo, Saanya Jain, Mr. Giovanni Melina, Ms. Pritha Mitra, Mr. Marcos Poplawski Ribeiro and Solo Zerbo, 'Climate Change and Select Financial Instruments: An Overview of Opportunities and Challenges for Sub-Saharan Africa' (2022) IMF Staff Climate Note 2022/009, https://www.imf.org/-/media/Files/Publications/Staff-Climate-Notes/2022/English/CLNEA2022009.ashx, accessed 31 December 2024.

critical tool in advancing CCUS globally by creating avenues for access to the technologies needed to combat climate change effectively.

4.4 Capacity Building Obligations

Capacity building is a fundamental component of the international climate change agenda. The Paris Agreement, aimed at enhancing the abilities of countries, especially developing ones, through effective means of addressing climate change by enhancing capacity building.[55] The requirements of capacity building in the agreement have significant implications for CCUS, giving that the technology involved is complex and resource-intensive. The essence of capacity building in the context of the Paris Agreement revolves around empowering countries to plan for, access, manage, deploy, and sustain climate technologies and practices such as the use of the CCUS.[56] For the effective realisation it means that CCUS developed countries should help developing countries develop the necessary skills, knowledge, and institutional frameworks to implement the technology.

Given that CCUS involves intricate processes and advanced technology, capacity building is critical for ensuring that countries can not only adopt technologies but also efficient operation and maintenance of the systems. In practical terms, capacity building for CCUS under the Paris Agreement can take various forms such as training personnel in the latest CCUS technologies and sharing best practice.[57] It is also about developing regulatory and legal frameworks that support CCUS initiatives and putting in place institutions and mechanisms for monitoring and verifying carbon capture and storage. Putting in place such measures are crucial in countries that are lacking in existing infrastructure or expertise. For developing countries, this capacity building is particularly important. Moreover, capacity building for CCUS extends beyond just technical aspects and can include public awareness and engagement, ensuring that

[55] Article 11 Paris Agreement.

[56] Meinhard Doelle, 'The Paris Agreement: Historic Breakthrough or High Stakes Experiment?' (2016) Schulich Law Scholars, Schulich School of Law, Dalhousie University.

[57] Mizan R Khan, J Timmons Roberts, Saleemul Huq and Victoria Hoffmeister, *The Paris Framework for Climate Change Capacity Building* (Routledge, 1st edn, 2018).

communities understand the benefits and potential risks of CCUS.[58] This is essential for gaining public support and addressing any social or environmental concerns that are associated with CCUS projects.[59] Capacity building therefore can lead to a broader adoption of CCUS globally, particularly in regions that have traditionally lagged in advanced climate technologies. This wider deployment can accelerate the global reduction of greenhouse gas emissions, helping countries to meet the targets set in the Paris Agreement, while spurring innovation and improvements in CCUS technologies, as more countries continue to engage in research and development.

Capacity building obligations of the Paris Agreement are likely to have a significant positive impact on the global advancement and implementation of CCUS projects. By equipping countries with the necessary tools, knowledge, and systems, there is the possibility that the success rate of the efforts put in place for CCUS will create avenue for it to be more accessible and becoming more effective solution for combating climate change worldwide.

5 The Future of CCUS and Net Zero

The future of CCUS in the context of achieving net zero emissions is dynamic. With projected growth and trends indicating an increasing role in global climate strategies, the CCUS offers much promise. The development of CCUS technologies is expected to escalate overtime and driven by a growing recognition of its indispensability in tackling hard-to-abate sectors like heavy industry and power generation. With countries

[58] For example, war may adversely affect the environment leading to destruction. See Muyiwa Adigun, 'Boko Haram's Radical Ideology and Islamic Jurisprudence' in J-M Iyi and H Strydom (eds) *Boko Haram and International Law* (Springer, 2018) 229. By implication, wars may lead to negative impact on CCUS project thereby causing environmental risks.

[59] Social security laws for example may be use to offer general protection. See Kehinde Anifalaje, 'Emerging Trends in the Regulation of Social Security' (2016) 43(2) Journal of Malaysian and Comparative Law 1, 1–30; Kehinde Anifalaje, 'A Social Security Perspective of Employees' Compensation Law in Nigeria' (2021) 10(2) The Journal of Corporate and Commercial Law and Practice 45, 45–82.

continuing to strive to meet ambitious climate targets in the Paris Agreement and Net Zero, the CCUS is seen as a crucial bridge in the energy transition.[60]

Given the enabling environment and cooperation, the evolution of CCUS technologies is likely to see improvements in capture efficiency and a decrease in operational costs, making them more accessible and appealing to a broader range of industries overtime. Furthermore, advancements in carbon utilisation techniques, such as converting captured carbon into useful products like building materials, fuels, carbon fibre, chemicals, plastics, are expected to create new markets and economic opportunities.[61] For example, captured carbon can be converted into polymers in the production of plastics and adhesives; it can also be used as synthetic fuel in the production of liquid fuel.[62] This shift could transform CCUS from a cost centred to a value-creating venture, thereby enhancing its appeal and adoption. The integration of CCUS with other technologies is a critical aspect of its future role in meeting the net zero transition target. There is the prospect that there can also be synergy of the CCUS technologies with renewable energy sources. International regime is a starting point for policy and regulation of CCUS.[63] With the need to keep obligations at international levels, nations will consider effective policy implementations for CCUS integration and provide a framework for financial incentives such as tax credits or carbon pricing mechanisms, which can significantly lower the economic barriers to CCUS adoption. Governments can, in furtherance of their

[60] IEA, Special Report on Carbon Capture Utilisation and Storage: CCUS in Clean Energy Transitions, Energy Technology Perspectives 2020, 13, https://iea.blob.core.windows.net/assets/181b48b4-323f-454d-96fb-0bb1889d96a9/CCUS_in_clean_energy_transitions.pdf.

[61] Patricia Larkin, Monica Gattinger and Stephen Bird, 'Carbon Capture, Utlization, and Storage: Public Confidence in Risk Decision-Making' in M Gattinger (ed) *Democratizing Risk Governance* (Palgrave Macmillan, 2023, 161–187). https://doi.org/10.1007/978-3-031-24271-7_7.

[62] Fatima Bello, 'Towards a Legal Framework for Deploying Carbon Technologies to Address the Environmental Impacts of Gas Flaring in Nigeria' (2022) 13(2) The Journal of Sustainable Development, Law and Policy 102–124, 109. https://doi.org/10.4314/jsdlp.v13i2.5.

[63] The implementation of effective policies has been identified as a key factor in the development of renewable energy, which includes CCUS. See Oniemola PK (2011), 29–38. Thus, a policy on CCUS introduced by international organisations will likely be duplicated and implemented by many member countries.

obligations, support research and development and establish clear regulatory frameworks and guidelines for the safe and effective implementation of CCUS.

The Net Zero target is not the personal project of individual nations but a global pact. At the 2021 Glasgow Climate Pact, 197 nations committed to the Net Zero target by 2050.[64] It is a huge project that requires a lot of plans, schemes and technologies to achieve. The schemes and technology needed to achieve the Net Zero target are expensive; this includes CCUS technologies. However, the Intercontinental Panel on Climate Change (IPPC) has estimated that achieving the 2 °C target of the Paris Agreement will cost double the cost without the use of CCUS.[65] While CCUS technologies may be expensive, it is actually cost saving. There is no doubt that CCUS plays an important role in meeting the Net Zero target.[66]

It would be an easier burden to bear if the private and public sectors jointly and actively participate in adopting CCUS technologies in their respective businesses. The oil companies have already taken vital steps to do this. Companies in other concerned sectors can also be made to adopt CCUS technologies through the use of incentives, policies and regulations. The Energy sector is the largest contributors to GHG emissions[67]

[64] Daniel C Etsy and Nathan de Arriba-Sellier, 'Zeroing in on Net-Zero: From Soft Law to Hard Law in Corporate Climate Change Pledges' (2023) 94(3) University of Colorado Law Review 635–679, 635.

[65] National Petroleum Council, 'The Role of CCUS in the Future Energy Mix' in *Meeting the Dual Challenge: A Roadmap to At-Scale Deployment of Carbon Capture, Use, and Storage* (Chapter 1, 2021, 1–19), https://www.energy.gov/sites/default/files/2022-10/CCUS-Chap_1-030521.pdf.

[66] International Energy Agency, 'Legal and Regulatory Frameworks for CCUS: An IEA CCUS Handbook', 3, https://iea.blob.core.windows.net/assets/bda8c2b2-2b9c-4010-ab56-b941dc8d0635/LegalandRegulatoryFrameworksforCCUS-AnIEACCUSHan dbook.pdf.

[67] T Bruckner, IA Bashmakov, Y Mulugetta, H Chum, A de la Vega Navarro, J Edmonds, A Faaij, B Fungtammasan, A Garg, E Hertwich, D Honnery, D Infield, M Kainuma, S Khennas, S Kim, HB Nimir, K Riahi, N Strachan, R Wiser, and X Zhang, 'Energy Systems' in: O Edenhofer, R Pichs-Madruga, Y Sokona, E Farahani, S Kadner, K Seyboth, A Adler, I Baum, S Brunner, P Eickemeier, B Kriemann, J Savolainen, S Schlömer, C von Stechow, T Zwickel and JC Minx (eds), *Climate Change 2014: Mitigation of Climate Change*. Contribution of Working Group III to the Fifth Assessment Report of the Intergovernmental Panel on Climate Change (Cambridge University Press, Cambridge, United Kingdom and New York, NY, USA, 2014), 511–597, 516.

and a large part of this comes from oil companies. It is absolutely necessary for these oil companies to participate in the mitigation schemes being adopted all around the world.[68] Shell, one of the leading international oil companies in the world,[69] regularly publishes its Sustainability Report where it outlines the steps the company has taken to transition its operations to Net Zero emissions as well as to ensure a safe and sustainable environment.[70] In its 2022 Sustainability Report, Shell stated that it has a 2050 target to achieve net zero emissions operations.[71] The Report also stated that Shell has actively been using carbon capture and storage technology and had invested about $220 million in operating costs for CCS.[72] ExxonMobil, another leading international oil company, has indicated in its 2023 Report that its operations align with the goals of the Paris Agreement and the International Energy Agency (IEA); and that it is working towards the 2050 Net Zero target.[73] ExxonMobil has adopted CCUS in its operations[74] and also has plans to commercialise carbon capture and storage.[75] Yet another of the largest oil companies in the world, Chevron, has published a report that shows its inclusion of CCUS

[68] Already, the Paris Agreement encourages the participation of the private sector which includes the oil companies.

[69] Susan Ariel Aaronson and David Deese, 'The Big Five International Oil Companies as Responsible Stakeholders in the Global Economy' (2009) 5(1) Transnational Corporations and Global Governance 27–54, 27.

[70] Shell Global, 'Shell Publishes Reports on Sustainability, Climate & Energy Transition Lobbying and Payments to Governments' (28 March 2023), https://www.shell.com/media/news-and-media-releases/2023/shell-publishes-reports-on-sustainability-climate-and-energy-transition-lobbying-and-payments-to-governments.html, accessed 24 November 2023.

[71] Shell Plc Sustainability Report 2022, 22, https://reports.shell.com/sustainability-report/2022/_assets/downloads/shell-sustainability-report-2022.pdf.

[72] Shell Sustainability Report 2022, 30.

[73] ExxonMobil, *Advancing Climate Solutions: Progress Report* (2023) 7, https://corporate.exxonmobil.com/-/media/global/files/advancing-climate-solutions-progress-report/2023/2023-advancing-climate-solutions-progress-report.pdf.

[74] Ibid. at 12.

[75] Ibid. at 6.

in its operations.[76] The report also indicates that CCUS will account for the removal of 6.2 gigatonnes (Gt) of CO_2 by 2050.[77]

International and regional institutions offer avenues for projection of net zero and other climate change mitigations measures.[78] The UN systems within the current framework and international coalitions could establish common standards and encourage cooperative efforts in CCUS research and projects, fostering a globally coordinated approach to carbon management. As the world moves towards net zero emissions, the potential of CCUS to contribute significantly to this transition is becoming increasingly evident in the light of the number of projects upcoming and implemented by both the private and public sector at domestic levels. Largely, it is the role of domestic law to give effect to a treaty.[79] The future of CCUS is likely to be characterised by technological innovation, increased integration with renewable energy and energy efficiency measures, and a strong foundation in supportive policies and international collaboration, which are instrumental in the global quest for a sustainable and carbon-neutral future.

6 Conclusion

In conclusion, the exploration of Carbon Capture, Utilisation, and Storage (CCUS) in the context of achieving Net Zero emissions reveals it has a pivotal role to play and substantial potential. The utilisation of the CCUS emerges not just as a technological solution, but constitute an important component in the avalanche of approaches required for addressing climate change that meets the objective of the Paris Agreement. Its ability to capture emissions from carbon intensive sectors, which are among the most challenging sources to achieve the decarbonisation

[76] Chevron Corporation, *Climate Change Resilience Report* (2023), https://www.chevron.com/-/media/chevron/sustainability/documents/climate-change-resilience-report.pdf.

[77] Ibid. at 18.

[78] Muyiwa Adigun, 'Legal Remedies for Energy Injustice in the ECOWAS Sub-Region: The Role of the ECOWAS Court (2024) 42(3) Journal of Energy & Natural Resources Law 363, 363–380; Muyiwa Adigun, 'A Human Rights Approach to Climate Litigation before the ECOWAS Court' (2024) 26(1) Environmental Law Review 16, 16–32.

[79] See Muyiwa Adigun, 'Enforcing ECOWAS Judgments in Nigeria Through the Common Law Rule on the Enforcement of Foreign Judgments' (2019) 15 Journal of Private International Law 130, 130–161.

goals, positions it as an indispensable tool in the global effort to reduce greenhouse gases. The potential of CCUS extends beyond mere carbon capture. Its utilisation aspect, which involves converting captured carbon into useful products, presents an innovative approach in carbon management. This not only mitigates emissions from anthropogenic activities but also fosters a circular carbon economy, creation of new economic opportunities while reducing the impacts of carbon footprints. In the same token, the storage component of CCUS, ensuring the safe and long-term sequestration of carbon dioxide, further solidifies its role in achieving the gaols of Net Zero.

Reflecting on the future of CCUS, it is clear that its success and widespread adoption depend on a combination of technological advancement, supportive policies, economic viability, and international collaboration, which can be achieved within the framework of the Paris Agreement. The role of international agreements such as the Paris Agreement in shaping climate policies and laws underscores the importance of global cooperation in climate action actualising Net Zero. Therefore, the CCUS, in this global context, is not just a national or regional solution but a key component of the international effort to mitigate climate change. As the world progresses towards a net zero future, CCUS stands as a testament to human ingenuity and our capacity to confront global challenges, most especially climate change. It represents a bridge between our current realities and a more sustainable, carbon-neutral future as envisioned by Paris Agreement and push for Net Zero. The journey towards Net Zero is undoubtedly complex and multifaceted, but with technologies like CCUS, combined with global cooperation and a commitment to sustainable development, it is a goal that remains within reach as the world continue to search for an optimal solution for resolving the global menace of climate change.

CHAPTER 6

CCU-Based Products and Their Technological Routes

Alexandre de Barros Gallo, Alberto J. Fossa, Cylon Liaw, and Edmilson Moutinho dos Santos

1 Introduction

Currently, fossil fuels are often used as energy and carbon sources in industrial processes. Especially in organic chemistry, carbon is an essential and indispensable component of manufactured products. In the future, CO_2 (carbon dioxide) is to be used as a key carbon source, thereby placing basic organic chemistry and certain areas of the energy and transport sectors on a new raw material basis that is independent of fossil

A. de Barros Gallo (✉) · A. J. Fossa · C. Liaw · E. Moutinho dos Santos
Research Centre for Greenhouse Gas Innovation (RCGI), University of São Paulo, São Paulo, Brazil
e-mail: alexandre.gallo@usp.br

Institute of Energy and Environment of the University of São Paulo (IEE-USP), São Paulo, Brazil

A. J. Fossa
e-mail: afossa@newencreative.com.br

E. G. Pereira et al. (eds.), *Carbon Capture Utilization and Storage*, Sustainable Development Goals Series,
https://doi.org/10.1007/978-3-031-81272-9_6

raw materials (crude oil, natural gas, and coal). The core of this is the capture, transport, and utilization of CO_2 and ultimately the realization of an industrial carbon (dioxide) cycle, which is currently being developed. Carbon cycles can be closed via many different chemical compounds; here the focus is on the direct and indirect use of CO_2. It can be separated directly in combustion processes, for example, or indirectly from the atmosphere.

Carbon Dioxide Capture and Utilization (CCU) is part of the set of technologies for climate change mitigation, which includes also carbon capture and storage (CCS). In this case, while the CCS provides for the final destination of the captured CO_2 to geological sites or depleted wells for oil and natural gas exploration, the CCU uses carbon dioxide as a raw material for conversion into products.

Possible routes of CO_2 use include those with chemical use (with direct conversion) and physical use (without conversion) of carbon dioxide. For the purposes of mitigating climate change, it is essential to highlight that each product from these CO_2 utilization routes contributes differently, depending on the time of storage of the carbon in its composition. In this sense, the category of construction materials (aggregates, cement, and concrete) has a notable advantage in conserving CO_2 in the long term in buildings, for example.

This chapter presents an overview of Carbon Capture and Utilization (CCU) basic concepts and a review of CCU experiences, followed by a depiction of possible technological routes, including examples and its maturity level.

2 Carbon Capture and Utilization (CCU) Overview

This section presents an overview of Carbon Capture and Utilization (CCU) by first discussing and presenting some basic concepts, particularly clarifying some overlaps and differences between CCU, CCS, and CCUS. This terminology approach is then succeeded by a review of CCU experiences, presenting some numbers regarding the production of CCU-based products and the status of CCUS projects around the globe.

2.1 Basic Concepts

In this context, a multitude of new terms, designations, and acronyms have emerged in recent years. The development and upscaling of corresponding technologies requires inter-disciplinary cooperation between companies from different sectors, as well as institutions from different fields of work. However, the use of terminology still varies from country to country and company to company. Terms often do not have a clear definition but are used synonymously despite having different meanings. Uniform definitions and a common understanding of terminology are, however, necessary to enable effective and efficient cooperation. This section aims to discuss definitions for terms in the field of CO_2 utilization and related fields of work (e.g., CO_2 storage or sector coupling). This will help to facilitate cross-national and cross-disciplinary communication between research institutes, companies, and users, as well as in political and public discourse.

The basic concepts can be discussed in four main contexts. The first is the explanation of CCU, CCS, and CCUS terms and how they differ from each other. The second involves disentangling CO_2 storage and CO_2 sequestration. The third involves highlighting the differences between CO_2 emission reduction, CO_2 reduction, and CO_2 removal. Finally, the last context details the definition of CO_2 utilization routes.

The terminology used in this section and throughout this chapter is mainly based on ISO[1] and DIN,[2] as well as considers information from the remaining references.

2.1.1 CCU, CCS, and CCUS

CO_2 capture and utilization (CCU) is a process comprising the separation and extraction of CO_2 from a CO_2-containing medium and the use of the CO_2 in a production process for the manufacture of a carbonaceous product or other use of the CO_2. CCU is sometimes also referred to as carbon recycling, CO_2 recycling, or upcycling, or CO_2 valorization. CCU covers all the terms, but the other terms do not cover each in themselves.

CO_2 capture, utilization, and storage (CCUS) is a process consisting of the separation and extraction of CO_2 from a CO_2-containing medium

[1] ISO 27917 Carbon dioxide capture, transportation, and geological storage—Vocabulary—Cross-cutting terms [2017].

[2] DIN SPEC 91458 Utilization of carbon dioxide—Terms [2023].

and its use in a production process, whereby the carbon is bound long-term in a product and will not be released again as CO_2. CCUS is often used to combine the terms CCU and CCS. This approach is not recommended, as it can lead to unnecessary confusion between the very different CCU and CCS methods. The term CCUS is recommended only when both processes are coupled.

CO_2 capture and storage (CCS) is a process consisting of the separation and extraction of CO_2 from a CO_2-containing medium, transportation, and long-term[3] storage of CO_2 or the carbon contained therein. CCS can help to prevent CO_2 from entering the atmosphere (CO_2 emission reduction) as well as to remove CO_2 from the atmosphere (CO_2 removal).

The delimitation of boundaries for CCU involves processes common to CCS, such as the capture, conditioning, and transport of carbon dioxide, as illustrated in Fig. 1. In the case of CCU, the subsequent processes depend on the CO_2 utilization route. In the case of CCS, the final process involves CO_2 storage, for example, in geological reservoirs.

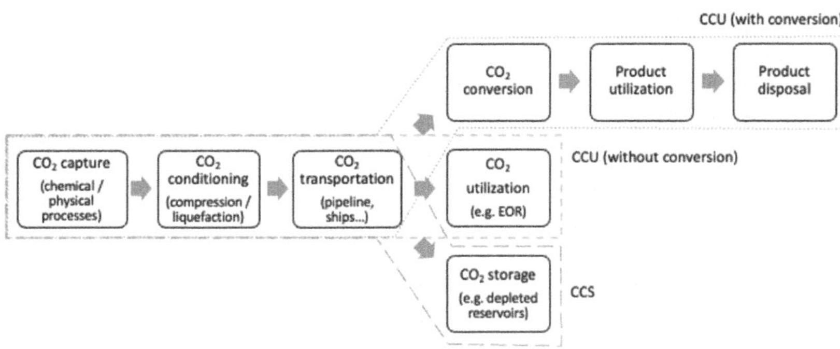

Fig. 1 CCS and CCU schematic processes[4]

[3] Long-term is defined as a minimum period of time required for CO_2 storage to be considered an effective option for mitigating climate change.

[4] Eduardo G. Pereira and others, 'Advancing Sustainability Through Carbon Capture, Utilization and Storage (CCUS) Technologies: The Hydrogel Case Study' [2023] 14 Journal of Sustainable Development Law and Policy (The), https://doi.org/10.4314/jsdlp.v14i1.11s.

2.1.2 CO_2 Storage and CO_2 Sequestration

CO_2 storage is a process in which CO_2, or the carbon contained therein, is stored long-term. CO_2 sequestration is often used as an alternative to CO_2 storage, but the latter is preferred in CCS and CCUS contexts as "sequestration" is more general and can also refer to non-anthropogenic processes (e.g., non-long-term uptake of carbon by living organisms).

In fact, **CO_2 sequestration** is an uptake of CO_2 in reservoirs other than the atmosphere. It can be achieved, for example, through land-use change, reforestation, and mineralization, as well as through various technologies of underground CO_2 storage.

2.1.3 CO_2 Emission Reduction, CO_2 Reduction, and CO_2 Removal

CO_2 emission reduction is a calculated net decrease of CO_2 emissions between a baseline scenario and an alternative process or scenario with comparable accounting frameworks. In most cases, a CO_2 emission reduction can be referred to as "avoided CO_2". CO_2 emission reduction is often referred to as CO_2 reduction. This terminology is not recommended in the context of CO_2 emission reduction.

CO_2 reduction is a chemical reaction in which CO_2 serves as an oxidant, takes up electrons, and is converted into a reduced form. This term is not recommended in the context of CO_2 emission reduction, as it is ambiguous to the definition of CO_2 emission reduction. The terminology is often used for electrochemical processes but is also applicable for CCU processes with chemical use.

CO_2 removal is an anthropogenic activity of long-term storage of CO_2 in anthropogenically developed or enhanced sinks that reduces the level of CO_2 in the Earth's atmosphere. CO_2 removal includes anthropogenic development and enhancement of existing and potential sinks (e.g., biological and geochemical) but does not include natural CO_2 uptake that is not directly triggered by human activity.

2.1.4 CO_2 Utilization and CO_2 Utilization Routes

CO_2 utilization is the use of CO_2 in a production process for the manufacture of a carbon-containing product or other use by exploiting its physical properties. The mere storage of CO_2 explicitly does not fall under CO_2 utilization. Two main routes of CO_2 utilization can be outlined, namely chemical CO_2 utilization and physical CO_2 utilization.

Chemical CO_2 utilization is a CO_2 utilization route in which CO_2 is used as a reactant in a chemical reaction. Chemical CO_2 utilization is often also referred to as material use, but this terminology is not recommended, because the term material use is not clearly defined.

Physical CO_2 utilization is a CO_2 utilization route mainly with regard to its physical properties and without changes to the chemical bonds.

2.2 *A Review of CCU Experiences*

Between 2019 and 2022, global CO_2 emissions grew continuously, apart from 2020, and reached a total of 36.80 gigatons (Gt),[5] Of this value, four sectors stand out, whose sum of emissions resulted in 34.75 $GtCO_2$ (2022), with the highest contribution from electricity generation, followed by the industry and transport sectors (see Table 1). Although there was a small variation in the period considered, the four sectors remained practically constant in their respective contributions.

Table 1 Global CO_2 emissions (2019–2022)[6]

Sector	*2019*	*2020*	*2021*	*2022*
Power generation (%)	41	42	42	42
Industry (%)	26	27	27	26
Transport (%)	24	22	22	23
Buildings (%)	9	9	9	9
Total (%)	100	100	100	100
Total ($GtCO_2$)	34.00	32.22	34.44	34.75

[5] IEA, 'Global CO_2 Emissions by Sector, 2019–2022' [2023] OECD/IEA, https://www.iea.org/data-and-statistics/charts/global-co2-emissions-by-sector-2019-2022.

[6] Ibid.

From a demand perspective, global CO_2 consumption jumped from 150 $MtCO_2$/year to 230 $MtCO_2$/year between 2000 and 2015.[7] By 2050, 370 $MtCO_2$/year is expected to be consumed according to the International Energy Agency's projection.[8] Currently, the main uses of carbon dioxide are in the manufacture of fertilizers and for the application of the enhanced oil recovery (EOR) technique (Fig. 2).

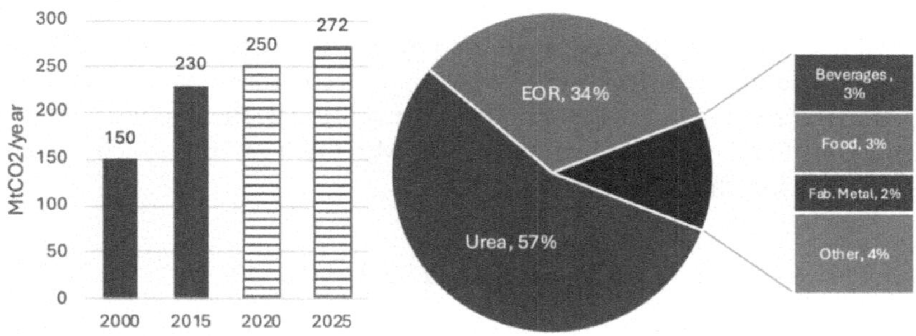

Fig. 2 Growth in global CO_2 demand over the years (left) and demand description in 2015 (right)[9]

Three factors are considered paramount to the successful implementation of CCU technology, namely: (1) the length of time the product retains CO_2; (2) the amount of CO_2 fixed in the product; and (3) the value of the product in the market.[10] In addition, in the view of the IEA,[11] five other key factors are decisive in assessing the benefits to the climate of the application of the CCU: (1) the origin of the CO_2 (natural

[7] B. Dziejarski, R. Krzyżyńska and K. Andersson, 'Current Status of Carbon Capture, Utilization, and Storage Technologies in the Global Economy: A Survey of Technical Assessment' [2023] 342 Fuel , https://doi.org/10.1016/j.fuel.2023.127776.

[8] IEA, 'Energy Technology Perspectives 2020: Special Report on Carbon Capture, Utilisation and Storage' [2020] OECD/IEA, https://www.iea.org/reports/ccus-in-clean-energy-transitions.

[9] Ibid.

[10] A.-M. O. Mohamed and others, 'Carbon Capture and Utilization' in A.-M. O. Mohamed and others, *Sustainable Utilization of Carbon Dioxide in Waste Management* (Elsevier, 2023), https://doi.org/10.1016/b978-0-12-823418-1.00001-9.

[11] IEA, 'Putting CO_2 to Use: Creating Value from Emissions' [2019] OECD/IEA, https://www.iea.org/reports/putting-co2-to-use.

deposits, fossil fuels, biomass, or atmospheric); (2) the product or service that the CO_2-based product is replacing; (3) how much and what form of energy is used to convert CO_2; (4) how long carbon is retained in the product; (5) the scale of opportunity for the use of CO_2.

Table 2 present an outlook regarding production figures for novel CO_2-based products and, in summary, it is observed a production of 1.3 Mt/a in total in 2022 growing to an expected production of more than 6 Mt/a in 2030.

Table 3 lists the projects associated with the CCU and/or EOR in operation, under construction, planned, or decommissioned. According to the International Energy Agency,[12] the list has the highest capacities in EOR projects and can exceed 8 $MtCO_2$/year of capacity, such as Petrobras in the Santos Basin.

3 CO_2 Utilization Routes

As presented in 2.1.4, CO_2 utilization can be allocated into two main routes considering if a CO_2 conversion process is present. Therefore, there is a route with direct use without CO_2 conversion, the physical CO_2 utilization, and a route with indirect use with CO_2 conversion route, the chemical CO_2 utilization. The main categories within each of these routes are presented in Fig. 3.

In the first route, physical CO_2 utilization, are included several applications that harness several distinct properties of CO_2, such as its high capacity for heat absorption, its stable and non-reactive nature, and its ability to act as a solvent. Through its direct use, carbon dioxide can be applied as a working fluid, solvent, among others, such as to improve yield in greenhouses and plant algae.

In the second route, chemical CO_2 utilization, CO_2 is used in conversion processes, which can be thermal, biological, electrochemical, photochemical, and other processes that alter the chemical bonds of CO_2. Some examples of the indirect use of CO_2 involve the production of chemicals, fuels, and building materials.

In the following Sects. 3.1 and 3.2 are presented, respectively, Physical and Chemical CO_2 utilization routes, including its main categories and

[12] IEA, 'CCUS Projects Database' [2023] OECD/IEA, https://www.iea.org/data-and-statistics/data-product/ccus-projects-database.

Table 2 Outlook of novel CO_2-based products in 2022 and 2030[13]

Product	*CO_2-based carbon content (%)*	*Production capacity in 2022*	*2030 Outlooks*
Aromatic polycarbonate (PC)	5	900,500 t/a	1.2 Mt/a
Ethanol	100	138,000 t/a	700,000 t/a for advanced fuel, chemical and polymer
Aliphatic polycarbonate (APC)	11–12	120,000 t/a	300,000 t/a mostly PPC, PEC, high molecular weight
Methanol	100	ca. 115,000 t/a	1 Mt/a mostly by CO_2 hydrogenation, some electrochemical pathways in development
Polycarbonate polyols	5–6	50,000 t/a	Increasing capacity by 2030 Low molecular APC, used in polyurethane synthesis
Polyhydroxy-alkanoates (PHA)	100	5000–10,000 t/a	ca. 30,000 t/a
Minerals	100	3 commercial plants in activity, several pilot and demonstration plants	Several commercial plants mostly used for cement applications

(continued)

[13] P. Ruiz and others, 'Carbon Dioxide (CO_2) as Feedstock for Chemicals, Advanced Fuels, Polymers, Proteins and Minerals—Short Version' [2023] nova-Institut GmbH (Ed.), Hürth, https://nova-institute.eu/news/pr/?id=428.

Table 2 (continued)

Product	*CO_2-based carbon content (%)*	*Production capacity in 2022*	*2030 Outlooks*
Methane	100	Several pilot plants	ca. 320,000 t/a fed into the gas grid
Hydrocarbons include kerosene, diesel, gasoline, naphtha, waxes	100	ca. 700 t/a	Due to the ReFuel Aviation EU proposal synthetic sustainable aviation fuels (SAF) should reach a share of 5% by 2035, this would mean about 3 Mt/a, automatic additional capacity will be achieved for the other fractions
Proteins	100%	Pilot plants	2 commercial plants, first in 2030 mostly for feed applications

Table 3 Projects associated to CCU applications[14]

Project name	*Country*	*Project type*	*Announced*	*Operation*	*Project status*	*Announced capacity ($MtCO_2/yr$)*	*Estimated capacity by IEA ($MtCO_2/yr$)*	*Sector*	*Fate of carbon*
3D DMX ArcelorMittal and IFPEN Dunkirk 'REUZE'	France	CCU	2022	2025	Planned	0.3	0.3	Iron and steel	Use
AEB Amsterdam	Netherlands	CCU	2021	2028	Planned	0.44	0.44	Power and heat	Use
AirCapture DACU at Nutrien Kennewick Fertilizer (WA)	United States	CCU	2022		Planned			DAC	Use
Airhive Alpha Lab pilot	Canada	CCU	2023	2024	Planned	0.001	0.001	DAC	Use
Aittaluoto power (Pori)	Finland	CCU	2022	2027	Planned	0.1	0.1	Power and heat	Use
Aker CC WtE e-fuels	Finland	CCU	2023		Planned	0.1	0.1	Power and heat	Use
Alberta Carbon Trunk Line (ACTL) (ALB)	Canada	Transport	2011	2020	Operational	14.6	14.6	Transport	EOR
Arcelor LanzaTech Carbalyst (Steelanol) Ghent	Belgium	CCU	2011	2022	Operational	0.125	0.125	Iron and steel	Use

(continued)

[14] IEA (n 13).

Table 3 (continued)

Project name	*Country*	*Project type*	*Announced*	*Operation*	*Project status*	*Announced capacity ($MtCO_2/yr$)*	*Estimated capacity by IEA ($MtCO_2/yr$)*	*Sector*	*Fate of carbon*
Arkalon CO_2 Compression Facility (KS)	United States	Full chain		2009	Operational	0.19–0.31	0.31	Biofuels	EOR
Avedøre Power Station CCU (Green Fuels for Denmark)	Denmark	CCU	2021	2026	Planned			Power and heat	Use
Balikpapan Refinery	Indonesia	Full chain	2022	2030	Planned	1.4	1.4	Other fuel transformation	Mixed
Bell Bay Powerfuels project (TAS)	Australia	CCU	2020	2027	Planned		0.411	Power and heat	Use
Berrima cement plant (NSW)	Australia	CCU	2022		Planned	0.1	0.1	Cement	Use
Bonanza BioEnergy CCUS (KS)	United States	Full chain	2010	2012	Operational	0.1–0.16	0.16	Biofuels	EOR
Boundary Dam CCS (SASK)	Canada	Capture	2008	2014	Operational	1	1	Power and heat	EOR
Bridgeport Energy Moonie CCUS Project	Australia	T&S	2018	2024	Under construction	0.12	0.12	T&S	EOR
CalCapture (Elk Hills) (CA)	United States	Capture	2019	2026	Planned	1.4–1.5	1.4	Power and heat	EOR

Project name	*Country*	*Project type*	*Announced*	*Operation*	*Project status*	*Announced capacity ($MtCO_2/yr$)*	*Estimated capacity by IEA ($MtCO_2/yr$)*	*Sector*	*Fate of carbon*
Carbon Clean CEMEX Rudersdorf plant phase 1	Germany	CCU	2021	2026	Planned	0.11	0.11	Cement	Use
Carbon Clean CEMEX Rudersdorf plant phase 2	Germany	CCU	2021		Planned	0.584	0.584	Cement	Use
Carbon TerraVault 1: Lone Cypress Hydrogen project (CA)	United States	Capture	2022	2025	Planned	0.1	0.1	Hydrogen or ammonia	EOR
Carbon2Business LafargeHolcim Laegerdorf (Westkuste100/ Hyscale100)	Germany	CCU	2020	2027	Planned	1	1	Cement	Use
Carbon2ProductAustria (C2PAT)—Lafarge Mannersdorf	Austria	CCU	2020	2030	Planned	0.7	0.7	Cement	Use
CarbonCapture DACU Carbon Cure at Gary Works (IN)	United States	CCU	2022		Planned	0.005	0.005	DAC	Use

(continued)

Table 3 (continued)

Project name	*Country*	*Project type*	*Announced*	*Operation*	*Project status*	*Announced capacity ($MtCO_2/yr$)*	*Estimated capacity by IEA ($MtCO_2/yr$)*	*Sector*	*Fate of carbon*
CCUS at Red Hills natural gas processing complex (NM) EOR Phase 2	United States	Full chain	2022	2024	Under construction	0.36	0.36	Natural gas processing/ LNG	EOR
CCUS at Red Hills natural gas processing complex (NM) EOR Phase 3	United States	Full chain	2022	2024	Planned	0.26	0.26	Natural gas processing/ LNG	EOR
CEIC-CNPC CCUS-EOR demonstration (Ningxia) Phase 1	China	Full chain	2023	2024	Under construction	0.4	0.4	Other fuel transformation	EOR
CEIC-CNPC CCUS-EOR demonstration (Ningxia) Phase 2	China	Full chain	2023	2025	Under construction	0.5	0.5	Chemicals	EOR
CEIC-CNPC CCUS-EOR demonstration (Ningxia) Phase 3	China	Full chain	2023	2027	Planned	1.5	1.5	Chemicals	EOR
Century plant (TX)	United States	Full chain	2008	2010	Operational	4.32–5	4.32	Natural gas processing/ LNG	EOR

Project name	*Country*	*Project type*	*Announced*	*Operation*	*Project status*	*Announced capacity ($MtCO_2/yr$)*	*Estimated capacity by IEA ($MtCO_2/yr$)*	*Sector*	*Fate of carbon*
Changling Gas plant/ Jilin Oil Field CO_2-EOR Full-scale (Jilin)	China	Full chain	2006	2018	Operational	0.43–0.6	0.6	Natural gas processing/ LNG	EOR
China Energy Guohua Jinjie Power (Shaanxi)	China	Full chain	2018	2021	Operational	0.15	0.15	Power and heat	Mixed
China Energy Taizhou power (Jiangsu)	China	Full chain	2021	2023	Operational	0.5	0.5	Power and heat	Mixed
Climeworks Norsk e fuel phase 1	Norway	CCU	2020	2025	Planned	0.039	0.039	DAC	Use
Climeworks Norsk e fuel phase 2	Norway	CCU	2020	2026	Planned	0.035	0.035	DAC	Use
Climeworks Norsk e fuel phase 3	Norway	CCU	2020	2029	Planned	0.184	0.184	DAC	Use
CNPC China Northwest hub phase 1 (Xinjiang)	China	T&S	2019	2025	Planned	1	1	T&S	EOR
CNPC China Northwest hub phase 2 (Xinjiang)	China	T&S	2019	2030	Planned	2	2	T&S	EOR

(continued)

Table 3 (continued)

Project name	*Country*	*Project type*	*Announced*	*Operation*	*Project status*	*Announced capacity ($MtCO_2/yr$)*	*Estimated capacity by IEA ($MtCO_2/yr$)*	*Sector*	*Fate of carbon*
CNPC China Northwest hub refinery hydrogen (Xinjiang)	China	Capture	2019	2025	Planned	1	1	Other fuel transformation	EOR
CO_2 capture at Duke Energy's East Bend Station (KY)	United States	Capture	2018		Planned			Power and heat	EOR
CO_2 TO METHANOL Finnfjord Ferrosilicon	Norway	CCU	2017		Planned	0.3	0.3	Iron and steel	Use
Coffeyville fertilizer Plant (KS)	United States	Full chain	2011	2013	Operational	0.7–0.9	0.9	Chemicals	EOR
Core Energy CO_2-EOR South Chester plant (MI)	United States	Full chain		2003	Operational	0.35	0.35	Natural gas processing/LNG	EOR
CPNC coal power Baijiantan Karamay (Xinjiang)	China	Full chain	2023	2027	Under construction	2	2	Power and heat	EOR
Crawford Nickel	Canada	CCU	2023		Planned	1.5	1.5	Other industry	Use
e-Alto	United Kingdom	CCU	2022		Planned			Biofuels	Use

Project name	*Country*	*Project type*	*Announced*	*Operation*	*Project status*	*Announced capacity ($MtCO_2/yr$)*	*Estimated capacity by IEA ($MtCO_2/yr$)*	*Sector*	*Fate of carbon*
ECOPLANTA (Tarragona)	Spain	CCU	2021	2027	Planned	0.34	0.34	Power and heat	Use
Emirates Steel Industries (Abu Dhabi CCS Phase 1)	United Arab Emirates	Full chain	2007	2016	Operational	0.8	0.8	Iron and steel	EOR
EM-Rhone	France	CCU	2023	2028	Planned	0.195	0.195	Cement	Use
Enhance Clive Sequestration Facility (ACTL) (ALB)	Canada	Storage	2011	2020	Operational	1.1–1.5	1.1	Storage	EOR
Enid fertilizer (OK)	United States	Full chain	1982	1982	Operational	0.68	0.68	Chemicals	EOR
FlagshipTHREE Umea Dava plant	Sweden	CCU	2023	2026	Planned		0.136	Power and heat	Use
FlagshipTWO Sundsvall Korsta plant	Sweden	CCU	2023	2026	Planned		0.177	Power and heat	Use
Gemah field EOR	Indonesia	Storage	2022	2031	Planned			Storage	EOR
Gerald Gentleman Station Carbon Capture (NE)	United States	Capture	2013	2025	Planned	3.8	3.8	Power and heat	EOR
Grandpuits biorefinery	France	CCU	2022	2025	Under construction	0.11	0.11	Other fuel transformation	Use

(continued)

Table 3 (continued)

Project name	*Country*	*Project type*	*Announced*	*Operation*	*Project status*	*Announced capacity ($MtCO_2$/yr)*	*Estimated capacity by IEA ($MtCO_2$/yr)*	*Sector*	*Fate of carbon*
Great Plains Synfuel Plant (ND) Weyburn-Midale (SK)	United States	Full chain	1997	2000	Operational	3	3	Other fuel transformation	EOR
Guanghui Energy CCUS integration project Phase 1 (Xinjiang)	China	Full chain	2021	2023	Operational	0.1	0.1	Chemicals	EOR
Gundih CCUS EGR	Indonesia	Full chain	2020	2030	Planned	0.3	0.3	Natural gas processing/LNG	EOR
Habshan-Bab gas plant (Abu Dhabi CCS Phase 2)	United Arab Emirates	Full chain	2020	2026	Under construction	1.5	1.5	Natural gas processing/LNG	EOR
HIF Cabo Negro eFuels Facility	Chile	CCU	2021		Planned	0.23	0.23	Biofuels	Use
HIF Paysandú eFuels Facility Alur bioethanol plant	Uruguay	CCU	2023	2025	Planned	0.15	0.15	Biofuels	Use
HIF Tasmania eFuels Facility (TAS)	Australia	CCU	2022	2028	Planned	0.3	0.3	Power and heat	Use
HIF USA eFuels Matagorda County (TX)	United States	CCU	2021	2027	Planned	2	2	DAC	Use

Project name	*Country*	*Project type*	*Announced*	*Operation*	*Project status*	*Announced capacity ($MtCO_2/yr$)*	*Estimated capacity by IEA ($MtCO_2/yr$)*	*Sector*	*Fate of carbon*
Horizon mine tailings (ALB)	Canada	CCU		2017	Operational	0.4	0.4	Other fuel transformation	Use
Huaneng Zhengning coal power plant (Gansu)	China	CCU	2021	2024	Under construction	1.5	1.5	Power and heat	Use
Hynovera	France	CCU	2023	2028	Planned		0.123	Power and heat	Use
Hynovi project (Vicat Montalieu plant)	France	CCU	2021	2026	Planned	0.5	0.5	Cement	Use
Hyskies Vattenfall CHP Lanzatech	Sweden	CCU	2021	2028	Planned	0.207	0.207	Power and heat	Use
Indian Oil Corporation Koyali refinery	India	Full chain	2021	2025	Planned	0.7	0.7	Other fuel transformation	Mixed
Jiling Petrochemical CCUS (Nanjing refinery) (Jiangsu) phase 1	China	Full chain	2021	2023	Operational	0.1	0.1	Other fuel transformation	EOR
Karamay Dunhua methanol plant (Xinjiang)	China	Full chain		2015	Operational	0.1	0.1	Chemicals	EOR

(continued)

Table 3 (continued)

Project name	*Country*	*Project type*	*Announced*	*Operation*	*Project status*	*Announced capacity ($MtCO_2$/yr)*	*Estimated capacity by IEA ($MtCO_2$/yr)*	*Sector*	*Fate of carbon*
Kemper county CCUS (MS)	United States	Full chain		2017	Suspended/canceled/decommissioned	3	3	Power and heat	EOR
Kotka power (Kotka)	Finland	CCU	2021	2026	Planned	0.11	0.11	Power and heat	Use
Kymijärvi power (Lahti)	Finland	CCU	2021	2026	Planned	0.11	0.11	Power and heat	Use
Labarge Shute Creek Gas Processing Plant 2010 expansion (WY)	United States	Full chain	2008	2010	Operational	3.5	3.5	Natural gas processing/LNG	EOR
Labarge Shute Creek Gas Processing Plant original (WY)	United States	Full chain	1984	1986	Operational	3.5	3.5	Natural gas processing/LNG	EOR
Lake Charles Methanol (LA)	United States	Capture	2009	2027	Planned	4–4.2	4.2	Chemicals	Mixed
Linde Clear Lake HyCO Plant (TX)	United States	CCU	2022	2023	Operational	0.18	0.18	Chemicals	Use
LIPOR Maia Energy Recovery Plant PtL facility	Portugal	CCU	2022		Planned	0.1	0.1	Power and heat	Use
Lost Cabin Gas Plant (WY)	United States	Full chain	2010	2013	Operational	0.9	0.9	Natural gas processing/LNG	EOR

Project name	*Country*	*Project type*	*Announced*	*Operation*	*Project status*	*Announced capacity ($MtCO_2/yr$)*	*Estimated capacity by IEA ($MtCO_2/yr$)*	*Sector*	*Fate of carbon*
Merritt Electrofuels Project DAC (BC)	Canada	CCU	2021	2026	Planned	0.25	0.25	DAC	Use
Mineral carbonation international Carbon Plant demonstrator	Australia	CCU	2021	2024	Under construction	0.3	0.3	Chemicals	Use
MOL Szank field CO_2 EOR	Hungary	Full chain		1992	Operational	0.16	0.16	Natural gas processing/LNG	EOR
Montauk Energy biogas plants (TX)	United States	CCU	2023		Planned	0.14	0.14	Biofuels	Use
Mote biomass-to-hydrogen plant (CA)	United States	Capture	2021	2025	Planned	0.14	0.14	Hydrogen or ammonia	Mixed
North East Alliance blue ammonia Yakutia Phase 1	Russia	Full chain	2021	2026	Planned			Hydrogen or ammonia	EOR
North East Alliance blue ammonia Yakutia Phase 2	Russia	Full chain	2021	2030	Planned			Hydrogen or ammonia	EOR
Nutrien Redwater Fertilizer (ALB) phase 1	Canada	Capture	2011	2020	Operational	0.3	0.3	Chemicals	EOR

(continued)

Table 3 (continued)

Project name	*Country*	*Project type*	*Announced*	*Operation*	*Project status*	*Announced capacity ($MtCO_2$/yr)*	*Estimated capacity by IEA ($MtCO_2$/yr)*	*Sector*	*Fate of carbon*
Nutrien Redwater Fertilizer (ALB) phase 2	Canada	Capture	2022		Planned			Chemicals	EOR
NWR Sturgeon Refinery (ALB)	Canada	Capture	2011	2020	Operational	1.3	1.3	Other fuel transformation	EOR
OCAP	Netherlands	Transport		2005	Operational	0.5	0.5	Transport	Use
PCS Nitrogen-Geismar plant (LA)	United States	Full chain	2011	2013	Operational	0.2–0.3	0.3	Chemicals	EOR
Pecos County Ultra Clean Fuels Refining (TX)	United States	Capture	2021		Planned			Other fuel transformation	EOR
Petra Nova Carbon Capture (TX)	United States	Full chain	2009	2023	Operational	1.4	1.4	Power and heat	EOR
Petrobras Santos Basin pre-salt oilfield CCS	Brazil	Full chain	2011	2013	Operational	3–10.6	10.6	Natural gas processing/ LNG	EOR
Project Air Perstorp Oxo capture	Sweden	CCU	2020	2027	Planned	0.406	0.406	Chemicals	Use
Project Interseqt—Hereford Ethanol Plant (TX)	United States	Full chain	2018	2024	Planned	0.35	0.35	Biofuels	EOR

Project name	*Country*	*Project type*	*Announced*	*Operation*	*Project status*	*Announced capacity ($MtCO_2/yr$)*	*Estimated capacity by IEA ($MtCO_2/yr$)*	*Sector*	*Fate of carbon*
Project Interseqt—Plainview Ethanol Plant (TX)	United States	Full chain	2018	2024	Planned	0.35	0.35	Biofuels	EOR
Ramba CCUS/EOR	Indonesia	Full chain	2022	2032	Planned			Natural gas processing/ LNG	EOR
Rohm chemical plant (×2)	Germany	CCU	2023		Planned	0.5	0.5	Chemicals	Use
Sailboat Green Methanol plant Shengong petrochemical (Jiangsu)	China	CCU	2021		Planned	0.15–0.16	0.16	Chemicals	Use
Shah gas plant (Abu Dhabi CCS Phase 2)	United Arab Emirates	Full chain	2020	2030	Planned	2.3	2.3	Natural gas processing/ LNG	EOR
Shand	Canada	Capture	2018		Planned	2	2	Power and heat	EOR
Shell Netherlands Refineries heavy residue gasification CCU	Netherlands	CCU		1997	Operational	0.75	0.75	Other fuel transformation	Use

(continued)

Table 3 (continued)

Project name	*Country*	*Project type*	*Announced*	*Operation*	*Project status*	*Announced capacity ($MtCO_2/yr$)*	*Estimated capacity by IEA ($MtCO_2/yr$)*	*Sector*	*Fate of carbon*
Sinopec Nanjing Chemical Industries CCUS Cooperation Project (Jiangsu)	China	Full chain	2015	2021	Operational	0.2	0.2	Chemicals	EOR
Sinopec Qilu Petrochemical Shengli (Shandong)	China	Full chain	2012	2022	Operational	0.7–1	1	Chemicals	EOR
Sinopec Shengli Power Plant CCS (Shandong)	China	Full chain	2011	2030	Planned	1–2	2	Power and heat	EOR
Stade facility Green Methanol	Germany	CCU	2019		Planned		0.274	Chemicals	Use
STRATOS (Ector County) (TX)	United States	Full chain	2019	2025	Under construction	0.5	0.5	DAC	Mixed
Suek Blue Ammonia Borodinsky	Russia	Full chain	2021	2027	Planned			Hydrogen or ammonia	EOR
Sukowati CO_2-EOR (Jambaran Tiung Biru gas processing)	Indonesia	Full chain	2020	2025	Planned	0.47 –0.93	0.93	Natural gas processing/ LNG	EOR
Taiwan Cement Hoping/Heping plant	Taiwan	CCU	2019	2030	Planned	0.1	0.1	Cement	Use

Project name	*Country*	*Project type*	*Announced*	*Operation*	*Project status*	*Announced capacity ($MtCO_2$/yr)*	*Estimated capacity by IEA ($MtCO_2$/yr)*	*Sector*	*Fate of carbon*
Tangguh LNG EGR	Indonesia	Full chain	2021	2027	Planned	2.5–3.3	3.3	Natural gas processing/LNG	EOR
Tarastenjärvi power (Tampere)	Finland	CCU	2021	2026	Planned	0.11	0.11	Power and heat	Use
Terrell Natural Gas Processing Plant (former Val Verde) (TX)	United States	Full chain		1972	Operational	0.4–0.5	0.5	Natural gas processing/LNG	EOR
The ZEROS project Chambers & Liberty County (Jefferson) (TX)	United States	Capture	2019	2024	Planned	1.5	1.5	Power and heat	EOR
Tracy DAC facility	United States	CCU	2021	2023	Operational	0.001	0.001	DAC	Use
Tumbler Ridge e-methanol project (BC)	Canada	CCU	2015	2028	Planned		0.233	Other industry	Use
Twence Hengelo facility	Netherlands	CCU	2019	2024	Under construction	0.1	0.1	Power and heat	Use
Uthmaniyah CO_2-EOR demonstration	Saudi Arabia	Full chain	2013	2015	Operational	0.8	0.8	Natural gas processing/LNG	Mixed

(continued)

Table 3 (continued)

Project name	*Country*	*Project type*	*Announced*	*Operation*	*Project status*	*Announced capacity ($MtCO_2/yr$)*	*Estimated capacity by IEA ($MtCO_2/yr$)*	*Sector*	*Fate of carbon*
Valero Port Arthur Refinery (TX)	United States	Full chain	2009	2013	Operational	0.9	0.9	Other fuel transformation	EOR
Velocys Bayou Fuels (MS)	United States	Capture	2019	2026	Planned	0.3–0.5	0.5	Biofuels	EOR
Veolia EfW to fuels	United Kingdom	CCU	2024	2030	Planned			Power and heat	Use
Waste to energy CCU to TES Wilhemshaven (unknown facility)	Germany	CCU	2023		Planned	0.4	0.4	Power and heat	Use
Waste to hydrogen Boson Energy (BEH2X LAVNT)	Latvia	CCU	2022	2024	Planned			Hydrogen or ammonia	Use
Weyburn-Midale storage	Canada	Storage		2020	Operational	3	3	Storage	EOR
Yanchang integrated CCUS Yulin Coal Chemical (Shaanxi) phase 1	China	Full chain	2015	2022	Operational	0.3	0.3	Other fuel transformation	EOR
Yanchang integrated CCUS Yulin Coal Chemical (Shaanxi) phase 2	China	Full chain	2022		Planned	2.4	2.4	Other fuel transformation	EOR

Project name	*Country*	*Project type*	*Announced Operation*	*Project status*	*Announced capacity ($MtCO_2$/yr)*	*Estimated capacity by IEA ($MtCO_2$/yr)*	*Sector*	*Fate of carbon*
Yanchang integrated CCUS Yulin Coal Chemical (Shaanxi) phase 3	China	Full chain	2022	Planned	2.3	2.3	Other fuel transformation	EOR
Zutica and Ivanic grad EOR extension	Croatia	Storage		Planned			Storage	EOR

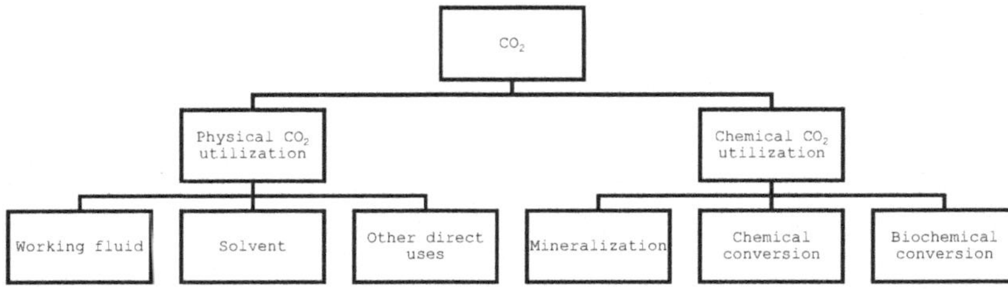

Fig. 3 CO_2 utilization routes[15]

examples. Finally, in Sect. 3.3 are presented the maturity level of CO_2 utilization routes.

3.1 *Physical CO_2 Utilization*

3.1.1 *Working Fluid*

The physicochemical characteristics of CO_2 allow its application as a working fluid in power cycles and heat transfer processes. For example, one application of CO_2 as a working fluid is in supercritical power cycles, in which CO_2 would replace water, promoting an increase in the efficiency of electricity generation.[16,17]

3.1.2 *Solvent*

Among the more traditional uses of CO_2 as a solvent, enhanced oil recovery (EOR) is a well-established commercial technology that has been applied since the 1970s. The technology involves injecting CO_2 into oil fields, which increases the overall pressure in the reservoir and improves the mobility of the oil, resulting in a greater flow of oil toward the production wells. Analogous to EOR, CO_2 can also be used in the enhanced coal-seamed methane recovery (ECBM) process.[18]

Other uses of CO_2 as a solvent involve extraction, cleaning, and deposition processes. Characteristics of supercritical CO_2 such as dissolving

[15] Pereira (n 4).

[16] V. Srinivasan and others, 'CO_2 Utilisation Roadmap' [2021] CSIRO (Commonwealth Scientific and Industrial Research Organisation).

[17] Ruiz (n 12).

[18] Srinivasan (n 16); Ruiz (n 12).

non-polar molecules and rapid penetration into solid materials, due to low viscosity and high diffusivity, are essential for these processes. Regarding extraction, CO_2 can act as a solvent, for example, in coffee and tea decaffeination processes, cocoa defatting, production of hop extracts, and sesame seed oil, among others. For cleaning, CO_2 can act as a solvent, e.g., for drying or cleaning aerogels, semiconductors, and pharmaceuticals. Finally, with respect to deposition, CO_2 can act as a solvent, for example in the deposition of dyes on polyester, fungicides on wood, and pharmaceuticals on polymers.[19]

3.1.3 Other Physical Uses

Most existing commercial applications involve the direct use of CO_2 (without conversion), traversing different sectors such as the food and beverage industry, metal manufacturing, refrigeration (dry ice), dry cleaning, health care, water treatment, and fire suppression.[20]

The case of the food and beverage industry is an example of multiple physical uses, from the use of food-grade CO_2 to produce carbonated beverages and dairy products, through the production of carbonated beverages, as well as in food preservation, for example. Furthermore, there is an increase in the use of CO_2 to enhance yields in industrial processes, such as the production of methanol and the cultivation of crops in greenhouses.[21]

3.2 Chemical CO_2 Utilization

3.2.1 Mineral Carbonation

Carbonation is an inherent geological process in weathering, whereby carbon dioxide (CO_2) combines with minerals in the Earth's crust to create carbonates. Thermal and chemical engineering can expedite this process to generate carbonates. Carbonation can be advantageous for the concrete industry since it allows for the integration of CO_2 into the production process of concrete. Carbonation occurring during the production process can enhance the durability of concrete, resulting in a

[19] Ibid.

[20] Ibid.

[21] Ibid.

reduction in the amount of cement needed and thereby lowering both the carbon footprint and the expenses associated with raw materials.[22]

Concrete is generally composed by cement, water, and aggregates like sand and gravel. CO_2 can substitute water in a process known as CO_2 curing, where the components are combined. Additionally, it can be utilized as a primary material for creating aggregates and in the manufacturing of cement (Fig. 4). These uses all entail the reaction of CO_2 with calcium or magnesium minerals, leading to the creation of carbonates, which are the specific type of carbon found in concrete. Carbonates possess a lower energy state compared to CO_2. Aggregates can be generated by chemically combining CO_2 with waste materials from power plants or industrial operations, such as iron slag and coal fly ash, which would otherwise be kept or disposed of in landfills.[23]

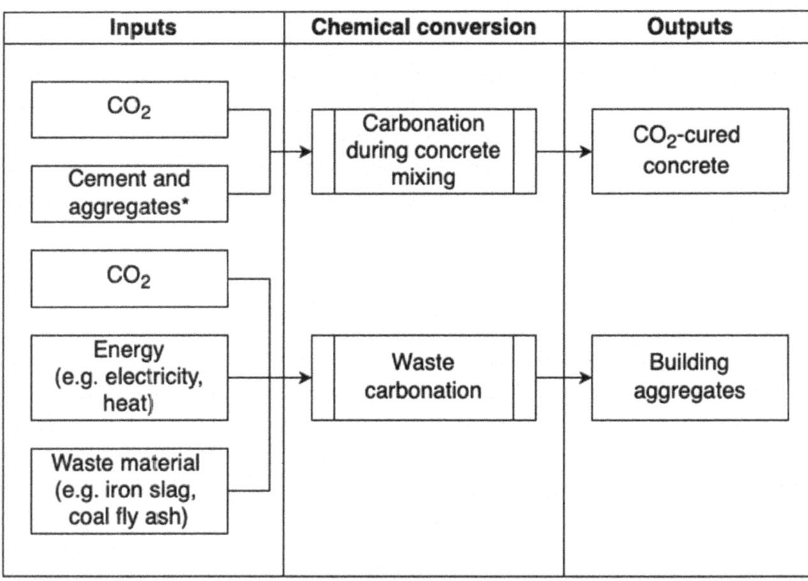

Fig. 4 Mature conversion routes for CO_2-derived building materials[24]

[22] IEA (n 11); Srinivasan (n 16).

[23] Ibid.

[24] IEA (n 11).

3.2.2 *Chemical Conversion*

Chemical compounds originating from CO_2 include a diverse range of carbon-based molecules, including polymers, textiles, solvents, and synthetic rubber. Carbon dioxide (CO_2) serves as a raw material to create various chemical intermediates (Fig. 5), which can then be transformed into a wide range of more intricate chemical compounds. These intermediate products, including ethylene, propylene, and methanol, are generated in significant quantities and can also serve as final products independently. The production of methanol and methane from CO_2 is the most technologically advanced method among the various chemical intermediates that may be derived from CO_2. Additionally, this process can also act as a precursor for several other chemical intermediates.[25]

The conversion of CO_2 into methanol and methane is considered the most technologically advanced method, similar to the production of fuels produced from CO_2. Methanol has the potential to be transformed into valuable carbon-based chemical intermediates, such as olefins, which are

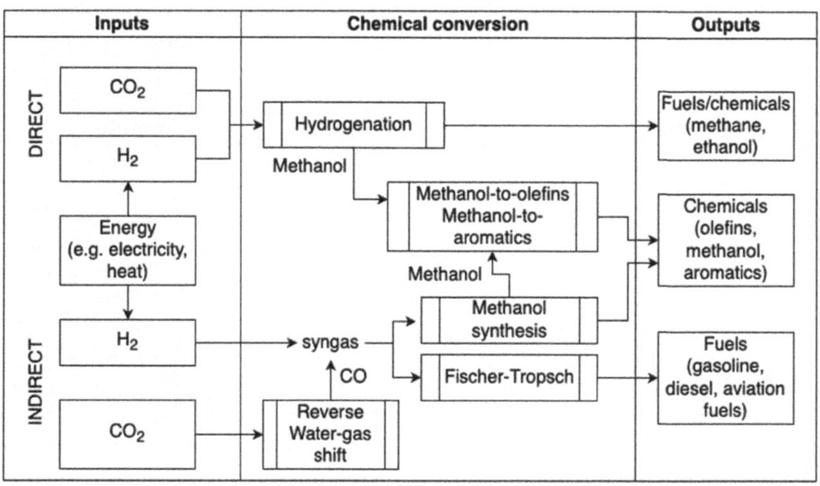

Fig. 5 Mature conversion route for CO_2-derived fuels and chemical intermediates[26]

[25] IEA (n 11); Srinivasan (n 16).

[26] IEA (n 11).

essential for producing plastics, and aromatics, which have various applications in industries including health and hygiene, food manufacturing, and processing.[27]

Polymers, a distinct category of chemicals, are utilized in the manufacturing process of plastics, foams, and resins. The carbon included in carbon dioxide (CO_2) can be utilized in the synthesis of polymers, thereby substituting a portion of the raw materials obtained from fossil fuels throughout the manufacturing procedure (Fig. 6). Contrary to the conversion of CO_2 into fuels and chemical intermediates, the process of using CO_2 to manufacture polymers requires minimal energy input. This is because the CO_2 is transformed into a molecule called carbonate, which has an even lower energy state. At now, multiple companies have polymer production facilities that utilize CO_2 as a raw material.[28]

3.2.3 Biological Conversion

The process of biological conversion of CO_2, facilitated by synthetic biology, involves the utilization of microorganisms to produce a range of different products. Considering that numerous specialized and valuable goods cater to the specific needs of the food and agriculture industries,

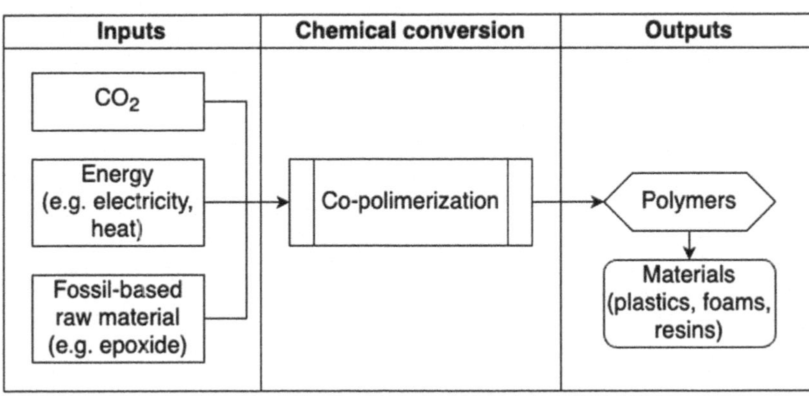

Fig. 6 Mature conversion route for CO_2-derived polymers[29]

[27] Ibid.

[28] IEA (n 11); Srinivasan (n 16).

[29] IEA (n 11).

such as alternative livestock feed, it is likely that the initial emphasis of CO_2 biological conversion will be on exploiting global food export prospects.[30]

During the process of biological CO_2 conversion, microorganisms such as algae, cyanobacteria, acetogens, and methanogens assimilate CO_2 and transform it into a diverse range of valuable compounds. Some of these products may include compounds of significant magnitude, such as ethylene and ethanol. Additionally, it is feasible to generate specialized and valuable chemicals, such as medications, nutrition, cosmetics, and fragrances. Despite being produced in smaller quantities, these goods can provide a more economically advantageous alternative compared to traditional methods of industrial synthesis. Microorganisms exhibit variable operating conditions, scalability, and application across various environments. Biological systems have the capability to be directly linked to exhaust gas streams or to directly harvest CO_2 from the atmosphere. Modular biological systems are more cost-effective and environmentally friendly than industrial chemical processes. They can operate efficiently on smaller scales and significantly reduce the production of toxic waste, including spent catalysts. In the future, it is possible that biological systems could generate a range of valuable compounds in significant quantities to accommodate evolving needs. These products would need to contend with other processes that utilize CO_2, such as thermochemical synthesis.[31]

3.3 CO_2 Utilization Routes' Maturity Level

The evaluation of the maturity level of different technologies of CCU-based products is still a challenge particularly for chemical CO_2 utilization, while most technologies within physical CO_2 utilization route have achieved a true Technology Readiness Level (TRL) of 9. Chemical CO_2 utilization encompasses a wide range of technologies, many of which have been proven to be conceptually viable through initial commercial projects. Thus, it can be inferred that almost all chemical CO_2 utilization strategies

[30] Srinivasan (n 16).

[31] Ibid.

have a TRL of 6. However, given the advanced stage of development, it is possible to identify specific CCU-based products using TRL scale.[32]

Table 4 presents a non-exhaustive list comprising several CCU-based products within the main categories of chemical CO_2 utilization, alongside with the main categories of physical CO_2 utilization route.

4 Conclusions

In summary, the development and implementation of Carbon Capture and Utilization (CCU) technologies are pivotal in mitigating climate change and transitioning toward a sustainable industrial future. The distinction between CCU, CCS (Carbon Capture and Storage), and CCUS (Carbon Capture, Utilization, and Storage) is crucial for understanding the different roles each process plays in managing carbon emissions.

CCU stands out by transforming CO_2 into valuable products, thereby creating a circular carbon economy. Two major technological routes, physical and chemical, offer promising applications, from producing fuels and chemicals to manufacturing construction materials. Each product's contribution to climate mitigation depends significantly on the duration of carbon storage, with construction materials offering long-term sequestration benefits.

From a technology readiness perspective, the maturity of CCU technologies varies widely. Some applications, such as the use of CO_2 in enhanced oil recovery (EOR) and the production of urea and methanol, are relatively mature and commercially deployed. Others, like the synthesis of advanced polymers or fuels through electrochemical pathways, are still in the development or pilot stages. The technology readiness levels (TRLs) for these technologies range from early research and development (TRL 1–3) to demonstration and deployment (TRL 7–9). Achieving widespread adoption and scaling up these technologies will require significant investment in research, development, and infrastructure.

In conclusion, the potential of CCU technologies to convert CO_2 into useful products presents an opportunity to reduce carbon footprints, drive economic growth, and support sustainable development. Ongoing

[32] Dziejarski (n 7); Ruiz (n 12).

Table 4 Maturity level for different technologies of CCU-based products[33]

CO_2 utilization route	*Utilization main categories*	*CCU-based products*	*TRL*
Physical CO_2 utilization	Working fluid	–	9
	Solvent	–	
	Other direct uses	–	
Chemical CO_2 utilization	Mineralization	Magnesium carbonate	4
		Calcium carbonate, sodium carbonate	7
		Sodium bicarbonate, concrete curing	8
	Chemical conversion	Malates	1
		Carbamates, formaldehyde, isocyanates, fine chemicals (e.g., alkanes, aromatics, olefins)	2
		Acetic acid, dimethyl ether, acrylic acid	3
		Ethylene glycol, lactones, oxalic acid	4
		CO_2-based Fischer–Tropsch products	5
		Ethanol, formic acid, syngas	6
		CO_2-based polymers	7
		Methane, dimethyl carbonate, cyclic carbonates	8
		Methanol, polyurethane, salicylic acid, polyols, CO_2-based polycarbonates, urea	9
	Biochemical conversion	CO_2-based enzymatic and microbial products	3
		Polymer, proteins	5–7
		Dry algae powder	8
		Methane, ethanol, isobutanol	9

research, technological innovation, and policy support are essential to fully realize the benefits of CCU and integrate it into broader climate action frameworks. By advancing the technology readiness of various CCU routes, we can ensure a more sustainable and resilient future.

[33] Ibid.

CHAPTER 7

Social Perception and Acceptance of CCS and CCU Technologies: A Comparison Through a Socio-Technical Approach

Karen Louise Mascarenhas, Guilherme Baccari, and Nathalia Weber

1 Introduction

Carbon Capture, Utilization, and Storage (CCUS) technologies hold significant promise in mitigating greenhouse gas emissions and addressing climate change. These technologies, which encompass Carbon Capture and Storage (CCS), Carbon Capture and Utilization (CCU), or both together (CCUS), contribute to the portfolio of alternatives needed to lead the energy transition towards a low-carbon society. However, their successful implementation is contingent upon understanding and addressing the social perception and acceptance issues surrounding them.

K. L. Mascarenhas (✉) · G. Baccari · N. Weber
Research Centre for Greenhouse Gas Innovation (RCGI), University of São Paulo, São Paulo, Brazil
e-mail: karenmascarenhas@usp.br

E. G. Pereira et al. (eds.), *Carbon Capture Utilization and Storage*, Sustainable Development Goals Series,
https://doi.org/10.1007/978-3-031-81272-9_7

It is important to study social perception and acceptance not just because many CCS projects have been cancelled due to public rejection, as is the case in Barendrecht in the Netherlands, Beeskow in Germany and Carson in the United States, leading to massive economic losses.[1,2,3] But also, engaging with the local communities, consumers, policymakers, academics, industry experts, investors, and all relevant stakeholders, considering their opinions, and adjusting the projects to better fit their necessities is paramount. Furthermore, as these technologies are implemented, people could draw their risk assessment on CCU based on their evaluation of CCS, as occurs with fracking interference in the risk perception of other carbon capture technologies in the United Kingdom.[4]

This chapter explores the multifaceted factors influencing how various stakeholders, including government entities, industry players, regulatory bodies, industry agents, specialists, and the general public, perceive CCS and CCU technologies globally.

The chapter begins by examining the diverse perceptions and attitudes towards CCUS technologies across distinct societal segments, specifying when differences are pointed out upon CCS or CCU technologies. When

[1] Brunsting S and others, 'Stakeholder Participation Practices and Onshore CCS: Lessons from the Dutch CCS Case Barendrecht' (2011) 4 Energy Procedia 6376.

[2] Dütschke E, 'What Drives Local Public Acceptance—Comparing Two Cases from Germany' (2011) 4 Energy Procedia 6234.

[3] Bradbury J and Wade S, 'Case Study of the Carson CCS Project' (Global CCS Institute 2010)

[4] Cox E, Pidgeon N and Spence E, 'But They Told Us It Was Safe! Carbon Dioxide Removal, Fracking, and Ripple Effects in Risk Perceptions' (2021) 42 Risk Analysis 1472.

G. Baccari
e-mail: guibaccari@alumni.usp.br

N. Weber
e-mail: webermasulino@usp.br

K. L. Mascarenhas
Institute of Energy and Environment, University of São Paulo, São Paulo, Brazil

G. Baccari
Institute of Psychology, University of São Paulo, São Paulo, Brazil

N. Weber
CCS Brasil, São Paulo, Brazil

these differences are not under analysis, we will adopt the term CCUS to refer to both technologies indistinctively. In this chapter, we delve into the underlying reasons for scepticism or acceptance, considering factors such as cultural norms, values, and prior experiences with industrial projects or environmental initiatives. By understanding these nuances, stakeholders can tailor communication strategies and engagement efforts to effectively address concerns and garner support.

The proposed discussion about social perception and acceptance is based on a socio-technical approach for analysing the impact of the uncertainties related to CCS.[5] Moreover, the chapter explores the impact of regulatory frameworks and government policies on social acceptance and investment confidence in CCUS projects. Clear and consistent regulations, coupled with supportive policies such as financial incentives or carbon pricing mechanisms, can incentivise industry adoption, and foster public trust in the technology's efficacy and safety.

A critical focus within this chapter is the examination of the CCUS value chain to evaluate potential social perception implications across distinct technological processes. This analysis facilitates a comprehensive understanding of how societal attitudes may fluctuate throughout various stages of CCS implementation, including capture, transportation, utilisation, injection, and storage activities. By deconstructing the CCS value chain into its elemental components, scholars can delineate the social repercussions and acceptance impediments associated with each phase, thereby guiding tailored approaches for engaging stakeholders and communicating effectively.

In conclusion, this chapter underscores the critical importance of considering social perception issues in planning and implementing CCUS projects. By proactively addressing stakeholders' concerns, fostering transparent communication, and promoting inclusive decision-making processes, the pathway to widespread acceptance and successful deployment of CCS and CCU technologies can be paved.

[5] Markusson N and others, 'A Socio-Technical Framework for Assessing the Viability of Carbon Capture and Storage Technology' (2012) 79 Technological Forecasting and Social Change 903.

2 Carbon Capture Storage and Carbon Capture Utilisation Perspectives in Social Perception

The social perception of carbon capture projects is a complex phenomenon. The technology itself assembles under similar nomenclatures a variety of different systems, such as Carbon Capture and Storage (CCS), industrial Carbon Capture and Storage (iCCS), Carbon Capture and Utilisation (CCU), Carbon Capture, Utilisation and Storage (CCUS), Carbon Capture, Utilisation and Storage with Enhanced Oil Recovery (CCUS-EOR), BioEnergy with Carbon Capture and Storage (BECCS) and/or Utilisation (BECCUS), among others. Such variety interferes with public and social perception in ways that are not yet fully known to engineers and researchers, thus possibly rendering CCUS projects deployment more difficult.

Scientific articles on the social perception of carbon capture technologies point out that laypeople don't have much knowledge or awareness on the subject, whether they were populations from developed countries like Germany, Japan, and France, or developing countries like Brazil, China, and Romania.[6] Therefore, they usually form pseudo or semi-informed opinions on the subject, which influences their acceptance behaviour.[7] Furthermore, people derive their perception based on their representations of other social objects, those which are more familiar and closer to them.[8] In other words, we could say that people tend to use the mental representation of known objects to judge the technology and form their opinions and behaviours to accept or reject CCUS. Such formulation agrees with the literature as Linzenich found that people evaluate CCS and CCU in a similar way, despite showing a higher CCU acceptance.[9]

[6] Lima PR and others, 'Environmental Awareness and Public Perception on Carbon Capture and Storage (CCS) in Brazil' (2021) 111 International Journal of Greenhouse Gas Control 103467.

[7] De Best-Waldhober M and others, 'Informed Public Opinion in the Netherlands: Evaluation of CO_2 Capture and Storage Technologies in Comparison with Other CO_2 Mitigation Options' (2012) 10 International Journal of Greenhouse Gas Control 169.

[8] Batel S and Devine-Wright P, 'Towards a Better Understanding of People's Responses to Renewable Energy Technologies: Insights from Social Representations Theory' (2014) 24 Public Understanding of Science 311.

[9] Linzenich A and others, 'Uncovering Attitudes Towards Carbon Capture Storage and Utilization Technologies in Germany: Insights into Affective-Cognitive Evaluations of Benefits and Risks' (2019) 48 Energy Research & Social Science 205.

In this book chapter, we will explore the differences in the social perception regarding CCS and CCU technologies when they are significant, considering them as complex socio-technical processes, discuss what literature says about the influence of one perception over the other, and finally bring an example from our own research.

To begin, let's define what we mean when we say social perception. First of all, research on social perception is treated as an element of a broader research field, namely the social acceptance field. The prevalent way science understood the matter has shifted over the years since its twofold start.[10] On the one side stand Wolsink studies alongside Pasqualetti and Butler and Bosley and Bosley (1988) on wind turbines placement, acceptance, and the Not In My Back Yard (NIMBY) effect.[11,12,13] On the other approach, Slovic, Flynn, Layman, and Renn studies focused on risk perception, especially regarding nuclear energy power plants.[14,15] Both lines of research began in the 1980s and comprise what Wolsink calls 'the take-off of social acceptance research'.[16] This phase focused on individuals' attitudes and beliefs, in other words in public instead of social acceptance, which the latter is a more modern, refined and embracing concept. Research was focused on the individual and their degree of acceptance of projects or technologies. Also, research would often carry misconceptions and presumptions no longer accepted as productive, such as the belief that opposition to wind power or other technologies like CCUS were deviant, that objectors were ignorant or misinformed, and that studies aimed to debunk opposition. Wolsink

[10] Wolsink M, 'Social Acceptance Revisited: Gaps, Questionable Trends, and an Auspicious Perspective' (2018) 46 Energy Research & Social Science 287.

[11] Wolsink M, 'The Social Impact of a Large Wind Turbine' (1988) 8 Environmental Impact Assessment Review 323.

[12] Pasqualetti MJ and Butler E, 'Public Reaction to Wind Development in California' (1987) 8 International Journal of Ambient Energy 83.

[13] Bosley P and Bosley K, 'Public Acceptability of California's Wind Energy Developments: Three Studies' (1988) 12 Wind Engineering 311.

[14] Slovic P, Flynn JH and Layman M, 'Perceived Risk, Trust, and the Politics of Nuclear Waste' (1991) 254 Science 1603.

[15] Renn O, 'Three Decades of Risk Research: Accomplishments and New Challenges' (1998) 1 Journal of Risk Research 49.

[16] Wolsink M, 'Social Acceptance Revisited: Gaps, Questionable Trends, and an Auspicious Perspective' (2018) 46 Energy Research & Social Science 287.

points to a major conceptual turn in the field in the 2000s; he argues that research shifted from focusing on the individual to encompassing the institutional aspects. The object of acceptance was changing to include more than cognitive factors and site characteristics; it went to the web of relationships related to the technologies, the communities, and other social actors such as policymakers and companies. Public acceptance was being replaced by social acceptance, which can be understood as 'a set of activities unfolding over time in complex, multilayered, polycentric processes that contain countless research questions'.[17] In other words, acceptance research was advancing towards a more socio-technical approach.

In the 2000s and 2010s, some of the most prominent theories and frameworks that aim to understand the social acceptance of CCS and other renewable technologies and innovations were developed. The most relevant are the concept of acceptance proposed by Wüstenhagen in 2007,[18] the Technology Acceptance Framework (TAF) proposed by Huijts in 2012,[19] and the socio-technical approach proposed by Markusson in 2012.[20]

In their seminal paper from 2007, Wüstenhagen and colleagues formulate a concept of acceptance for renewable innovations. Namely, they define social acceptance as composed of three factors: socio-political acceptance, community acceptance, and market acceptance. The first refers to the acceptance of technologies or projects by key stakeholders, policymakers, and the general public. Historically, it has been the most studied element of acceptance and represents a global level of attitudes regarding certain technologies, which is mostly depicted as scores in surveys, polls, and questionnaires. Community acceptance refers to the

[17] Wolsink M, 'Social Acceptance Revisited: Gaps, Questionable Trends, and an Auspicious Perspective' (2018) 46 Energy Research & Social Science 287.

[18] Wüstenhagen R, Wolsink M and Bürer MJ, 'Social Acceptance of Renewable Energy Innovation: An Introduction to the Concept' (2007) 35 Energy Policy 2683.

[19] Huijts NMA, Molin EJE and Steg L, 'Psychological Factors Influencing Sustainable Energy Technology Acceptance: A Review-Based Comprehensive Framework' (2012) 16 Renewable and Sustainable Energy Reviews 525.

[20] Markusson N and others, 'A Socio-Technical Framework for Assessing the Viability of Carbon Capture and Storage Technology' (2012) 79 Technological Forecasting and Social Change 903.

behaviours of the communities directly affected by the project sitting process, such as local stakeholders, residents, and authorities. Such a factor is much linked to the communication between the project deployers and the local communities of importance, as well as their perception of justice and trust. At last, market acceptance is the one relative to consumers, investors, and intra-industry stakeholders' perceptions and behaviours regarding certain technologies. It can be comprehended as the process where an innovation is adopted or not by the market, considering that different actors play different roles in the spread of a product or innovation.[21]

Building upon Wüstenhagen's and other works on the acceptance of renewables, as well as renowned psychological theories, Huijts, and colleagues create a comprehensive framework for sustainable energy technologies (SETs) acceptance in order to facilitate further research on the acceptance behaviours regarding these technologies and the psychological traits that influence this process. They draw from the theory of planned behaviour, Schwartz's norm activation model, and the hedonic motives and affect theories in order to formulate a model where acceptance results from a process involving the evaluation of costs and benefits, moral evaluation, and the feelings associated with the technology. This process is influenced by many variables, as reinforced by the earlier studies, such as previous experience, knowledge, trust, sense of fairness, affects, perceived costs, benefits and risks, perception of the problem and the outcome, social and personal norms, external behaviours controls, attitudes, and intention.[22] Huijts work allowed for a more cohesive understanding of the psychological factors involved in the acceptance behaviour and, despite their focus on individual acceptance, they tried to link the individual sphere with the social context and norms. This way, the authors exemplified a tendency in research that was better elaborated in the socio-technical approach for renewables and sustainable innovations.

[21] Wüstenhagen R, Wolsink M and Bürer MJ, 'Social Acceptance of Renewable Energy Innovation: An Introduction to the Concept' (2007) 35 Energy Policy 2683.

[22] Huijts NMA, Molin EJE and Steg L, 'Psychological Factors Influencing Sustainable Energy Technology Acceptance: A Review-Based Comprehensive Framework' (2012) 16 Renewable and Sustainable Energy Reviews 525.

Finally, we draw from Markusson work that approached our field of interest, CCS/CCUS, from a socio-technical point of view.[23] Such an approach is set up around the relationship between society and technology innovation, it conceptualises that both are intricated into each other as a system and, thus, as a society and its many agents build, develop, accept, or reject a technology. Also, this technology changes society in various ways, being spatial or social. They co-evolve. In other words, the socio-technical approach tries to include the institutional element in the psychological and social models of acceptance, as well as the engineering and technological innovations. Therefore, one of the main contributions of such an approach is to formulate and analyse the most relevant uncertainties in the fields of interest. In the CCS case, for example, Markusson and colleagues proposed 7 fields of uncertainty: (1) variety of pathways, (2) safe storage, (3) scaling up and speed of development and deployment, (4) integration of CCS systems, (5) economic and financial viability, (6) policy, political, and regulatory uncertainty, and (7) public acceptance.

Exploring the social-technical approach, we will concentrate on the main aspects that influence social perception and acceptance of CCUS technologies. As so, we will begin analysing the influence of the variety of pathways through the CCUS value chain, followed by considerations of three areas of uncertainties: the safe storage issues, the scaling up and speed of development and deployment aspects, and the integration of CCS systems. Next, we approach the government policies and regulations and the investors' confidence in CCUS projects together with the economic and financial viability and finalise with the social perception and acceptance aspects.

[23] Markusson N and others, 'A Socio-Technical Framework for Assessing the Viability of Carbon Capture and Storage Technology' (2012) 79 Technological Forecasting and Social Change 903.

3 Analysing Social Perceptions Across the Technology Pathways Through the CCUS Value Chain

Understanding the intricacies of the CCS and CCU value chains is paramount for evaluating the potential implications of social perception across each macro stage or technological process. This comprehension allows for a nuanced analysis of how societal attitudes and responses may vary depending on the specific stages of CCUS deployment, encompassing capture, transportation, utilisation, injection, and storage activities. By dissecting the CCUS value chain into its constituent components, researchers can discern the social impacts and acceptance challenges associated with each stage, thereby recommending targeted strategies for communication and engagement with diverse stakeholders.

CCUS encompasses a diversified spectrum of technologies aimed at capturing CO_2 emissions from various point sources such as power plants, cement factories, and industrial facilities, subsequently either storing it underground, utilising it for industrial processes, or combining both. The CCUS framework comprises three primary applications: Carbon Capture and Storage (CCS), Carbon Capture and Utilisation (CCU), and the integrated CCUS approach, which incorporates, for example, CCS and Enhanced Oil Recovery (EOR), in which processes of injecting water, chemicals or gases as CO_2 into a reservoir are implemented to increase the ability to extract oil from the well.

Within the CCUS paradigm, CCS primarily involves the geological storage of captured CO_2, emphasising its role in long-term carbon sequestration. Conversely, CCU focuses on utilising captured CO_2 from industrial processes, thereby reducing emissions and potentially creating value-added products. In this sense, CCU considers CO_2 as a raw material for soft drink beverages or to create value by designing integrated CO_2 capture and conversion processes for producing chemicals or fuels such as methanol. In this concept, CO_2 is considered a valuable chemical building block for fuels and materials. Ultimately, the distinction between CCS and CCU is critical, as it delineates between storage-centric and utilisation-centric approaches to carbon management.

Moreover, the terms 'CCS' and 'CCUS' are often used interchangeably in discourse, reflecting a broader understanding of carbon capture and storage technologies. Variations such as Bioenergy with Carbon Capture and Storage (BECCS) and Direct Air Carbon Capture and Storage (DACCS) introduce additional nuances to the CCUS landscape, highlighting the diversity of applications and deployment strategies within the field.

Despite the technical distinctions, the social perceptions and impacts of CCUS projects remain understudied and inadequately addressed. Deploying carbon capture technologies, whether for storage or utilisation purposes, can elicit varied responses from local communities and stakeholders, influenced by factors such as project proximity, perceived environmental risks, and socio-economic considerations.

For instance, implementing CCS infrastructure, including CO_2 transport pipelines and injection sites, may trigger heightened community attention, particularly in onshore settings where human activities are more concentrated. Conversely, CCU projects may encounter less resistance, especially when integrated with existing industrial facilities, potentially garnering positive perceptions due to their perceived contribution to carbon recycling and emission reduction.

In this context, the societal acceptance of CCUS applications varies, with EOR-focused projects facing greater scrutiny compared to BECCS and DACCS initiatives, which offer additional carbon removal benefits. However, challenges persist, particularly regarding the public's comprehension of complex issues such as the nexus between bioenergy and deforestation. There is a strong misperception, mainly from Europeans and Americans, that Brazil's deforestation issue is due to sugarcane plantations and the competition between food and energy production, which the latter is seen as privileged over the first.[24] This highlights the need for targeted communication and education efforts to address misinformation and promote informed decision-making.

Other aspects that challenge the perception and acceptance of CCUS activities are the uncertainty about storage safety, the speed and scaling up of the development and deployment of projects, and the integration of CCUS systems as a whole.

[24] Coelho ST and Goldemberg J, 'Sustainability and Environmental Impacts of Sugarcane Biofuels' (2019) In Khan MT and Khan IA, 'Sugarcane Biofuels: Status, Potential, and Prospects of the Sweet Crop to Fuel the World' 409 Springer.

4 Perceptions About Safe Storage, Scaling up and Speed of Deployment, and System Integration of CCS Technologies

One of the fundamental aspects of CCUS technologies is storage safety. Despite the many uncertainties often discussed, such as those highlighted by Markusson et al. (2012), the safety of geological CO_2 storage is well established. Extensive research and practical implementation, as supported by the open letter 'CCS is needed and ready to address climate change',[25] demonstrate that the long-term security of CO_2 storage has been validated through decades of projects like Sleipner and Snohvit in the North Sea, which have operated safely for 27 and 15 years, respectively. These projects, along with the Boundary Dam and Quest facilities in Canada, underline that geological CO_2 storage is a proven, reliable solution. This confidence is supported by rigorous monitoring practices and advanced technologies ensuring site integrity. Additionally, a study published in Nature, 'Estimating geological CO_2 storage security to deliver on climate mitigation',[26] reinforces that geological storage is a secure and effective climate mitigation strategy. Therefore, while we continue to study and improve process efficiencies, the safety of geological storage is well understood and firmly established.

Scaling up and deploying CCUS technologies depend on the supply chain and the building up of capacity on the technology and the diverse competing pathways. Developing such infrastructure is challenging and takes time to evolve. During the process, if the public is not involved in the discussion and decision-making, it can compromise the understanding and acceptance of CCUS technologies. CCUS are complex systems that involve many components, are not set in a linear process, and imply many uncertainties, which is not easy for the diverse stakeholders to comprehend and follow.

Even though there is an impression that information is the main element capable of working to gain the public's acceptance, that is not the whole truth. The social perception of diverse stakeholders and the

[25] Fennell P and others to international leaders, 'CCS Is Needed and Ready to Address Climate Change—Open Letter to International Leaders' (10 June 2023).

[26] Alcalde J and others, 'Estimating Geological CO_2 Storage Security to Deliver on Climate Mitigation' (2018) 9 Nature Communications.

perception in different countries is related to differences in values and framings, and not only the result of a lack of scientific knowledge.

Nevertheless, the societal acceptance of CCUS technologies extends beyond local infrastructure development, encompassing broader conceptual challenges related to carbon sequestration and climate change mitigation. Public understanding of the urgency to address greenhouse gas emissions, coupled with concerns over social and economic implications, shapes perceptions of CCUS initiatives, influencing their feasibility and adoption.

5 Government Policies and Regulations on Social Acceptance and Investment Confidence in CCUS Projects

Government policies and regulatory frameworks exert significant influence on the social acceptance and the climate investment appeal for CCUS projects. Clear and consistent regulations provide stability and certainty for both industry stakeholders and the public, thereby fostering trust in the efficacy and safety of CCUS technologies. By establishing transparent guidelines and standards for deploying and operating CCUS facilities, governments can mitigate uncertainty and instil confidence in the viability of CCUS as a climate change mitigation strategy.

Supportive policies, such as financial incentives and carbon pricing mechanisms, are instrumental in incentivising industry adoption of CCUS technologies. Financial incentives, such as tax credits or grants for carbon capture projects, help offset the significant capital costs associated with implementing carbon capture infrastructure and make CCUS projects more economically viable for businesses. Additionally, carbon pricing mechanisms, such as carbon taxes or cap-and-trade systems, create a financial incentive for industries to reduce their carbon emissions by investing in CCUS technologies. By internalising the costs of carbon emissions, carbon pricing mechanisms encourage companies to explore low-carbon alternatives, thereby driving innovation and investment in CCUS projects.

Furthermore, government policies can facilitate stakeholder engagement in community consultation when planning and implementing CCUS projects. Mandating public consultations and environmental impact assessments for proposed CCUS facilities ensures that local communities have a voice in the decision-making process and that potential social and environmental concerns are adequately addressed. Engaging with local communities from the outset can help build trust and consensus around CCUS projects, reducing the risk of opposition or resistance from affected stakeholders, whether they are local governments or residents.

The impact of regulatory frameworks and government policies on social acceptance and investment confidence in CCUS projects is profound and multifaceted. Clear and consistent regulations, coupled with supportive policies such as financial incentives and community engagement requirements are essential for fostering public trust, encouraging industry investment, and driving the widespread adoption of CCUS technologies as a key tool in mitigating climate change.

Additionally, the social-technical approach also considers the social perception of the CCUS, which will be discussed further in the next section, wrapping up the concepts presented in Sect. 2.

6 Wrapping up Social Perception and Acceptance of CCUS

We have focused, therefore, on our definition of the concept of social perception as the process where social agents, for example, government representatives, policymakers, industry agents, academic members, investors, and the lay public view and judge any object, but especially technologies and innovations and the implications of such perceptions on the collective and institutional acceptance of these technologies.[27] In this sense, social perception is a formative element of social acceptance, a more complex and dynamic process, where psychological and individual factors interact with social and institutional factors in a mutual way in which the result is the acceptance behaviour in any form by the involved social actors. In other words, the behaviour in question can be the passive acceptance by the local community, the active acceptance of policymakers, or even the full rejection from activists. It also results from many variables such as the technology readiness level, the way the project was communicated, previous history of the local community, psychological factors, such as values and believes, social norms or place attachment, risks and

[27] Mascarenhas K, 'Compreensão Da Percepção Pública Da Ação Climática: Desafio Para as Lideranças Brasileiras' (dissertation, Universidade de São Paulo 2023) 1.

benefits perceptions, social factors like demographic or economics, and others. It is important to point out that this framework differentiates from the more usual studies that consider just the community/public perception/acceptance, which encompasses only local communities and eventually direct stakeholders.

So, we garner from the technology acceptance framework (TAF) to discuss the main similarities and differences between the social acceptance of CCS and CCUS. Specifically, the discussion revolves around 4 axes: (1) awareness and knowledge; (2) trust, fairness, and affects; (3) perception of costs, risks, and benefits; (4) other factors, like outcome and efficacy perception, social norms, and behavioural control.

Two of the most researched factors in the field are people's level of awareness regarding climate change and CCS/CCUS, and their level of knowledge of the problem and the technology. The first noticeable point is that CCS and CCUS awareness is quite low in most countries,[28] including Brazil.[29] In relation to knowledge, Dowd et al. found, through an online survey in Japan, Australia and the Netherlands, that most people have only a basic understanding of CO_2 along with many misconceptions regarding, for example, flammability and potential health effects.[30] This lack of education on CO_2 and its properties has a significant impact on people's understanding of CCS and its role in tackling climate change. Such results are reinforced by findings from focus groups in the UK, Belgium, the Netherlands, Germany, Spain, and Poland where people's little knowledge of CO_2 affected their perception of CCS.[31,32] An example of misconceptions about carbon capture technologies and

[28] L'Orange Seigo S, Dohle S and Siegrist M, 'Public Perception of Carbon Capture and Storage (CCS): A Review' (2014) 38 Renewable and Sustainable Energy Reviews 848.

[29] Netto AL and others, 'A First Look at Social Factors Driving CCS Perception in Brazil: A Case Study in the Recôncavo Basin' (2020) 98 International Journal of Greenhouse Gas Control 103053.

[30] Dowd A-M and others, 'Investigating the Link Between Knowledge and Perception of CO_2 and CCS: An International Study' (2014) 28 International Journal of Greenhouse Gas Control 79.

[31] Oltra C and others, 'Lay Perceptions of Carbon Capture and Storage Technology' (2010) 4 International Journal of Greenhouse Gas Control 698.

[32] Upham P and Roberts T, 'Public Perceptions of CCS in Context: Results of NearCO_2 Focus Groups in the UK, Belgium, the Netherlands, Germany, Spain and Poland' (2011) 4 Energy Procedia 6338.

CO_2 molecules is view on Wallquist's paper showing that people thought that CO_2 was stored as a gas in large cavernous spaces, and like a balloon in which the applied pressure could lead to an explosion or blow-up.[33]

Tcvetkov and colleagues review that knowledge plays a role in social acceptance behaviour, but it isn't always a major one.[34] Furthermore, Wallquist's study suggests that such a relation between knowledge and acceptance is mediated through the risk and benefits perception.[32] We believe that such finding extrapolates to the CCUS field as Liu et al. found evidence of a positive relationship between cognition and acceptance of CCUS in China mediated by risk/benefit perception.[35] This is also present in van Heek et al. as laypersons evaluated CCU products as capable of CO_2 leakage during contact and after disposal, increasing their perceived risks for human health and the environment.[36]

Our second proposed axis, trust, fairness, and affects, relates mostly to people's emotional responses towards CCU. Affective and emotional responses are important factors in the construction of social perception, even because it has been found that offering more information and correcting misconceptions on CCS, CCU, or CO_2, thus improving knowledge, has a limited impact on acceptance.[37] Likewise, the major reviews on CCS perception take trust as the leading emotional state involved in the process, followed by fairness perception.[38,39] Trust

[33] Wallquist L, Visschers VH and Siegrist M, 'Impact of Knowledge and Misconceptions on Benefit and Risk Perception of CCS' (2010) 44 Environmental Science & Technology 6557.

[34] Tcvetkov P, Cherepovitsyn A and Fedoseev S, 'Public Perception of Carbon Capture and Storage: A State-of-the-Art Overview' (2019) 5 Heliyon.

[35] Liu B and others, 'How Public Cognition Influences Public Acceptance of CCUS in China: Based on the ABC (Affect, Behavior, and Cognition) Model of Attitudes' (2021) 156 Energy Policy 112390.

[36] Van Heek J, Arning K and Ziefle M, 'Reduce, Reuse, Recycle: Acceptance of CO_2-Utilization for Plastic Products' (2017) 105 Energy Policy 53.

[37] Brunsting S, de Best-Waldhober M and Terwel BW, '"I Reject Your Reality and Substitute My Own!" Why More Knowledge About CO_2 Storage Hardly Improves Public Attitudes' (2013) 37 Energy Procedia 7419.

[38] L'Orange Seigo S, Dohle S and Siegrist M, 'Public Perception of Carbon Capture and Storage (CCS): A Review' (2014) 38 Renewable and Sustainable Energy Reviews 848.

[39] Tcvetkov P, Cherepovitsyn A and Fedoseev S, 'Public Perception of Carbon Capture and Storage: A State-of-the-Art Overview' (2019) 5 Heliyon.

also has an important mediating effect on the effective communication between communities and project deployers, specifically as Academia and NGO stakeholders, are viewed as more trustworthy than government or industrial stakeholders.[40,41]

In the CCU field, the impact of trust and distrust, in its relation to fairness, affect, and emotion, is translated mostly directly in product perception and acceptance. Van Heek et al. found that such connection is mediated by trust or distrust regarding the operative company. Furthermore, she points out that, on the one hand, the construction of trust is the result of 4 factors: customer relationship, the company's transparency, moral values, and the company's reputation. On the other hand, distrust results from customer deception, bad customer orientation, bad working conditions, and negative image.[42] One example is that the environmental motives for carbon dioxide utilisation (CDU) were viewed with scepticism when proposed by the industry in the UK and in Germany, the participants resorted to other more trusted stakeholders as government regulation, or called for a better communication framing in order to know why they should trust the process and believe in the ecological benefits of CDU.[43]

Also the relation between trust and risk/benefit perception suggested in CCS literature is present in the CCU technology.[44] Their research points out that institutions which can gather support from trusted stakeholders and other institutions for the implementation of their own CCU projects are better seen by the public. Therefore, in the case of CCU

[40] Li Q and others, 'Public Awareness of the Environmental Impact and Management of Carbon Dioxide Capture, Utilization and Storage Technology: The Views of Educated People in China' (2017) 19 Clean Technologies and Environmental Policy 2041.

[41] Offermann-van Heek J and others, 'Trust and Distrust in Carbon Capture and Utilization Industry as Relevant Factors for the Acceptance of Carbon-Based Products' (2018) 6 Frontiers in Energy Research.

[42] Offermann-van Heek J and others, 'Trust and Distrust in Carbon Capture and Utilization Industry as Relevant Factors for the Acceptance of Carbon-Based Products' (2018) 6 Frontiers in Energy Research.

[43] Jones CR, Olfe-Kräutlein B and Kaklamanou D, 'Lay Perceptions of Carbon Dioxide Utilisation Technologies in the United Kingdom and Germany: An Exploratory Qualitative Interview Study' (2017) 34 Energy Research & Social Science 283.

[44] Liu B and others, 'How Public Cognition Influences Public Acceptance of CCUS in China: Based on the ABC (Affect, Behavior, and Cognition) Model of Attitudes' (2021) 156 Energy Policy 112390.

projects and products, trust behaves as the ground where every action and information shared between community and industry stakeholders takes place, as one of the pillars with which perception and acceptance are built. So, in order to gain the trust of relevant communities and stakeholders, project deployers should engage them in the decision-making process, disclose the available information regarding the project, including economic gains, offer compensations in case of disasters, invest in the community, aim for a fair process by implementing and distributing gains and losses.

Risks (economic risks included) and benefits are often considered counterparts in economics, but in the social perception literature, they are treated as different constructs. Despite being the most researched factors in the literature and considered what 'underlies approval/disapproval of the CCS technology',[45] risks and benefits have different weights in influencing acceptance behaviour.[46] On the one side, in addition to generic local benefits such as job creation, infrastructure investment, and economic growth, CCS's benefit would be the ability to keep using fossil fuels while diminishing emissions, allowing a more gradual transition. On the other, carbon capture technologies relate to risks such as the lock-in of fossil fuels usage and a downgrade in investments in renewable energy technologies. Other risks like CO_2 leakage, effects on human and animal health, water and soil contamination, earthquakes, and explosions are also important cognitions driving the acceptance behaviour.

Variant from CCS papers, CCU papers most often focus only on product perception, however Linzenich studied its overall perception in comparison with other technologies. They found CCU to stand in between renewable energy (perceived as the least risky) at one side, and fossil fuel and nuclear (perceived as riskier) at the other.[47] It corroborates the findings from Arning et al. where CCU was more positively

[45] Tcvetkov P, Cherepovitsyn A and Fedoseev S, 'Public Perception of Carbon Capture and Storage: A State-of-the-Art Overview' (2019) 5 Heliyon.

[46] L'Orange Seigo S, Dohle S and Siegrist M, 'Public Perception of Carbon Capture and Storage (CCS): A Review' (2014) 38 Renewable and Sustainable Energy Reviews 848.

[47] Linzenich A, Arning K and Ziefle M, 'Acceptance of Energy Technologies in Context: Comparing Laypeople's Risk Perceptions Across Eight Infrastructure Technologies in Germany' (2021) 152 Energy Policy 112071.

ranked than CCS.[48] Such a difference is explained as both have different perceived risks and benefits. If CCS main perceived risks are leakages and the production of a lock-in state with fossil fuels, CCU perceived risks revolve around product usage and handling, and product disposal, which could lead primarily, to health and environmental damage, respectively. An example is seen in the evaluation of CO_2-derived thermal insulation boards by German consumers, where they presented a positive perception of CCU and CCU products despite being concerned with health and environmental issues, and even if such health and environmental risks are non-existent or minimal according to specialists.[49]

Finally, we present the most relevant aspects regarding the other factors influential on social perception and acceptance in the fourth axis. First it is important to consider the *Not in my back yard* (NIMBY) effect controversy. Theoretically the concept of NIMBY represents a behaviour where people present a general acceptance of a technology but reject it when placed near their homes. The controversy resides in the fact that some studies confirm the effect while others do not, arguing that other concepts explain better the technology rejection behaviour.[50]

These other aspects are socio-demographic aspects, willingness to pay for CCS/CCU, acceptance between technologies, the effects of culture, and the influence of government policy and stakeholders interactions. To provide a deep explanation on each of these factors runs out the aim of this chapter. But such lack is minimised as each of them is influenced by one or another of the summarised factors, awareness, knowledge, perception of risks, benefits and economics, and trust, in both frameworks proposed in the field, namely Huijts from 2012 and Tcvetkovs from 2019, already cited in this chapter. In order to illustrate such a web of relationships between factors, some excerpts from our empirical research are presented.

[48] Arning K and others, 'Same or Different? Insights on Public Perception and Acceptance of Carbon Capture and Storage or Utilization in Germany' (2019) 125 Energy Policy 235.

[49] Arning K, Offermann-van Heek J and Ziefle M, 'What Drives Public Acceptance of Sustainable CO_2-Derived Building Materials? A Conjoint-Analysis of Eco-Benefits vs. Health Concerns' (2021) 144 Renewable and Sustainable Energy Reviews 110873.

[50] Tcvetkov P, Cherepovitsyn A and Fedoseev S, 'Public Perception of Carbon Capture and Storage: A State-of-the-Art Overview' (2019) 5 Heliyon.

In 2023, we studied the perception of specialists in CCS technology and non-specialists through a Q-methodology research model, we used a forced choice instrument and explanatory semi-structured interviews.[51] During the interviews four participants spontaneously brought up the matter of CCU or CCUS, illustrating some points discussed in the present chapter. For example, participant 11, a carbon capture specialist, viewed CCU as a necessity to reach net-zero and placed the technology above CCS, arguing that CCU generates products and value and is more economically viable. Such speech demonstrates the influence of his perception on the risks and benefits of such technologies into his acceptance behaviour, as well his awareness regarding the whole global scene and the importance of reaching net-zero in order to mitigate climate change. For him the benefits surpass the risks.

A more direct example, from participant 14, another carbon capture specialist, discusses the matter of trust. When asked if she would trust the government for injecting CO_2 into reservoirs, she answered:

> ID_14: Government... I would say that the industry, the oil industry, absolutely! They make it with their hands down. It lasts a year to make an agreement to know [in the government], they [industry] made it within a month, they have got everything modelled, estimated, everything is okay. So, I wouldn't say that it is the government that would do it, the government doesn't have the technical corpus for that, but the oil industry, certainly. I don't see any problem with them. In truth, they already do CCUS, which is the injection inside an oil and gas reservoir increasing the pressure in order for oil and gas extraction.

Participant 20, a specialist, talked about greenwashing in her interview and demonstrated her thoughts on whether CCUS would create the lock-in status of dependency on fossil fuels and who would finance the energy transition:

> ID_20: [Greenwashing] from the oil industry, mainly. I think that today inside CCS, and correct me if I say wrong, one of the biggest question marks relates to the chain with use. CCUS, for example, is a technology that can be considered for mitigating the climate crisis, but once it increases fossil fuels production, what is the balance, right? There has been an important discussion with teacher Cristina where she spoke about it. Who has

[51] McKeown B and Thomas D, 'Q Methodology' (1988) SAGE Publications Inc.

> the money? Who has the money is the oil industry. Who is going to finance the energy transition? Who has got money. So, it is possible for us to say "we can't do it [CCUS] because it is greenwashing. Because we could be investing all this money only in renewable energy". Would we invest all? Who would invest all this money? The ones who would invest in renewable energy don't have money. How are we going to do it? I believe it is an important discussion.

This last excerpt shows participant 23, a specialist, view on what is the role of CCS and CCUS inside the energy transition context:

> ID_23: Yes, I believe it has a very important role, actually all the technologies that are being studied, so much in the part of storage as in utilisation also, which is CCUS. According to the data, mainly from what I follow from Paulo Artaxo, who works with the impact on the forests, the harbours, monsoons and everything else. And really, if you do not implement all the technologies that you've got until now, even further thinking on population growth, which is still increasing, and that the energy demand will continue to rise, if all the possibilities aren't implemented... Obviously to also stop deforestation, in collaboration with that.

The texts above illustrate what literature on the social perception of CCUS considers most: awareness and knowledge, risk and benefits, trust and fairness. However, the factors aren't absolute categories, they interconnect with each other and are present together in the participant's discourse. For example, participant 23 has a positive perception of CCUS, even showing acceptance behaviours, such perception is derived in part from her knowledge obtained with a trustworthy source, Paulo Artaxo, and from her awareness that climate change is an important issue. For her, the risk of climate change surpasses the possible risks related to CCUS. In the same way, if we draw from the socio-technical approach (Markusson), we could say that the field of uncertainty of public acceptance relates to all other fields, and all fields relate to each other.

Finally, we summarise the CCUS acceptance research made in Brazil, so far. Despite operating one of the biggest CCUS projects, the Petrobras EOR Santos Basin,[52] our country doesn't have consistent research on the acceptance aspects of the technology and the scarce studies on

[52] Global CCS Institute, 'The Global Status of CCS: 2023' (2023) 1.

the theme point for CCS acceptance instead of CCUS and indicate a low level of knowledge regarding carbon capture technologies.[53] From Netto et al. it is noteworthy to indicate that from 57 interviewed residents from the Recôncavo Basin in Bahia, a region with an important history with the oil and gas industry, 23 knew what EOR was, despite not knowing carbon capture for strictly storage purposes.[54] At last, we mark a paper from Mascarenhas and colleagues where the main challenges for the implementation of CCS in Brazil are discussed from a socio-technical approach, as we believe that many of their insights for CCS may apply in the case of CCUS.[55]

7 Final Remarks and Limitations

As discussed in this chapter, CCUS systems are complex and composed of many parts that may interact with each other in ways that are not linear. These parts affect each other, transforming their results into forms that are difficult to track down. Here we based our analysis of social perception and acceptance on the seven uncertainties regarding CCS deployment to discuss their effect on people and their thoughts.[56]

We also presented a few studies on public or social perception in the world and in Brazil, and illustrated them with our most recent research on the subject, sharing transcripts of CCS specialists and non-specialists. We are aware of the limitations of this discussion and invite other scholars to discuss further this so important matter of the social perception of the many stakeholders about technologies such as CCUS.

Advancing the societal acceptance and understanding of CCUS technologies requires concerted efforts to bridge knowledge gaps and address

[53] Lima PR and others, 'Environmental Awareness and Public Perception on Carbon Capture and Storage (CCS) in Brazil' (2021) 111 International Journal of Greenhouse Gas Control 103467.

[54] Netto AL and others, 'A First Look at Social Factors Driving CCS Perception in Brazil: A Case Study in the Recôncavo Basin' (2020) 98 International Journal of Greenhouse Gas Control 103053.

[55] Mascarenhas KL and others, 'Challenges for the Implementation of Carbon Capture and Storage (CCS) in Brazil: A Socio-Technical Approach' (2019) 2 Polytechnica 1.

[56] Markusson N and others, 'A Socio-Technical Framework for Assessing the Viability of Carbon Capture and Storage Technology' (2012) 79 Technological Forecasting and Social Change 903.

misconceptions at both local and global scales. By fostering interdisciplinary research and engaging stakeholders in dialogue, academia can play a pivotal role in elucidating the complexities of the CCUS value chain and promoting evidence-based decision-making towards a sustainable low-carbon future.

CHAPTER 8

CCUS: Legal Developments, Policies, and Challenges

Eduardo G. Pereira, Opeyemi Omotuyi, Tara Righetti, and Hannah Hylton Edwards

1 Introduction

The achievement of a climate-neutral society that provides sustainable green growth and economic development constitutes one of the major priorities on the global agenda in the twenty-first century.[1] Climate

[1] G.A. Res. 70/1, Transforming Our World: The 2030 Agenda for Sustainable Development, 21 October 21 2015.

E. G. Pereira (✉)
Institute of Energy and Environment, University of São Paulo (IEE-USP), São Paulo, Brazil
e-mail: egp.portugal@gmail.com

O. Omotuyi
Federal University Oye Ekiti, Oye, Nigeria

E. G. Pereira et al. (eds.), *Carbon Capture Utilization and Storage*, Sustainable Development Goals Series,
https://doi.org/10.1007/978-3-031-81272-9_8

change can have devastating impacts on human health, ecosystems, agriculture, water, and energy.[2] These impacts are antithetical to the realization of fundamental human rights and the global sustainable development agenda. While efficiency and conservation measures and a transition to zero-carbon generation and fuels will be necessary, there is a growing consensus that these measures alone will not be sufficient.[3] Therefore, the utilization of low-carbon and negative-emission technologies that help reduce atmospheric carbon concentrations and decarbonize the economy are essential. One of the evolving technologies in this regard is carbon capture, utilization, and storage (CCUS).

CCUS is identified as one of the most advanced technologies available for the reduction of emissions and the reduction of carbon levels in the atmosphere.[4] CCUS involves the capture of carbon from emitting sources or direct carbon capture from the atmosphere, and subsequent compression and transportation of captured carbon for use in a range of applications, or injection into deep geological formations for permanent storage.[5] The mitigation pathways identified by the IPCCC 6th assessment report rely extensively on CCS and to a lesser extent, direct air capture. They acknowledge that CCS is critical to decarbonizing otherwise hard-to-abate sectors and to achieving net-zero industrial emissions.[6] As succinctly put by the IEA, CCUS can play four critical roles in the

[2] Intergovernmental Panel on Climate Change, "Climate Change Widespread, Rapid, and Intensifying", https://www.ipcc.ch/2021/08/09/ar6-wg1-20210809-pr/, accessed 13 April 2023.

[3] IPCC, Climate Change 2022 Mitigation of Climate Change Working Group III Contribution to the Sixth Assessment Report of the Intergovernmental Panel on Climate Change, 4 April 2022 [hereinafter IPCC 6th Assessment Report).

[4] Ibid.

[5] International Energy Agency, "Carbon Capture, Utilisation and Storage", 17 October 2022, https://www.iea.org/fuels-and-technologies/carbon-capture-utilisation-and-storage, accessed 7 April 2023.

[6] IPCC 6th Assessment Report, supra note 3.

T. Righetti
School of Energy Resources, University of Wyoming, Laramie, WY, USA

H. H. Edwards
University of the West Indies - St. Augustine Campus, St. Augustine, Trinidad and Tobago

transition to net zero including; tackling emissions from existing energy assets; posing as a solution for sectors where emissions are hard to abate; posing as a platform for clean hydrogen production; and removing carbon from the atmosphere to balance emissions that cannot be directly abated or avoided.[7]

The deployment of CCUS can help counties reach emissions reductions targets while still pursuing economic growth and a cleaner environment, thus advancing energy justice. Global energy demand is expected to grow between now and 2050 and to outpace the construction of new renewable sources.[8] Decarbonizing fossil production can assure that affordable energy is available without compromising climate goals. Promoting access to advanced and cleaner fossil-fuel technology positively impacts the achievement of a just energy transition and ensures energy justice and security. CCUS may further sustain economic activity in areas dependent on fossil energy production and generation while concurrently improving local air quality in frontline communities by reducing air pollution. Finally, CCUS may provide economic benefits in point source and sequestration communities through the creation of clean energy jobs and positioning businesses to compete in markets that increasingly demand low-carbon products and sustainable practices.[9]

According to a forecast by the International Energy Agency (IEA), CCUS is expected to account for almost 15% of the cumulative emissions reduction, and its contribution to the net-zero agenda increases as the technology improves and the related costs fall.[10] Similarly, an analysis by Mckinsey & Company posits that over 25,000 industrial carbon emitters across the globe can be decarbonized through the adoption of

[7] International Energy Agency, "CCUS in Clean Energy Transitions: CCUS in the Transition to Net-Zero Emissions", op. cit.

[8] Daniel Raimi et al., "Global Energy Outlook 2023: Sowing the Seeds of an Energy Transition", Resources for the Future, March 2023.

[9] United States Council on Environmental Quality, "Carbon Capture, Utilization, and Sequestration Guidance", US Federal Register, vol. 87, no. 32, 16 February 2022, https://www.federalregister.gov/documents/2022/02/16/2022-03205/carbon-capture-utilization-and-sequestration-guidance, accessed 14 April 2023.

[10] International Energy Agency, "CCUS in Clean Energy Transitions: CCUS in the Transition to Net-Zero Emissions", https://www.iea.org/reports/ccus-in-clean-energy-transitions/ccus-in-the-transition-to-net-zero-emissions, accessed 13 April 2023.

CCUS technologies.[11] Although estimates for the level of growth necessary differ, it is clear that a massive expansion of CCUS is needed to meet Paris agreement commitments.[12]

Expanding CCUS will however require governments to address current barriers to the scalable deployment of CCUS techniques. These include incomplete or absence of regulatory and policy frameworks, lack of emissions reduction mandates such as carbon pricing, insufficient incentives for investment in CCUS technologies, and challenges with social license, justice, and public acceptance of CCUS applications. These barriers inhibit the industrial implementation and up-scaling of CCUS applications. Likewise, the successful development and deployment of CCUS relies on the establishment of legal and regulatory frameworks to ensure the effective stewardship of CCUS activities and the safe and secured storage of carbon.[13] Although regulatory approaches differ, a commitment to scaling CCU necessitates significant public investments in innovation, outreach and education, and regulation.[14] IEA identifies 25 priority legal and regulatory issues for CCUS legislative frameworks including permitting and environmental reviews, enabling first mover projects, addressing liability and transboundary issues, and ensuring safe and secure storage. In addition, legislation should integrate CCUS with other legal frameworks for emissions regulation or reduction, including emissions trading and reporting programs and carbon taxes, and with regulations related to climate disclosures and greenwashing. In some cases, legislative frameworks may also include policies providing specific support to cover for the high costs of CCUS operations.[15] This may

[11] Mckinsey & Company, "Scaling the CCUS Industry to Achieve Net-Zero Emissions", 28 October 2022, https://www.mckinsey.com/industries/oil-and-gas/our-insights/scaling-the-ccus-industry-to-achieve-net-zero-emissions#/, accessed 13 April 2023.

[12] Ibid. See also Int'l Energy Agency, Net Zero by 2050 A Roadmap for the Global Energy Sector, at 79–80; 94–99, October 2021, https://iea.blob.core.windows.net/assets/deebef5d-0c34-4539-9d0c-10b13d840027/NetZeroby2050-ARoadmapfortheGlobalEnergySector_CORR.pdf.

[13] International Energy Agency, Legal and Regulatory Frameworks for CCUS: An IEA CCUS Handbook (IEA Publications, 2022) 3.

[14] Sonja Thielges et al., "Committed to Implementing CCU? A Comparison of the Policy Mix in the US and the EU", Frontiers in Climate, vol. 4, November 2022.

[15] Elisabeth Dutschke and Vicki Duscha, "Is There a Future for CCUS in Europe? An Analysis of the Policy Framework and Societal Support", 16th International Conference on Greenhouse Gas Control Technologies (GHGT-16), 23–27 October 2022, Lyon, France.

include provisions on direct investments in innovation, and research and development, or in loan guarantees, as through the US programs authorized by the Inflation Reduction Act and the EU's Net Zero Industry Act, or through tax incentives such as those in place in the United States and Canada.

This chapter explores legal trends, policies, and norms emerging with regard to CCUS. The chapter further addresses regulatory barriers and opportunities concerning the implementation of public policies and laws on the development and deployment of CCUS technology.

2 Legal Challenges and Concerns

The successful global deployment of CCUS is largely dependent upon the formation and implementation of robust legal and regulatory frameworks, geared toward efficacious stewardship and oversight of CCUS activities. Such frameworks will serve numerous objectives, with the foremost being to ensure safe, secure, and permanent CO_2 storage in geological foundations.[16] CCUS laws and regulations must also ensure the protection of the environment and public health, clarify the rights and responsibilities of CCUS stakeholders, and provide a legal foundation for the development, operation, and long-term management of CO_2 storage resources.[17]

There are a number of salient goals underpinning the establishment of these legal and regulatory frameworks. A comprehensive regulatory regime will promote CO_2 storage as a strategy to combat climate change while also ensuring the protection of public health and the environment.[18] Regulation must provide CCUS project operators with clarity and certainty while also comprehensively regulating injection activities to prevent and protect against losses of containment and damages to health, safety, or the environment. These goals can be accomplished through permitting, monitoring, and verification requirements, and establishment of clear rules that establish rights and responsibilities relating to

[16] International Energy Agency, Legal and Regulatory Frameworks for CCUS: An IEA CCUS Handbook (IEA Publications, 2022) 7.

[17] Ibid.

[18] International Energy Agency, "Legal Aspects of Storing CO_2—Update and Recommendations", 2007, https://www.iea.org/reports/legal-aspects-of-storing-co2, accessed 28 April 2023.

civil liability and property ownership and access, including specific guidelines on intellectual property.[19] Additionally, CCUS regulatory frameworks should establish clear systems of administrative oversight, including enforcement and emergency response, and should facilitate transparency through information disclosure and public participation.[20]

All stages in the process of CCUS must be adequately regulated, including capture, transport, and use. Many of these components of the CCUS process will already be covered to some extent by existing regulations, particularly in areas where there is existing oil and gas development or GHG regulation. Notably, most of these pre-injection activities occur on the surface and thus may fit within existing regulatory and liability frameworks applicable to industrial activities. For instance, chemical waste produced by capture equipment may be regulated under hazardous waste frameworks, process emissions, and containment losses from pipelines regulated by air quality rules, and transportation safety regulated by applicable rules related to rail, pipelines, or vessels. CO_2 sequestration, however, is largely novel and presents complex new issues. Even countries that have authorized injections of CO_2 as part of waste disposal or enhanced recovery, may not have regulatory frameworks to address its unique risks and potential legal harms. Consistent with this analysis, as demonstrated through a systemic review of CCUS policies and regulations within key countries and under international conventions, CO_2 storage has been the focus of most new CCUS regulations.[21]

The following section details legal challenges with CCUS related to safety, property, and liability, and drawing from current examples globally, presents possible solutions for the mitigation of same. Additionally, the paper describes resources, including the ISO and IEA frameworks, that are available to jurisdictions looking to adopt CCUS regulatory frameworks and adapt them consistently with local laws. While these frameworks are still evolving and may not be fully comprehensive, these examples constitute a valuable knowledge base for the increasing number of jurisdictions that have identified the usefulness of CCUS but are yet to establish legal frameworks for CO_2 storage.

[19] Ibid.

[20] L. Huang, "The Legal Framework of Carbon Capture and Storage (CCS) Technology", Journal of Political Science and Law, vol. 36, 2014, 10–19.

[21] CCUS Legal and Regulatory Database—Data Tools—IEA.

2.1 *Preventing Leakage*

The primary objective of CCUS regulation is the prevention of leakage into freshwater aquifers or the atmosphere. This is notwithstanding that the safety and security of geological storage is essential to the successful widespread adoption and sustained public acceptance of CCUS, and its effectiveness as a carbon mitigation measure.[22] Loss of containment into the atmosphere could create significant health impacts[23] thereby negating any emission reduction benefits of carbon storage, and resulting in contribution to climate change through process emissions associated with CCUS. Large releases of CO_2 could also result in ecological damage, including soil acidification.[24] Moreover, migration of CO_2 within the surface could adversely impact drinking water resources or mineral production activities in the subsurface.[25]

Detailed legal frameworks for site characterization, well construction, and monitoring verification and storage can mitigate these potential public health and environmental damage associated with CO_2 storage. For instance, the United States regulates injection activities through Class II and Class VI of the Underground Injection Control Program of the Safe Drinking Water Act.[26] This program is administered by the US Environmental Protection Agency or by State agencies which have been granted primacy.[27] Currently, all States with significant oil and gas

[22] International Energy Agency, Legal and Regulatory Frameworks for CCUS: An IEA CCUS Handbook (IEA Publications, 2022) 54.

[23] See, e.g., John Fogarty and Michael McCally, "Health and Safety Risks of Carbon Capture and Storage", Journal of the American Medical Association, vol. 303, 2010, 67, 68.

[24] Hao Zhang, "Regulations for Carbon Capture, Utilisation and Storage: Comparative Analysis for Development in Europe, China, and the Middle East", Resources, Conservation & Recycling, The Chinese University of Hong Kong Faculty of Law Research Paper No. 2021-38, 2021, 73, available at https://www.law.ox.ac.uk/sites/default/files/migrated/oscola_4th_edn_hart_2012quickreferenceguide.pdf, accessed March 20th 2023.

[25] Hyun-Kwon Do et al., "Evaluation of Long-Term Impacts of CO_2 Leakage on Groundwater Quality Using Hydrochemical Data from a Natural Analogue Site in South Korea", Water, vol. 12, no. 5, May 2020, 1457.

[26] 42 U.S.C. § 300h-9(b).

[27] 42 U.S.C. § 300g-3(a)(2).

production activities have primacy over Class II, which regulates injections of oil and gas waste and for enhanced hydrocarbon recovery.[28] Thus far, only North Dakota, Wyoming, Louisiana, and West Virginia have primacy over the Class VI program.[29] The Class VI regulatory program requires extensive site characterization so as to ensure long-term containment, corrective action for existing subsurface penetrations, and well construction and mechanical integrity requirements.[30] Additionally, the Class VI program includes long-term monitoring, reporting, and verification requirements to validate plume migration models and identify indications of loss of containment.[31] Finally, extensive financial assurances are required to protect the public and property owners against potential loss or damage.[32]

In the European Union and European Economic Zone, prevention of leakage is accomplished via the Geologic Storage Directive.[33] Like the Class VI program, the geologic storage directive sets requirements for site characterization and injection.[34] These requirements apply to any carbon project exceeding 100,000 tons.[35] These requirements have been transposed into national frameworks within member States, the UK, and States who are parties to the European Economic Area Agreement.[36] Additional regulatory requirements may arise under memoranda of understanding

[28] The United States Environmental Protection Agency (US EPA), "Primary Enforcement Authority for the Underg4round Injection Control Program", https://www.epa.gov/uic/primary-enforcement-authority-underground-injection-control-program-0.

[29] Ibid.

[30] 40 C.F.R. part 146 subpart H (2023).

[31] Ibid.

[32] Ibid.

[33] DIRECTIVE 2009/31/EC OF THE EUROPEAN PARLIAMENT AND OF THE COUNCIL of 23 April 2009 on the geological storage of carbon dioxide and amending Council Directive 85/337/EEC, European Parliament and Council Directives 2000/60/EC, 2001/80/EC, 2004/35/EC, 2006/12/EC, 2008/1/EC and Regulation (EC) No 1013/2006, 2009 O.J. (L 140/114).

[34] Ibid.

[35] CCS Directive, supra note 38 at Article 2.

[36] https://climate.ec.europa.eu/eu-action/carbon-capture-use-and-storage; the report from the Commission to the European Parliament and the Council on the implementation of the CCS Directive from 2014: https://eur-lex.europa.eu/legal-content/EN/ALL/?uri=CELEX:52014DC0099.

developed under the London Protocol for offshore projects,[37] or under member States laws. Norway, for example, has developed a detailed framework for geologic storage within its petroleum and pollution regulatory frameworks.[38]

For project operators in States without existing regulatory regimes for subsurface injections or States in the process of designing regulatory programs, technical standards prepared by the International Standardization Organization may provide an essential resource. ISO standard 27914 sets details technical requirements relative to projects for geologic storage including site characterization, well construction, management, monitoring, and verification.[39] Notably, the ISO standard is not a regulation. Instead, it sets forth best practices as developed through consensus processes with technical experts from member countries and organizations.[40] Associated technical reports developed by the technical committees may provide additional guidance and information. These guidelines and reports may be helpful when developing regulations. Additionally, in countries without a CCUS regulator or with incomplete regulatory frameworks, the ISO process of independent verification may be used to validate the technical sufficiency of projects.[41]

[37] Tim Dixon and Andrew Birchenough, "Exporting CO_2 for Offshore Storage—The London Protocol's Export Amendment", Proceedings of the 15th Greenhouse Gas Control Technologies Conference 15–18 March 2021, Available at SSRN: https://ssrn.com/abstract=3820609 or http://dx.doi.org/10.2139/ssrn.3820609.

[38] Silvia Andrea Cupertino et al., *Relevant Aspects of Carbon Storage Activities' Liability in Paradigmatic Countries: Australia, Brazil, Canada, European Union, Japan, Norway, United Kingdom, and United States*, in Carbon Capture and Storage in International Energy Policy and Law, Hirdan Katarina de Medeiros Costa and Carolina Arlota, Eds., Elsevier (2021).

[39] Int'l Org. for Standardization, ISO/TC 265 Carbon Dioxide Capture, Transportation, and Geological Storage.

[40] https://www.iso.org/developing-standards.html.

[41] J. Banks and A. Miranda-Gonzalez, "A Methane Champion: Colombia Becomes First South American Country to Regulate Methane from Oil and Gas", Clean Air Task Force, 2022; Commission Delegated Regulation (EU) 2021/2139 of 4 June 2021 supplementing Regulation (EU) 2020/852 of the European Parliament and of the Council by establishing the technical screening criteria for determining the conditions under which an economic activity qualifies as contributing substantially to climate change mitigation or climate change adaptation and for determining whether that economic activity causes no significant harm to any of the other environmental objectives.

2.1.1 Permitting

As with any large-scale industrial project, a rigorous and transparent permitting process is crucial for CCUS projects. As a minimum, comprehensive regulatory programs typically require permits for exploration, well construction and injection, and creation of monitoring, measurement, and verification plans. Additionally, governments must delegate regulatory authority to an appropriate permitting authority or regulatory body and ensure that these authorities possess the requisite resources and capacity to process and evaluate permitting applications. Land use regulations and industrial siting processes may further impact construction and operation of surface facilities.

Permitting processes for CO_2 storage typically involve two primary phases[42]:

- Authorization for site exploration: Authorization for site exploration may differ significantly according to local law. Site exploration activities may include seismic surveys, baseline testing, and drilling geologic characterization wells. Where subsurface storage resources are owned or controlled by the State, this may require exploration licenses from the government. Characterization wells may be permitted under frameworks related to drilling of oil and gas or geothermal wells, if any.[43] Because these characterization wells could later operate as conduits for CO_2 leakage, the construction requirements for characterization wells are critical. Additional rights and authorizations, including for seismic surveys, may be needed from private surface or subsurface owners.[44]
- Authorization for well construction and injection: As part of the permitting process for well construction and injection, the regulator will review the project proponents' models related to site characterization to assess risks of leakage and damage to human health and the environment. In addition, the regulator will evaluate the financial capabilities of the project operator and the sufficiency

[42] International Energy Agency, Legal and Regulatory Frameworks for CCUS: An IEA CCUS Handbook (IEA Publications, 2022) 40.

[43] Wyoming oil and gas conservation commission rules and regulations, Chapter 3, Section 9.

[44] Joseph A. Schremmer, "Pore Space Property", Utah Law Review, vol. 2021, no. 1, 2021.

of monitoring, decommissioning, and rehabilitation plans prior to authorizing construction of an injection well.[45] Once a well has been constructed, the regulator may review additional data and the results of mechanical integrity test prior to granting permission to inject.

2.1.2 Monitoring, Measurement, and Verification (MMV)

Requirements for Monitoring, Measurement, and Verification are essential to validating the models generated during the site characterization process and assuring processes to identify and respond to leakage indicators. Monitoring techniques may include collection of data from soils and monitoring wells, seismic studies, and pressure monitoring. Though the requirements of each MMV plan will vary according to the scope, size, and location of the project, generally these plans ought to include monitoring programs and methods sufficient to: establish and maintain a baseline survey for the storage site until injection commences, monitor the injection facilities, the storage site (including the CO_2 plume), and the surrounding environment, compare the ongoing monitoring results against the baseline survey for the storage site, compare the actual behavior of the storage site against its anticipated behavior based on the results of the site characterization process and monitoring results, detect, attribute, and assess significant leakage, unintended migration or other irregularity in the storage site, quantify, as required by the relevant authority, the volumes of CO_2 associated with significant leakage or unintended migration, detect migration of CO_2, detect significant adverse effects on the surrounding environment, and assess the effectiveness of any corrective measures taken.[46] Plans should be modified as necessary throughout the project lifecycle.

2.2 Environmental Impact Assessment ("EIA")

Environmental Impact Assessment regulations are important tools to facilitate public participation processes and assure regulators have adequately considered the potential social, economic, and environmental impacts of sequestration projects. The EIA is identified in Principles 14

[45] IEA, Legal and Regulatory Framework.

[46] International Energy Agency, Legal and Regulatory Frameworks for CCUS: An IEA CCUS Handbook (IEA Publications, 2022) 60.

and 15 of the Stockholm Declaration as an "essential tool" for reconciling conflicts between development and the environment in order to obtain "maximum social, economic and environmental benefits for all."[47] An EIA is a standard practice for major infrastructure projects in many jurisdictions and would typically be a requirement for any projects which receive multilateral and donor finance.[48] In addition, an EIA may be a required precursor to grants of government authorizations for land use or permit. These processes may require re-evaluation and updates over the course of the project lifecycle. Judge Weeramantry indicated in his view that the EIA is not meant to be a mere assessment prior to the commencement of the project but a continuing assessment and evaluation that takes place as long as the project continues in operation.[49] While the EIA process varies depending on the jurisdiction, the main steps tend to include the following: screening, scoping, analysis of alternative options, project description, environmental baseline review, legislative review, impact identification, prediction, significance and mitigation, environmental management plan and monitoring program, reporting, review, project implementation and operation.[50]

Numerous countries and States have laws which require environmental assessments. The National Environmental Policy Act (NEPA) of the United States requires an assessment of the environmental impacts of "significantly affecting the quality of the human environment."[51] While Courts have found that grants of a Class VI permit or 45Q credits would not trigger NEPA, the grant of injection rights for CCUS on publicly owned property likely would. This could provide an avenue for legal

[47] United Nations, "Declaration of the United Nations Conference on the Human Environment" (adopted 16 June 1972) U.N. Doc. A/Conf.48/14/Rev. 1(1973). Art. 15.

[48] International Energy Agency, Legal and Regulatory Frameworks for CCUS: An IEA CCUS Handbook (IEA Publications, 2022) 38.

[49] *Gabcikovo-Nagymaros Project (Hungary/Slovakia)*, Judgment, I.C.J. Reports 1997, p. 7 [68].

[50] International Energy Agency, Legal and Regulatory Frameworks for CCUS: An IEA CCUS Handbook (IEA Publications, 2022) 38.

[51] 42 U.S.C. § 4332(2)(C).

challenges to project authorizations.[52] In other cases, federal involvement in projects such as those constructed under demonstration programs could trigger a separate NEPA analysis. In the EU, the Environmental Impact Assessment Directive requires that environmental assessments are undertaken for significant projects.[53] Additionally, the Strategic Environmental Assessment Directive requires preparation of environmental reports assessing plans or programs.[54]

2.3 *Property Rights*

Property rights are essential to obtaining access to land and subsurface storage space for CCUS. CCUS projects will require property rights for pipelines, injection and monitoring wells, access roads, and compressors, and for the conduct of seismic surveys. Ownership of these above-ground surface rights may be well defined within existing property law systems. However, CCUS projects also require extensive property rights within the subsurface for the available storage capacity within the pore spaces within rocks.[55] Legal rules to clarify real property ownership and access to the subsurface, the relationship to other subsoil resources, and title to injected CO_2 are essential parts of the governance of CCUS projects.[56]

2.3.1 Access

Determining ownership, access, and the nature of entitlements in these subsurface spaces presents novel questions of real property law. The

[52] State Sequestration: Federal Policy Accelerates Carbon Storage, But Leaves Full Climate, Equity Protections to States, https://papers.ssrn.com/sol3/papers.cfm?abstract_id=4269719, Righetti, T., et al., "The Carbon Storage Future of Public Lands", Pace Environmental Law Review, vol. 38, 2021, 181–220.

[53] Directive 2011/92/EU of the European Parliament and of the Council of 13 December 2011 on the assessment of the effects of certain public and private projects on the environment.

[54] Directive 2001/42/EC of the European Parliament and of the Council of 27 June 2001 on the assessment of the effects of certain plans and programs on the environment.

[55] Tara Righetti, "Correlative Rights and Limited Common Property in the Pore Space: A Response to the Challenge of Subsurface Trespass in Carbon Capture and Sequestration", Environmental Law Reporter, vol. 4, 2017, 10420.

[56] Id.

ownership of the geological formation varies between jurisdictions.[57] In the United States, rights to subsurface property are fragmented across state, federal, and private landowners, and may be further divided between both private and public surface and mineral owners.[58] Even in jurisdictions where mineral resources are generally understood to belong to the state, there may be some questions as to the extent of private ownership and control of the non-mineral subsurface vacuity necessary for geologic sequestration. Governments that have declared ownership of valuable subsurface substances (e.g., coal and hydrocarbons) have allowed ownership of other subsurface substances to remain with the surface estate.[59] To provide certainty and allow contracting around property rights, the law must evolve to determine ownership of these resources, either through the enactment of new codes or common law rules. Additionally, laws related to compensation, expropriation, and amalgamation of subsurface property will be needed.

Likewise, the law will need to define the extent of rights to exclude and the application of associated claims of trespass and nuisance.[60] As CO_2 is injected, it will displace tremendous volumes of brine—pushing these formation fluids into adjoining properties and increasing pressurization, which in turn may limit the suitability of those formations for CCUS or other uses. Legal rules related to indirect migration and subsurface trespass are currently ill-defined and have not yet been applied in the context of CCUS.

2.3.2 Subsurface Rights—Competing Claims

Operations may further require brine withdrawal for pressure managements or recompletion and plugging of legacy or operating oil and gas wells. As such, the law must determine the relative rights and interrelationship between storage property and property interests in minerals and water.[61] As a result of these different systems of regulation of CO_2 storage

[57] UCL Carbon Capture Legal Programme, "Property Rights", available at, https://www.ucl.ac.uk/cclp/ccspropertyrights.php, accessed 15 May 2023.

[58] Righetti et al., Supra note 59.

[59] UCL Carbon Capture Legal Programme.

[60] Joseph A. Schremmer, "A Unifying Doctrine of Subsurface Property Rights", Harvard Environmental Law Review, vol. 46, 2022, 525.

[61] Id.; UCL Carbon Capture Legal Programme, "Property Rights", available at https://www.ucl.ac.uk/cclp/ccspropertyrights.php, accessed 15 May 2023.

in different jurisdictions, reservoirs and injection sites may be subject to competing claims. In this context, the interaction between CO_2 storage proponents and petroleum and mineral explorers and producers needs to be carefully balanced.[62]

2.3.3 Ownership of Injected CO_2

The ownership of CO_2 must be clearly defined throughout the life cycle of CCUS projects. Ownership of CO_2 carries with it responsibility for stewardship, monitoring, and verification, and potential liability for CO_2 releases. Determination of ownership will also be critical if CO_2 is later withdrawn for utilization. Additionally, defining and recognizing a property right in injected CO_2 will be important in determining the application of legal doctrines including those of conversion and abandonment.

Like other aspects of property law, CO_2 ownership is governed by specific legal regimes of national and sub-national jurisdictions, and may be governed by a combination of common law, civil law, statute, and contract.[63] Ownership may vary over the lifecycle of a project. For instance, Wyoming law clarifies that title to injected CO_2 remains with the injector.[64] However, the law also provides a mechanism whereby the title to the CO_2 and certain liabilities associated therewith may be transferred to the state upon satisfaction of certain conditions.[65]

Yet even with such frameworks, ownership of a specific molecule of CO_2 may not be clear. Maintaining clear ownership distinctions, however, will become increasingly difficult as CCUS projects reach scale and the possibility of subsurface pressure and plume interactions increases. Complications also arise given the dual classification of CO_2 as both a commodity and waste. Where there are overlapping mineral and surface property interests in the subsurface, there is a possibility that injected CO_2 would be owned by the injector, whereby the mineral owner could have rights to any native CO_2 present in the formation fluids and gasses. Where the injector only holds rights to the pore spaces, ownership may transfer as CO_2 mineralizes within the surrounding subsurface formations.

[62] Ibid.

[63] Ibid.

[64] Wyo. Stat. 35-11-318.

[65] Wyo. Stat. §§ 35-11-319(b), (c).

2.4 *Liability Concerns*

Liability for CCUS projects may arise under public or private law.[66] Public law liabilities may arise under environmental laws governing water and air quality, disposal of hazardous wastes, protection of cultural resources and wildlife, employee and workplace safety, and rules related to climate disclosures. Private law claims arising in tort may include trespass, nuisance, interference with business advantage, and conversion. Additionally, commercial arrangements related to the CO_2 projects, including agreements with landowners, may give rise to contractual liability. To address some of these concerns, many commercial agreements and regulatory programs include extensive insurance, bonding, and surety requirements.[67] These mechanisms are designed to protect the public from assuming unfunded liabilities in the event of operator insolvency.

The most challenging and complex issue related to liability for CO_2 storage activities relates to long-term liability for losses of containment and releases of CO_2. Storage activities and stewardship obligations are likely to extend beyond the life of commercial project operators. As a result, some US States and the European Union have put in place mechanisms which allow transfer of liability for plume releases to the public upon satisfaction of certain conditions including demonstrations of plume stability consistent with modeling and proper plugging, abandonment, and reclamation of injection facilities. These transfer mechanisms, however, are largely limited to future stewardship over sequestration projects and may not include private claims arising under contract or claims originating prior to transfer of stewardship obligations.

The government of Alberta assumes the long-term liability of licensed carbon sequestration projects in the province. Within the Mines and Mineral Act, the provincial government becomes the owner of any captured carbon and assumes all obligations of the lessee. The government also indemnifies a lessee against liability for damages attributable

[66] Elizabeth J. Wilson et al., "Assessing a Liability Regime for Carbon Capture and Storage", Energy Procedia, vol. 1, 2009, 4575, 4576 (internal citation omitted); Ian Havercroft and Richard Macrory, "Legal Liability and Carbon Capture and Storage: A Comparative Perspective", Global Carbon Capture and Storage Institute at 11 (2014); Michael Faure, "Liability and Compensation for Damage Resulting from CO_2 Storage Sites", William & Mary Environmental Law and Policy Review, vol. 40, 2016, 387.

[67] 40 CFR 146.85.

to such licensed carbon sequestration project.[68] These obligations are funded by carbon storage operators through per-tonne payments into the Post-closure Stewardship Fund.[69] These funds are collected pursuant to the Carbon Sequestration Tenure Regulations (CSTR) and can be used for ongoing monitoring, maintenance, and remediation associated with CCUS operations.[70]

3 CCUS: Global Trends, Legal and Regulatory Policies and Norms

The achievement of carbon reduction and negative emissions through the adoption of CCUS technology calls for the implementation of appropriate regulations and policies that support the development and deployment of such technology. As affirmed by the United Nations (UN) through its sustainable development agenda, the prevention of climate change requires the integration of preventive and mitigating measures into national policies, strategies, and planning.[71] Such regulatory and policy support may take the form of government regulation through the command-and-control approach, market-based regulation and incentives, and the adoption of principles consistent with the just and responsibility deployment of emissions reduction and carbon removal technologies.[72] To meet nationally determined contributions, governments around the world are implementing regulatory and policy support for the development and application of CCUS technology within their jurisdictions.

3.1 United States

The United States (US) Congress affirms that large-scale deployment of CCUS is significant for the achievement of the global climate goals,

[68] Mines and Minerals Act, Revised Statutes of Alberta 2000 op. cit., sec 121.

[69] Carbon Sequestration Tenure Regulation, Alberta Regulation 68/2011, sec 20.

[70] Ibid.

[71] United Nations, "Sustainable Development Goals", goal 13.

[72] Gabriel Weil, "The Carbon Price Equivalent: A Metric for Comparing Climate Change Mitigation Efforts Across Jurisdictions", Dickinson Law Review, vol. 125, 2024, 475.

while also relevant for driving economic development, technological innovation, and higher employment opportunities.[73] The US encourages CCUS principally through publicly funded tax credits and research and development.

The principal mechanism for incentivizing CCUS in the United States is the 45Q tax credit. 45Q provides an injector with a payment for each metric ton of qualified carbon oxide injected for permanent geologic storage. Although the 45Q tax credit for carbon sequestration has existed since 2008, the credit was underutilized. In 2022 the US Congress significantly enhanced the credit with passage of the Inflation Reduction Act (IRA) including by providing increases in the credit amount, an extension of the commence to construct deadline, and allowing direct pay.[74] In addition, the IRA added a credit specific for direct air capture (DAC) facilities.[75] Credit amounts differ depending on whether the project is developed for enhanced recovery, saline storage of industrial captured CO_2, or DAC projects. These changes have resulted in a rapid increase in private investment in CCUS projects. Since passage of the IRA, applications for Class VI injection permits have increased by more than 10x.

The US also provides support for CCUS through funding for research and development, demonstration projects, and loan guarantees.[76] The Infrastructure Investment and Jobs Act, Energy Act of 2020, and Inflation Reduction Act have provided billions of dollars in public funding across its CCUS value chain.[77] These funds are aimed at offering financial support for the development of CCUS in the US through policy mechanisms including research and development funding, loans, and permitting

[73] Inflation Reduction Act of 2022, H.R.5376, 117th Congress (2021–2022), sec 40301, https://www.congress.gov/bill/117th-congress/house-bill/5376, accessed 20 April 2023.

[74] Ibid., sec 13102.

[75] Ibid., sec 13104.

[76] Congressional Research Service, Carbon Capture and Sequestration (CCS) in the United States (Angela C. Jones & Ashley J. Lawson eds., 2022); Innovative Technologies Loan Guarantee Program, 87 Fed. Reg. 105 (1 June 2022).

[77] https://crsreports.congress.gov/product/pdf/R/R47034/1.

support.[78] Additionally, the IRA provided substantial new funding for demonstration projects, expansion of the CarbonSafe Program, front end engineering and design (FEED) studies, sequestration hubs, and CO_2 transport. CCUS may also be enabled by funds provided through the IRA for hydrogen projects.[79]

Pursuant to Biden Administration policies, projects receiving federal funding in the United States are subject to numerous requirements designed to address justice concerns associated with energy projects, including CCUS. These programs have two primary objectives: to address environmental justice by assuring that marginalized communities do not disproportionately bear the burdens of industrial projects; and assuring that the benefits of federal funding are equitably distributed consistent with restorative goals. The Justice 40 initiative, for example, assures that 40% of federal funding under certain programs be allocated to disadvantaged communities.[80] For any project receiving federal funding, the applicant must prepare an environmental justice analysis and plans for community benefits and engagement.[81] Projects are further be eligible for additional tax credit enhancements where projects are located in coal and fossil reliant "energy communities" that are disproportionately impacted by the energy transition and where certain wage and apprenticeship program requirements are met.[82]

A recent rule promulgated by the United States Environmental Protection Agency (EPA) mandates CCUS in some cases. The rule, promulgated under Sect. 111 of the Clean Air Act,[83] requires CCUS and/or hydrogen

[78] See generally the Infrastructure Investment and Jobs Act, H.R.3684, 117th Cong. (2021–2022), Title III – Fuels and Technology Infrastructure Investments, https://www.congress.gov/bill/117th-congress/house-bill/3684/text, accessed 20 April 2023.

[79] Matt Bright, "The Inflation Reduction Act Creates a Whole New Market for Carbon Capture", Clean Air Task Force, 22 August 2022.

[80] https://www.whitehouse.gov/environmentaljustice/justice40/.

[81] https://www.energy.gov/infrastructure/about-community-benefits-plans.

[82] Increased amounts of Credit of Deduction for Satisfying Certain Prevailing Wage and Registered Apprenticeship Requirements, 89 Fed. Reg. 53184 (June 25, 2024, https://www.federalregister.gov/documents/2024/06/25/2024-13331/increased-amounts-of-credit-ordeduction-for-satisfying-certain-prevailing-wage-and-registered; U.S. Internal Rev. Service, Notice 2024–30, Energy Community Bonus Credit Amounts Under the Inflation Reduction Act of 2022 (Mar. 22, 2024), https://www.irs.gov/pub/irs-drop/n-24-30.pdf.

[83] 42 U.S.C. § 7411(a)(1).

co-firing as a best system of emissions reduction at new natural gas generation and new or existing coal-fired generation facilities.[84] Existing plants must either install equipment with 90% capture efficiency by 2035 or to close.

3.2 Canada

The Canadian government encourages the development and deployment of CCUS technologies through a combination of tax incentives and carbon pricing. Between 2022 and 2030, the government has budgeted for investment tax credit for CCUS projects which include 60% tax credit for investments in direct air capture equipment, 50% tax credit for investment in other CCUS equipment, and 37.5% tax credit for investment in equipment for carbon transportation, storage and use.[85] This encourages investments in the development of CCUS technologies, as well as a rapid scale-up of CCUS deployment in Canada.

CCUS is also driven by stringent carbon pricing regulations that apply to many heavy-emitting industries. Over the same period covered by the tax credits, emissions charges assessed pursuant to the Greenhouse Gas Pollution Pricing Act will increase from $65 CAD per ton of GHG emissions to rise to $170 CAD in 2030.[86] Operators may therefore pursue CCUS to avoid the cost of emissions penalties. Operators investing in CCUS may further be encouraged by the ability to trade offset credits within Canada's Greenhouse Gas Offset Credit System, authorized under the Environmental Violations Administrative Monetary Penalties Act and the Greenhouse Gas Pollution Pricing Act.[87] While

[84] New Source Performance Standards for Greenhouse Gas Emissions From New, Modified, and Reconstructed Fossil Fuel-Fired Electric Generating Units; Emission Guidelines for Greenhouse Gas Emissions From Existing Fossil Fuel-Fired Electric Generating Units; and Repeal of the Affordable Clean Energy Rule, 89 FR 39798, 39812 (May 9, 2024).

[85] See the Government of Canada Budget 2022, https://www.budget.canada.ca/2022/report-rapport/toc-tdm-en.html, accessed 20 April 2023.

[86] Government of Canada, Update to the Pan-Canadian Approach to Carbon Pollution Pricing Plan 2023–2030 (August 2021).

[87] Canadian Greenhouse Gas Offset Credit System Regulations (SOR/2022-111).

not currently included, the Compendium of Federal Offset Protocols is expected to be expanded to cover direct air capture and sequestration.[88]

CCUS policy may be further supported through provincial policies. For instance, Alberta has developed a framework of regulatory support for CCUS through amendments made to the existing provincial oil and gas laws such as the Mines and Minerals Act, as well as the implementation of the Carbon Sequestration Tenure Regulations, and the Carbon Capture and Storage Funding Act. Pursuant to the revised Mines and Minerals Act, the Minister responsible for mines and minerals is empowered to enter into an agreement with a lessee, pursuant to which the lessee is granted the right to inject captured carbon into a subsurface reservoir for sequestration.[89] To ensure an effective CCUS system in Alberta, the government has also enacted the Carbon Capture and Storage Funding Act, aimed at encouraging and expediting the design, construction and operation of CCUS projects in the province.[90] The Minister of Energy is empowered to make available funds for the expenses incurred by the Department of Energy to develop or refine a regulatory system, further education, or acquire information or studies regarding CCUS projects in Alberta.[91] Earlier in 2023, Newfoundland announced the net-zero project with the objective of advancing technical, economic, and policy frameworks to drive offshore CO_2 storage.[92]

3.3 United Kingdom

Pursuant to the declaration of a green industrial revolution agenda, the UK government has announced a ten-point plan which set out measures that will help in the achievement of the revolution.[93] One of the listed

[88] Government of Canada, News Release, "Canada Launches Greenhouse Gas Offset Credit System to Support a Clean, Green Economy", 8 June 2022.

[89] Mines and Minerals Act, Revised Statutes of Alberta 2000, Chapter M-17, sec 116(1).

[90] Carbon Capture and Storage Funding Act, Statutes of Alberta 2009, Chapter C-25.

[91] Carbon Capture and Storage Funding Regulation, Alberta Regulation 64/2010.

[92] https://energynl.ca/wp-content/uploads/2023/03/NZP-CCUS-White-paper-Main-Report-Ver-09.pdf.

[93] "The Ten Point Plan for a Green Industrial Revolution", 2020, https://assets.publishing.service.gov.uk/government/uploads/system/uploads/attachment_data/file/936567/10_POINT_PLAN_BOOKLET.pdf, accessed 21 April 2023.

measures is the adoption of CCUS solutions for the decarbonization of the industrial sector, with a target to capture 10metric tons of carbon annually by 2030.[94] Among measures intended to encourage CCUS, the plan included a commitment to raise the UKs research and development investment to 2.4% of its national GDP by 2027.[95] Subsequently, a White Paper on Energy was implemented to put in place strategies for the transformation of the UK energy system pursuant to the green industrial revolution. The White Paper emphasizes the acceleration of UK's transition to clean energy system, through the creation of investment opportunities that enable a smart and flexible energy system, and harnesses the power of competition and innovation.[96] The report acknowledges that the UK cannot meet its emissions reduction targets through efficiency measures and generation switching alone, and thus identifies the deployment of CCUS and clean hydrogen as a significant strategy for the successful decarbonization of UK's industrial sector.[97] These conclusions are carried through in the UK's net-zero and industrial decarbonization strategies which endeavor to decarbonize the UK economy while maintaining competition and avoiding carbon leakage.

To fund the development of CCUS infrastructure and facilities, the UK has implemented green finance programs to encourage innovation and reduce the cost of the net-zero transition. In 2020, the UK government announced the Carbon Capture and Storage Infrastructure Fund (CIF) with budgetary allocations of up to one billion pounds.[98] While not specific to CCUS, the UK has additionally implemented a Net-Zero Innovation Portfolio which provides funding for such low-carbon

[94] Ibid.

[95] "The Ten Point Plan for a Green Industrial Revolution", op. cit.

[96] See "The Energy White Paper: Powering Our Net Zero Future" 2020, https://assets.publishing.service.gov.uk/government/uploads/system/uploads/attachment_data/file/945899/201216_BEIS_EWP_Command_Paper_Accessible.pdf, accessed 21 April 2023.

[97] Ibid.

[98] "The Carbon Capture and Storage Infrastructure Fund: An Update on Its Design", 2021, https://www.gov.uk/government/publications/design-of-the-carbon-capture-and-storage-ccs-infrastructure-fund/the-carbon-capture-and-storage-infrastructure-fund-an-update-on-its-design-accessible-webpage, accessed 21 April 2023.

technologies and systems thereby decreasing the costs of decarbonization in the country.[99] One of the primary objectives of the portfolio is to bring down the cost of capturing and sequestering carbon, and helping the industrial sector to understand the opportunity for developing and deploying novel carbon capture technologies.[100] The portfolio also supports the research and development of direct air capture technologies in the UK.[101] In addition, a report by the UK CCS Research Community suggests that the UK government is committed to an improved regulatory stability and transparency by the establishment of an independent economic regulation for carbon transport and storage networks thereby, enabling investors to earn a fair return on their investments under a predictable and stable regulatory framework.[102] The UK also benefits from a well-developed research community that is driving innovation and policy. The UK Carbon Capture and Storage Research Centre includes more than three hundred academic members collaborating with CCUS organizations across the world.[103]

The UK is on the cusp of passing a regulatory framework for CCS. The Energy Bill is currently undergoing relevant legislative processes in the UK Parliament. Upon Royal Assent, the bill will establish a legislative framework for the regulation of the licensing of carbon transport and storage, and commercial arrangements for industrial carbon capture and storage and for hydrogen production in the UK. The regulatory framework provided is designed to attract private finance and remove market barriers to CCUS investments, and provide long-term revenue certainty essential to develop and scale-up the CCUS industry in the UK.[104]

[99] See the Government of UK, "Net Zero Innovation Portfolio", https://www.gov.uk/government/collections/net-zero-innovation-portfolio, accessed 20 April 2023.

[100] Ibid.

[101] Ibid.

[102] See the UK Department for Business & Trade, "Carbon Capture, Usage and Storage", https://www.great.gov.uk/international/content/investment/sectors/carbon-capture-usage-and-storage/, accessed 21 April 2023.

[103] UK Carbon Capture and Storage Research Centre (UKCCSRC), https://ukccsrc.ac.uk/, accessed 21 April 2023.

[104] UK Parliament, "Energy Bill (HL) 2022–23", https://bills.parliament.uk/bills/3311, accessed 21 April 2023.

3.4 *Europe*

CCUS is primarily regulated alongside other low-carbon and net-zero technologies in Europe using market-based or economic policies linked with the achievement of Europe's climate neutrality by 2050. The 'European Green Deal' is aimed at transforming the European Union (EU) into a modern, resource-efficient, and competitive economy with targets such as ensuring negative greenhouse gases emissions by 2050, among others.[105] This requires the creation of effective regulatory environment for clean technologies and enhancement of the competitiveness of the net-zero industry. To achieve this, the 'Green Deal Industrial Plan' was proposed to create a supportive environment for scaling-up the EU's manufacturing capacity for the net-zero technologies and products required to meet Europe's climate targets. Some of the pillars of the plan include predictable and simplified regulatory environment, and faster access to funding, etc.[106] This led to a proposed Net-Zero Industry Act, which is designed to strengthen the development of EU's net-zero industry through measures and strategies that facilitate investments in, and incentivize demand for, net-zero technologies. Likewise, the Act is designed to support large-scale technological capacity building for Europe's workforce in the development of net-zero technologies. Some of the proposed measures and strategies include the simplification of the regulatory framework for the development of such technologies, and a scale-up of the capacity to store carbon in Europe.[107] Particularly, the Net-Zero Industry Act proposes to facilitate investments in clean technologies by supporting large-scale technological capacity building to enable the development of net-zero technologies in Europe. To achieve

[105] The European Commission, "A European Green Deal", https://commission.europa.eu/strategy-and-policy/priorities-2019-2024/european-green-deal_en, accessed 21 April 2023.

[106] The European Commission, "The Green Deal Industrial Plan", https://commission.europa.eu/strategy-and-policy/priorities-2019-2024/european-green-deal/green-deal-industrial-plan_en, accessed 21 April 2023.

[107] The European Commission, Proposal for a Regulation of the European Parliament and of the Council on Establishing a Framework of Measures for Strengthening Europe's Net-Zero Technology Products Manufacturing Ecosystem "Net Zero Industry Act", 2023, https://commission.europa.eu/strategy-and-policy/priorities-2019-2024/european-green-deal/green-deal-industrial-plan/net-zero-industry-act_en, accessed 21 April 2023.

this, the Act proposes the coordination of efforts to ensure the availability of a skilled workforce required for the net-zero industry through the setting up of specialized Skills Academies in Europe. It also proposes to foster the set-up of net-zero manufacturing projects by streamlining administrative and permit-granting processes. It requires EU member States to ensure that permits for net-zero manufacturing projects are granted through faster and less cumbersome procedures. This expedites the implementation of new CCUS projects, and also cut costs and other administrative burden for businesses intending to undertake CCUS projects. More importantly, the Act aims at attracting investments and creating better conditions and market access for clean technologies in the EU.[108] These will increase the competitiveness of Europe's net-zero industry. Although a wide range of net-zero technologies is intended to benefit from the Net-Zero Industry Act, the Act specifically proposes to facilitate and enable CCUS projects by enhancing the availability of carbon storage sites in the EU.[109]

To address the challenge of funding such novel and capital-intensive technologies in the EU, the European Commission has implemented several policies to promote the development of low-carbon technologies including CCUS. The 'Innovation Fund' is dedicated to the development and deployment of innovative low-carbon technologies in the EU. Pursuant to the EU's carbon pricing system, polluters pay for their greenhouse gases emissions through the Emissions Trading System (ETS). Funds generated from the ETS are invested into the Innovation Fund for the financing of innovative low-carbon technologies including CCUS.[110] Additionally, Horizon Europe, which succeeded the Horizon 2020 program of the EU, is the EU's primary funding program for research and innovation in the EU with a budget of over ninety-five billion euro until 2027. The fund facilitates collaboration and knowledge sharing and strengthens the impact of research

[108] Ibid.

[109] Ibid.

[110] The European Commission, "Climate Action: Innovation Fund", https://climate.ec.europa.eu/eu-action/funding-climate-action/innovation-fund_en, accessed 21 April 2023.

and innovation in developing, supporting, and implementing EU's policies particularly on climate change and sustainable development.[111] For cross-board projects, the TEN-E policy encourages the development of cross-border carbon networks including carbon transport and storage facilities between EU member States, and with neighboring third countries. These carbon infrastructure projects benefit from financial support under the Connecting Europe Facility (CEF) as projects of common interest (PCIs).[112] The CEF is a primary EU funding policy for promoting growth, jobs, and competitiveness through targeted infrastructure investment.[113]

The EU also encourages multi-stakeholder engagement in support of the development and deployment of CCUS technologies through various multi-stakeholder initiatives including, the Strategic Energy Technology Plan Working Group on CCUS, and the European Technology and Innovation Platform 'Zero Emissions Platform'.[114] The European Commission also established the CCUS Forum in 2021 as a robust stakeholder consultation platform bringing together representatives of EU institutions, the EU, and third countries, and the private sector so as to facilitate the deployment of CCUS technologies in Europe. Likewise, the EU's CCUS Zero Emission Network (ZEN) was created to enable knowledge sharing between different stakeholders and between European regions in order to support full chain development of CCUS in the EU. The CCUS ZEN aims at building a coherent ecosystem of CCUS actors capable of decarbonizing the European industrial sector and delivering the EU's climate targets.[115]

[111] The European Commission, "Research and Innovation: Horizon Europe", https://research-and-innovation.ec.europa.eu/funding/funding-opportunities/funding-programmes-and-open-calls/horizon-europe_en, accessed 21 April 2023.

[112] Trans-European Networks for Energy Regulation (TEN-E), Regulation (EU) 2022/869, https://eur-lex.europa.eu/eli/reg/2022/869/oj, accessed 21 April 2023.

[113] See the Connecting Europe Facility, https://commission.europa.eu/funding-tenders/find-funding/eu-funding-programmes/connecting-europe-facility_en#:~:text=The%20Connecting%20Europe%20Facility%20(CEF,infrastructure%20investment%20at%20European%20level, accessed 21 April 2023.

[114] The European Commission, "Carbon Capture, Storage and Utilisation", https://energy.ec.europa.eu/topics/oil-gas-and-coal/carbon-capture-storage-and-utilisation_en, accessed 21 April 2023.

[115] Ibid.

3.5 International Cooperation

Countries across the globe are also engaging in bilateral agreements to support the development of CCUS technologies in their different jurisdictions. For instance, Japan and Indonesia agreed to cooperate on the development and deployment of decarbonization technologies in order for both countries to transition to clean energy. A memorandum of cooperation which was signed to this effect in 2022 aims at a collaboration between both countries in the development and deployment of clean technologies that enhance energy transition including hydrogen, CCS, and CCUS, among others.[116] This is in line with the UN sustainable development agenda which affirms global partnership and collaboration as a veritable means of implementing the sustainable development goals. The agenda emphasizes the significance of collaboration and international cooperation on issues such as science, technology, and innovation, among others, as well as knowledge sharing on mutually agreed terms, as a means for the achievement of sustainable development globally.[117]

The United States has engaged with numerous countries to support development of stable and transparent regulatory and investment regimes for CCUS through the Commercial Law Development Program in the Department of Commerce. Focused on green energy and energy transition, the program helps facilitate regulatory development by organizing workshops and providing technical expert support. Additionally, the US Department of Energy has partnered in several international demonstration projects.[118] The International Strategy, Policy, and Implementation Group within the National Renewable Energy Lab has further supported technology deployment and policy support for new technologies abroad. These and similar programs in other countries are compatible with the capacity building, finance, and technology transfer mechanisms of the Paris Agreement.

International collaboration can also be facilitated through participation in international coalitions supporting CCUS policy and governance. Among these are the International Standardization Organization, the

[116] See "Japan, Indonesia to Cooperate on Hydrogen, Ammonia and CCS", Reuters, 10 January 2022, https://www.reuters.com/business/sustainable-business/japan-indonesia-cooperate-hydrogen-ammonia-ccs-2022-01-10/, accessed 21 April 2023.

[117] United Nations, "Sustainable Development Goals", goal 17.

[118] https://netl.doe.gov/carbon-management/carbon-storage/globalcollaborations.

Association for International Energy Law, the Energy Bar Association, and the European Federation of Energy Law Associations. These organizations and coalitions support development of consensus regulations that will support compatibility and stability across jurisdictions.

4 Recommendations and Conclusion

Deployment of CCUS is indispensable to the achievement of the climate targets endorsed by the Paris Agreement. Factors, including the absence of or incomplete regulatory and policy framework, lack of emissions reduction mandates, insufficient incentives for CCUS investments, however pose significant barriers to the scalable deployment of CCUS technologies.

In light of the foregoing considerations, and in order to ascertain and determine the way forward regarding the deployment of CCUS, recommendations must be proffered. It is undeniable that the process geared toward implementing CCUS is an exceedingly lengthy and technical one, which requires a great deal of expert instruction and international negotiation. This painstaking process may motivate the question as to whether the global deployment of CCUS is a fruitful, or even feasible, endeavor.

These barriers may be effectively addressed through the adoption of robust legal and regulatory frameworks for CCUS both at national and international levels. Moreover, to ensure sustainable deployment of CCUS, States must enact economic policies to encourage project development, establish and enforce legal frameworks to encourage transparency and protect public interests in health, safety, the environment, and property. Even with appropriate regulatory regimes, CCUS will not achieve its commercial potential without public support. This is required for research and development related to technology demonstration and site characterization. Additionally, to achieve social license and address concerns of distributive justice, deep community engagement, outreach, and education are essential. Furthermore, States can access grants to assist with the effective implementation of CCUS. These efforts are underway to varying extents in jurisdictions worldwide. Communication and knowledge sharing can however help accelerate these developments.

A salient step, in order for States to realize the foregoing recommendations, is for States to invest in research and development of CCUS. Investing in research and development of CCUS is crucial for advancing this technology. Research and development efforts will be necessary to improve the efficiency and cost-effectiveness of CCUS, exploring new methods of capture, and finding innovative and economically beneficial uses for captured carbon dioxide. Many States renege on this technology, owing to their lack of knowledge on the processes, or lack of funds to finance its effective implementation.

As stated above, one cardinal recommendation, though admittedly a formidable proposition, is the creation by States and international organizations of legal and regulatory frameworks. All the elements, considerations, and processes in CCUS are interlinked, meaning that once the cumbersome task of formulating standardized legal frameworks is established, all elements and stages, being so interlinked, can be regulated together, and assist States in understanding their respective obligations and rights during each stage of the process. Therefore, the creation of set policies and regulations may encourage States to invest in the CCUS technology and will assist in ensuring its safe and responsible implementation. These legal instruments will assist in promoting transparency, regulating liability, monitoring, and long-term storage.

Further, in order to ensure that States abide by these policies, these policies can be formulated to require the implementation of technical committees to oversee each stage of the process. The proponents of carbon capture, storage, and utilization constitute an integrated system, and the regulation and governance thereof should ideally not be separated. Each stage requires specific materials, risk management, monitoring, and quantification challenges, which, as part of an integrated system, must be considered in accordance with the prior stages. These technical committees may be necessary to ensure that properly qualified and skilled experts are implemented to oversee the various functions of the processes. These committees would oversee the full process of CCUS system, from capturing the carbon dioxide, transporting it to a final location, storing it, and finally utilizing the carbon for specific products. The committees would contemplate and provide specific technology, terminology, environmental considerations, risk management, GHG quantification and verification, health and safety, and other related CCUS activities. These technical committees will assist in enforcement and interpretation of the policies and regulations.

Additionally, promoting public awareness is key to its successful and widespread implementation. By raising awareness about the benefits of CCUS, States can encourage individuals and communities to support and renege in sustainable practices. Through educational campaigns, public forums, and media outreach, States and International organizations, can empower people make informed choices and understand the benefits of CCUS. When the public is made aware of CCUS, it encourages support for its implementation and advocacy for sustainable practices. Raising public awareness about CCUS assists States in implementing same by generating support among its population. This, in turn, results in more widespread advocacy for its inclusion in government policies, and funding initiatives.

States, especially developing States, are also encouraged to seek and access grants to implement the CCUS technology within their State, by exploring various funding opportunities. States can start by researching government programs, international organizations, and private foundations that offer grants, specifically for CCUS projects. States can then collaborate with these international organizations such the Global CCS Institute, the International Energy Agency, and the United Nations Framework on Climate Change to provide funding, as well as expertise and guidance. Additionally, States can collaborate with research institutions, industry partners, and non-profit organization to strengthen their grant applications and increase their chances of securing funding. Pursuant to proper research and development, States would be better equipped to find the right resources and make a completing case for the implementation of CCUS within their State.

In light of the above, it is evident that in order to achieve our climate mandates, States must take steps to conform with the recommendations stated herein. While some recommendations provide obvious challenges, with the correct application, the implementation of CCUS can be feasible and effectual.

CHAPTER 9

Standardization Issues Related to CCUS

Alberto J. Fossa, Alexandre de Barros Gallo, and Edmilson Moutinho dos Santos

1 Introduction

Carbon capture, utilization, and storage (CCUS) technologies have the potential to play a significant role in reducing greenhouse gas emissions. Standardization is crucial for ensuring that CCUS projects can be accurately monitored, verified, and reported, which is essential for building trust among stakeholders and investors. Without standardized protocols, it becomes difficult to compare and assess the performance of different projects, hindering the scalability and cost-effectiveness of CCUS technologies.

A. J. Fossa (✉) · A. de Barros Gallo · E. M. dos Santos
Research Centre for Greenhouse Gas Innovation (RCGI), University of São Paulo, São Paulo, Brazil
e-mail: afossa.ms@outlook.com

E. M. dos Santos
Institute of Energy and Environment of the University of São Paulo (IEE-USP), São Paulo, Brazil

E. G. Pereira et al. (eds.), *Carbon Capture Utilization and Storage*, Sustainable Development Goals Series,
https://doi.org/10.1007/978-3-031-81272-9_9

Additionally, standardization is necessary for ensuring compliance with regulatory requirements and for facilitating the development of a robust carbon market. By addressing these standardization issues, we can unlock the full potential of CCUS technology in mitigating climate change and transitioning to a low-carbon economy.

In this chapter, we will explore the importance of addressing standardization issues, considering its strategic role, and then discuss it in the context of CCUS technologies.

2 The Strategic Role of Standardization

Within the technical regulatory framework, standardization (national or international) has played a relevant role in technological development and the promotion of innovation in various sectors. It is understood that the development of mechanisms for the exploitation of natural resources will benefit from the adequate development of consistent technical bases and good practices consolidated through the consensus of experts. In particular, normalization can play a prominent role in a sustainable process of unconventional gas development, as it acts in the control, minimization, and mitigation of any impacts caused by its exploration and production. This chapter recalls the most important principles about aspects of standardization in order to clarify its main functions and connections with the development of sustainable markets.

2.1 Background on Standardization Strategies

Due to economic expansion, newly industrialized economies like Taiwan and Singapore, as well as growing East Asian countries like South Korea and Japan, faced new industrial conditions in the 1980s. The United States and Europe were forced to restructure their competitiveness standards in order to introduce new industrial strategies, safeguard intellectual property rights, and loosen antitrust laws in response to the development of these rising nations. These regulations were put in place with an emphasis on international competition to strengthen their innovation systems.

The increasing application of consensus-based technical standards is closely linked to the loosening of antitrust regulations to encourage the establishment of business consortiums for the purpose of collaborative research. Consortia were considered primarily for two reasons. One of them has to do with working with strategic partners to develop new technologies. The second is the creation of industry-applicable technical standards to address issues like product compatibility and customer safety. The development of technical standards within a consortium has a greater impact on the industrial universe compared to the development of new technologies. This is because standards have significant pro-competitive effects, such as facilitating price competition by making it easier to compare standardized products. Furthermore, nations leading the way in the development of technical standards enjoy a competitive edge, and globally, the adoption of particular standards is essential to maintaining competitiveness.

Consequently, as will be explained below, the governments of the United States and the European Union have been actively involved in monitoring, regulating, and advancing the process of defining technical standards.

We begin with US industrial strategies that were actively pursued in the 1980s and discuss how normalization and R&D promotion are related. In the early 1980s, there was a growing trend in industrial policy to support R&D. This trend was cemented with the implementation of the National Cooperative Research Act (NCRA) in 1984. Additionally, the President's Commission on Industrial Competitiveness's 1985 publication of the Young Report served as a metaphor for US industrial policies that supported research and development in order to foster innovation. According to this analysis, there are other factors at play in the US's diminishing global competitiveness besides the country's currency's appreciation. The report advised the public and private sectors to implement new initiatives to promote research and development (R&D) in order to fully address the upgrading of industry competitiveness. In 1993, the NCRA was redrafted as the National Law for Cooperation in Research and Production, which expanded the scope of research and permitted the promotion of production. As a result, the US adopted industrial policies that prioritized R&D through business consortiums.

It is common for standardization initiatives and R&D encouragement to be taken into different considerations. But in actuality, there is a close connection between these two phenomena. It is easy to see how these two events are related when antitrust laws are being discussed. Antitrust regulations hindered research and development in the US prior to 1980. Several businesses work together to produce technology through cooperative research. Participating businesses will, therefore, probably monopolize the technology. When businesses engage in unethical collaborations, it can inhibit or even completely eradicate competition, which raises antitrust issues. Because of this, antitrust laws were used to regulate R&D financing, and cooperative research that broke these regulations was subject to harsh penalties. Companies with a particular degree of market share collaborating to develop regulatory standards were regarded as having corporate behavior that might be distinct from that of other market participants.[1]

Nevertheless, the impact of antitrust laws was lessened following the introduction of new industrial strategies for research incentives in the 1980s, which involved multiple corporations. Even if several companies had a certain amount of market share, these guidelines said that they could come up with normative specifications through joint research and development as long as third parties could access the normative specifications (i.e., the results of the joint search). The new law made it feasible to create technological standards that have consequences akin to the formation of "de facto standards" through consensus standards, like agreements formed by groups of people before market competition really began.

Similar to the United States, the EU started promoting industry-wide standardization in the 1980s to restructure a worldwide competitive strategy. Furthermore, in 1992, the European Union (EU) sought to enhance trade harmonization inside the European Single Market and facilitate economic bloc integration, which prompted it to address industry standardization. In order to prevent regulations and specifications that could operate as trade barriers and be incompatible with each country's domestic markets, the EU's interconnected markets needed common standards.

The European Commission established a regulation known as No. 418/85 in 1984 that increased the antitrust protection of R&D. This regulation offers broad exemptions for horizontal R&D partnerships to businesses whose total market share is not above 20%. Additionally, the Commission may provide exemptions for cooperative efforts, such as research and development, manufacturing, and sales, in accordance with Article 85(3). Establishing a consortium to work together on the creation of any industry standard is a simple process for businesses.

[1] H. Hirabayashi, 'Antitrust law guidelines concerning joint research and development' [1993] Commercial Law Standardization, International Division of Labor, and Platform Business Center.

The European Commission (EC) released another significant decision in 1985, titled "A New Approach to Technical Harmonization and the Use of Standardization." By the end of 1992, the European Commission had suggested reworking processes and procedures in order to strengthen its internal market. At the time, national standards were established independently in each nation, but universal standardization throughout the region was required. Because of this, standardization efforts have been promoted throughout Europe in the unique form of regional standards, or CEN, CENELEC, and ETSI, comprising national and international standards. The European Commission (EC) pushed European firms to align their processes and products with new regional standards based on this new methodology. This process sped up the restructuring of industrial capacities and the establishment of regional standardization. In Europe, standardization has advanced toward the goal of achieving the integration of regional economies. Through this approach, European businesses constructed their business models that leveraged standardization as the cornerstone for corporate development and improved their capacity to use standardization as a strategic instrument.

The literature provides numerous examples of the advantages of standard development as a tactic for creating new firms and technologies.[2] For instance, GSM was the most widely used mobile phone communication technology worldwide in the 2010s. One successful standard that was adopted through the new European standardization procedure is the GSM standard. The European Conference of Postal and Telecommunications Administrations (CEPT) GSM Working Group started developing standards for telecommunications in 1982, but it was found that this procedure took a long time.

The European Telecommunications Standards Institute (ETSI) replaced the former CEPT as the body in charge of developing standards when it was founded in 1988. This action signaled a significant shift in protocol. The people in charge of telecommunications policy and the postal and telegraph services (Postal Telecom and Telegraph, or PPT) in each nation took the lead in developing the standards independently prior to the creation of ETSI in the CEPT standardization process. Nevertheless, consensus standardization was followed with the involvement of telecom equipment and installation businesses, in addition to the officially

[2] H. Tatsumoto, K. Ogawa and J. Shintaku, 'Strategic standardization: Platform business and the effect on the international division of labor' [2011] 10 Annals of Business Administrative Science ISSN 1347-4456.

responsible agents previously involved, once the standardized process was moved to ETSI. The process was accelerated, and more useful consensus standards for the market were produced as a result of this inclusion and method expansion.

The telecom equipment and installation sector in Europe trailed well behind its counterparts in the US and Japanese markets in terms of the analog cellular phones that were popular at the time. During the process of standardizing digital phones, Europe was concerned that the foreign telecom sectors would get access to a sizable integrated market that would arise from normalization. As a result, telecommunications equipment and installation businesses actively participated in the standardized process for developing new models, in contrast to traditional standardization processes. As a result, a standardized process was created, enabling the European telecom sector to compete on the world market.

The telecommunications equipment and installation business played a significant role in the completion of the GSM technology standards. The GSM system's standards and criteria, for instance, omit certain features in areas like base station-base station controller interfaces. Utilizing their technical know-how, telecom equipment manufacturers can more easily set themselves apart in non-standard domains. On the other hand, because international roaming is necessary, there are now stricter standards in areas like billing protocols that rely on the knowledge and skills of operators. This made it possible for operators with fewer technological resources to adopt and run a GSM system. By implementing the basic GSM systems offered by European telecom equipment firms, even new operators—particularly those in developing nations who are unfamiliar with GSM technology—can quickly begin offering their services. GSM therefore made it easier for European producers of telecom gear, like Ericsson and Nokia, to enter international markets.

Europe's and America's industrial policies have changed, as was previously mentioned. Research projects and regional economic cooperation were approved alongside consensus normalization. Many businesses use initiatives like consortia to forge agreements and develop industry-wide standards in a flexible manner when it comes to consensus standardization. The new kind of normalization is notable in the context of business plans because it allows organizations to develop a set of consensus requirements that align with their own business models and strategies. This is due to the fact that consensus normalization, which also becomes a powerful tool for strategies, has a significant impact on corporate rivalry situations.

2.2 Standardization as a Business Platform

We can start by thinking about "open areas," which provide access to information about a particular process, product, or technology, and "closed areas," which protect or restrict knowledge and related technologies. Because these areas haven't been normalized, it should be noted that the extra value doesn't always converge for all "closed areas." When a technological architecture is opened and a "closed area" stays the same, it only realizes the added value it had prior to the opening of the technological architecture. Following the conversion of some technical architecture components into "open areas" through standardization, modifications are required in order to gather value-added that has been scattered throughout the "open area" and place it in the "closed area." While being a "closed area" is necessary, it is not a sufficient criterion for value addition.

Standardization distributes the "open area's" extra value. Reorganizing the technical architecture is required to gather and obtain distributed added value in the "closed area" that has not been standardized. The term "strategic positioning of standardization" refers to this approach.[3] The most important normalizing positioning strategy is the trading platform. A platform leader is a business that has a business platform.[4] As a platform for producing the final product, the platform leader provides its components. The platform has standardized interfaces and consists of a central component along with additional peripherals. The platform leader intentionally uses standardization to create a global industrial ecosystem.

For instance, Intel spearheaded the effort to standardize each PC interface in the middle of the 1990s. Simultaneously, Intel combined peripheral circuits as a collection of parts to create a platform, concentrating on the CPU. It then made this platform widely available to Taiwanese laptop and motherboard manufacturers. Thus, Intel not only democratized technology by standardizing processes, but also changed the nature of its business from manufacturing CPUs alone to manufacturing platforms that include component sets (chipsets). So, rather than offering these platforms to already-established businesses, it chose to do

[3] H. Tatsumoto and C. Takanashi, 'Competition for consensus standards' in J. Shintaku and M. Eto, *Consensus Standard Strategies: All About Business Utilization* (Nikkei Publishing, 2008).

[4] A. Gawer and M. Cusumano 'Platform leadership: how Intel, Microsoft, and Cisco drive industry innovation' [2002] Harvard Business School Press.

so for emerging-market businesses. As a result, Intel was able to gain from the value addition that standardization brought about in the larger product market.[5]

Comparable instances can also be found in the DVD player industry. In the mid-1990s, DVD players' primary component, the autonomous optical pickup (OPU), was standardized, and Sanyo Electric adopted this technology. Sanyo Electric established a company to offer this integrated platform by integrating components including mattresses, base decks, and actuators. This platform was primarily made available to up-and-coming businesses in nations like China by Sanyo Electric. Like Intel, Sanyo Electric restructured its OPU division following normalization, which allowed it to standardize and get more added value in the larger product market.

The fundamental idea of the platform—controlling the "open area" from the perspective of the "closed area"—is illustrated by these corporate actions.[6] As was previously said, it is simpler to distinguish between the "closed area" and the "open areas." Nevertheless, the "closed area" is not a sustainable source of profit if it has no control over the "open area." To gain control over the "open area," a firm creates a platform to combine a central component with additional peripherals and creates a unique business model that captures value-added distributed throughout the "open area."

Intel gained control over motherboards and notebooks by managing a platform that included peripherals, or the "chipset," in addition to the CPU, which is the main component. Using Intel's platform, numerous Taiwanese start-ups have effectively entered the market and rapidly increased their output. Here is an example of how knowledge integration into the platform helps new businesses enter the "open area." Even businesses with no technological experience, like startup enterprises in Taiwan, were able to produce goods by utilizing the platform that Intel offered.

The aforementioned examples can be examined from the perspective of the financial gain connected to technological advancement. The two main components of controlling the "open area" from the "closed area" are partner company selection and exclusive ownership of the "closed area."

[5] H. Tatsumoto, K. Ogawa and T. Fujimoto, 'The effect of technological platforms on the international division of labor: A case study on Intel's platform business in the PC industry' in A. Gawer, *Platforms, Markets, and Innovation* (Edward Elgar, 2009).

[6] K. Ogawa 'Platform formation mechanism in the electronics industry in Japan' [2008] 7(6) Akamon Management Review.

As a domain of the company's business, exclusive ownership of the "closed area" serves to bar any other companies from entering the area of specific and protected information. Exclusive control over the "closed area" is largely dependent on implicit knowledge and intellectual property.[7] To increase the implicit area of knowledge, Intel and Sanyo Electric, in the aforementioned cases, have integrated peripheral areas in addition to the main components. Explicit information is the initial form of interface information between the main and peripheral components. Nevertheless, this explicit interface information becomes implicit when peripheral components are incorporated into primary components. In addition, there is interdependence among the pieces following integration. Additionally, the platform provider provides implicit data that helps gather this dependency information. Another company can only get this type of implicit knowledge if it pays a significant fee to the company that provides the platform. Through the integration of core and peripheral components, platform providers can create an entrance barrier into the "closed area".

Similarly, patents and other intellectual property amassed on the platform act as a barrier to entry for new businesses. For instance, Qualcomm holds a number of crucial patents and has significantly contributed to the development of the CDMA digital mobile phone system. Qualcomm has created integrated circuits that incorporate the baseband, power management, and encoding processing necessary for the CDMA technique. This integrated circuit is provided as a platform by Qualcomm. The platform that is constructed using Qualcomm's "chipset" is home to numerous important patents and has effectively blocked other businesses from competing to provide CDMA System chipsets. Otherwise, the likelihood of proprietary ownership has significantly increased due to intellectual property.

To choose partner companies in the "open area," demand for this kind of well-established platform must be generated. Put differently, the platform building business needs to locate a new partner company in the "open area" that will utilize the platform that the platform leader is providing. Rather than being established businesses, partner companies in the "open area" are typically start-ups, notably those in developing nations.

[7] D. Teece, 'Profiting from technological innovation: Implications for integration, collaboration, licensing, and public policy' [1986] 15 Research Policy.

The rationale behind this conduct takes into account the fact that businesses with advanced technology typically don't use a platform. Companies with less technological experience, on the other hand, tend to adopt platforms that are already accessible. Businesses with advanced technology dislike having a platform provided to them. Ingenious businesses typically detest black-box components—like platforms—that are rich in implicit knowledge and will not utilize platforms offered by other businesses. This is because they run the danger of losing their competitive advantages if they are ignorant of the contents or components. New businesses, on the other hand, lack this notion because they have not accumulated much technology. These businesses view the platform as a practical means of utilizing cutting-edge technology. Because of this, young businesses with less technological experience frequently join forces with platform companies operating in the "open area." Businesses in developing nations typically take full advantage of this chance.

Furthermore, one could argue that the "open area" has little profitability. Because it is simpler to enter an open market than a closed one, profitability is frequently poor. Because they can handle poor profitability through low-cost operations, new enterprises in emerging countries are typically the platform provider's key partners. Due to their low operating costs, businesses in emerging nations typically serve as the primary platform partners. Additionally, institutional promotion strategies are put into place in a lot of developing nations to make platforms easier to access,[8] which keeps operations viable. Because of these factors, alliances between businesses in developed and developing nations were formed via the adoption of platforms in a variety of industries, including CDMA, PCs, and DVDs.

2.3 *Functions of Standardization in Technological Development*

The "innovation systems functions approach" is employed as a heuristic framework to comprehensively and holistically analyze the roles and functions of standardization in technological innovation. Walz[9] utilized this

[8] H. Tatsumoto, 'The impact of country-specific advantages on global competitiveness: A case study on taxation in the semiconductor industry' [2009] 2(2) International Business Studies.

[9] R. Walz, 'The role of regulation for sustainable infrastructure innovations: the case of wind energy' [2007] 2 International Journal of Public Policy.

approach to analyze the impact of regulation on the many functions of innovation systems. This method establishes strong links between the various roles of standardization in innovation. This section introduces the concept of the function approach, followed by a discussion on certain functions that are particularly relevant to standardization.

2.4 Approach to the Functions of Innovation Systems

Roles, also known as "determinants of innovations," refer to the specific actions and variables that have a significant impact on the creation, dissemination, and application of innovations.[10] By systematically identifying and mapping out these roles, one can focus on understanding the dynamics of what is actually happening in the system and gain insights into how different components and forces interact to shape the complex and progressive development of innovation systems.[11] Moreover, unlike other approaches to studying innovation systems that focus on specific regions or sectors, the functions approach provides a more consistent perspective for making meaningful comparisons across countries and technological domains because it emphasizes the key activities carried out within the overall innovation system.

Numerous theoretical and empirical investigations were conducted in order to pinpoint a set of primary roles that impact the functioning of innovation systems. These roles are connected to various problems and actions that occur throughout the innovation process.[12,13] After reviewing these earlier studies, Bergek and others[14] suggest the following list of seven functions, which are then further refined as follows and combined with other approaches:

[10] C. Edquist, 'The systems of innovation approach and innovation policy: An account of the state of the art' [2001] DRUID Conference.

[11] M. P. Hekkert and others, 'Functions of innovation systems: A new approach for analyzing technological change' [2007] 74(4) Technological Forecasting and Social Change.

[12] C. Edquist and B. Johnson, 'Institutions and organizations in systems of innovation' in C. Edquist, *Systems of Innovation: Technologies, Institutions, and Organizations* (Pinter, 1997).

[13] R. Smits and S. Kuhlmann, S., 'The rise of systemic instruments in innovation policy' [2004] 1 International Journal of Foresight and Innovation Policy.

[14] A. Bergek and others, 'Analyzing the functional dynamics of technological innovation systems: A scheme of analysis' [2008] 37(3) Research Policy.

a. The process of creating new information from the innovation system's existing knowledge base and spreading it to other activities is known as knowledge development and dissemination. We will refer to this process as "new knowledge creation and transfer" to avoid misunderstanding the broader role that innovation plays in influencing knowledge development and dispersion.
b. Mechanisms that affect the path of research within the innovation system between various rival technologies, applications, markets, business models, etc. are included in the category of influence on research direction.
c. Investigating novel applications, markets, and technology is a key component of entrepreneurial experimentation as it fosters social learning and lowers uncertainty around the development of innovation systems.
d. Market creation encompasses actions that facilitate the market's progression, starting with the opening of a learning space in the nursing market, increasing volume in the bridge market, and ending with the mass market.
e. In order for the new technology to be deemed acceptable and desirable by the relevant players, resources must be raised, demand must be created, and actors must gain political clout. This is accomplished through social acceptability and compliance with the appropriate institutions.
f. The mobilization of necessary resources, such as expertise, financial capital, human capital, and complementary assets, is a component of resource mobilization.
g. The creation of positive external economies through political power, information and knowledge flows, pooled labor markets, combinatorial opportunities, and uncertainty resolution is a necessary step in the development of positive externalities.

2.5 Functions of Standardization in Supporting Technological Development

We may investigate normalization's role in and interactions with the overall innovation system from a more comprehensive and integrated standpoint by examining how normalization is viewed in the literature that uses the function method. You can also look at other writings that talk about normalization in the context of innovation. These writings are in line with the functions approach, which lets you see how the different roles and functions of normalization help innovation systems.

It can be seen that standardization generally gives innovation systems a generic role that affects the creation and transfer of knowledge. Standards publications facilitate the dissemination of technical knowledge created at different stages of innovation systems by coding and transferring state-of-the-art technology and best practices among a range of actors.[15,16] Furthermore, the process of normalization itself advances knowledge development by offering a platform for group cognitive processes where actors from diverse backgrounds—researchers, business, government, and users—come together to generate novel concepts and solutions.[17,18,19] The standardization process, which incorporates conversations from a range of participants, is also a useful avenue for the transfer and dissemination of information since it allows them to exchange cutting-edge research and best practices based on their experiences.[20,21,22]

The five purposes of innovation systems that are expressly supported by standardization are legitimacy, influence on the direction of the research,

[15] R. H. Allen and R. D. Sriram, 'The role of standards in innovation' [2000] 64 Technological Forecasting and Social Change.

[16] G. M. P. Swann, 'The Economics of Standardization: An Update' [2010], https://assets.publishing.service.gov.uk/government/uploads/system/uploads/attachment_data/file/32444/10-1135-economics-of-standardization-update.pdf.

[17] W. Bergholz, B. Weiss and C. Lee, 'Benefits of standardization in the microelectronics industry and their implications for nanotechnology and other innovative industries' [2006], http://www.iec.ch/about/globalreach/academia/pdf/bergholz-1.pdf.

[18] K. Blind, 'Standardization as a catalyst for innovation' [2009] Inaugural Address: Research in Management Series, http://papers.ssrn.com/sol3/papers.cfm?abstract_id=1527333.

[19] O. Hogan, C. Sheehy, and R. Jayasuriya, 'The economic contribution of standards to the UK economy' [2015].

[20] OCST, 'Voluntary industry standards and their relationship to government programs' [1993], http://www.strategicstandards.com/files/GovernmentStandards.pdf.

[21] G. Tassey, 'Standardization in technology-based markets' [2000] 29(4) Research Policy.

[22] CIE, 'Standards and the economy' [2006], http://www.thecie.com.au/content/publications/Economic_impact_of_standards.pdf.

development of positive externalities, creation, and transfer of new knowledge and entrepreneurial experimentation. These functions have been covered in a number of academic and practical literatures.

Legitimacy is translated into reducing uncertainty by providing the necessary information and increasing social acceptance and managing conflicts by literature on innovation systems, while in other literature on normalization is described as encouraging innovators to participate in innovation by reducing future uncertainty and increasing consumer confidence, facilitating market access and enabling subsequent innovation.

Influence on the direction of the research is perceived in literature on innovation systems as performance specifications and criteria to guide the direction of learning and powerful mechanisms for selecting dominant designs or specific technology. Other literature on normalization recognizes as quality and performance standards to provide guidance on how to achieve a goal, articulating demands from leading clients and selection and prioritization of certain technologies.

Development of positive externalities are presented in literature on innovation systems as increasing attractiveness to customers through network effects and economy of scale savings, while on other literature on normalization it is seen as variety reduction standards for economies of scale and compatibility and interface standards to generate positive network effects.

Creation and transfer of new knowledge is mainly present on other literature on normalization as providing an essential platform on which new innovations can emerge, as a measurement and testing standards as an infrastructure system to support R&D cooperation and knowledge transfer channel codifying R&D results and facilitating efficient communication.

Entrepreneurial experimentation is mainly present on other literature on normalization as promoting competition and leveling the playing field, quality, and performance standards to improve skills building, and interoperability standards to promote product integration and their interactions.

2.6 Incentive Policies for Standardization

Numerous experts and policy studies attest to the growing recognition of the significance of standardization and the need for a methodical, forward-looking approach to it in order to foster technological progress.

Strong frameworks and forecasting procedures, in particular, have the potential to be valuable tools for standards organizations and politicians in predicting future standards demands and connecting them to research and other innovation efforts within a larger strategic framework.[23,24] In order to effectively anticipate and manage standardization, numerous attempts have been made to conduct such strategic, forward-looking evaluations in practice, particularly in areas of new technology.

For instance, to coordinate standardization efforts in a number of domains with opportunities related to technical standards, such as additive manufacturing, cloud computing, and smart grid, the US National Institute of Standards and Technology (NIST) has created a number of conceptual frameworks and roadmaps.[25,26,27] The roadmapping approach is used by German standardization bodies to predict future standardization requirements in a range of interdisciplinary fields, including intelligent manufacturing and electric vehicles.[28,29,30] Additionally, a standardization roadmap for additive manufacturing has been developed, with an emphasis on identifying the technical standards that are required

[23] European Commission, 'A strategic vision for European standards: Moving forward to enhance and accelerate the sustainable growth of the European economy by 2020' [2011], http://eur-lex.europa.eu/LexUriServ/LexUriServ.do?uri=COM:2011:0311:FIN:EN:PDF.

[24] F. Scapolo and others, 'How will standards facilitate new production systems in the context of EU innovation and competitiveness in 2025?' [2014], https://ec.europa.eu/jrc/sites/default/files/jrc-foresight-study-web_en.pdf.

[25] NIST, 'NIST framework and roadmap for smart grid interoperability standards, release 1.0' [2010], http://www.nist.gov/public_affairs/releases/upload/smartgrid_interoperability_final.pdf.

[26] M. Hogan and others, 'NIST cloud computing standards roadmap, version 1.0' [2011], http://www.nist.gov/itl/cloud/upload/NIST_SP-500-291_Jul5A.pdf.

[27] NIST, 'Measurement science roadmap for metal-based additive manufacturing' [2013], https://www.nist.gov/system/files/documents/el/isd/NISTAdd_Mfg_Report_FINAL-2.pdf.

[28] NPE, 'The German standardization roadmap for electromobility, version 3.0' [2012], https://www.din.de/resource/blob/77456/4407b11ea3d0ad1ac1a8889eab8e6b15/nr-3-0-en-data.pdf.

[29] DKE, 'The German standardization roadmap Industrie 4.0' [2014], https://www.din.de/resource/blob/65354/1bed7e8d800cd4712d7d1786584a7a3a/roadmap-i4-0-e-data.pdf.

[30] SASAM, 'Additive manufacturing: Sasam standardization roadmap' [2014], The European Technology Platform Manufuture.

and formulating development strategies to support this industry through standardization management. Strategic standardization is given particular consideration in the EU project "Towards a European Strategy in Synthetic Biology" (TESSY) as an efficient policy support for a developing field of synthetic biology.[31]

Based on the analysis of current procedures, it seems that one of the most extensively used policy instruments for standardization forecasting and strategic management is the standardization roadmap. It offers organized, forward-looking viewpoints for organizing and creating standardization plans in a timely and relevant way. Many nations' standards bodies have created standardized roadmaps in a variety of technological and industrial fields where efficient standardization management is crucial for the strategic development of the country, particularly in newly emerging multidisciplinary fields. It is believed that roadmap activities result in more coordinated and harmonized standardization efforts. They also offer a forum for interaction and cooperation among many stakeholders, which is necessary for efficient and timely standardization.

Even though they are becoming more and more popular, policymakers and standards bodies still face significant obstacles when attempting to implement systematic forecasting exercises for standardization, which includes implementing roadmaps. Actually, from the standpoint of public policy, there hasn't been much scholarly writing on the subject of anticipating or forecasting the strategic management of normalization. The work of Goluchowicz and Blind,[32] who describe their experiences using Delphi studies to propose potential future domains of normalization, is a noteworthy exception. So, there is a lack of a thorough and methodical look at the different factors that need to be taken into account in these forecasting exercises in order to figure out what normalization needs will be in the future and come up with the right strategies.

[31] TESSY, 'Final report: TESSY achievements and future perspectives in synthetic biology' [2008].

[32] K. Goluchowicz and K. Blind, 'Identification of future fields of standardization: An explorative application of the Delphi methodology' [2011] 78(9) Technological Forecasting and Social Change.

3 CCUS Standardization

The inconsistent use of acronyms, technical abbreviations, and terminology adds to the technical complexity of CCUS. This uncertainty can impair public perception and communication, which can affect how policies are formulated and how projects are financed and resourced. In order to improve transparency in conversations concerning CCUS solutions, stakeholders must engage in objective, knowledge-based discourse. This requires accurate and consistent communication. Standardized consensus is required for more than just effective communication; it also applies to technical aspects. A lot of technical solutions, from GHG streams to CCU-based products, depend on cooperation and component interchange across stakeholders, who have to operate within reasonable specification limits. A unified foundation must be established for accounting for carbon emissions, life cycle evaluation, and technical–economic assessment—all crucial for calculating the practical outcomes of applying various CCUS solutions. Standardization for CCUS is essential in this situation because it removes transactional obstacles and spreads technical knowledge, enabling everything from project execution to the creation of public policy.[33,34,35]

3.1 Main CCUS Technical Issues for Standardization

Carbon capture, utilization, and storage (CCUS) is a suite of technologies aimed at reducing carbon dioxide (CO_2) emissions from industrial processes and power generation. CCUS technical aspects involve the four main steps of the CCUS process, namely capture, transportation, utilization, and storage. Two other technical are also relevant for CCUS: monitoring and verification; and safety and environment. The main technical issues for standardization considering the six previously presented aspects from CCUS are the following:

[33] R. Neerup and others, 'A call for standards in the CO_2 value chain' [2022] Environmental Science and Technology, https://doi.org/10.1021/acs.est.2c08119.

[34] B. Olfe-Krautlein and others, 'Why terminology matters for successful rollout of carbon dioxide utilization technologies' [2022] 4 Frontiers in Climate, https://doi.org/10.3389/fclim.2022.830660.

[35] V. Sick and others, 'The need for and path to harmonized life cycle assessment and techno-economic assessment for carbon dioxide capture and utilization' [2020] 8 Energy Technology, https://doi.org/10.1002/ente.201901034.

1.1. Capture Technologies:

1.1.1. Performance Metrics: Standardize metrics for capture efficiency, energy consumption, and operational reliability across different capture technologies (post-combustion, pre-combustion, and oxyfuel combustion).

1.1.2. Design and Operation: Establish guidelines for the design, operation, and maintenance of capture facilities to ensure optimal performance and safety.

1.2. Transportation:

1.2.1. Pipeline Specifications: Standardize the design, materials, construction, and maintenance practices for CO_2 pipelines to ensure safety and efficiency.

1.2.2. Transport Methods: Develop standards for CO_2 transport via ships and trucks, including container specifications, handling procedures, and safety protocols.

1.3. Storage:

1.3.1. Site Selection Criteria: Define criteria for selecting suitable geological storage sites, including geological stability, capacity, and proximity to CO_2 sources.

1.3.2. Injection Protocols: Standardize CO_2 injection methods, pressures, and rates to ensure safe and effective storage.

1.3.3. Monitoring and Verification: Establish uniform methods for monitoring and verifying stored CO_2, including seismic imaging, pressure monitoring, and geochemical analysis to detect leaks and ensure long-term containment.

1.4. Monitoring and Verification:

1.4.1. Monitoring Techniques: Standardize the use of technologies and methodologies for continuous monitoring of storage sites.

1.4.2. Data Reporting: Create standardized protocols for data collection, reporting, and sharing to ensure transparency and facilitate regulatory compliance.

1.5. CO_2 Utilization:

1.5.1. Product Quality: Develop standards for products derived from CO_2 utilization, such as chemicals, fuels, and building materials, to ensure consistent quality and safety.
1.5.2. Process Efficiency: Establish benchmarks for the efficiency and environmental impact of CO_2 utilization processes.

1.6. Safety and Environmental Standards:

1.6.1. Risk Assessment: Standardize risk assessment procedures for CCUS projects to identify and mitigate potential hazards.
1.6.2. Environmental Impact: Develop guidelines for assessing and minimizing the environmental impact of CCUS operations, including potential impacts on air, water, and soil quality.
1.6.3. GHG accounting: Develop guidelines for assessing and quantification of CO_2 captured and CO_2 stored either in geological formations (CCS) or in a product (CCU).

By standardizing these technical issues, the CCUS industry can achieve greater efficiency, safety, and public acceptance, thereby accelerating the deployment of these critical technologies for mitigating climate change. In the next section is discussed where these technical issues could be addressed by international standardization forums.

3.2 Forums for CCUS standardization

The first reference as an international forum for CCUS standardization is ISO/TC 265, Carbon Dioxide Capture, Transportation, and Geological Storage. The ISO/TC 265 was created in 2011 with the clear scope of dealing with the standardization of design, construction, operation, planning and environmental management, risk management, quantification, monitoring and verification, and related activities in the field of capture, transport, and storage of carbon dioxide (CCS).

The initial structure contemplated the creation of six working groups (WGs) to focus on specific CCS themes: Capture (WG1), Transportation (WG2), Storage (WG3), Quantification and verification (WG4), Cross-cutting issues (WG5), and EOR issues (WG6). This work structure contemplating these WGs was responsible for publishing 13 standards between the years of 2016 and 2024. All published standards primarily cover the issues of carbon capture, transport, and storage in geological formations.

Since 2018, the possibility of considering aspects of carbon use has emerged in the discussions of ISO/TC 265, which would be underestimated despite the great potential interest in mitigating the effects of climate change. An ad hoc group created under ISO/TC 265 has assessed some CCU possibilities and drawn some conclusions:

- CO_2 geo-utilization has a high potential for CO_2 emissions reduction, and other types of enhanced hydrocarbon and water recovery should be discussed.
- CO_2 to mineralization has a high potential for CO_2 emissions reduction but needs a business case and assessment of CO_2 emissions reduction.
- CO_2 for chemicals, fuels, and bioconversion market size is too small for a high impact on CO_2 emissions savings and needs a LCA of all environmental impacts, particularly due to high energy consumption, where renewable energy is essential.

This ad hoc group also identified some conclusions regarding standardization procedures:

- CO_2-based products belong to other technical committees (e.g., product-specific standards).
- There is a need for harmonized guidelines to evaluate GHG emission savings since the work of ISO/TC 207, Environmental Management, is too generic for CCU.
- There is a need for a gap assessment between other technical committees and ISO/TC 265 for the need for standards for CCU and CCS.

Therefore, a second relevant standardization forum emerges, the ISO/TC 207, more specifically its subcommittee SC 7, Greenhouse gas and climate change management and related activities. Despite being more focused on developing more comprehensive standards, their standards are essential references to be considered in CCUS standardization.

The conclusions also refer to other technical committees for product-specific standards, which is a specificity of CCU standardization. Each

CCU route, as depicted in the previous section, might need standards to specify their products, their production, and/or application.

For instance, ISO/TC 61, Plastics, in its subcommittee SC 14, Environmental Aspects, has worked on standards for biobased plastics. It was defined and specified a way to measure the amount (mass) of CO_2 taken from the air and put into plastic (ISO 22526). Other examples encompass hard-to-abate industry sectors such as Iron and Steel, where ISO/TC 17, Steel, is developing standards for calculating carbon emission intensity; and Concrete and Cement, where CEN/TC 104, Concrete, has developed a technical report to assist on the carbonation and carbon dioxide (CO_2) uptake in concrete (CEN/TR 17310).

Besides product-specific standards, CCU standardization also needs cross-cutting standards, connecting relevant subjects from different references. This type of standard might fit well within ISO/TC 265, as discussed by Gallo and others.[36] This approach has been taken by the Brazilian standardization body (ABNT) when creating a committee to CCUS (*Captura, Transporte, Utilização e Armazenamento de Dióxido de Carbono*) in April 2022 and, more recently, by the European standardization body (CEN) when creating the new CEN/TC 474, Carbon dioxide Capture, Transportation, Utilization, and Storage, in June 2024.[37]

CEN/TC 474 will work on existing ISO/TC 265 standards, but it will also consider new documents customized to the European stakeholders needs and experiences. This new European TC has currently six working groups (WGs) to focus on specific CCUS themes: Pipeline Transportation (WG1), CO_2 Stream Composition and Quality (WG2), Non-pipeline Transportation (e.g., Shipping, Rail, Trucking) (WG3), Utilization (WG4), Storage (WG5), and CO_2 Accounting (WG6). From this work structure, WG1, WG2, WG5, and WG6 have already started development of new working item proposals (NWIP), and some of them are to be balloted soon.

[36] A. Gallo and others, 'Towards the International standardization of carbon dioxide capture, Transportation, Utilization and Storage (CCUS) technologies: Current challenges and future directions' [2021] 12 Journal of Sustainable Development Law and Policy, https://doi.org/10.4314/jsdlp.v12i2.2.

[37] CEN, 'Kick-off CEN/TC 474. Carbon dioxide capture, transportation, utilisation and storage. Presentation at ISO/TC 265. Jakarta, Indonesia' [2024].

Finally, forums for CCUS standardization mentioned throughout this chapter are summarized in Table 1, while some of their CCUS relevant standards are reported in Table 2.

4 Lessons Learned and Conclusions

One of the key lessons learned is the critical role of standardization in the successful deployment of CCUS technologies. Standardized metrics, guidelines, and protocols are essential to ensure efficiency, safety, and public acceptance. Standardization facilitates transparency, comparability, and regulatory compliance, which are crucial for scaling up CCUS projects and integrating them into broader climate strategies.

Another important lesson is the technical complexity introduced by the inconsistent use of technical abbreviations and terminology. This

Table 1 Forums for CCUS standardization[38]

Std. body	*Forum*	*Title*	*Main CCUS-related topics*
ISO	ISO/TC 265	Carbon dioxide capture, transportation, and geological storage	Capture, transportation, storage, quantification, EOR
ISO	ISO/TC 207/SC 7	Environmental management—Greenhouse gas and climate change management and related activities	GHG accounting
ISO	ISO/TC 17	Steel	CCU accounting
ISO	ISO/TC 61	Plastics	CCU accounting
CEN	CEN/TC 474	Carbon dioxide capture, transportation, utilization and storage	Capture, transportation, utilization, storage, accounting
CEN	CEN/TC 104	Concrete	CCU accounting
ABNT	ABNT/CB-038/ CE-038:007.002	Captura, Transporte, Utilização e Armazenamento de Dióxido de Carbono	Capture, transportation, utilization, storage

[38] Non-exhaustive list based on findings presented in this chapter.

Table 2 CCUS-related standards[39]

Forum	*Identifier*	*Title*	*CCUS-related topic*
ISO TC 265	ISO/TR 27912:2016	Carbon dioxide capture—Carbon dioxide capture systems, technologies, and processes	Technology overview
ISO TC 265	ISO 27913:2016	Carbon dioxide capture, transportation, and geological storage—Pipeline transportation systems	Transportation
ISO TC 265	ISO 27,914:2017	Carbon dioxide capture, transportation, and geological storage—Geological storage	Storage
ISO TC 265	ISO/TR 27915:2017	Carbon dioxide capture, transportation, and geological storage—Quantification and verification	Quantification
ISO TC 265	ISO 27916:2019	Carbon dioxide capture, transportation, and geological storage—Carbon dioxide storage using enhanced oil recovery (CO_2-EOR)	EOR
ISO TC 265	ISO 27917:2017	Carbon dioxide capture, transportation, and geological storage—Vocabulary—Cross-cutting terms	CCS terminology
ISO TC 265	ISO/TR 27918:2018	Lifecycle risk management for integrated CCS projects	Risk management
ISO TC 265	ISO 27919-1:2018	Carbon dioxide capture—Part 1: Performance evaluation methods for post-combustion CO_2 capture integrated with a power plant	Capture
ISO TC 265	ISO 27919-2:2021	Carbon dioxide capture—Part 2: Evaluation procedure to assure and maintain stable performance of post-combustion CO_2 capture plant integrated with a power plant	Capture
ISO TC 265	ISO/TR 27921:2020	Carbon dioxide capture, transportation, and geological storage—Cross-Cutting Issues—CO_2 stream composition	CO_2 stream composition
ISO TC 265	ISO/TR 27922:2021	Carbon dioxide capture—Overview of carbon dioxide capture technologies in the cement industry	Specific sector CCS details

(continued)

[39] Non-exhaustive list based on findings presented in this chapter.

Table 2 (continued)

Forum	*Identifier*	*Title*	*CCUS-related topic*
ISO TC 265	ISO/TR 27923:2022	Carbon dioxide capture, transportation, and geological storage—Injection operations, infrastructure, and monitoring	Storage
ISO TC 265	ISO/TR 27925:2023	Carbon dioxide capture, transportation, and geological storage—Cross-cutting issues—Flow assurance	Storage
ISO TC 265	ISO/DTR 27926	Carbon dioxide capture, transportation, and geological storage—Carbon dioxide enhanced oil recovery (CO_2-EOR)—Transitioning from EOR to storage	EOR/storage
ISO TC 265	ISO/DIS 27927	Carbon dioxide capture—Key performance parameters and characterization methods of absorption liquids for post-combustion CO_2 capture	Capture
ISO TC 265	ISO/CD 27928	Carbon dioxide capture, transportation, and geological storage—Performance evaluation methods for CO_2 capture plants connected with CO_2 intensive plants	Capture
ISO TC 265	ISO/DTR 27929	Transportation of CO_2 by ship	Transportation
CEN TC 104	CEN/TR 17310:2019	Carbonation and CO_2 uptake in concrete	CCU accounting
DIN	DIN SPEC 91508	Requirements for the life cycle assessment of CCU- and CCS-technologies	CCUS accounting
DIN	DIN SPEC 91458	Utilization of carbon dioxide—Terms	CCU terminology
ABNT CE 038:007.002	Project under development	Capture and utilization of carbon dioxide (CCU)—Part 1—CO_2 utilization routes (*Captura e utilização de dióxido de carbono (CCU)—Parte 1 - Rotas de utilização de CO_2*)	CCU terminology/ characterization

inconsistency can impair communication, public perception, and policy formulation. Establishing a unified foundation for technical terms, performance metrics, and operational guidelines is therefore vital to streamline processes and enhance understanding among stakeholders.

Carbon capture and utilization (CCU) presents unique standardization challenges that differ from those of carbon capture and storage (CCS). Each CCU route might require specific standards for product quality, process efficiency, and environmental impact assessment. Addressing these challenges necessitates a concerted effort from various technical committees and standardization bodies.

International cooperation through standardization forums such as ISO/TC 265 and CEN/TC 474 is instrumental in developing and harmonizing CCUS standards. These forums provide a platform for discussing and addressing standardization issues, ensuring that CCUS technologies can be effectively integrated into global efforts to combat climate change.

By learning from these lessons and focusing on comprehensive standardization, the CCUS industry can overcome current barriers and accelerate the deployment of technologies essential for achieving climate goals.

CHAPTER 10

CCUS in Australia

Tina Soliman Hunter

1 Introduction

As a petroleum producing nation, to date Australia has relied on both oil and gas, but particularly gas as a fuel for both generation of electricity and the production of heat for industry. As Australia becomes part of the global effort to reduce mean temperature below 2 °C, and even restrict to 1 °C, there is a continual focus on the use of gas as a fuel. Recognising that Australia has ample reserves of reserves guess in the central and western parts of the country, although there is a shortage on the East Coast, there is a debate about whether Australia should use the natural gas for domestic energy and heat use, and to also export the gas as liquefied natural gas (LNG) to Asian states.

The importance of gas to Australian industry and electricity, as well as the export market is enormous. At present natural gas comprises around

T. Soliman Hunter (✉)
Transforming Energy and Markets Research Centre, Macquarie University, Sydney, NSW, Australia
e-mail: tina.solimanhunter@mq.edu.au

E. G. Pereira et al. (eds.), *Carbon Capture Utilization and Storage*, Sustainable Development Goals Series,
https://doi.org/10.1007/978-3-031-81272-9_10

27% of energy consumption,[1] with coal and oil comprising an additional 64%.[2] As Australia seeks to wean itself off oil and coal, there is a strong likelihood that there will be a shift to natural gas use, at least in the interim. Although natural gas emits significantly less carbon dioxide then oil and coal, it is nevertheless CO_2 intensive, requiring attention in removing such CO_2 before it reaches the atmosphere.

Carbon capture, utilisation, and storage (CCUS) is of enormous interest as a tool in reducing CO_2 emissions and assisting a country in meeting its nationally determined contribution under the Paris agreement. Indeed, the IEA notes that (CCUS) as critical for the energy transition, since it is the only group of technologies that contributes to both reducing emissions in key sectors directly and removing CO_2 to balance emissions in 'hard to abate' sectors such as concrete and mining.[3]

Australia has taken the role of technology to heart, with the 2020 Low Emissions Technology Statement providing a strong indication that Australia will be utilising 'technology not taxes' in its quest for net zero by 2050.[4] As such, there is a strong reliance on CCUS and hydrogen as the magic silver bullets in reducing CO_2 emissions. Given that hydrogen is yet to be produced at commercial scale and application, then the technological tool at hand for Australia, particularly given its excellent geological basins for carbon sequestration, is CCUS.

The development of a CCUS industry in Australia might be seen as pointless given the strong emphasis on developing renewable energy as the primary energy source for the national electricity market (NEM), however there is a huge need for heat in many of Australia's industries, particularly as Australia has been set on the path to a manufacturing renaissance under the modern manufacturing strategy.[5] Given that fossil

[1] Department of Climate Change, Energy, the Environment and Water, *Australian Energy Update, 2023* (2024), https://www.energy.gov.au/sites/default/files/Australian%20Energy%20Update%202023_0.pdf.

[2] Department of Climate Change, Energy, the Environment and Water, *Australian Energy Update, 2023* (2024), https://www.energy.gov.au/sites/default/files/Australian%20Energy%20Update%202023_0.pdf.

[3] IEA, Net Zero by 2050: A Roadmap for the Global Energy Sector (4th revision, 2021).

[4] Steven Geroe, '"Technology Not Taxes': A Viable Australian Path to Net Zero Emissions?' (2022) 165 *Energy Policy* 112945.

[5] Department of Industry, Science, and Resources, Making It Happen: The Australian Government's Modern Manufacturing Strategy (2020).

fuels, especially gas, are likely to be used in Australia for decades to come as both a primary fuel for the modern manufacturing strategy,[6] and as a transition fuel,[7] CCUS is essential for emissions reduction.

2 CCUS Projects, Challenges, and Opportunities

There has been a strong history of investigating the possible use of CCUS in Australia for CO_2 emissions abatement, and historically there has been a focus on research and development (R&D).[8] Although there are still few commercial projects operating to date, Australia's history of CCUS R&D since the mid-2000s has enabled CCUS to be readily adopted. Many, although not all jurisdictions, have either initiated or welcomed CCUS projects. These projects are outlined in Fig. 1 and will be considered in greater detail below. As can be seen in Fig. 1, all but two of Australia's CCUS projects are in various stages of planning, with Gorgon and Moomba the only commercial CCUS projects in Australia.

2.1 Commonwealth Offshore Waters

The Australian Government (the Commonwealth) holds jurisdiction for any CCUS activities in waters from 3 nm seaward. Although its legislative competence includes CCUS activities, very little activities have been undertaken to date. At present there are two CCUS activities in Commonwealth waters. Both are still in the planning stage, and if approved, will be the first commercial offshore CCUS projects in Australia.[9]

[6] Tina Soliman-Hunter, 'Covid-19 and the Shift Towards Net-Zero Emissions in Australia: Accelerating the Energy Transition' (2021) 19(5) *Oil, Gas, and Energy Law* (OGEL), https://www.ogel.org/, 14.

[7] Tina Soliman-Hunter, 'Covid-19 and the Shift Towards Net-Zero Emissions in Australia: Accelerating the Energy Transition' (2021) 19(5) *Oil, Gas, and Energy Law* (OGEL), https://www.ogel.org/, 14.

[8] Reference here the CO_2 CRC.

[9] Commercial CCUS facilities are defined by the *Global CCS Institute* as a facility where CO_2 is captured and transported for permanent storage as part of an ongoing commercial operation, the facilities have economic longevity similar to the facility whose CO_2 they capture and must support a commercial return and meet a regulatory requirement while operating. Conversely, a pilot and demonstration facility are one where CO_2 is captured for testing, developing, or demonstrating CCUS technologies or processes. In

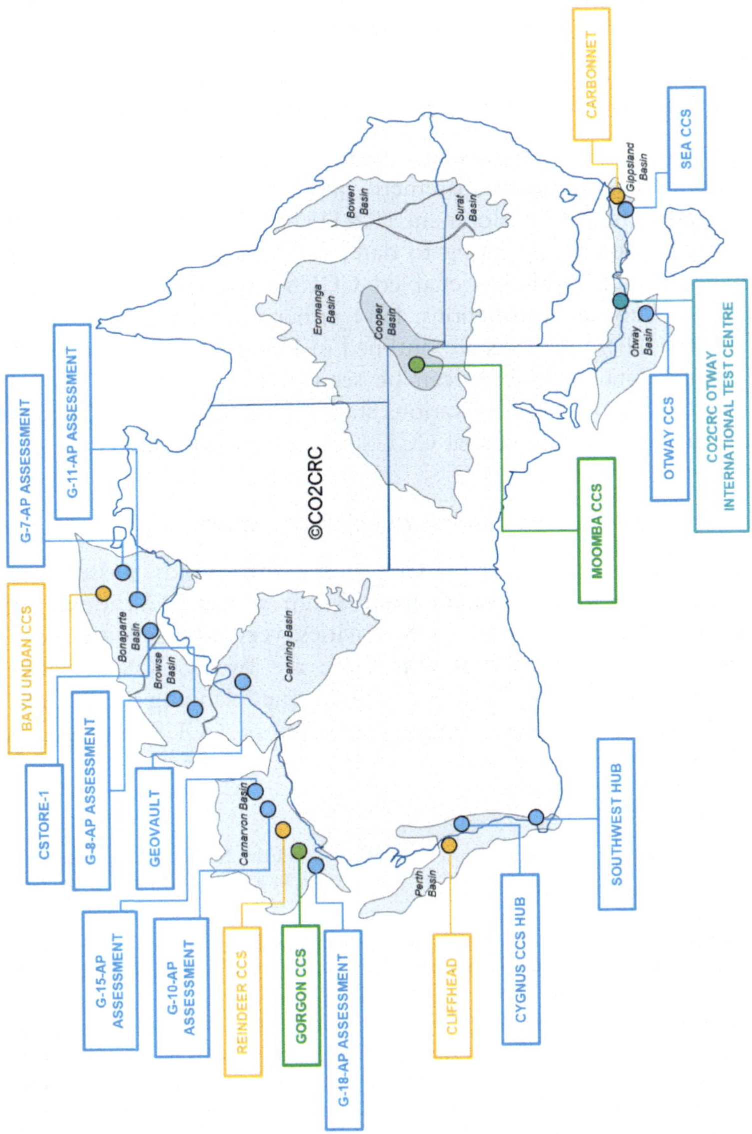

Fig. 1 CCUS projects in Australia (*Source* https://co2crc.com.au/)

2.1.1 *Pilot Energy—Cliff Head Oil Basin—Conversion from Production Field to CO_2 Storage*

The Cliff Head CCUS Project aims to be the first commercial offshore CCUS/UCS in Australian Commonwealth Waters. The project is based on the conversion of the late-stage, mature Cliff Head Oil Field, and its associated offshore and onshore infrastructure, into a permanent CO_2 storage facility.

Due to reservoir properties which include thick and competent overlying geological seals, as well as existing petroleum infrastructure, Pilot Energy (Pilot) identified the potential to convert the depleted Cliff Head oil field (Cliff Head), into a site for UCS of CO_2. Located in Production Permit WA-31-L in the northern Perth Basin, offshore Western Australia, Cliff Head oil field is located in Commonwealth waters approximately 12 kms offshore, and 20 km south-southwest of Port Denison and Dongara. Water depth is on the order of 15–20 m. The project is scheduled to commence capturing and storing CO_2 from 2025 to 2026, with a storage potential of 50 Mt of CO_2, injected at a rate of 665,000–1 million t/pa for 15 years.

Should this project succeed, it is likely to provide and exemplar for other CCS projects in Commonwealth Waters. However, concerns arise given that the injection is into a depleted petroleum reservoir, which is not an ideal reservoir for injection—rather injection for CO_2 storage is more in associated saline aquifers.[10]

2.2 *UCS in Commonwealth Waters*

In August 2022 two CCS exploration permits were awarded in Commonwealth waters off the coast of the Northern Territory and Western Australia. The permits are for area G-7-AP over GHG21-1 in the Bonaparte Basin, a joint venture between INPEX, Woodside Energy, and

the pilot/test project the captured CO_2 may not be permanently stored, and it has a short lifespan compared to commercial facilities; and a commercial return is not expected from the facility. See Global CCS Institute, *Facilities Database* (2023), https://co2re.co/FacilityData.

[10] Tina Soliman-Hunter, 'Role of Underground Carbon Storage to Assist Reaching Net Zero by 2050: Perspectives on Petroleum Reservoirs' (2024) *International Journal of Petroleum Technology* 1–15.

TotalEnergies, and for area G-8-AP over GHG21-3 in the Browse Basin for Woodside Energy.

The award of G-7-AP off the coast of Darwin (Northern Territory) will assist the Northern Territory in realising its ambition to be a CCUS Hub a reality,[11] with Resources Minister King reiterating that 'Australia has the capacity to continue to be an energy export leader, at the same time as developing a domestic offshore UCS industry.'[12]

2.2.1 Bayu-Undan UCS Field

The Bayu-Undan CCUS project is located offshore Timor Leste, with Darwin is set to be an important regional hub for CO_2 capture, transfer, and storage at Bayu-Undan, as well as to other potential CCUS locations offshore the Northern Territory that are still in the exploration and feasibility stages. However, previous tensions regarding maritime boundaries[13] and Australia's conduct towards Timor Leste[14] may influence progression of the project.

In March 2022, Santos announced that it had entered the front-end engineering and design (FEED) phase for the proposed Bayu-Undan UCS project. This project involves the export of carbon dioxide from Australia via pipeline to the Bayu-Undan gas field, located across the Australian/Timor Leste maritime boundary, and therefore regulated under a special regime.

In May 2023, Santos announced the signing of four Memoranda of Understanding (MOU) for the proposed storage of CO_2 emissions from third parties to underpin the initial development of the Bayu-Undan CCUS project, where front-end engineering design (FEED) work is

[11] The Hon Madeleine King MP, *New Offshore Greenhouse Gas Storage Acreage to Help Lower Emissions* (Media Release, 24 August 2022), https://www.minister.industry.gov.au/ministers/king/media-releases/new-offshore-greenhouse-gas-storage-acreage-help-lower-emissions, 1.

[12] The Hon Madeleine King MP, *New Offshore Greenhouse Gas Storage Acreage to Help Lower Emissions* (Media Release, 24 August 2022), https://www.minister.industry.gov.au/ministers/king/media-releases/new-offshore-greenhouse-gas-storage-acreage-help-lower-emissions, 1.

[13] Dai Tamada, 'The Timor Sea Conciliation: The Unique Mechanism of Dispute Settlement' (2020) 31(1) *European Journal of International Law* 321–344.

[14] Bernard Collaery, *Oil Under Troubled Water: Australia's Timor Sea Intrigue* (Melbourne University Press, 2020).

nearing completion.[15] The four non-binding MOUs for CO_2 supply to Bayu-Undan CCUS have been executed with potential upstream gas and LNG projects offshore the Northern Territory, in Darwin, and an energy and industrial conglomerate in Korea.

These MOUs indicate that demand for Bayu-Undan storage at Bayu-Undan CCUS could exceed 10 million. Tonnes per annum (Mtpa). It is expected that the Bayu-Undan CCUS project with CO_2 piped offshore to the reservoir from Darwin, will offer a relatively low-cost carbon solution, staying within the Australian Government's proposed price cap on Australian Carbon Credit Units.

The importance of the Bayu-Undan CCUS project is that it has the potential to reduce the absolute emissions and emissions intensity of Australian gas and LNG projects and other industrial activities in the Northern Territory. It achieves this by providing permanent CO_2 storage and providing a potential Scope 3 emissions solution for Australia's exports to Asia. Furthermore, large customers in countries such as Korea are also considering Bayu-Undan as a location for storage of its captured energy and industrial emissions in a manner similar to Norway's *Longship Project*.[16]

2.2.2 *Other Reservoirs*

According to a study in 2020 by the Global CCS Institute, it is highly likely that the Petrel Sub-basin (Bonaparte Basin) in the Joseph Bonaparte Gulf, approximately 180 km west of Darwin, is suitable for CO_2 sequestration.[17] Reviews of the Petrel Sub-basin completed over the past two decades indicate that it is highly prospective for geological storage, with a capacity to receive up to 20 million tonnes per year of CO_2 and containing several hundred million tonnes of total storage resources.[18]

[15] As at April 2024.

[16] For an overview of the Norwegian Longship project see Tina Soliman Hunter, 'Role of Underground Carbon Storage to Assist Reaching Net Zero by 2050: Perspectives on Petroleum Reservoirs' (2023) *International Journal of Petroleum Technology* 164–179.

[17] Global CCS Institute, *Carbon Capture and Storage Hub Study: Prepared for the Government of the Northern Territory of Australia* (2020), https://cmc.nt.gov.au/__data/assets/pdf_file/0006/1052898/q20-0114-gccsi-nt-css-hub-study-final-report.pdf.

[18] Global CCS Institute, *Carbon Capture and Storage Hub Study: Prepared for the Government of the Northern Territory of Australia* (2020), https://cmc.nt.gov.au/__data/assets/pdf_file/0006/1052898/q20-0114-gccsi-nt-css-hub-study-final-report.pdf.

Such storage well exceeds the storage capacity of the Norwegian Longship Project.

2.3 Western Australia

2.3.1 Gorgon

The only fully commercial Australian CCUS project is Chevron's Barrow Island in Western Australia (WA). This project falls under Western Australian jurisdiction and is regulated under the Barrow Island State Agreement, since the pre-combustion carbon capture was part of the overall Barrow Island CCS project.[19]

The Gorgon project captures CO_2 at the LNG processing facility on Barrow Island, storing the captured CO_2 by injection into the Dupuy Formation, a deep saline reservoir 2 km beneath Barrow Island.[20] The Dupuy Reservoir characterisation was initially utilised to select the target formation, with further seismic studies, seismic modelling, and a comprehensive appraisal including the drilling of an appraisal well, utilised to confirm the Dupuy formation as the target reservoir. In addition, a suite of robust subsurface modelling was undertaken to gauge the impact of CO_2 injection. However, a lack of knowledge meant that a flexible development plan for CO_2 disposal was required to ensure CO_2 sequestration occurred.

The target Dupuy Formation is characterised by fine to medium-grained blocky sandstone, capped by a fining upward unit at the top, the upper massive sand, thought to be a slope deposit, and containing important intra-reservoir siltstone baffles such as the *Perforans* 'Shale,' a very low permeability siltstone.[21] Comprehensive geomechanical modelling of Dupuy Formation based on numerous data sources: pore pressures, modified leak-off tests, mini-frac data, image log data from the Gorgon CO_2 data well, fault geometric mapping from seismic data, and rock mechanics data from core analysis. This modelling was significant in the evaluation

[19] See Sect. 3 for a comprehensive overview of the regulatory framework for CCUS in Western Australia.

[20] M. Flett et al., 'Subsurface Development of CO_2 Disposal for the Gorgon Project' (2009) 1 *Energy Procedia* 3031–3038. Note that the captured CO_2 is NOT injected into the existing petroleum-producing formation.

[21] M. Flett et al., 'Subsurface Development of CO_2 Disposal for the Gorgon Project' (2009) 1 *Energy Procedia* 3031–3038.

of the Dupuy Formation for subsurface disposals of CO_2. In addition, the modelling significantly influenced well placement and well trajectory, identified constraints on formation containment, and provided well operability limits for injection pressure.[22]

Hydrodynamic studies were also undertaken and were an important component of the containment assessment process. These studies detailed formation faults and their ability to seal or be part of a communication network and confirmed the integrity of the top seal.[23] The pressure and salinity data acquired demonstrated that aquifer pressure separation and salinity differences of at least 20,000 parts per million total dissolved fluids (PPM TDS), equivalent between the Dupuy and the overlying aquifers existed, indicating the existence of hydraulic separation between the aquifers over the production time span, and demonstrating that at least one, and possibly two, competent regional seals were present above the Dupuy Formation.[24]

Although extensive data acquisition, geomechanical modelling and hydrological studies were undertaken to identify and assess the target formation, the Gorgon CCS has not been entirely successful. Since the commencement of CO_2 injection at Barrow Island in August 2019, Chevron has injected 5,000,000 tonnes of greenhouse gases. Although this volume of injection is large, surpassing the annual injection rates of the Norwegian Sleipner commercial CCS project, it has fallen well short of the estimated target of almost 10 million tonnes of CO_2 per annum. Independent analysis estimates that this shortfall of around 4.6 million tonnes will cost Chevron around $100 million to offset via carbon credits. An analysis of the cause of this shortfall in sequestration points to failure in engineering and technology applications and frequent inefficiencies,[25] all contributing to the ongoing failure to inject predicted volumes of CO_2.

[22] M. Flett et al., 'Subsurface Development of CO_2 Disposal for the Gorgon Project' (2009) 1 *Energy Procedia* 3031–3038.

[23] M. Flett et al., 'Subsurface Development of CO_2 Disposal for the Gorgon Project' (2009) 1 *Energy Procedia* 3031–3038.

[24] M. Flett et al., 'Subsurface Development of CO_2 Disposal for the Gorgon Project' (2009) 1 *Energy Procedia* 3031–3038.

[25] Bruce Robertson and Milad Mousavian, *Gorgon Carbon Capture and Storage: The Sting in the Tail* (2022), Institute for Energy Economics and Financial Analysis, https://ieefa.org/wp-content/uploads/2022/03/Gorgon-Carbon-Capture-and-Storage-The-Sting-in-the-Tail_April-2022.pdf.

Western Australia has several other CCUS projects in feasibility stage at present. Most notably, the Southwest Hub Project, centred around the industrial activities of Perth, is a government led initiative to capture CO_2 emissions from industry in the area and to store around 800,000 tpa in the local Lesueur Sandstone formation. Given the absence of a traditional caprock, successful use of the Lesueur Formation as a CO_2 reservoir depends on containment assurance by residual trapping and dissolution.[26] Confidence in the suitability of the Lesueur as a CO_2 storage reservoir has increased as a result of modelling[27] which shows that the Lesueur meets the requisite criteria to be considered a formation reservoir, since injected CO_2 stops migrating through the formation after 500 years.[28]

2.4 South Australia

2.4.1 Moomba Basin CCS Pilot

The Moomba CCS demonstration project, undertaken by Santos at its Cooper-Eromanga basin gas field which supplied significant gas volume to Australia's eastern states, seeks to store CO_2 as part of its Moomba carbon storage project. This project has the long-term objective to establish a large-scale carbon storage hub at Moomba, capable of storing more than 400 million tonnes of CO_2 by injecting the CO_2 into the depleted and/ or depleting oil and gas reservoirs of the Cooper basin. This injection will accomplish two main aims. Firstly, it will provide a secure storage solution for major carbon emitting activities relating to gas processing in Queensland NSW and SA, and potentially the NT. Secondly, it will enable Santos to utilise the CO_2 for EOR in depleting reservoirs as required. The source of CO_2 was Moomba gas plant's existing CO_2 separation facilities. After a successful pilot test, Santos has continued to invest in this project, particularly given the shift to blue hydrogen and CCS as part of the toolkit for shifting to net-zero emissions.

In June 2021, Santos and partner Beach Energy were awarded $15 million from the Australian Government's *Carbon Capture Use and Storage Development Fund*, part of the $162 million Moomba CCS

[26] Department of Energy, Mines, Industrial Regulation and Safety, *Southwest Hub Project* (2023).

[27] Odin Reservoir Consultants, *Dynamic Model of the Harvey Area* (2018).

[28] Department of Energy, Mines, Industrial Regulation and Safety, *Southwest Hub Project* (2023).

project, which will permanently store 1.7 million tonnes of CO_2 per year initially, scalable to store up to 20 million tonnes annually. The final investment decision (FID) on the project was made in November 2021, proceeding to FEED and first injection commenced on target in Q4 2024.

The geology of the Cooper-Eromanga basin where Santos' CCS project is to occur comprises a favourable environment, with geological thick layers of sandstone (Algebuckina Sandstone and Wallumbilla Formation), capped with mudstone formations (Allaru Mudstone).[29]

2.5 *Queensland*

A trial for carbon capture from coal-fired generation assets (ZeroGen project) was undertaken in the early 2010s. The Queensland (Qld) based ZeroGen project was to comprise of a 400 MW commercial scale power generation facility, with an integrated coal gasification power plant for the capture and storage of CO_2 emissions while generating low-emissions base-load electricity for the national electricity market.[30] Located in central Queensland, the geo-sequestration of the CO_2 was to be into the northern Denison Trough. As part of the project GHG exploration permits were awarded to the project proponent under the Greenhouse Gas Storage Act 2009 (Qld), which was implemented for the project to proceed.[31]

Depending on the literature consulted, the ZeroGen project was either an overwhelming success,[32] or a spectacular failure.[33] Although the project was due to progress to demonstration phase by 2015, with a full-scale plant by 2020, the ZeroGen project was placed in liquidation in

[29] M. Ahmad and T. Munson, Geology, and mineral resources of the Northern Territory: Chapter 41: Eromanga Basin. NT Geological Darwin (Northern Territory): 2013b.

[30] A. Garnett, C. Greig and M. Oettinger (eds) ZeroGen IGCC with CCS: A Case History (2014).

[31] A. Garnett, C. Greig and M. Oettinger (eds) ZeroGen IGCC with CCS: A Case History (2014).

[32] P. Ashworth et al., 'Public Acceptance of Carbon Dioxide Capture and Storage in a Proposed Demonstration Area' (2009) 223(3) *Proceedings of the Institution of Mechanical Engineers, Part A: Journal of Power and Energy* 299–304.

[33] B. Browne and T. Swann, *Money for Nothing*. AIIA Submission to the Clean Energy Finance Corporation Amendment Bill 2017 (2017).

2011, and officially cancelled. The full history of the ZeroGen project, including the failures, is an important lesson for carbon capture projects, and that full-scale deployment of a project is preferable to trial phase.[34]

Although the ZeroGen project made little contribution to advancing the development of CCS in Queensland, it did pave the way for other projects, especially by establishing a legal framework that enabled the establishment of other projects. At present feasibility projects are underway in the Surat Basin.[35]

2.6 Victoria

The Otway Project, developed and led by the Cooperative Research Centre for Greenhouse Gas Technologies (CO_2CRC) which was formed in 2003, was Australia's first demonstration project of deep geological storage of CO_2. The aim of the project was to provide technical information on storage, monitoring and verification processes and associated technologies to help inform the development and deployment of a commercial CCS industry. Located at Nirranda South in south-western Victoria, the project cost more than $40 million and injected more than 65,000 tonnes of CO_2 into a depleted gas reservoir about 2100 metres below the earth's surface (CO_2CRC, 2010a).

The CO_2 source was gas extracted from a natural gas reservoir about 2.25 km from the injection site, compressed, and piped to the injection well. Findings were consistent with predicted modelling, with no evidence of leakage of CO_2.[36]

The target formation was the depleted gas reservoir is the Waarre C (sandstone) geological formation, underlain by two additional Waarre formations (B and A), then underlain by the Eumeralla Formation. The Waarre C sandstone is capped by the Flaxman Formation, which is in turn overlain by the Belfast Mudstone and the Paaratte Formation (CO_2CRC,

[34] A. Garnett, C. Greig and M. Oettinger (eds) ZeroGen IGCC with CCS: A Case History (2014).

[35] Queensland Government, *Surat Basin Carbon Capture and Storage Project* (2023), https://www.qld.gov.au/environment/management/environmental/eis-process/projects/current-projects/surat-basin-carbon-capture-and-storage-project.

[36] Cooperative Research Centre for Greenhouse Gas Technologies, *CO_2CRC Otway Project: Overview* (2010). Retrieved from http://www.co2crc.com.au/otway/index.html.

2010b). As a pilot project, the CO_2CRC Otway Project was considerably smaller in its injection volume compared to commercial scale CCS projects, which aim to inject more than one million tonnes of CO_2 per year. However, the learnings of this study have been invaluable for commercial scale projects.[37] These include:

1. a carefully chosen site is critical for the success of the CCS project. Failure to choose the site is likely to result in CO_2 leakage,
2. The selected geology should comprise of a series of thick layers of porous sandstone. This will ensure the capacity of storage and even distribution of CCS into the formation,
3. Saline formations are preferred over non-saline,
4. Monitoring is critical for project success for not only assuring geology and storage, but also for social acceptance and social licence to operate,
5. The geological caprock overlaying the storage formation should be a mudstone (or similar) to prevent leakage of CO_2,
6. Learnings from the project are not just geological, but also regulatory, as the project serves as an excellent tool for understanding the complexity of the regulatory and approvals process under the planning regime.

The Otway Pilot CCS project undertaken by the CO_2CRC provided an excellent opportunity to explore social issues pertaining to CCS in the community. One of the CO_2CRC's key requirements for the project was community acceptance, therefore a community consultation plan was put in place early (2005) with a focus on building successful relationships with stakeholders, informing and educating the community about CCS, and transparent, timely communication (Ashworth et al., 2011). Research regarding the Otway Basin project demonstrated that a critical challenge for project proponents considering the deployment of CCS projects across the country is to identify the benefits that a project will bring to the local community. The specific value proposition for a CCS project will vary depending on perceived local benefits and can depend on issues such as the size and socio-economic status of the community (Ashworth, 2010).

[37] Santos, *Cooper/Eromanga Basins CCUS. DEM Oil and Gas Roundtable, November 2018* (2018).

In examining the community engagement process and activities that occurred during the CO_2CRC Otway Project, a number of successful strategies were identified (Ashworth et al., 2011), including the need for early proactive engagement with stakeholders to build trust and form working relationships; a need to undertake social research in the community to establish a baseline; informal face-to-face meetings with key landholders and stakeholders. to build relationships and trust between CO_2CRC and the community, and in discovering issues and concerns relating to compensation and access to land; the appointment of a trusted community member as a liaison officer to build trust between CO_2CRC and the community; the establishment of a community reference group, seen as integral in communicating the issues and concerns of the community to the project team; and an open day, providing the public with a tour of the site followed by lunch and an open formal discussion.

Major social issues regarding the project were also identified (Ashworth et al., 2011). OF primary importance was the lack of communication, respect, and regard for landholders' properties by a seismic crew when seismic testing, straining relationships between CO_2CRC and some landholders. Efforts to repair damage were seen by some community members as inadequate and the communication process poorly handled. Another critical issue was a landholder's refusal of access to his property unless adequate compensation was given. Negotiations between the landholder and the developers were unsuccessful and ultimately, the land was acquired under land acquisition legislation. This acquisition did not sit well with other landholders and caused delays in the project development.

These social issues experienced by the Otway Project are analogous to those experienced in the ZeroGen project in Queensland (Ashworth et al., 2009), as well as CSG project proponents (GISERA, 2013), and addressing these issues are critical to the success of any project that involves CCS.

2.7 New South Wales (NSW)

Initially CCUS in NSW was assessed under the Delta CCS Demonstration Project, which sought to demonstrate the feasibility of Post

Carbon Capture technology in NSW.[38] Similar to the ZeroGen project in Queensland, the project attempted to capture the GHG emitted from the burning of coal, and if successful would enable the continued use of coal for electricity generation (clean coal technology).[39] Established in 2010, the project ceased in 2014 due to the withdrawal of Commonwealth Government funding.[40]

NSW continues its feasibility assessment of CCUS and is currently in stage 2 of assessment (determining the suitability of identified reservoirs), with no commercial activities as at March 2024.[41] In 2009 the NSW Government's Carbon Storage Taskforce identified the Darling Basin and the Sydney Basis as national priorities to determine their feasibility for CO_2 storage. Initial wells were drilled to define basin stratigraphy, identifying the Pondie Range Trough and the Poopelloe Lake Trough as possible suitable storage sites.[42] This exploration has now entered state 2 of assessment, examining the two identified basins for potential CO_2 storage.[43]

[38] NSW Department of Mining, Exploration and Geoscience, *Carbon Capture and Storage* (2016), https://meg.resourcesregulator.nsw.gov.au/invest-nsw/coal-innovation-nsw/carbon-capture-and-storage.

[39] NSW Department of Mining, Exploration and Geoscience, *Carbon Capture and Storage* (2016), https://meg.resourcesregulator.nsw.gov.au/invest-nsw/coal-innovation-nsw/carbon-capture-and-storage.

[40] NSW Department of Trade and Investment and Resources and Energy, *Report to NSW Parliament: Coal Innovation NSW Fund 2013–14* (2014), https://www.parliament.nsw.gov.au/tp/files/8039/Report%20of%20the%20Coal%20Innovation%20NSW%20Fund%20for%202013-14.pdf.

[41] NSW Government, *NSW CO_2 Storage Assessment Program: Stage 2* (2022), https://meg.resourcesregulator.nsw.gov.au/sites/default/files/2022-11/co2-storage-assess-stage2-flyer.pdf.

[42] NSW Department of Mining, Exploration and Geoscience, *NSW CO_2 Storage Assessment Program* (2023), https://meg.resourcesregulator.nsw.gov.au/invest-nsw/coal-innovation-nsw/nsw-co2-storage-assessment-program.

[43] NSW Government, *NSW CO_2 Storage Assessment Program: Stage 2* (2022), https://meg.resourcesregulator.nsw.gov.au/sites/default/files/2022-11/co2-storage-assess-stage2-flyer.pdf.

3 Legal Framework, Policies, and Norms in Australia—Description and Analysis

3.1 Policy and Government Strategy

The IEA's perspective on the role of technology in achieving NZE by 2050 was iterated in Australia's Low-Emissions Technology Roadmap in 2020 and updated in the annual Low Emissions Technology Statement.[44] The cornerstone of Australia's policy for carbon emissions reduction, the Low-Emissions Technology Statement incorporates the Technology Investment Roadmap (TIR), which reviews and refines the Australian Government's technology-led approach to reducing emissions.[45] The Australia's Government policy and strategy regarding low emissions investment is articulated through the annual TIC, with the 2022 TIC's goal of supporting the commercial uptake of low emissions technologies, and to develop new technologies to assist with the reduction of CO_2 emissions. At the forefront of the technologies is the development of a strong CCUS industry in Australia.

The role and importance of CCUS was reiterated in 2022 when Commonwealth Minister for Resources and Northern Australia Madeleine King noted that 'carbon capture and storage has a vital role to play to help Australia meet its net zero targets. Australia is ideally placed to become a world leader in this emerging industry, with large, stable, offshore geological formations for greenhouse gas storage.'[46]

Australia's climate policy, set by the Australian Government as required under international law commitments,[47] was finally committed to legally binding greenhouse gas emissions reduction targets in 2023. The targets are set out in the Commonwealth *Climate Change Act 2022* (CCA).

[44] Australia Government Department of Industry, Science, Energy and Resources, *Technology Investment Roadmap: First Low Emissions Technology Statement 2020*, https://webarchive.nla.gov.au/awa/20220603080623/https://www.industry.gov.au/data-and-publications/technology-investment-roadmap-first-low-emissions-technology-statement-2020.

[45] Department of Climate Change, Energy, the Environment and Water, *Technology Investment Roadmap* (2020), https://www.dcceew.gov.au/climate-change/publications/technology-investment-roadmap.

[46] The Hon Madeleine King MP, *New Offshore Greenhouse Gas Storage Acreage to Help Lower Emissions* (Media Release, 24 August 2022) 1.

[47] The Paris Agreement and the accompanying Nationally Determined Contributions.

Section 10(1) of the CCA sets two targets: a reduction in Australia's net greenhouse gas emissions to 43% below 2005 levels by 2030, and a reduction in Australia's net greenhouse gas emissions to zero by 2050.

3.2 Indigenous Interests

A critical consideration in any CCUS project in Australia is the interests of the first nation peoples. These interests are addressed at commonwealth level under the *Native Title Act 1993* (Cth), with additional requirements under Northern Territory Legislation.

3.2.1 Native Title Act—Future Acts

Section 24AA(1) of the Native Title Act (NTA) sets out a regime under which actions of the Commonwealth or state and territories *after* the commencement of the Native Title Act (1993), which may affect native title rights, can be done lawfully, and are known as *future acts*. The future acts regime is procedural since it creates a right for native title holders (NTH) or native title claimants (NTC) to negotiate an agreement. Unlike the rights conferred under the Aboriginal Lands Act 1976 (NT), it does not give NTC or NTH the right to veto the proposed future act.

Where a proposed future act attracts the right to negotiate (RTN), such as the grant of a right to explore for petroleum, the registered NTH or NTC is entitled to negotiate with the relevant government (Commonwealth or state), and the third party in whose favour the future act will be made. Since the right to negotiate extends to both determined NTH, and registered NTC, it is likely that any agreement negotiated will need to be reached with multiple Aboriginal parties.

3.2.2 CCS Titles and the Right to Negotiate

Often, the obligations under the NTA are dealt with concurrently with the process to obtain the requisite title (e.g. injection licence). Regardless of the process followed, a licence may not be granted until native title issues have been addressed. The NTA creates procedures to be followed during the right to negotiate process. The NTA establishes the right to negotiate procedure for the grant petroleum tenements, requiring a prospective applicant for a petroleum tenement (grantee) applies to the relevant government for the grant of the tenement, who is required to notify registered or potential native title parties of the proposed act, under s29 of the NTA. The title applicant is also required to give public notification of the application by publication in a newspaper, thereby creating an opportunity for unregistered native title claimants to register their claims. New native title claims must be lodged within three months and accepted by the National Native Title Tribunal (NNTT) for registration within four months of the notification date if the new native title claimant is to be granted the right to negotiate.

After the notification requirement has been met, negotiations, which are conducted in 'good faith,'[48] may commence. There is a wide scope for the negotiations, which may include agreement on profit-sharing or compensation, uses of the land, etc., with the negotiations geared towards the native title party granting its consent to the future act. Although the State is considered a party, negotiations generally take part between the licence holder (or applicant) and the native title party. The parties are required to negotiate in good faith.[49]

If an agreement is reached, the parties (including the Government) sign an agreement, and the right to negotiate process is complete. The requisite title can then be granted (subject to any other requirements, such as environmental approval). If agreement is not reached, either party may request that the NNTT mediate (at any time) or arbitrate (after six months) the right to negotiate process. In the case of arbitration, the NNTT determines whether the future act will be allowed, and if so, under what conditions. Such a determination must not provide for profit-sharing arrangements.

Appeals regarding future act determinations lie in the Federal Court. The renewal or re-grant of a mining and petroleum tenement does not attract the right to negotiate process if the tenement was granted before 23 December 1996 or under the future act regime, and the renewal or re-grant does not extend the rights or actions allowable under the tenement.

[48] As required under s31(1)(b) of the NTA. For the content of the requirement to negotiate in good faith, see Western Australia v Taylor (1996) 134 FLR 211 at 224–5: the 'Njamal Indicia'.

[49] As required under s31(1)(b) of the NTA. For the content of the requirement to negotiate in good faith, see Western Australia v Taylor (1996) 134 FLR 211 at 224–5: the 'Njamal Indicia'.

3.2.3 *Indigenous Land Use Agreements Under the Native Title Act*

The NTA establishes a system of registered indigenous land use agreements (ILUAs) under which the party seeking permission for future acts may reach an agreed and binding position with the native title holders. ILUAs may cover a broad range of subject matter, including validation and conditioning of previous and proposed future acts (including the retrospective validation of previously invalid acts), compensation or benefits sharing for commercial activities on native title land, the exercise of native title rights in an area, and the extinguishment of native title rights, if required.

There are significant procedural hurdles to the registration of an ILUA, including long notification periods before registration. The alternative is the right to negotiate process, outlined in s3.3.2 above, which is regarded as considerably less onerous. However, there are several reasons why an ILUA is concluded for petroleum activities, as outlined in Table 1.

3.3 *Regulatory Framework for CCS*

The legislative and regulatory framework for CCS needs to cover both onshore sequestration which is primarily a state matter, and offshore sequestration, which is a Commonwealth matter. Currently, state and Commonwealth legislation is patchy, with comprehensive and uniform regulation still lacking in some jurisdictions. Some jurisdictions do not have a legislative framework in place to cover issues in various stages of CCUS, such as transport, injection, monitoring, and financial liability. At state/territory levels there are glaring gaps in the regulatory framework, with some jurisdictions (such as NSW) not yet having regulation in place, while other jurisdictions such as Victoria, establishing a comprehensive greenhouse gas Act. Still others, such as the Northern Territory, Western Australia and South Australia, have incorporated CCS into existing legislation.

3.3.1 *Liability*

Given the lack of CCS legislation, there is no specified liability attribution for CCS projects in the NT at present. However, existing Australian

Table 1 Analysis of use of ILUAs for activities in the NT

Indigenous land use agreements	
Overview	*Conditions and/or features*
ILUA process can commence at any time—there is no need to wait for a s29 notice	The obligation to identify native title parties to the ILUA lies on the project proponent (compare this to the right to negotiate, under which native title claimants are responsible for responding to a s29 notice)
The ILUA process may be faster than a litigated outcome	Due to the registration process, there is an opportunity for objections to be lodged against the ILUA. Refusal to register an ILUA may itself result in litigation
An ILUA can be broad covering any matter which may affect native title. Compare this to the RTN, which only deals with issues in s29 notice	No time frame in which agreement must be reached. Under the RTN, after six months the matter could be referred to the NNTT for arbitration
An ILUA is binding on all future native title claimants	ILUAs are binding by registration—if it is not registered, it is not effective (and there is no process to ensure registration, unlike the RTN, under which parties can force an outcome by arbitration)
Since an ILUA can cover multiple tenements under a project, there is an economy of scale for ILUAs over large projects	The process for obtaining a registered ILUA is lengthy and costly

Source Compiled by Author

legislation[50] requires that proponents of CCUS projects 'demonstrate to the satisfaction of the relevant Commonwealth or state regulators that the formations they plan to use can safely contain the target volume of the GHG to be injected and stored in certain formations, without leaking or compromising the formation's geotechnical integrity.'[51] Furthermore, this legislation empowers regulators to take administrative action against titleholders when greenhouse gas substances leak or have the potential

[50] *Greenhouse Gas Storage Act 2009* (Qld), and *Greenhouse Gas Geological Sequestration Act 2008* (Vic).

[51] Tim Power, *How Australian Laws and Regulations Affect Carbon Capture and Storage* (2021), https://www.whitecase.com/insight-our-thinking/how-australian-laws-and-regulations-affect-carbon-capture-and-storage.

to leak following injection. These powers include the right to direct the proponent to suspend or cease injection, cancel the title, and/or take reasonable steps to remedy the situation.

3.4 Commonwealth Regulation of CCS

In Australia, the injection of CO_2 in offshore reservoirs is regulated under several commonwealth instruments. However, the capture and transport of CO_2 is regulated by the jurisdiction where the activity takes place. Seaward of three nautical miles, the Commonwealth will also regulate the pipelines and infrastructure construction. The area from Baseline seaward to 3 nautical miles is known as commonwealth waters and falls under state jurisdiction following the offshore constitutional settlement.[52]

When a project proponent intends to dispose of CO_2 by underground CO_2 sequestration (UCS) in marine areas, they are required to obtain a dumping permit under the Commonwealth's *Environment Protection Sea Dumping Act 1981* (EPSDA). To grant a sea dumping permit for the purpose of UCS, the minister or delegate must be satisfied that the material for dumping meets the criteria set out in the *1996 London Protocol to the London Convention*. This particularly pertains to the composition of the CO_2 stream and any contaminants found within that stream. Permits will also be required for the storage loading, transportation, and sequestering at appropriate offshore sites to ensure that there are no significant adverse impacts on human health or the marine environment.

Commonwealth regulation of CCUS activities is confined to Commonwealth Waters (three nautical miles seaward from baseline) and is regulated under the *Offshore Petroleum and Greenhouse Gas Storage Act 2006* (Cth) *(OPGGSA)*. Individual states/NT regulate access to and extraction of gas. However, for Matters of National Environmental Significance (MNES),[53] the Commonwealth regulates activities through the *Environment Protection and Biodiversity Conservation Act 1999* (Cth).

[52] For a detailed consideration of jurisdictions and regulation of these jurisdictions, refer to Tina Soliman Hunter and John Chandler, *Offshore Petroleum Law in Australia* (Lexis Nexis, 2nd ed., 2021).

[53] Discussed in Sect. 3.4.2 below.

3.4.1 *Offshore Petroleum and Greenhouse Gas Storage Act 2009 (Cth)*

To inject into a reservoir in Commonwealth Waters,[54] the project proponent will need to hold a greenhouse gas injection licence under the Commonwealth's *Offshore Petroleum and Greenhouse Gas Storage Act* 2006 (OPGGSA) and its regulations.

Typically, a project proponent commences with a greenhouse gas (GHG) assessment permit, which enables a permit holder to drill a well and inject air, water, or petroleum to determine if the reservoir is suitable for CO_2 injection. An assessment permit is granted either by work-programme bidding (ss296–300 OPGGSA), or via cash bid (ss303–307 OPGGSA).

After examining the results of assessment wells and seismic modelling (both two- and three-dimensional) the assessment permit holder, if believing the assessed licence area can hold CO_2, will provide a *declaration of identified greenhouse gas storage formation* as required under s312 of OPGGSA. As in the case of a declaration of a petroleum discovery, the title holder is obligated to apply for a greenhouse gas injection licence under ss361–364 of OPGGSA.

As specified in s318 of OPGGSA, an assessment title holder, an injection title holder who has not carried on any injection or storage of CO_2, an unsuccessful applicant for a GHG injection lease, or the holder of a petroleum retention lease can apply for a GHG holding lease, which is akin to a petroleum retention lease under ss141–146 of OPGGSA. To obtain a GHG holding lease, the identified GHG storage formation must be wholly situated in the permit area and the applicant is currently 'not in a position' to inject and permanently store a GHG substance but is likely to be in such a position within 15 years. These generous terms for retaining the right to inject GHG provide the project proponent with the ability to plan sequestration projects, as well as utilising the

[54] For the division of waters in the Australian offshore zone refer to Michael White, 'Australia's Offshore Legal Jurisdiction Part 1: History & Development' (2011) 25(1) *Australian and New Zealand Maritime Law Journal* 3–18; Michael White, 'Australia's Offshore Legal Jurisdiction Part 2: Current Situation' (2011) 25(1) *Australian and New Zealand Maritime Law Journal* 19–34.

current *method* for CCUS to obtain Australian Carbon Credit Units[55] while also producing petroleum (although not when the CO_2 is injected for enhanced oil recovery).

The OPGGSA regulates greenhouse gas storage in Australian Commonwealth waters,[56] which commence from three nautical miles from baseline and extend to the boundary of Australia's exclusive economic zone or the outer margins of the continental shelf, whichever is greater. The extent of Commonwealth waters is outlined in the Sea and Submerged Lands Act 1973 (Cth). The OPGGSA also regulates offshore early gas activities in Australian Commonwealth waters. Under OPGGSA the Australian Government awards greenhouse gas storage titles in Commonwealth waters. The Act is also responsible for the administration of greenhouse gas injection and storage. It is administered by the responsible Commonwealth minister (RCM), who makes major policy decisions regarding CCS, including the granting greenhouse gas titles, the imposing GHG title conditions, GHG title cancellations, the release of offshore GHG acreage for storage, and formulating and enforcing conditions for the award of a GHG storage permit.

The GHG titles are administered by the National Offshore Petroleum Titles Administrator (NOPTA), and safety is regulated by the National Offshore Petroleum Safety and Environmental Management Authority (NOPSEMA). All offshore titles for greenhouse gas storage are tracked through the National Electronic Approvals Tracking System (NEATS). Safety well integrity and petroleum operations including environmental management are regulated by NOPSEMA, who also assesses and accepts environmental plans in relation to offshore greenhouse gas storage. Where EOR is undertaken greenhouse gas storage and petroleum titles may coexist, and NOPSEMA will regulate both activities. Under OPGGSA requirements a project proponent can nominate identified storage prospects, although the proposed storage formation must fall wholly within that nominated area.

[55] Australian Clean Energy Regulator, *Carbon Capture and Storage Method* (Web Page, 31 August 2022), https://www.cleanenergyregulator.gov.au/ERF/Choosing-a-project-type/Opportunities-for-industry/carbon-capture-and-storage-method#:~:text=The%20crediting%20period%20is%20the,starting%20than%20many%20other%20methods.

[56] Australian Commonwealth Waters, Coastal Waters and State Waters are defined in s6 of the *Offshore Petroleum and Greenhouse Gas Storage Act 2006* (Cth).

The Commonwealth Government, as a matter of policy, does not offer greenhouse gas storage titles within licencing rounds and instead relies on application or nomination by an interested party.[57] It is important to note that native title interests, as well as other marine activities such as petroleum, fishing, aquaculture, or tourism may have an interest in areas where CCUS activities may occur in Commonwealth waters. When greenhouse gas storage areas allocated during formal bidding rounds, applications will be made under the normal offshore application procedures set out in the OPGGSA, after the release areas are advertised in the Australian Government Gazette. Criteria for the award of a title in such will be provided by NOPTA and will include geological information about offshore basins. Applications will be assessed against the application criteria set out in OPGGSA and will vary depending on whether the allocation of a title will be by bid (work Programme or Cash bid), or through discretionary criteria.[58]

At present, there are only two active offshore greenhouse gas assessment permits, both located in Commonwealth waters off the coast of Victoria. The titleholder is the Crown in right of Victoria, under the Otway project as part of the CO_2CRC's research agenda.

3.4.2 Environmental Protection and Biodiversity Act 1999

Offshore CCUS is regulated by the Commonwealth, with onshore CCUS under the regulatory purview of States and territories. However, as a function of the Commonwealth constitution, the Commonwealth still has regulatory control over onshore CCUS in accordance with the *Environment Protection and Biodiversity Conservation Act 1999* (Cth) (EPBCA), which provides for the protection of the environment in some circumstances, as well as the protection of biodiversity including some habitats, in accordance with international treaties and conventions. Therefore, in most instances, there will be an additional layer of commonwealth regulatory approval for CCUS activity under the EPBCA.

[57] Department of Industry, Science and Resources, *Offshore Greenhouse Gas Storage Acreage Release Process* (2023), https://www.industry.gov.au/mining-oil-and-gas/offshore-greenhouse-gas-storage/offshore-greenhouse-gas-storage-acreage-release-process.

[58] For a detailed examination of the allocation of titles under bid or discretionary methods, see Tina Hunter, 'Access to Petroleum Under the Licencing and Concession System' (chapter 2) in Tina Hunter (ed), *Regulation of the Upstream Petroleum Sector: A Comparative Study of Licencing and Concession Systems* (Edward Elgar, 2015).

While the EPBCA is Commonwealth legislation and has as its ambit environmental protection, it does not apply to all CCUS activities. Rather, it applies only where the activity falls under an area where referral for assessment is required under the EPBCA. Day-to-day environmental management falls under the ambit of the states and territories onshore. Under the EPBCA an action will require approval from the environment minister if *the action has, will have, or is likely to have a significant impact on a MNES*. The eight MNES are listed threatened species and ecological communities, migratory species protected under international agreements, Ramsar wetlands of international importance, Commonwealth marine parks, World Heritage properties, the Great Barrier Reef Marine Park, any nuclear actions, and coal seam gas (CSG) and large coal mining developments—impact on water resources (the 'water trigger').[59]

Therefore, while the EPBCA is not 'core' environmental legislation, it is nonetheless important and needs to be considered when examining regulation of CCUS activities. Under s523 of the EPBCA, an action is defined broadly to include a project, a development, an undertaking, an activity or a series of activities, or an alteration of any of these things. A significant impact is defined as 'an impact which is important, notable, or of consequence, having regard to its context or intensity. Whether or not an action is likely to have a significant impact depends on the sensitivity, value, and quality of the environment, which is impacted, and on the intensity, duration, magnitude, and geographic extent of the impacts.'[60]

3.4.3 Environment Protection (Sea Dumping) Act 1981 (the Sea Dumping Act)

An offshore UCS permit application form under the EPSDA is currently being developed, and the formal consider Australia's international obligations under the London convention in practical and reflect the criteria set out in Annex 2 of the London Protocol; Revised 2012 Specific Guidelines for Assessment of Platforms or Other Man-Made Structures at Sea; and Risk Assessment and Management Framework for CO_2 Sequestration in Sub-Seabed Geological Structures.

[59] Department of Agriculture, *Water and the Environment, Significant Impact Guidelines 1.1*, http://www.environment.gov.au/epbc/publications/significant-impact-guidelines-11-matters-national-environmental-significance.

[60] Department of Agriculture, *Water and the Environment, Significant Impact Guidelines 1.1*, http://www.environment.gov.au/epbc/publications/significant-impact-guidelines-11-matters-national-environmental-significance.

3.5 *Northern Territory Regulation of CCS*

3.5.1 Constitutional Limitations

Under s122 of the Australian Constitution, the Parliament may make laws for the government of any territory surrendered by any State to and accepted by the Commonwealth, or of any territory placed by the Queen under the authority of and accepted by the Commonwealth, or otherwise acquired by the Commonwealth, and may allow the representation of such territory in either House of the Parliament to the extent and on the terms which it thinks fit. This means that the Commonwealth can make any laws in relation to any activity in the Northern Territory. Whereas the NT Government can make laws in relation two carbon capture and storage, BRS needs to be aware that any legislation made by the NT Parliament can be overturned by the Commonwealth and replaced by Commonwealth legislation at will. Previous examples of such action have included NT legislation for assisted dying and the mining of uranium in Jabiluka.

Land title and tenure in the Northern Territory is more complex compared to other Australian jurisdictions, with three types of land to tenure in the NT: freehold, Aboriginal freehold, and pastoral leasehold.

Northern Territory freehold land, also known as fee simple estate, means that the government has passed all interest in that land to the owner, except for ownership of resources such as petroleum and minerals. Freehold land is primarily found in cities and towns. This land is not affected by native title, which is extinguished under s22 of the Native Title Act 1993 (Cth).

Aboriginal freehold land is a type of landholding that is unique to the NT. It was established in 1976 under the *Aboriginal Land Rights Act 1977* (NT) (ALRA), which converted former aboriginal reserves into permanent aboriginal freehold land. This type of freehold is inalienable, meaning it cannot be sold or gifted (although it may be leased), and is formally held by Aboriginal land trusts which sign agreements on behalf of Traditional Owners (TOs). In mining and petroleum activities on aboriginal freehold land there are additional application requirements under the ALRA. It is important to understand that when exploration activities are to occur on aboriginal freehold land, consent MUST be obtained from the TOs and the relevant federal minister, and an agreement must be reached by all parties.

The TOs have the right to refuse or suspend access to their land for exploration, and once exploration is refused this can be permanent. Part IV of the ALRA sets out the process for obtaining permission on aboriginal freehold land, and the process is illustrated in Fig. 2. The relevant Land Council (in this instance the Central Land Council) has the legal power to assist indigenous people negotiate with governments and private companies over projects on aboriginal lands. They also support Aboriginal people in managing their traditional lands and seas, including issuing permits for entry fishing filming and other activities.

The third type of land tenure in the NT is *pastoral leases*. Under the *Pastoral Land Act* activities on these lands must implement the sustainable use of natural resources and remain economically viable. These leases primarily pertain to agricultural activities especially grazing.

3.5.2 *Regulation of Petroleum Operations, Environment, and Infrastructure*

As of January 2024, there is no regulatory framework for the regulation of GHG/CO_2 sequestration or storage in underground formations. However, petroleum activities in the NT are regulated under the *Petroleum Act 1984* (NT), the Petroleum Regulations 2020, the 2019 Code of Practice for Onshore Activities, and the 2019 Schedule of Onshore Petroleum Exploration and Production Requirements 2019. Such a regulatory regime is messy, with the use of both a code of practice (albeit enforceable under the operation of s58(ab) of the Petroleum Act), and a Schedule. However, the *Petroleum Act 1984* (NT) only applies to petroleum titles and does not include CO_2 storage.

Environmental management, protection and the regulation of environmental processes are undertaken under the Environment Protection Act 2019 (EPA) and its Regulations, as well as the Petroleum (Environment) Regulations 2016, as outlined in Table 2.

3.5.3 *NT Major Projects Regulation*

In the NT, major projects are defined as those projects that are significant contributors to the economic development of the NT, either through the creation of jobs or business opportunities. Major projects can either be government initiated and/or facilitated projects and private sector-initiated projects. Project proponents are encouraged to utilise the major projects status policy framework to develop projects such as shell gas development in CCS and to take advantage of the benefits of the

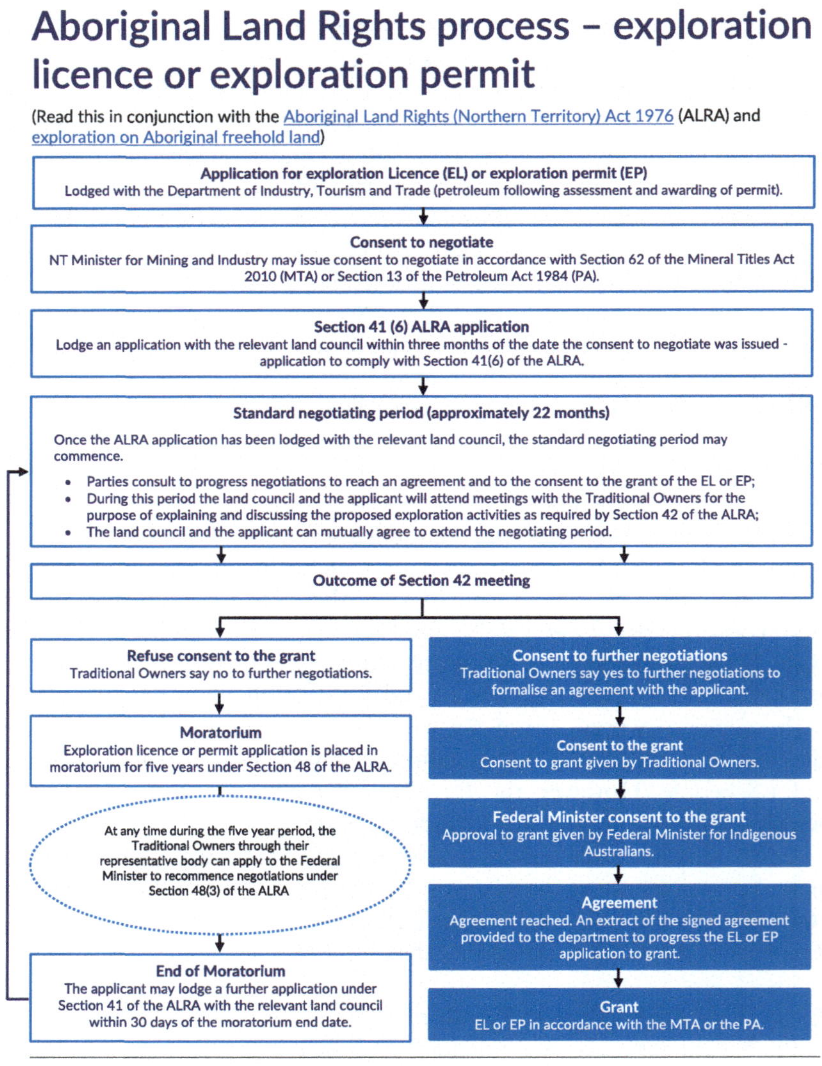

Fig. 2 Flowchart of process to obtain permission to undertake exploration on Aboriginal Freehold Land (*Source* NT Government)

Table 2 Overview of regulatory framework for Petroleum operations and infrastructure, NT

PETROLEUM ACT 1984 and PETROLEUM REGULATIONS 2020 ENVIRONMENT PROTECTION ACT 2019 and ENVIRONMENT PROTECTION REGULATION 2019 *Final Report of The Scientific Inquiry into Hydraulic Fracturing in the Northern Territory* (Pepper Report)				
General Codes, Policies and Legislation	**Granting of titles, assessment and approvals**	**Land use, planning and protection**	**Titles/activities environment regulation**	**Community**

• Code of Practice for Onshore Petroleum Activities 2019 (minimum standards)
  • Legally enforceable (s58 of petroleum Act 1984)
• Schedule of Onshore Petroleum Exploration and Production Requirements 2019
  • No legal enforceability
	• Petroleum Act 1984
	• Administered by Department of Industry, Tourism and Trade
	• NT Planning Scheme
	• -Gas Pipelines Access (Northern Territory) Act 1998
	• Energy Pipelines Act 2009
		• NT Planning Scheme
		• Environment Protection Act 2019
			• Code of Practice for Onshore Petroleum Activities 2019
			• Schedule of Onshore Petroleum Exploration and Production Requirements 2019
			• Environment Protection Regulations 2020
			• Waste Management and Pollution Control Act 1998
			• Waste Management and Pollution Control Regulations 2011
			• Environmental Offences and Penalties Act 1996
			• Water Acts
				• Land access arrangements (pastoral or Crown lease, freehold)
				  • Petroleum Act 1984
				  • Petroleum Regulations 2020
				• Aboriginal Land Councils
				• Aboriginal Land (Northern Territory) Act 1976 (no Land Access Agreement)
				• Native Title Act 1993 (ILUA)

Source Compiled by Author

major project status. These benefits include an acknowledgement of the significance of the project to the NT, identification, and mapping of project-related government approvals, whole of government coordination and facilitation of the project and project-related government approvals, and opportunities to use the NT Government investment attraction facilities and activities, including overseas trade missions, access to markets and potential sources of finance. If major project status is conferred the project will have a dedicated case manager who will be the single point of contact until project delivery occurs.

At present the NT's CCUS hub, as part of the *Middle Arm Sustainable Development Precinct*, is progressing, on the back of the 2020 Global CCS Institute Study to implement a Darwin-based CCS hub.[61] However, to implement this project, regulatory reform for capture, storage and transport activities prior to final injection still requires development, as does regulation for storage in Northern Territory state and coastal waters.

[61] Global CCS Institute, *Carbon Capture and Storage Hub Study: Prepared for the Government of the Northern Territory of Australia* (2020), https://cmc.nt.gov.au/__data/assets/pdf_file/0006/1052898/q20-0114-gccsi-nt-css-hub-study-final-report.pdf.

3.6 Regulation of CCS in Australian States

Any activity undertaken onshore in Australian states are subject to existing environmental and planning laws, including assessments and permits. What follows is an overview of CO_2 storage specific regulation on top of existing environmental and planning regulation.

3.6.1 Western Australia

Carbon capture and storage on Barrow Island as part of the Gorgon Project is regulated under the Barrow Island Agreement (BIA), in accordance the *Government Agreements Act 1979* (WA). In conjunction with the environmental impact assessment requirements under the *Environmental Protection Act 1986* (WA).

The BIA is an Act to ratify and authorise the State Agreement between the West Australian Government and the Gorgon joint venture relating to undertake offshore production of natural gas, process the gas onshore for LNG export, construct the necessary infrastructure to undertake the gas processing on Barro Island, and to make provisions for the conveyance and underground disposal of CO_2 recovered during gas processing on Barrow Island.

The BIA only applies to the activities undertaken on Barrow Island, and only those activities undertaken by the Gorgon joint venture. It does not apply to any other CCS activity in any other part of WA, or by any other project proponent. It is important to note that although all of the activities take place under the BIA, the *Environment Protection Act 1986* (WA) also applies to activities under the BIA.

Part 4 of the BIA pertains to the conveyance and underground disposal of CO_2. Under s13 of the BIA, CO_2 injection on Barrow Island requires the Minister's approval, which will be granted upon application as required in s13(2), and includes the rate, volume, and composition of the proposed CO_2 injected, the expected duration of the proposed disposal, the method for injection and disposal of the CO_2, the capability of the underground reservoir to confine the disposed CO_2, and the technical advice and data available to support the disposal proposal.

Under s14A of the BIA, the WA Government indemnifies the Gorgon joint venture from post-closure common liability claims for either loss or damage caused by CO_2 injection. Section 14B of the BIA empowers the WA Government to ask the Commonwealth to give notice that it is satisfied with the storage of CO_2 in the formation, and that there is no significant risk of leakage, or significant risk of adverse impacts to the geotechnical integrity of the formation, the environment, or human health and safety. This right of request to the Commonwealth commences 15 years after injection activities have ceased.

Section 14D of the BIA indemnifies the Gorgon joint venture for their common law liability to independent third parties, arising after the liability assumption date, for loss or damage caused by the injection of CO_2 into the formation where that loss or damage is attributable to van act or omission in the carrying out of CO_2 injection operations where approval for the operation is obtained under s13 of the BIA. Furthermore, s17A of the BIA protects the Gorgon joint venture for an action in tort for any wrongdoing where the action has been undertaken in good faith.

In the event of new CCUS activities occurring onshore and in state waters in Western Australia, regulatory reform of the existing *Petroleum And Geothermal Energy Resources Act 1967 (Western Australia)* (PGERA) has occurred to incorporate CO_2 injection—s5 of the PGERA incorporates 'the injection of carbon dioxide… into an underground reservoir or other subsurface formation'[62] into the definition of a petroleum operation. As such, the injection of CO_2 is now subject to regulation under the PGERA.

3.6.2 Victoria

In 2008 the *Greenhouse Gas Geological Sequestration Act 2008* (Vic) (GGGSA) entered into force. As set out in s1, the purpose of the GGGSA is to facilitate and regulate the injection of greenhouse gas substances into Victorian onshore underground geological formations for the purpose of permanent storage of those gases, including to facilitate and regulate the exploration for suitable underground geological storage formations.

This act was established by the Victorian Parliament to enable a pilot project for CO_2 storage in the Otway basin to be undertaken. Comprehensive in its scope, the GGGSA utilises the principles of sustainable development and intergenerational equity and recognises that biological diversity and ecological integrity must be maintained. This Act applies for the onshore sequestration of CO_2, with offshore sequestration in Victoria's coastal waters regulated under the *Offshore Petroleum and Greenhouse Gas Storage Act 2010* Vic. Under the GGGSA the minister

[62] As defined in section 2 of the Barrow Island Act 2003.

has the capacity to make a declaration for the protection of land for significant environmental reasons, for economic operations or activities, or for any other reason the minister considers appropriate. Section 14 of the GGGSA stipulates that geological storage formations underground are the property of the Victorian Crown, who retain Crown land rights over these formations. Upon injection of CO_2, and where an injection and monitoring licence is cancelled or surrendered, the Crown becomes the owner of any substance injected into the underground geological formation under that licence.

Part 3 Division 1 addresses the rights and obligations relating to GHG sequestration exploration permits (EP), outlining the procedures for obtaining and renewing a permit, and variations of work programmes. Under sections 37–55, an injection testing plan must be prepared and approved by the EP holder. Upon discovery of an underground geological storage formation the relevant minister must be notified, who has the right to give directions regarding the formation discovered. The GGGSA also establishes the requirements for retention leases and injection and monitoring licences.

The GGSA is extremely comprehensive and is likely to serve as a model act for other jurisdictions. It is the only onshore CCS-specific act to have been applied to underground injection activities.

3.7 South Australia

South Australia has not established a stand-alone CO_2 injection-specific regulatory framework for CCUS and is unlikely to do so in the near future. Rather, in a manner similar to Western Australia, in 2023 South Australia amended its existing petroleum legislation, the *Petroleum and Geothermal Energy Act 2000* (SA) (PGEA) to include CO_2 as a regulated resource and a regulated substance under the meaning in s5. In addition, the definition of a regulated activity in s10(1)a of the PGEA includes the 'utillisation of a natural reservoir to store petroleum or another regulated substance,' including includes CO_2 since it is defined as a regulated substance under s5 of the PGEA.

3.8 Queensland

The first CCUS project in Queensland, the ZeroGen Project, was regulated under the *Greenhouse Gas Storage Act 2009* (Qld) (GGSA), which was established specifically for the ZeroGen trial. The GGSA remains current and will be utilised to regulate further CO_2 storage projects.

Like the Northern Territory, large CCUS projects are likely to be regulated under the Queensland State Development and Public Works Organisation Act 1971 (Qld) (SDA), since the project was declared to be a significant project under section 26(1)(a) of that act. The SDA is utilised for major oil and gas development projects in Qld, and therefore large CO_2 injection projects are likely to fall within its ambit. Where a project is proclaimed as a significant project under the SDA, all approvals, including environmental approvals will be undertaken SDA, rather than individual pieces of legislation, to enable coordination and fast-tracking of the project.

3.9 NSW

As noted in section 2 above, NSW is currently engaged in a programme of assessing potential storage basins. To date, CCUS-specific legislation has not been implemented. In 2010 the Greenhouse Gas Storage Bill was introduced to the NSW Parliament, seeking to establish a regime for the permanent underground storage of GHG. The Bill was quashed, and no new Bill has been introduced to date.

Unlike Western Australia and South Australia, NSW has not altered their existing petroleum legislation, the *Onshore Petroleum Act 1991* (NSW) (OPA) to incorporate the injection of CO_2. Rather, the OPA refers only to CO_2 existing in a natural state (s6), and to the recovery of CO_2 (s138).

4 Conclusion

Given its geological age, reservoirs, and accumulated data from existing petroleum activities, Australia has vast capacity to store CO_2, and its location in the Asia Pacific region makes it a CO_2 storage destination. There is only one commercial CO_2 storage project at present, that of Barrow Island in Western Australia. This project has been plagued with under performance, however it has clearly demonstrated that CO_2 capturing storage in Australia is feasible. The CO_2 CRC in the Otway Basin has made great strides in research into C2 storage in Australia and provided invaluable information regarding opportunities and challenges in CCUS in Australia. By virtue of its location and existing facilities at Middle Arm,

The Northern Territory seeks to leverage these to become a regional CO_2 CCS hub. Onshore, the Moomba CCUS project in South Australia will provide a plethora of data that can be utilised for the development of other commercial projects in Australia.

The regulation of CCUS in Australia is complex, given the large number of jurisdictions, the diversity of CO_2 reservoirs, the interaction between state and Commonwealth law, and approaches by prevailing regulators. Offshore from three nautical miles, in commonwealth waters, the Commonwealth regulates CCUS under the OPGGSA. Few jurisdictions have sought to emulate the Commonwealth approach to regulating CCUS. Western Australia and South Australia are keen to develop CCUS projects and have taken the approach of amending existing petroleum legislation to regulate these activities. On the other hand, Victoria and Queensland, because of their early mover status in CCUS activities, have implemented CCUS-specific legislation. The NT, as it seeks to become a CCUS hub for both Australia and imported CO_2, is developing its facilities yet to date a bespoke greenhouse gas storage legislative regime is yet to materialise. Like western Northern Territory, NSW is yet to develop CCUS legislation, nor has it amended existing petroleum legislation. This is likely to be the result of the stage that NSW currently is at, assessing its reservoirs for the potential to store CO_2.

CHAPTER 11

CCUS in Brazil

Eduardo G. Pereira, Juliana Costa, and Edmilson Moutinho dos Santos

1 Introduction

The growing concern surrounding the global warming and the increase in the temperature require an urge for comprehensive and collaborative efforts to address the challenges posed by climate chance. The global

E. G. Pereira (✉)
Institute of Energy and Environment, University of São Paulo (IEE-USP), São Paulo, Brazil, São Paulo, Brazil
e-mail: egp.portugal@gmail.com

J. Costa
Federal University of Rio de Janeiro, Rio de Janeiro, Brazil

E. M. dos Santos
Institute of Energy and Environment, University of São Paulo (IEE-USP), São Paulo, Brazil

E. G. Pereira et al. (eds.), *Carbon Capture Utilization and Storage*, Sustainable Development Goals Series,
https://doi.org/10.1007/978-3-031-81272-9_11

warming is a phenomenon provoked specially by the increase of greenhouse gas emissions in the atmosphere with increasing adverse effects in natural and human systems.[1]

In order to combat this challenge several countries adopted the Paris agreement in 2015 during COP-21[2] which imposed key objectives and joint obligations for countries to limit its increase in temperature in 1,5°C, in addition to promoting changes in its financial flows in order to invest in low-carbon emissions development strategies aiming at achieving net zero emissions.

However, the initiatives taken by the countries proved insufficient to achieve the objectives of the Paris agreement by the end of the century. According to the Intergovernmental Panel on Climate Change ("IPCC") humanity will have to remove carbon emissions from the atmosphere, which can occur naturally by means of reforestation, or using technologies such as the carbon capture utilization and sequestration method ("CCUS").[3] According to the International Energy Agency ("IEA") 2023 World Energy Outlook edition the pipeline of CCUS projects around the world expects to capture around 400 million tonnes of carbon dioxide ("CO2") in coal plants, in the oil and gas and intensive industry sectors.[4]

Under this context, CCUS projects stand out as a promising choice to nations and companies pursuing reduction of their CO2 emissions. Investing in this business could be an important alternative for the

[1] Brazilian Petroleum Industry—IBP. Technical Note regarding carbon market. Available at: https://www.ibp.org.br/personalizado/uploads/2023/07/nt-mercados-de-carbono.pdf.

[2] Decree No. 9.073/2017 enacted the Paris Agreement. Article1(a) provides that the Paris Agreement aims at maintaining the increase in the global average temperature below 2°C compared to pre-industrial levels and making efforts to limit this temperature increased to 1,5°C compared to pre-industrial levels, recognizing that this would significantly reduce the risks and impacts of climate change.

[3] Intergovernmental Panel on Climate Change—IPCC. "Climate Change 2022—Mitigation of Climate Change". Page 166.

[4] International Energy Agency—IEA. World Energy Outlook 2023. "*Pathways for the Energy Mix*".

decarbonization of industries specially in the segments of gas, oil and coal.[5]

Brazil plays a significant role in the development of CCUS projects. The Brazilian International Chamber of Commerce recommends for the private sector to reduce its emissions not only by means of reforestation projects, but also in projects that apply technologies for storing and capturing CO2 in industrial process.[6] Brazil's location in a geologically stable region combines with the country's vast territorial extension indicates the huge capacity of the country's underground gas storage, with lower level of risk when compared to countries that suffer from natural disasters such as earthquakes and volcanic eruptions.

First, this work aims at presenting in item 2 a general overview regarding the technology and infrastructure required for the implementation of CCUS projects, considering that the high-level technology plays a key role in the development of the projects and directly impacts the investment decision making process. In item 3 this paper will approach the specific and applicable regulation of CCUS in Brazil, aiming at providing readers with legal information regarding current and under development normative frameworks. At last, item IV will provide for an overall analysis of challenges faced by Brazil for the successful implementation of CCUS projects, which will hopefully be overcome in a near future.

2 CCUS Technologies and Required Infrastructure

The potential of CCUS activities is directly related to the access and development of the technology its implementation requires. Therefore, this chapter shall provide for a brief resume of the technology applied in CCUS projects and the corresponding operational process.

CCUS technology comprises the capture of CO2 emissions produced mainly in large point sources such as industrial facilities, power plants and refineries that use either fossil fuels or biomass as fuel. The hole CCUS

[5] According to IEA the industry segments are liable for the following quantity of CO2 emissions: (i) 9% for the gas sector; (ii) 2% of the oil sector; and (iii) 29% for the coal sector. Information obtained in the IEA statistic chart from July, 2020.

[6] Brazilian International Chamber of Commerce. "*Opportunities for Brazil in Carbon Markets*". 2022 Report.

backbone structure encompasses the stages of separation and capture, transport and storage (or usage) of CO2.

The first stage of separation and capture consists of dissolving other gases produced during the combustion or industrial processes. There are three leading technologies methods to capture the CO2, which are (i) pre-combustion (ii) post-combustion; and (ii) oxy-fuel combustion. With an in-depth technical language, the pre-combustion mechanism involves transforming the fuel into a gas mixture consisting of hydrogen and CO2 before it is burnt. As the post-combustion process, it removes CO2 from the flue gas, by using a chemical solvent for instance, after the combustion process of fossil fuels, and the remaining hydrogen-rich mixture can be used as fuel. At last, the oxy-fuel system involves burning a fuel with almost pure oxygen to produce CO2 and steam, and the released CO2 can be subsequently captured.[7]

Once the CO2 is captured, it can be either used onsite, or compressed and transported to a suitable storage location, by pipeline, rail, truck or ship. Destination of such compressed CO2 can be for creation of new products or valuable services—reason why the word "Utilization" is included in CCUS—or for permanent injection and storage in geological formations in the subsoil—reason why the word "Storage" is included in CCUS.

Among the main types of underground storage area, the following formations could be used to store CO2: oil and gas reservoirs, in active or depleted fields, geological formations containing highly saline water, unmineable coal deposits, shale rich in organic matter and basalt. Oil and gas reservoirs and saline reservoirs are those bearing the most advanced projects in the world, since they can support higher volumes of CO2 comparing to the other geological formations.[8]

Despite the technological complexity of the methods described above, the energy industry already detains developed expertise in the technology methods required for the performance of CCUS activities. The oil and gas exploration and production activities involve technologies designed to

[7] The London School of Economics and Political Science. "*What is carbon, capture, usage and storage (CCUS) and what role can it play in tackling climate change*". March 2023. Available at: https://www.lse.ac.uk/granthaminstitute/explainers/what-is-carbon-capture-and-storage-and-what-role-can-it-play-in-tackling-climate-change/.

[8] Brazilian Energy Company - EPE. "*Carbon capture and storage: A brief guide on one of the key alternatives for transformation the oil oil and gas sector in Brazil*". August 2023.

locate and map CO2 storage areas, drill wells, pressurize and inject CO2 and monitor the process and possible leaks. These technologies are largely used to improve the oil recovery factor, namely in Brazil the enhanced oil technique, which has been applied by Petrobras for years in the pre-salt fields. Petrobras developed the first CO2 separation process associated with natural gas in ultra-deep waters, injecting the separated CO2 in a water depth of 2,200 m, which is the record for the deepest CO2 gas injection ever made in the world.[9]

3 CCUS Policy and Regulatory Framework Under Development in Brazil

Even though the applicable legislation regarding CCUS projects in Brazil is still under development, Brazil has already taken the first steps to the construction of a solid regulatory framework directed to the mitigation of climate change and combat of carbon emissions.

3.1 *Current Applicable Legal Provisions*

In 2021, the Brazilian executive power published Federal Law 14.134/2021 ("New Gas Law"), which provides for activities relating to the transport of natural gas,[10] and for the flow, treatment, processing, underground storage, packaging, liquefaction, regasification and commercialization of natural gas.

Paragraph 2 of article 3 of the New Gas Law allows for the equivalent applicability of the aforementioned legislation to other gases which do not fall within the definition of natural gas[11] established by the New Gas Law, but which adhere to specifications established by the ANP for an alternative gas. Therefore, the New Gas Law indicates that if the CO2 is duly specified by ANP all provisions regarding transportation and

[9] Brazilian Energy Company—EPE. "*Carbon capture and storage: A brief guide on one of the key alternatives for transformation the oil oil and gas sector in Brazil*". August 2023.

[10] In accordance with article 177 of the Brazilian Federal Constitution.

[11] Item XXI of article 2° provides for the following definition of Natural Gas: "*any hydrocarbon that remains in a gaseous state under normal atmospheric conditions, extracted directly from oil or gas reservoirs, the composition of which may contain wet, dry and residual gases*".

underground storage of CO2 could be applicable to the CCUS activities, enabling execution of CCUS activities in Brazil with legal security and clarity regarding the enforceability of legal rules. Until now, ANP has not enacted any CO2 specifications. For the purpose of Chapter 4, item of this paper, all references to "*natural gas*" shall be interpreted as to include any other gas specified by ANP, including CO2 if specified in the future.

Chapter 5 of the New Gas Law provides specifically for the underground storage of natural gas. The chapter establishes that ANP is the competent authority to grant the authorization for companies or consortium of companies incorporated under Brazilian law to carry out the activity of underground storage of natural gas, at sole risk of the interested party.[12] As the ANP systematic for offering areas and granting exploratory rights for oil and natural gas fields by means of bidding process and open acreage cycles, the New Gas Law provides that ANP shall be responsible for defining the geological formations for underground storage and the applicable rules for the grant of the authorization.[13]

Interested parties may request from the ANP information on geological data relating to areas with potential for underground storage of natural gas, upon payment of a fee as established by the ANP.[14] Interested parties may also carry out exploratory research activities to confirm the suitability or accurateness of such data, after authorization by the ANP is granted.[15] The data obtained by the private party during exploratory activities must be forwarded to the ANP free of charge.[16] It should be noted that ANP oil and natural gas exploration and production contracts—whether in concession or production sharing regime—provides that the first stage of the contracts is the exploratory stage, which differs from the possibility to execute exploratory activities provided for in the New Gas Law for the CCUS market, to the extent

[12] In this provision it is possible to identify similarities to the applicable framework for the exploration and production activities under the concession regime, as the most recent version of ANP concession contract provides in clause 2.2 that the concessionaire shall be responsible for all costs, risks and consequences related to the execution of the operations, according to the most recert version of the concession contract available in ANP website.

[13] Article 20 caput and paragraph 1st of the New Gas Law.

[14] Article 21, caput of the New Gas Law.

[15] Article 21, paragraph 1 of the New Gas Law.

[16] Article 21, paragraph 2 of the New Gas Law.

that in the oil and gas industry companies under contract are not allowed to execute exploratory activities without being under contract with just an authorization granted by ANP.

Article 22 of the New Gas Law establishes that third parties other than the holder of the authorization shall have access to underground storage facilities, under the terms established by ANP.[17] ANP shall also define the periods which the access to such facilities shall not be mandatory, considering the course of the investments for these facilities.[18]

Even though the applicable legislation regarding CCUS projects in Brazil is still under development, Brazil has already taken the first steps to the construction of a solid regulatory framework directed to the mitigation of climate change and combat of carbon emissions.

The Brazilian Federal Constitution provides in article 20 that the Brazilian natural resources, including those disposed in the subsoil, belong to the Federal Government.[19] Article 23 of the New Gas Law provides that this constitutional provision is not applicable for the natural gas storage in geological formations, but the company responsible for the storage of the gas shall not withdraw of gas superior of the injected amount.[20] Noncompliance with this provision shall subject the company to penalties of cancelling the authorization and other penalties provided for in article 2 of Law No. 8.176/1991.[21]

The ANP may cancel the authorization for geological storage of naturel gas in the following cases: (i) liquidation or bankruptcy of the holder of the authorization; (i) request by the holders of the authorization; (iii) complete and definitive deactivation of the storage facility; (iv) major non-compliance with the obligations arising from the New Gas Law, applicable

[17] Article 22, caput of the New Gas Law.

[18] Article 22, paragraph 1 of the New Gas Law.

[19] Article 3 of Federal Law No. 9.478/1997 also provides that Deposits of oil, natural gas and other fluid hydrocarbons in the national territory, including the land, territorial sea, continental shelf and exclusive economic zone, belong to the Union.

[20] Article 23, caput and paragraph 1 of the New Gas Law.

[21] Article 23, paragraph 2 of the New Gas Law. Article 2 of Law No. 8.176/1991 provides that it is a crime against property to produce goods or exploit raw materials belonging to the Federal Government, without legal authorization, which penalty is imprisonment from one to five years and payment of a fine.

regulations and storage services contract, under ANP regulations; and (v) failure to comply with independence and autonomy requirements established by the New Gas Law.[22]

3.2 Bills

The legislative procedure in Brazil involves several steps from the introduction of a bill to its enactment as a law. Initially, the bill is introduced and subsequently directed to various committees for in-depth debate, which might undergo significant amendments or might even be recommended for rejection. Following the committee, the bill is then brought to the house floor for further discussion. Upon its approval, it is then sent to a review by the other house (from the House to the Senate, or vice versa). Once both houses of Congress have approved the bill, it is sent for presential approval. Finally, with the President's approval, the bill is published in the official gazette for its enactment.

The key goal of the bills discussed in this work is to contribute to the emissions reductions and create a suitable environment to the development of CCUS activities, establishing safe legal conditions for companies to decide whether to invest in such projects. It should be noted that the bills were also designed aiming at achieving plurality of industry agents acting in the CCUS industry, meaning that its terms would be applicable not only for the oil and gas industry, but also for the entire energy business and its industrial processes.

As CCUS relevance in the commitment to combating climate change and promoting sustainable growth increases, multiple bills addressing its regulations have been introduced in Brazil. This multiplicity introduces a complex scenario where overlapping frameworks could lead to delays and overly complication of its legislative procedure.

Upon the multiplicity of bills, the Brazilian legislative procedure has a practice of attaching related bills to one another, aiming at consolidating the discussion, analysis and voting processes (in Portuguese, the term of attaching a bill is "*apenso*"). Once bills are attached, they are reviewed together through the committee stages and, if approved, proceed as a unified package to subsequent stages of the legislative process.

[22] Article 10, items I to IV and Article 23, paragraph 2.

Expanding upon the brief overview of the legislative procedure above, there are currently two Bills under approval by the National Congress providing for guidelines related to the implementation of CCUS projects in Brazil.

3.3 1. No. 528/2020 ("Future Fuels Bill")

The former Future Fuels Bill n. 528/2020 was finally approved in October 2024 as President Luiz Inácio Lula da Silva signed the Fuel of the Future Law (Law No. 14.648/2024), which marked a significant step toward a more sustainable energy future. This law builds on previous ideas, such as the Future Fuels Bill, with the primary goal of putting Brazil as a leader in the global energy transition. The legislation establishes a comprehensive framework that encourages low-carbon fuels and strengthens the regulatory environment for carbon capture, utilization and storage (CCUS) projects, emphasizing their vital role in meeting climate objectives and reaching carbon neutrality.[23]

The law emphasizes producing and using renewable fuels such as biodiesel, green diesel, biogas and biomethane. It specifies financial incentives and regulatory support to speed the deployment of biofuel infrastructure and encourage biomethane production. These efforts attempt to diversify Brazil's energy mix while lowering its carbon impact. Importantly, the law promotes technical innovation, promoting public-private partnerships to create sophisticated CCUS systems. R&D hubs for biofuels and carbon capture will be built to assist this innovation, ensuring Brazil'''s competitiveness in the developing global energy environment.[24]

The law considerably improves the regulatory environment for CCUS in Brazil by empowering the National Petroleum Agency (ANP) to monitor and regulate all elements of carbon capture and storage projects.

[23] 'Lula Enacts Fuel of the Future Law: "Brazil Will Drive the World's Largest Energy Revolution"' (Planalto) <https://www.gov.br/planalto/en/latest-news/2024/10/lula-enacts-fuel-of-the-future-law-201cbrazil-will-drive-the-worlds-largest-energy-revolution201d#:~:text=OF%20THE%20FUTURE-,Lula%20enacts%20Fuel%20of%20the%20Future%20law%3A%20%E2%80%9CBrazil%20will%20drive,for%20biomethane%2C%20among%20other%20initiatives> accessed 30 November 2024

[24] Santos D, '"fuel of the Future" Law Is Signed into Law' (Demarest, 17 October 2024) <https://www.demarest.com.br/en/fuel-of-the-future-law-is-signed-into-law/> accessed 30 November 2024

ANP is entrusted with defining recommendations for technical and environmental requirements, operating safety, and long-term monitoring of storage facilities. The law also tackles possible conflicts between CCUS activities and oil and gas operations, providing defined structures for collaboration and resource sharing. This integration is viewed as critical, given Brazil's significant experience in oil and gas exploration, which may be used to enhance CCUS infrastructure and knowledge.[25]

One of the law's most significant elements is its emphasis on environmental protection. All CCUS projects must undergo comprehensive environmental impact studies and follow tight monitoring and mitigation measures. Operators must provide complete reports outlining the safety procedures in place to prevent CO2 leaks and other possible dangers. Transparency is an important element, with data from these evaluations made publicly available to encourage community confidence and responsibility. This emphasis on environmental stewardship is meant to strike a balance between industrial development and ecological conservation, a vital factor in winning public support for CCUS programs.

The Fuel of the Future Law also includes methods for integrating CCUS projects into Brazil's overall carbon market plan. Under this structure, CCUS projects can create carbon credits that can be exchanged both locally and globally. This market-based strategy provides a financial incentive for businesses to participate in carbon capture systems, possibly opening up new revenue streams and building a stronger carbon reduction ecosystem. By combining CCUS with carbon trading, the law prepares Brazil to attract global investment and engage more actively in international efforts to combat climate change.[26]

Despite its ambitious breadth, the law's proper implementation will need overcoming various obstacles. Infrastructure development remains a significant challenge, particularly in rural places where delivering and storing collected CO2 can be logistically difficult and expensive. Significant investment will be required to construct pipelines, storage facilities

[25] Demirkol S, 'ANP Releases Study Report on Carbon Capture, Usage, and Storage (CCUS)' (Brazilian NR, 27 April 2024) <https://braziliannr.com/2024/04/27/anp-releases-study-report-on-carbon-capture-usage-and-storage-ccus/> accessed 30 November 2024

[26] ——, 'Brazil Leads Green Energy with Law on Fuels of the Future' (Brazilian NR, 14 October 2024) <https://braziliannr.com/2024/10/14/brazil-leads-green-energy-with-law-on-fuels-of-the-future/> accessed 30 November 2024

and other critical infrastructure. Another possible problem is regulatory coordination, which would need good collaboration across numerous government agencies to speed permitting procedures and assure compliance with environmental and safety regulations. Capacity building will also be critical, with training programs and knowledge-sharing activities needed to provide industry stakeholders with the skills and experience required to effectively manage CCUS projects.

The Fuel of the Future Law also establishes a framework for public-private collaborations, highlighting collaboration as a crucial success method. The law encourages corporations to move into CCUS projects by using Brazil's current oil and gas expertise, which includes understanding of geological formations and drilling methods, to assist carbon storage. The regulation further emphasizes the significance of data sharing and collaboration by requiring ANP to give interested parties with access to technical data on sedimentary basins. This option is designed to speed up project development by allowing corporations to build on previous geological investigations rather than completing fresh surveys from the start.[27]

Another distinguishing feature of the legislation is its emphasis on long-term surveillance and responsibility. Operators are held accountable for safe CO2 storage, with severe consequences for noncompliance including fines and termination of operating licenses. The government is also required to set clear guidelines for decommissioning storage facilities and handling post-closure duties, ensuring that stored CO2 is securely retained for the foreseeable future. This emphasis on long-term supervision is critical to preserving public trust in CCUS as a safe and effective climate mitigation solution.

Overall, the Fuel of the Future Law represents a watershed moment in Brazil's approach to sustainable energy and climate policy. By establishing a clear and supportive legal framework for CCUS, the law lowers investment risks and makes the environment more appealing to domestic and foreign stakeholders. It establishes Brazil as a global leader in the transition to low-carbon energy, with the potential to become a centre

[27] Santos D, '"fuel of the Future" Law Is Signed into Law' (Demarest, 17 October 2024) <https://www.demarest.com.br/en/fuel-of-the-future-law-is-signed-into-law/> accessed 30 November 2024

for CCUS research and implementation. However, fulfilling this potential will need continued collaboration among government, industry and civil society to confront the difficulties ahead and capitalize on the possibilities provided by this groundbreaking law.[28]

3.4 2. No. 1.425/2022

Bill No. 1.425/2022 ("CCUS Bill") outlines an extensive regulatory framework for the capture of CO2 and subsequent reutilization aimed at reducing greenhouse gas emission, as well as advancing Brazil's decarbonization initiatives according to its economic strategy concerning climate change and sustainable development.[29]

CCUS Bill introduces that the capture of CO2 shall come from various stationary sources, including direct air capture.[30] Injection and permanent storage shall occur within the geological formations located in Brazil's sedimentary basins of the national territory, in the exclusive economic zone or on the continental shelf under Brazilian jurisdiction. The geological storage shall only occur on formations that upon previous assessment do not show any kind of environmental impact and risk, especially regarding the scape of CO2 storage.[31] The legislation project establishes that its CO2 injection rules are not applicable for hydrocarbon recovery technique.[32] This exclusion represents a clear intention to separate the regulation for the CCUS activity and the oil production recovery technique, even with similar operational natures.

According to the CCUS Bill, the exploration of geologic reservoirs for the permanent storage of CO2 shall be executed by a private party upon the grant of a 30-year authorization by the Executive Power.[33] The competent regulator shall further provide for guidelines regarding the

[28] Santos D, '"fuel of the Future" Law Is Signed into Law' (Demarest, 17 October 2024) <https://www.demarest.com.br/en/fuel-of-the-future-law-is-signed-into-law/> accessed 30 November 2024

[29] Article 1 of CCUS Bill.

[30] Article 1, paragraph 1 of CCUS Bill.

[31] Article 1, paragraph 2 and 3 of CCUS Bill.

[32] Article 1, paragraph 5 of CCUS Bill.

[33] Article 5 caput and paragraph 1 of CCUS Bill.

decommissioning and closure of CO_2 injection infrastructure.[34] The draft legislation states that the CCUS activity may be developed simultaneously with mineral exploration, hydrocarbon exploration and production and water resource exploration activities in a block under previously governmental contract or authorization, but that in the event of the project's coexistence impossibility, execution of the CCUS project will depend on the consent of the holder of the pre-existing right, in accordance with regulations.[35]

The CCUS project authorization shall be cancelled in the following cases: (i) non-compliance with the conditions of the authorization; (ii) if once the project starts, the level of operation is lower than the minimum stipulated in the authorization for 3 (three) consecutive years, or for 10 (ten) years added together during the term of the authorization; (iii) if cancellation is required to prevent or mitigate serious environmental degradation; (iv) if cancellation is required to attend priority uses of the geological reservoir, to protect collective interest, which are technically incompatible with the activity of permanent storage of CO_2.[36]

The CCUS Bill provides that the Executive Power shall disclose the geological reservoirs for which the private parties may request authorization to develop CCUS projects, describing the CO_2 storage capacity.[37] The Executive Power shall design an entity responsible for the studies regarding the geological reservoirs and its CO_2 storage capacity. The competent governmental body—most likely the ANP—may organize a public consultation to receive expressions of interest in storage blocks in previously identified reservoirs, and private companies that acquire the authorization right of the storage blocks they submitted expression of interest shall reimburse the entity that executed the studies.[38]

Private parties may request a qualified authorization for CO_2 storage in reservoirs that have not been disclosed by the Executive Power, provided that they submit studies regarding such reservoir capacity and meet the

[34] Article 5, paragraph 2 of CCUS Bill.

[35] Article 5, paragraph 3 of CCUS Bill.

[36] Article, items I to IV of the CCUS Bill.

[37] Article 8, caput of the CCUS Bill.

[38] Article 8, paragraph 2 of the CCUS Bill.

safety standards and requirements.[39] As of the conditions for the authorization request, the CCUS Bill establishes the following requirements (i) indication of the operator responsible for the operation of the injection infrastructure, without prejudice to the joint and several joint and several liability of other consortium members; (ii) the block requested for storage; (iii) proof of fiscal and legal regularity and technical and economic-financial qualification to support the activity; (iv) an operating plan for the development of the activity; (v) a monitoring proposal for the period during which the CO2 storage activity will be carried out; (vi) contingency plan, under the terms of the regulations; (vii) financial guarantees compatible with the burdens and risks involved; (viii) documents proving the storage capacity of the geological block subject of the authorization request.[40]

It should be noted that once an expression of interest has been received for a geological storage block, the competent governmental body shall hold a public call for 30 (thirty) days to allow other companies to express their interest in such geological storage block.[41] If after 15 days from the beginning of the public call there are not any other companies interested in the geological storage block, the authorization will be granted to the company that submitted the initial authorization request.[42] If there are more than one company interested in the geological storage block, the competent governmental authority shall try to make the demands of all parties compatible. If unsuccessful, the competent governmental authority shall grant the authorization to the most advantage proposal according to the following criteria: (i) capacity to decarbonize their activities; (ii) feasibility of capture, transport and storage activities; and (iii) project implementation capacity.[43]

With focus on the environmental, social and public health accountability, the CCUS Bill determines that damages from CCUS' operation are held by the operator (its consortium members or third party with shared responsibility).[44]

[39] Article 8, paragraph 4 of the CCUS Bill.

[40] Article 9, paragraph 1, items I to VIII of the CCUS Bill.

[41] Article 9, paragraph 3 of the CCUS Bill.

[42] Article 9, paragraph 4 of the CCUS Bill.

[43] Article 9, paragraph 5 and 6 of the CCUS Bill.

[44] Articles 15, 16, 17 and 18 of the CCUS Bill.

Lastly, to consolidate the CCUS initiative into the Brazilian legal and regulatory framework, it introduces amendments to the Petroleum Law. We highlight the objective into promoting research and development of renewable energy and decarbonization initiatives, guidelines to promote carbon capture, and establishes ANP as the regulatory entity responsible for overseeing carbon capture and storage activities.

4 Next Steps for Implementing CCUS Projects in Brazil

In Brazil there are a few initiatives regarding CCUS projects which indicates the industry excitement regarding Brazil as a potential for the development of CCUS.

Company Repsol Sinopec Brazil announced in the end of 2022 a partnership with University Pontíficia Católica do Rio Grande do Sul (PUCRS) and the German startup DACMa MnbH to develop a pioneering project in Latin America, for the development of CO2 capture technology directly from the air, through the construction of an experimental plant capable of capture around 300 tonnes of CO2.[45]

Petróleo Brasileiro S.A. ("Petrobras") has signed an agreement with the government of Rio de Janeiro for a pilot carbon capture project, which is estimated to capture up to 30% of the state's emissions. The company has reserved a budget of R$ 25 billion for investments in renewable energy for the next five years. According to Petrobras' CCUS project, the gas will be taken to the Barra do Furado station, in the municipality of Quissamã (RJ), around 60 kilometres away. It will then be injected and stored in a salt reservoir. The project will have the capacity to store 100,000 tonnes of CO2 per year.[46]

Despite the CCUS projects ongoing in Brazil, the country still suffers from lack of accurate mapping of the geological storage areas and its potential, or a solid legal framework to allow the implementation of these

[45] News provided for EPBR. "*CCS: Opportunities and challenges for Brail to expand carbon capture*". August 2023. Available at: https://epbr.com.br/ccs-as-oportunidades-e-desafios-do-brasil-para-a-expansao-da-captura-de-carbono/.

[46] News provided by Brazilian journal. Available at: https://www1.folha.uol.com.br/ambiente/2023/12/petrobras-tem-projeto-piloto-para-captura-de-carbono-no-rio-de-janeiro.shtml.

projects with legal security. The lack of finance credit lines, public policies, public investments and tax benefits for the development of CCUS projects are also an obstacle for Brazil to attract investors, either for the construction of the required infrastructure or for the development of the geological storage of CO2 stage. According to the Research Center for Greenhouse Gas Innovation ("RCGGI") the barriers for the implementation of CCUS projects are no longer the high-level required technology, but overcoming the high costs, political barriers and management challenges.[47]

As countries around the world promote global initiatives to reduce greenhouse gas emissions, the proposed Bills discussed throughout this work represents a significant step for Brazil into achieving carbon neutrality and reducing its carbon footprint. Although its proactive approach, the encouragement for developing technologies for the capture, transport, storage and exploration of low-carbon-emission energy sources highly depends on multiple external factors, such as, availability of financial resources, including significant investment in technology and infrastructure. Besides the CO2 emissions reduction, regulating CCUS and creating governmental benefits for the development of CCUS projects may attract new business opportunities for Brazil, which has all geological factors at its favour to turn the country a huge CCUS hub.

The development of a consolidated carbon market is also a powerful tool to support CCUS project. This is because the carbon market operates on the principle of emissions trading, by means of which nations or companies can buy and sell carbon credits as they quantify the reduction or removal of greenhouse gas emissions. This would be a financial incentive for the investment in CCUS projects, whereby companies could monetize their emissions reduction. The combination of the CCUS market and the carbon credit trade market could help developing countries to attract funding from global entities seeking to offset their own emissions.

It is also worth analysing that government authorities must adapt themselves to the new activities intrinsic to the renewable energy market,

[47] Research Centre for Greenhouse Gas Innovation—RCGI. "*Barriers to implementing carbon capture and storage are political, social and commercial, say experts*". 2023. Available at: Barreiras para implementação de captura e armazenamento de carbono são políticas, sociais e comerciais, afirmam especialistas - Só mais um site Universidade de São Paulo - Sites (usp.br).

beyond from the construction of regulatory rules point of view. Under the Brazilian reality, the Ministry of Mines and Energy and the ANP, as the government agency competent to deal with the matter, would need to develop robust monitoring, reporting and verification mechanisms to accurate the emissions reductions. These systems must be based on the principle of transparency and efficiency, which enhances trust among market participants and ensures the integrity of the carbon credit system.

5 Conclusão

This paper has provided a comprehensive overview of CCUS activities in Brazil, focusing on the existing regulatory framework and the legal provisions under construction by the Fuel of the Future Law and CCUS Bill, which ultimate goal is to shape the landscape of CCUS implementation.

The discussion on the current and under construction regulatory framework with an analysis of the applicable provisions of the New Gas Law, the Fuel of the future Law and the CCUS Bill highlighted both strengths and weaknesses. The existing regulatory landscape, although indicative of Brazil's commitment to addressing climate change, has shown certain gaps and ambiguities, particularly concerning the absence of a dedicated and comprehensive legislative structure specifically tailored for CCUS activities.

In light of the insights gleaned from this examination, it is evident that a collaborative effort between government, industry and environmental stakeholders is imperative to navigate the complexities of CCUS implementation successfully. As Brazil continues to develop its regulatory framework and refine its legislative approach, it is crucial to strike a balance that not only encourages technological innovation and investment but also ensures environmental sustainability and social acceptance.

CHAPTER 12

CCUS in Canada

Allan Ingelson and Carter Czaikowski

1 Introduction

Fossil fuel combustion in the oil and gas industry is a major source of carbon dioxide (CO_2) emissions, and in light of its commitments under the Paris Agreement,[1] the Government of Canada has taken steps to significantly reduce Canada's national greenhouse gas (GHG) emissions. Two basic types of CO_2 emissions reduction projects deployed in Canada are enhanced oil recovery (EOR) projects and carbon capture and storage (CCS). EOR projects that emit 82% less emissions than traditional oil production operations are the more common type of project, as they recover otherwise uneconomic oil from subsurface reservoirs. The additional revenue from EOR projects makes them a more attractive

[1] *Paris Agreement to the United National Framework Convention on Climate Change*, 12 December 2015 (Report No. FCCC/CP/2015/L.9/Rev.1).

A. Ingelson (✉)
Faculty of Law, University of Calgary, Calgary, AB, Canada
e-mail: allan.ingelson@ucalgary.ca

C. Czaikowski
University of Calgary, Alberta, AB, Canada

E. G. Pereira et al. (eds.), *Carbon Capture Utilization and Storage*, Sustainable Development Goals Series,
https://doi.org/10.1007/978-3-031-81272-9_12

investment to oil and gas developers. CCS projects are strictly designed to capture and store CO_2 emissions into pore space in depleted oil and gas reservoirs. The goal of CCS technology is to prevent approximately 90% of the CO_2 emissions from major industrial emitters from being released into the atmosphere.[2] Initially oil, natural gas, helium, and CO_2 have been stored naturally in subsurface geological reservoirs around the world for hundreds of millions of years. Decades of Canadian petroleum reservoir engineering experience with EOR projects is now being used to mitigate the impacts of climate change.

Due to different types of gas injection, extraction and storage projects in the same area, there is the potential for multiple subsurface use disputes between hydrocarbon developers and CCS project operators. Alberta, the largest oil and gas producing province in Canada has amended its legislation and developed new regulations to provide increased certainty of tenure for CCS developers to prevent rights disputes among oil, gas, and CO_2 project developers, and thereby encourage more investment in CCS projects. This chapter discusses the legal framework for CCUS and CCS projects in Canada, including one of the world's largest transboundary EOR projects in Weyburn, Saskatchewan that provides for the transportation and injection of CO_2 from the United States and the Quest project in Alberta. Issues that remain to be addressed to provide a more effective and efficient CCS regulatory framework are discussed.

2 Canadian Projects

With both the provincial and federal governments identifying CCS as an important part of CO_2 reduction strategies, both levels of government have set emission reduction plans and goals. The Canadian government's goal to reduce CO_2 emission by 325 Mt/s by 2050 is set out through its Clean Energy Fund. Alberta released the Climate Change Strategy in 2008 with the goal of reducing emission by 200 megatons by 2050, with CCS projected to provide 70% of these savings

[2] Maucler, T., Campbell, W., Rujis, J. "Carbon capture and storage: assessing the economics" (2008) McKinsey & Company Inc., online: https://projects.mcrit.com/foresightlibrary/attachments/ccs_assessing_the_economics_mckinsey.pdf.

through the *Carbon Capture and Storage Funding Act* and its associated regulations.[3] In Canada the provincial governments have jurisdiction over energy resources and environmental matters within each province. As provincial governments have jurisdiction over energy development within provinces (oil, gas, wind, solar, hydroelectricity, geothermal), EOR and CCS projects are typically located on provincial lands and therefore provincially regulated. There are several different options for suitable underground storage of CO_2, with several selection criteria needed to be met to ensure that the CO_2 will be stored permanently. The natural onshore storage options available globally are depleted oil and gas fields, deep saline aquifers, coalbed methane reservoirs, and salt caverns. Offshore options are available as well, however, in Canada this is still in the research stage in Nova Scotia.[4] The stable rock formation of the Western Canadian Sedimentary Basin (WCSB), with depleted oil and gas reservoirs in the provinces of British Columbia, Alberta and Saskatchewan, provides desirable locations for CCS projects near large CO_2 emitting facilities located in the province of Alberta. The depleted reservoirs in the WCSB that once stored oil and gas resources are now used for CO_2 storage. Deep saline aquifers in the WCSB hold significant potential for CCS projects in Canada,[5] and the potential to eliminate 40% of Canada's GHG emissions by 2050.[6] The solution is supported by national and international experts who see this as a technology that can be applied to several applications such as power generation and oil sands.[7]

[3] Energy Resources Conservation Board Decision, ABERCB 008: Shell Canada Limited, Application for the Quest Carbon Capture and Storage Project, Radway Field, 10 July 2012.

[4] Natural Resources Canada, Clean fossil fuels—Carbon capture & storage, storage of CO2, 2008, June 4, 2010 http://canmetenergy-canmetenergie.nrcan-rncan.gc.ca/eng/clean_fossils+furls/carbon_capture_storage/storage.html.

[5] Natural resources Canada, Canada's Fossil Energy Future: The Way Forward on Carbon Capture and Storage, 4 June 2010, online: https://natural-resources.canada.ca/publications/fossil-energy-future/1167.

[6] *Ibid.*

[7] Bulut D., "The Canada CCS experience" {PowerPoint slides} (2007). Retrieved June 8, 2010, from http://www.netl.doe.gov/publications/proceedings/07/rcsp/pdfs?Bulut%20The%20Canadian%20CCS%20Experience_Dec07.pdf.

2.1 *Saskatchewan—An International EOR Project*

Canada is a proven leader in research and development of CCS projects. One of the world's largest EOR CCUS projects was launched in 2000 near Weyburn, Saskatchewan.[8] The project was designed as an international collaborative scientific study to assess the technical feasibility of CO_2 injection and storage in geological formations of depleted oilfields.[9] The CO_2 injection into the Weyburn field was initially around 5,000 tones/day, which otherwise would have been released into the atmosphere.[10] From this project, a Best Practices Manual was developed for the design, implementation, monitoring, and verification of CO_2 geological storage. The project is composed of two different CO_2 operations, with both purchasing CO_2 from North Dakota. It is one of the first cross-border transfers of CO_2 through the sale of physical quantities of CO_2 sold commercially for EOR with an additional benefit of carbon sequestration.[11] The Cenovus Energy storage project, one of the operations, is recognized as the worlds largest geological CO_2 storage project. The other project, run by Apache Canada, is using CO_2 injection for EOR to produce 2540 barrels of oil per day and anticipates that over 10 Mt of CO_2 will be stored over the lifespan of the project.[12] The Weyburn and Midale fields are expected to produce around 220 million additional barrels of incremental oil through displacement with CO_2, extending its life by approximately 20–25 years.[13] The Weyburn project verified the ability of an oil reservoir to securely store CO_2 for significant periods of time. The Weyburn project is unique in that it is still developing oil at this time, and several oil wells are drilled in the oil field to produce the oil. Therefore, the need for constant testing is crucial to ensuring that CO_2 does not escape. The surface, seismic, and chemical monitoring have repeatedly shown that these geological formations are securely storing the CO_2 which was injected for EOR activities. Further, the studies and

[8] Government of Saskatchewan, Saskatchewan Announces Carbon Capture and Utilization and Storage Priorities, 7 September 2021.

[9] Government of Canada, International Energy Agency Greenhouse Gas Weyburn-Midale CO2 Monitoring and Storage Project, 4 March 2013.

[10] Wikipedia, Weyburn-Midale Carbon Dioxide Project.

[11] *Supra* note 10 & 11.

[12] *Supra* note 10.

[13] *Supra* note 11.

surveys of the surrounding land and water done during the first phase confirmed that CO_2 had not leaked into the biosphere.[14] Thirty percent of the injected CO_2 could be released into the atmosphere, however, it is captured during production in order to be re-injected.[15] Even though research and performance has shown no signs of leaks from wells due to continued drilling in the Weyburn and Midale fields and that no CO_2 has escaped from the geological formations of the project, geologist caution that leaks may occur from certain types of geological formations where there is continued drilling.[16] Another project in Saskatchewan is SaskPower's Boundary Dam Unit 3 CCS Facility, which is the world's first CCS facility on a commercial powerplant.[17] It has captured more than 5 million tons of CO2 thus far.

2.2 *Alberta—The Shell Quest Project*

Since the 1970s CO_2 has been successfully injected and sequestered for the purpose of EOR in Alberta to increase the amount of petroleum that can be recovered from subsurface oil and gas reservoirs. Injecting CO_2 into some oil and gas fields can increase the reservoir pressure to facilitate additional production of hydrocarbons from the fields.[18] In light of the economic value of the hydrocarbons produced from EOR projects, there have been a much higher number of EOR projects in Alberta (at least 48 projects) compared to only two CCS projects. CO_2 storage sites in Alberta include empty pore spaces in depleted oil and natural gas reservoirs from which the hydrocarbons have been extracted, active oil and natural gas fields, and deep saline geological formations.[19] In Alberta, decades of gas injection and storage regulatory experience

[14] *Supra* note 11.

[15] McElligot, S., "Canada's PTRC finds No CO_2 seepage at Saskatchewan's EOR operations" (2007) 10:1 Gasification News, 10–10, 1p.

[16] Fogarty, J., McCally, M., "Health and safety risks of carbon capture and storage" (2010) 202:1 The Journal of the American Medical Association, 67–68.

[17] International CCS Knowledge Centre, Open Letter to International Leaders, 10 July 2023, online: https://ccsknowledge.com/news/ccs-is-needed-and-ready-to-address-climate-change---open-letter-to-international-leaders.

[18] *Supra* note 5.

[19] Orr, F.M., "Onshore geologic storage of CO_2" (2009) 325:5948 Science 1656–1658.

with EOR injection projects was applied to the review and approval of the first dedicated long-term CCS project approved in 2012, the Quest Carbon Capture and Storage Project.[20] The Quest project is designed to capture and store more than 1 million tons of CO2 annually in deep subsurface saline reservoirs.[21] The application for the project by Shell Canada Limited set out a pipeline to transport CO_2 from the Shell Scotford Upgrader to a proposed injection of CO_2 into the Basal Cambrian Sands. The lifespan of the Project is estimated to be more than 25 years, as per the lifespan of the Scotford Upgrader. The goal of the project is to reduce GHG emissions, as there are no other large scale commercial alternatives in Alberta to direct GHG reductions.[22] The project provides knowledge to help accelerate the implementation of other CCS projects as well as helping reduce the overall carbon footprint of oil sands, which is important to the security of the energy supply in North America. The CCS project is designed to capture around 80% of the CO_2 emissions being generated by the Scotford Upgrader, equaling to about 1.2 Mt/a of CO_2.[23] The Project required an Environmental Assessment as per the Canadian Environmental Assessment Act (CEAA) that identified mitigation measures to minimize adverse environmental and social impacts.[24]

The Government of Alberta has approved 25 CCS projects to proceed to the evaluation stage, as a part of their request for project proposals for carbon sequestration hubs. In the fall of 2021, the Alberta Government requested Expression of Interest from companies who would be interested in developing and operating a carbon sequestration hub in Alberta.[25] On June 26, 2024, ATCO EnPower announced that it had made a final investment decision together with its partner Shell Canada Limited to proceed with the first phase of an open-access Atlas Carbon Storage Hub, to store industrial emissions from Alberta's Industrial Heartland

[20] *Supra* note 4.

[21] *Supra* note 21.

[22] *Supra* note 20.

[23] *Supra* note 20.

[24] *Supra* note 20.

[25] Government of Alberta, Request for Full Project Proposals For Carbon Sequestration Hubs, 3 March 2022.

region.[26] ATCO noted the success of Shell Canada Limited in storing million 9 million tons of CO_2 at its Quest CCS facility in Alberta since 2015, as a factor in proceeding with the carbon storage hub in 2024. The Atlas Hub project will be anchored by CO_2 volumes from another Shell carbon capture project called Polaris. A certain regulatory framework in Alberta and funding to support CCS projects from the Canadian federal government[27] should promote more CCS projects in the province.

2.3 British Columbia

Northeastern British Columbia (B.C.) has a number of large natural gas processing plants. The Western Canada Sedimentary Basin (WCSB), extending from Alberta into northeastern BC, has porous and permeable reservoirs in the form of depleted natural gas pools and deep saline formations that are suitable for CO_2 storage. Legislation in B.C. has set greenhouse gases reduction targets for the province to reduce emissions levels to 33% below 2007 levels by 2020, and 80% by 2050.[28] To store CO_2, the gas must be captured. Sour gas and flu gas contain varying amounts of CO_2, that must be isolated before it can be transported via pipeline. CO_2 is removed from sour gas through the removal of H_2S, with B.C. having 12 approved sites for acid gas disposal in northeastern B.C. CO_2 is removed from the flu gas in three possible ways: (1) post-combustion using amine solvent, (2) pre-combustion—fuel mixed with steam, and (3) oxyfuel combustion—oxygen used in initiation combustion. After the CO_2 is captured, it must be transported via pipeline to a storage site, with northeastern B.C. currently having several short acid gas pipelines active.[29] Next, a porous and permeable reservoir is needed to store the injected CO_2. Potential storage reservoirs include

[26] ATCO EnPower, news release, June 26, 2024; atcocom/en-ca/about-us/news/2024/123013-atco-enpower.

[27] Government of Canada, Carbon Capture, Utilization, and Storage (CCUS) Investment Tax Credit (ITC), 21 June 2024, online: https://www.canada.ca/en/revenue-agency/services/tax/businesses/topics/corporations/business-tax-credits/clean-economy-itc/carbon-capture-itc.html. See also, Government of Alberta, Alberta Carbon Capture Incentive Program, online: https://www.alberta.ca/alberta-carbon-capture-incentive-program.

[28] Alf Hartling, "Carbon Capture and Storage in British Columbia", (2008) Ministry of Energy, Mines and Petroleum Resources, 25–31.

[29] *Ibid.*

depleted reservoirs, deep saline formations, non-mineable coal beds, salt caverns, and abandoned gas facilities. Further, oil pools that are near depletion can potentially utilize CO_2 injections for EOR. In B.C. CCS site monitoring is required to comply with provincial government regulations and can play an important role in establishing liability for damage that could be incurred from leakage. Monitoring programs begin with reservoir simulator models to identify areas of potential CO_2 migration pathways. In order for the storage space to be used for CO_2 sequestration, the gas pools must be depleted and production operations have ceased—something that could be decades away.[30]

2.4 *Newfoundland and Labrador*

Newfoundland and Labrador (NL) offshore oil and gas production is responsible for 1.2% of Canada's oil and gas sector emissions. In 2020 the Government of NL committed to achieve net zero GHGs by 2050. NL's Net Zero Project's objectives regarding CCS will be to provide (1) technical and economic analysis; (2) policy and regulatory analysis; and (3) development for a roadmap to advance CCS as an option for reducing GHG offshore emissions.[31] However, as CCS has not yet been applied in the Canadian offshore context, this will require concerted research and development. In NL offshore facilities space and weight are at a premium, meaning the combustion capture technology that has the least complicated process, requires less energy, and involves minimal equipment installation will be favored. Technologies are advancing but none that are fully commercialized to meet the challenge of implementing carbon capture on offshore NL. While EOR is also a recovery method that can utilize CO_2, accessing the required amounts for full EOR would be difficult offshore, with a more likely scenario for utilizing CO_2 offshore is alternating CO_2 and water injection.[32]

[30] *Ibid.*

[31] "The Net Zero Project: Carbon Capture, Utilization and Storage Offshore Newfoundland and Labrador" (February 2023), online: https://energynl.ca/wp-content/uploads/2023/03/NZP-CCUS-White-paper-Main-Report-Ver-09.pdf.

[32] *Ibid.*

3 Legal Framework for CCS

The province of Alberta is the largest oil and gas producing province and thus has the greatest opportunity for CCS and CCUS. Further as Alberta already has CCS and Enhanced Oil Recovery (EOR) projects in the province, it will be the focus of the discussion for the legal frameworks, policies, and best practices in relation to carbon sequestration. The provincial regulator called the Alberta Energy Regulator (AER) regulates both types of projects in the province. There are four stages in the CCS process: capture, transport, injection, and storage or disposal. In the first stage, CO_2 is captured and separated from the industrial production source, usually a facility where there is significant fossil fuel combustion of CO_2. A variety of technologies may be employed to separate CO_2 from other gases. While it is the provincial government that regulates the existing CCS and EOR operations located on provincial lands, recently there has been increased interest in promoting these projects by the federal government as part of a strategy to reduce national GHG emissions.

In light of the Government of Canada's *Net Zero Emissions Act,* adopted in 2021[33] recognizing that oil and gas production will continue during the Canadian energy transition, the federal government announced a refundable investment tax credit for corporations developing carbon capture, utilization, and storage projects (CCUS). The aim of the credit is to incentivize expansion of CCUS, with it offsetting the cost of purchase and installation for eligible equipment. The CCUS tax credit operates on a sliding cost scale with 60% for eligible capture equipment used in a direct air capture project", 50% for other eligible capture equipment, and 37.5% for eligible transportation, storage, and use equipment.[34]

[33] *Net Zero Emissions Act,* S.C. 2021, c. 22.

[34] Greg Johnson, Thomas McInerney, Brad Gilmour, Brendan Sigalet & Peter Ciechanowski, "Canadian Budget Proposed New Investment Tax Credit For Carbon Capture, Utilization and Storage" (April 12, 2022), online: https://www.bennettjones.com/Blogs-Section/Canadian-Budget-Proposes-New-Investment-Tax-Credit-For-Carbon-Capture-Utilization-and-Storage.

3.1 *Alberta's Legal & Regulatory Framework*

Carbon capture, utilization, and storage (CCUS) refers to a range of technologies and processes that capture carbon dioxide (CO_2) from sources such as industrial facilities, transport the CO_2 through pipelines, then inject it into deep subsurface geological formations (e.g., saline aquifers or depleted oil and gas reservoirs) for permanent storage. Legislation initially adopted to regulate oil and gas well drilling, injection, and withdrawal of natural gas and the construction of pipelines is also employed to regulate CCS projects. The main acts include the *Mines and Minerals Act*,,[35] *Oil and Gas Conservation Act* (OGCA)[36] and regulations, *Responsible Energy Development Act* (REDA)[37] and associated regulations, and the *Environmental Protection and Enhancement Act* (EPEA).[38]

The AER regulates two basic types of CCUS projects:

(1) Carbon Capture and Storage (CCS)

CCS projects are also referred to as CO_2 sequestration schemes or dedicated storage. A CO_2 sequestration scheme is the permanent storing and trapping of CO_2 in an approved subsurface formation. A carbon sequestration tenure or agreement from the Government of Alberta is required if a company wants to apply for a CCS project. An example of a CO_2 sequestration project that has received AER approval is the Shell Quest CCS project.

(2) Enhanced Oil Recovery (EOR)

Using CO_2 in EOR schemes improves the production of remaining oil in subsurface reservoirs, which also has the effect of permanently sequestering some CO_2. The AER has regulated EOR projects for more than 25 years.

In light of decades of oil and gas regulatory experience and related engineering expertise with the injection and withdrawal of oil and natural gas, the AER has the authority to regulate the oil and gas-related facilities

[35] Mines and Mineral Act, RSA 2000, c. M-17.

[36] *Oil and Gas Conservation Act*, RSA 2000, c O-6.

[37] *Responsible Energy Development Act*, SA 2012, c R-17.3.

[38] *Environmental Protection and Enhancement Act*, RSA 2000, c E-12.

that capture CO_2, the pipelines that transport the CO_2, and gas subsurface injection activities. The following AER directives outline specific requirements on how wells are to be used to inject CO_2 into subsurface formations so that CO_2 can be safely stored in subsurface hydrocarbon reservoirs.

- *Directive 020: Well Abandonment* contains the requirements for abandonments, casing removal, zonal abandonments, and plug backs.
- *Directive 051: Injection and Disposal Wells—Well Classifications, Completions, Logging, and Testing Requirements* contain the requirements for wellbore design, wellbore integrity logging, operational monitoring, and reporting of injection wells.
- *Directive 056: Energy Development Applications and Schedules* contains the requirements for facilities and pipelines that handle CO_2 and the requirements for wells that inject CO_2.
- *Directive 065: Resources Applications for Oil and Gas Reservoirs* contains the subsurface requirements for CCS and CCUS processes.
- *Directive 071: Emerging Preparedness and Response Requirements for the Petroleum Industry* contains the requirements for preparing and implementing an emergency response plan.
- *Directive 087: Well Integrity Management* contains the requirements for testing, reporting, and repairing isolation packers, surface casing vent flows, gas migration, and casing failures.

Recent Regulatory Developments. In March 2022, the AER moved to create increased regulatory certainty for additional CCUS projects in Alberta and to provide a clear regulatory framework for industry and internal procedures for AER staff.[39] The relevant regulatory instruments are AER Directives 020, 051, 056, 065, and 087. The AER's subsurface requirements for CO_2 schemes are set out in *Directive 065: Resources Applications for Oil and Gas Reservoirs*, Units 2 and 4.

Provincial Laws:

- The *Mines and Minerals Act* provides that the provincial government owns the pore space for CO_2 sequestration schemes. The

[39] Alberta Energy Regulator, Regulatory Change Report, 24 March 2022.

Alberta Department of Energy and Minerals (ADEM) issues the CO_2 tenure rights required to test depleted subsurface oil and gas reservoirs for their viability to store CO_2 indefinitely.

- Surface rights for CO_2 sequestration and CO_2 EOR and storage projects are managed the same as for oil and gas projects in the province, under the *Public Lands Act*, *Surface Rights Act*, and the *OGCA*. The government is accountable for the long-term liability if there are CO_2 leaks/releases from CCS reservoirs in the future.
- The *Carbon Sequestration Tenure Regulation* enables the Government of Alberta to issue evaluation permits, agreements, and leases for carbon sequestration in Alberta.
- The *Oil and Gas Conservation Act (OGCA)* permits the AER to approve CO_2 schemes if CO_2 injection will not interfere with the recovery or conservation of oil and gas, or an existing use of the underground formation for storing oil and gas.

In January 2024, the Alberta Department of Energy and Minerals (ADEM) released Ministerial Order 060/2023 (MO 060/2023) that delegates several technical decision-making powers concerning carbon capture and storage (CCS) from the Minister of Energy to specified officials within the AER.[40] This change is designed to improve the efficiency of the CCS regulatory process by distinguishing the respective responsibilities of the Department of Energy and Minerals for CCS development policy decisions versus CCS-related technical decisions between and the AER, respectively. MO 060/2023 clarifies the powers of the Minister overseeing the ADEM as under Part 9 of the *Mines and Minerals Act*, and the Carbon Sequestration Tenure Regulation, Alta Reg 68/2011 can now be exercised by officials within the AER holding particular titles or offices.[41] It enables the AER officials to manage various functions such as the approval of monitoring, measurement, and verification plans, and the issuance of closure certificates—critical elements for CCS operations. MO 060/2023 delegates to AER officials the following powers, functions, and designations of the Minister under the *MMA* and the Carbon.

[40] Release of the Text of a Ministerial Order Delegating Technical CCS-Related Decision-Making Authority to the Alberta Energy Regulator (29 January 2024).

[41] Such as: Director, Oil & Gas Subsurface; Waste & Storage, Director; In Situ, Manager; In Situ Applications Approvals, Manager; Oil & Gas Subsurface Applications, Manager; Geology & Hydrogeology, or Senior Reservoir Engineer.

Sequestration Tenure Regulation:

- Approval of MMV plans for Part 9 sequestration lessees[42];
- MMV compliance reports[43];
- Closure plans and applications for closure certificates[44];
- The authority to reject closure certificate applications if deemed incomplete or inaccurate[45];
- Verification that all preconditions for issuing a closure certificate are met, including stability and predictability of the sequestered CO_2, with no significant risk of future leakage[46];
- Approval of MMV plans that must be filed by permittees and lessees[47]; and
- Authority to approve and renew closure plans.[48]

The order provides for technical decisions to be made by the AER, a regulator that is independent from the Alberta Government, as the AER has the requisite expertise to improve the efficiency of the regulatory framework. However, the Minister of ADEM retains certain important powers such as the Crown's mineral proprietary rights, including the authority to issue permits and leases on behalf of the Crown. In addition, the Minister continues to oversee the Post-closure Stewardship Fund,[49] which includes the authority to set the contribution rates for the fund.[50]

With regard to regulating CCS and enhanced recovery projects, under Sect. 39 of the OGCA no company may commence an EOR scheme or the injection, disposal, or storage of any fluid into a subsurface geological formation through a well without the approval of the AER.[51] As CCS projects employ similar gas injection technology and processes, the AER

[42] *MMA*, s 116(3)(a).

[43] *MMA*, s 116(3)(c).

[44] *MMA*, ss 116(3)(e), 120(1).

[45] *MMA*, s 120(2).

[46] *MMA*, s 120(3)(f).

[47] CSTR, ss 7(1), 15.

[48] CSTR, ss 18–19.

[49] *MMA*, s 122.

[50] CSTR, s 20.

[51] *Oil and Gas Conservation Act*, RSA, 2000, c O-6.

with experience in regulating EOR well injections and monitoring, evaluates and approves CCS projects that satisfy the regulator's criteria.[52] The OGCA and the associated regulations and AER directives classify different types of wells for injection of CO_2 or other gases used in EOR or disposal or storage. *Directive 065: Resource Applications for Oil and Gas Reservoirs* contains the subsurface requirements for CCS and EOR processes. These wells are subject to increased levels of monitoring and higher standards for well cementing, casing, completions, and testing. EOR project applicants must notify companies which hold adjacent subsurface interests that may be affected by their activities. The plugging and abandonment of oil and gas wells, including EOR wells and those for acid gas disposal operations, to protect the environment and public health and safety is regulated by the AER to ensure that downhole and subsurface infrastructure is left in a stable and safe condition, including the isolation and protection of aquifers and groundwater.[53] *Directive 020: Well Abandonment* contains the requirements for abandonments, casing removal, zonal abandonments, and plug backs. Well completion reports and plug logs must be submitted to the AER. Much of this regulatory scheme can be applied directly to the abandonment phase of a CCUS project, with one exception; it does not provide for ongoing monitoring and verification of the well after abandonment.[54] This monitoring and verification, over a long period, is required for CCS projects. Regulation of EOR projects is aimed at optimizing worker and public safety, environmental protection, and equitable treatment of adjacent well licensees.[55] *Directive 087: Well Integrity Management* contains the requirements for testing, reporting, and repairing isolation packers, surface casing vent flows, gas migration, and casing failures. EOR projects have similar geographical scale to CCS projects, which makes them a useful analogy in terms of regulation, but on the other hand their objectives are different; the injection of CO_2 is both a means to increase hydrocarbon recovery and

[52] *Ibid.*

[53] Alberta Energy Regulator, AER Directive 020: Well Abandonment; effective September 5, 2023; released with AER Bulleting 2023-32.

[54] *Ibid.*

[55] Alberta Energy Regulator, AER Directive 065: Resources Applications for Oil and Gas Reservoirs; March 27, 2023; released with AER Bulletin 2023-12. Retrieved from: https://www.aer.ca/documents/directives/Directive065.pdf.

store CO_2 in the long term. When considering gas containment, reservoir characteristics, and hydraulic isolation, the AER requires a significant amount of information to demonstrate that the injected substances will be contained "within a defined area and geologic horizon", to protect the environment and public health and safety.[56] *Directive 051: Injection and Disposal Wells—Well Classification, Completions, Logging and Testing Requirements* contains the requirements for wellbore design, wellbore integrity logging, operational monitoring, and reporting of injection wells. The above environmental protection and health and safety issues are analogous considerations in the AER review of proposed CO_2 well injection and storage CCS projects. *Directive 071: Emerging Preparedness and Response Requirements for the Petroleum Industry* contains the requirements for preparing and implementing an emergency response plan.

To transport CO_2 for long-term injection and storage in Alberta, a proposal to develop a CO_2 pipeline must include a detailed plan. *Directive 056: Energy Development Applications and Schedules* contains the requirements for facilities and pipelines that handle CO_2 and the requirements for wells that inject CO_2. Efficient and reliable CO_2 pipelines have been approved by the provincial regulator.[57] For instance a 240-km pipeline called the Alberta Carbon Trunk Line was proposed in 2009 to transport CO_2 throughout the region.[58]

3.1.1 TIER Regulations

The Technology Innovation and Emissions Reduction Regulation (TIER) under the *Emissions Management and Climate Resilience Act*, sets out Alberta's approach to reducing emissions from industrial emitters. TIER sets emission benchmarks for large emitters in the province, defining them as "… a facility that has direct emissions of 100 000 CO2e tons or more in 2016 or a subsequent year". TIER's goal is to make operations more energy efficient by staying below emission standards. However if facilities are unable to meet the benchmark set by TIER then they can either (a)

[56] *Ibid*, Sect. 4.2.2.

[57] Aydin, G., Karakurt, I., Aydiner, K., "Evaluation of geologic storage options of CO2: applicability, cost, storage capacity and safety" (2010) 38:9 Energy Policy 5072–5080.

[58] Government of Alberta, News release: New Pipeline will Enhance Carbon Capture and Storage, 24 November 2009, online, https://www.alberta.ca/release.cfm?xID527386278A12C1-C3D7-722E-E591EC672F9FC009.

purchase a "fund credit" for each ton of excess emissions produced; (b) user emission performance credits from a TIER-regulated facility; or (c) submit emissions offsets generated under an approved offset protocol.[59] TIER, Alberta's output-based pricing system is deemed equivalent to the federal approach under the *Greenhouse Gas Pollution Pricing Act* (GGPPA).[60] In 2023, new changes were introduced to TIER including the creation of a variety of new TIER credits, such as for CCS. The new credits allow a facility to exchange an offset credit for a sequestration credit if the offset was created from the geological sequestration of carbon dioxide, and the credit can then be used to satisfy compliance obligations under TIER.[61] Further, TIER-regulated facilities are able to convert sequestration credits into capture recognition tons and then deduct them from the facility's net emissions.[62]

Section (1)(ee) of TIER sets out "net geological sequestration" as the quantity of carbon dioxide that is sequestered in a geological formation, less the quantity of specified gases, expressed in CO2e tons, released as a result of the sequestration activity". This sets out a separate definition from "net emissions" as "the net emission for a regulated facility determined under Sect. 13″ and "net sequestration" as "the quantity of carbon dioxide sequestered in a sink, other than a geological formation, less the quantity of specified gases, expressed in Co2e tons released as a result of the sequestration activity" (Sect. (1)(dd); (ff). Further, TIER Sect. 19(2) sets out that the net geological sequestration of carbon dioxide requirements for the geological sequestration to be recognized by one or more emission offsets. These requirements include the facility and geological formation being in Alberta, the geological sequestration must not be required by law, the construction of infrastructure and the sequestration must be after 2002, the sequestration must be quantifiable and measurable, the carbon dioxide sequestered must not have reduced a regulated facility's total regulated emissions, and the sequestration must be permanent. If TIER Regulations are not abided by, penalties are set out under

[59] DBH Law, "Amendments to TIER Syste Align Alberta with Federal Emissions Reduction Approach", 13 February 2023, online: https://dbhllp.com/amendments-to-tier-system-align-alberta-with-federal-emissions-reduction-approach/.

[60] *Greenhouse Gas Pollution Pricing Act,* S.C. 2018, c.12, s. 186.

[61] *Ibid.*

[62] *Technology Innovation and Emissions Reduction Regulation*, Alta Reg 133/2019.

Sect. 4 of the regulations, with a fine up to $50,000 per day for individuals and $500,000 per day for corporations.[63]

3.2 Property Rights in Alberta

There are different types of property rights holders in Alberta: Provincial (Crown lands), Private (freehold lands), Federal (Crown), and First Nation lands. If not regulated properly this can lead to disputes regarding rights and access. The legal system needs to balance competing commercial interests to provide fairness and certainty to encourage investment in new CCS projects.[64] Under the *Alberta Mines and Minerals Act*, 2000 the DEM previously the Alberta Department of Energy grants subsurface petroleum rights, such as petroleum and natural gas licenses and leases, and the rights for CCS projects. The Alberta Government added a ninth part to the *Mines and Minerals Act* (MMA) to clarify the rights of CCS project developers. Section 116(1) provides: "the Minister may enter into an agreement with a person that grants that person the right to inject captured CO_2 into a subsurface reservoir for sequestration", on the condition that "the lessee of an agreement referred to in subsection (1) shall obtain a well licence and the approval of the Regulator under the OGCA prior to drilling or using a well for the purposes of this section" (MMA, Sect. 116(2)). In addition, under the MMA the Alberta Government created a new regulation called the *Carbon Sequestration Tenure Regulation*[65] that contains clear definitions that provided increased certainty to CCS project developers:

1. In this Regulation,
(b) "carbon sequestration lease" means an agreement under section 116 of the Mines and Minerals Act issued in the form of a lease specified under section 9;
(c) "deep subsurface reservoir", in respect of a permit or lease, means the pore space within an underground formation that is deeper than 1000 m below the surface of the land within the location of that permit or lease.

[63] Ibid, Sect. 4.

[64] Bankes, N., Poschwatta, J., Shier, E.M., "The legal framework for carbon capture and storage in Alberta" (2008) 45:3 Alberta Law Review 585–630.

[65] *Mines and Minerals Act*, RSA 2000, c M-17.

CCS project developers must obtain CO_2 injection and storage rights from the Alberta Government, called an Evaluation Permit, and leases, which must comply with the Alberta MMA and the *Carbon Sequestration Tenure Regulation* (CSTR). A second important consideration is the establishment of a clear and certain legal framework for surface rights, including access to the land for CCS project infrastructure, and CO_2 pipelines that may be required to transport CO_2 from the capture site to the well injection site. Two key elements of the surface rights legislation are the requirement for mineral developers to make a genuine effort to negotiate reasonable compensation with surface rights owners for surface access and land disturbances or nuisance (such as noise) that may arise. In Alberta, if the surface rights owner and the project developer are unable to agree on the amount of compensation and sign a written surface lease agreement, a government tribunal called the Land and Property Rights Board will hold a hearing at the request of either party, where either party can make submissions and the tribunal will decide the amount of compensation. The legal certainty and fairness provided under this surface rights compensation system has encouraged oil and gas development in Alberta and will facilitate the development of new CCS projects as well.

In Alberta leases are granted to the CCS project applicant under the MMA to provide the right to drill wells, conduct evaluations, test, and inject CO_2 into deep subsurface reservoirs for a 15-year term, which in renewable in accordance with the Act and its regulations.[66] For the Quest project Shell Canada Limited made its application adhering to *Directive 065*,[67] for approval to dispose of CO_2 within the sequestration area of interest located in central Alberta northeast of the city of Edmonton.[68] The pore space tenure would not preclude companies that hold or acquire existing petroleum and natural gas rights within the area of interest for drilling wells. The provincial oil and gas regulator concluded that the site was appropriate for long-term CO_2 storage. When it approved the project the regulator stipulated 23 conditions in the project approval requiring additional data collection, analysis, and reporting and the company must secure separate approval for drilling additional injection wells.[69] Under

[66] Supra, note 35.

[67] Supra note, 54.

[68] Supra note, 4.

[69] Supra note 4.

the *Carbon Capture and Storage Statutes Amendment Act*,[70] long-term liability for CCS projects was an important issue that was addressed, that provided for the transfer of liability for the project such as CO_2 leaks from proponent to the Government of Alberta (GOA) after the operating phase of the project has ceased and the closure phase is completed. Shell must demonstrate that the sequestered CO_2 is securely contained within the storage complex. The approach to monitoring the site for leaks for years after the injection operations are complete parallels the approach adopted in the EU.

3.3 Environmental Protection Considerations

In Alberta regulation is a completely different process than the allocation of the tenure rights by the DEM that was discussed in the previous section, as it involves granting of approvals to drill wells and to inject substances. As such, proposed well drilling and operations are reviewed and considered by the independent AER, including the potential environment and public safety impacts of the proposed project.[71] The provincial oil and gas regulator has adapted its comprehensive and sophisticated oil and gas well regulatory framework for CCS projects. The regulation of petroleum operations in Alberta is under the jurisdiction of the AER, under the *Responsible Energy Development Act*[72] and the Alberta EPEA. Under EPEA, no person may discharge any contaminant from any industrial or trade premises onto or into land, or into water, unless the discharge is expressly allowed by a rule in a regional plan, by a resource consent, or by regulations. Project review and approval regulation by the AER is separate from ownership and the granting of dispositions by the DEM. The main instrument employed by the AER to regulate drilling and the operation and CCS injection wells is the approval process for well licenses under the OGCA to protect the environment and the health and safety of industry workers, residents, and the general public. Even if it holds a petroleum and natural gas license, a company cannot drill an oil or gas well without a license from the AER pursuant to the OGCA. The above-discussed environmental and health and safety considerations also

[70] *Carbon Capture and Storage Statutes Amendment Act, 2010*, SA 2010, c 14.

[71] Supra note 52.

[72] Supra note 36.

apply to CCS operations that may take place near oil and gas operations, especially if the target storage formation is a depleted oil or gas reservoir. One main issue is that the CO_2 that is injected in CCUS operations can migrate into adjoining formations, damaging petroleum resources, and making petroleum production more difficult. A second important issue is that CO_2 storage formations can be compromised by surrounding oil and gas wells. In cases when plugging and abandonment of an injection well has been inadequate, a CO_2 leak is possible. CO_2 is corrosive, and when combined with water forms acid that can attack steel and cement. In light of the potential impacts on oil and gas production from injecting CO_2 as part of CCS projects, oil and gas developers may be concerned about new restrictions on their operations and that is why regulations are important to respect the rights of oil and gas operators. A significant addition was made to the Alberta *Mines and Minerals Act* (Part 9) that contains new requirements specifically tailored to CCS projects that must be satisfied to drill CO_2 test wells into a potential subsurface reservoir and for long-term storage to reduce GHG emissions in response to climate change. An agreement with the provincial government is required to inject CO_2 and new terms were added to provide clear direction to CCS project developers.

The Alberta Government has set out under Sect. 115 of the MMA the process, information, reporting, and requirements to drill CCS injection test wells for the evaluation of potential CO_2 storage reservoirs.[73] As a condition of evaluating whether a subsurface reservoir would be suitable to store CO_2 for the long-term by drilling a CO_2 test well, as with oil and gas wells drilled on provincial lands in the province, the Alberta Government requires the lessee who has entered into an agreement to "obtain a well license and approval of the Regulator under the OGCA prior to drilling or using a well" for testing the suitability of a reservoir to hold CO_2 as part of a CCS project.[74] In addition to reviewing the proposed site testing program, the AER requires the following information for project review and consideration:

A lessee of an agreement under this section shall be in accordance with the regulations:

[73] Supra note 66, Sect. 115.

[74] Supra note 66, Sect. 116(2).

1. submit a monitoring, measurement and verification plan for approval;
2. comply with a monitoring, measurement and verification plan that has been approved;
3. provide reports with respect to the lessee's compliance with the monitoring, measurement and verification plan;
4. fulfill the work requirements with respect to the location of the agreement;
5. submit a closure plan for approval;
6. comply with a closure plan that has been approved.[75]

3.4 Consultation

Meaningful consultation is an essential element of an effective and efficient regulatory framework, and the Shell Quest project provides an example of how it is done in Alberta. In July 2012, the provincial oil and gas regulator considered and approved the first application for a CCS project in Alberta.[76] The CCS project proponent Shell Canada Limited stated at the hearing that "one of the principles of its sustainable development policy is its commitment to engage with stakeholders in a meaningful way, to identify issues, and to search for mutually agreed-upon solutions".[77] At the hearing, Shell reported that it had carried out an extensive and comprehensive stakeholder consultation program for the CCS Project.[78] The company indicated that it provided information about the proposed project to landowners, occupants, and residents within 5 km of the CO_2 capture site and within the CO_2 pipeline emergency planning zone, and to indigenous peoples and their organizations in the project area.[79] In addition the project proponent also notified leaseholders, petroleum and natural gas rights holders, government agencies,

[75] Supra note 66, Sect. 116(3).

[76] Supra note 4.

[77] Supra note 4, paragraph 1.

[78] Global CCS Institute, "The Quest for Less CO2: Learning from CCS Implementation in Canada. A Case Study on Shell's Quest CCS Project" 18 November 2015, online: https://www.globalccsinstitute.com/archive/hub/publications/196788/quest-less-co2-learning-ccs-implementation-canada.pdf (Accessed 6 January 2019).

[79] Supra note 4.

regional and municipal governments, special interest groups, nongovernmental organizations, industry participants, and industry associations about the proposed CCS project.[80] As part of meaningful consultation, the CCS project proponent distributed project information packages, organized and offered community information and project feedback sessions, prepared and distributed a CCS project site-specific newsletter, and create a dedicated toll-free telephone number to answer questions from residents and the general public about the proposed project. The company also carried out several information workshops in communities located in the project area to provide information, answer questions, and discuss potential concerns with residents and members of the general public.[81] The proponent met face-to-face with landowners along the pipeline right-of-way and within the emergency planning zone to provide project information, discuss questions and concerns, and get access to properties for seismic surveys, water well testing, and other environmental data collection. Shell stated that these meetings resulted in about 30 pipeline reroutes and confirmation of nonobjection from 109 of 111 landowners.[82] The proponent also established a website where stakeholders could find information, ask questions, and express concerns. The proponent had representatives at various community events to provide local residents with the opportunity to ask questions and get information about the project.[83] The proponent reported that it learned several things from its CCS project stakeholder consultation program. First, the importance of early consultation. The proponent reported that it had initiated community consultations at the start of the application process to secure feedback long before the final decision on the project.[84] Second, the importance of meaningful consultation. The proponent submitted that its engagement showed a meaningful concern for how people felt about the project, and it identified problems and potential solutions. Based on feedback from community people, project details were modified, as illustrated by the 30 reroutes of the project pipeline. The proponent indicated

[80] Supra note 70.

[81] Supra note 70.

[82] Supra note 4, paragraph 84.

[83] Supra note 4, paragraph 368.

[84] Supra note 4, paragraph 370.

that it was committed to maintaining open lines of communication with municipal authorities and government agencies.[85]

3.5 Project Approval

Following along with the Shell Quest Project as discussed in the section above, the regulator reported that in assessing the CCS project, it had to consider whether the project application was in the public interest generally and analyze the social, economic, and environmental impacts of the project.[86] The regulator also noted that it had the power, where necessary, to apply conditions to mitigate site-specific or local impacts. At the hearing the regulator concluded that the CCS project proponent's public consultation program initiatives were effective. The regulator noted a number of project modifications to the CCS project made by the company to address public concerns, and that the proponent was able to secure 109 nonobjections to the project with only two landowners continuing to object. The small number of parties appearing at the hearing in opposition to the project supports the regulator's assessment that participant involvement programs had been effective. The Board noted the apparent strong foundation of communication and engagement with area landowners and communities. The regulator concluded that the communication and public consultation program initiated exceeded the minimum oil and gas industry Participant Involvement Program requirements in AER Directive 056. The regulator commended the proponent on its meaningful communication and consultation to date. The Board strongly supported the proponent's plan to consider forming community advisory panels to help with the communication of complex monitoring data and developments. The Board noted that in determining whether or not there was a need for the project and its assessment of whether the project is in the public interest, the regulator must weigh the benefits against the risk factors that are present, given the nature of the development, the proposed location, and other factors associated with the specific situation. In addition, the regulator had to ensure that any site-specific or local impacts are mitigated to an appropriate and acceptable level. Should the regulator determine that the risks cannot be sufficiently mitigated and

[85] Supra note 4, paragraph 374.

[86] Supra note, 4 paragraph 390.

that the risk exceeds the potential benefit, the project would not be in the public interest and therefore the project would not be approved.[87] The regulator reported that in evaluating whether or not to approve the project it had considered the GHG emission reduction policies of both the governments of Canada and Alberta. The regulator noted that both the national and regional governments had adopted strategies and targets to reduce GHG emissions as part of their respective climate change strategies. Developing effective CCS technologies is an important part of both the Canadian and Alberta Government GHG reduction strategies. To this end, the Alberta Government enacted legislation to provide for the funding of CCS projects, to establish a framework to address long-term liability for stored or sequestered CO_2, and to address access and ownership of pore space for CO_2 storage. The Government of Canada also supports the reduction of GHG emissions through its Clean Energy Fund and is aiming to reduce CO_2 emissions by 325 Mt/a by 2050 through a number of approaches, including CCS.[88] The Canadian federal government has committed to supporting research into clean energy technologies and will also invest public funds into the CCS project. The regulator concluded that direct capture and storage of CO_2 emissions has the potential to be one of the leading available processes capable of significantly reducing the amount of GHGs released to the atmosphere in Alberta and elsewhere. The regulator also concluded that the project is at a suitable surface location to minimize impacts to existing land-use activities in the area, and that the project design and monitoring program were adequate to mitigate potential impacts of construction and operation of the project. The regulator also concluded that impacts to individual residents or landowners in the area will be minimal. The regulator, after reviewing the evidence presented at the hearing, decided that the project was in the public interest as there was a need for the project and after considering the environmental, social, and economic impacts of the project.[89] The regulator concluded that the potential for a CO_2 release was "very low" and that the overall public safety, environmental, and social risks and impacts associated with the capture of CO_2, and its subsequent transmission and injection were also very low. After approving

[87] Supra note, 4 paragraph 393.

[88] Supra note, 4 paragraph 60.

[89] Supra note, 4 paragraph 409.

the application for the project in light of the novelty of the CCS project and the uncertainty as to potential impacts from the CCS operations, the regulator imposed 14 conditions on the project. This information is required to be included in each annual report and presentations to the regulator.[90]

4 Concerns and Challenges

4.1 *Liability*

Potential liability for environmental and property damage from CO_2 leaks and human injuries was a major issue that had to be addressed in Alberta's legal framework before CCS site developers were prepared to invest in the development of CCS projects. Different types of liability that might arise include civil liability (e.g., under the torts of negligence, nuisance, trespass and occupiers' liability); criminal liability for acts or omissions causing serious harm to workers, residents, and members of the public; environmental liability for failing to comply with the Alberta EPEA and the associated regulations including the unauthorized release of substances; and gas emissions exceeding those allowed under the regulations. Consistent with its GHG emissions reduction policies and laws and to encourage private investment to fund the desired CCS projects, the Alberta Government decided to assume long-term liability for damage arising from CO_2 storage projects by allowing the facility operator to transfer liability to the government subject to certain conditions.[91] The Alberta Government recognized that corporations simply do not continue in perpetuity and de facto the government would inherit the risk of damage from CO_2 leakage over an extended period of time. However, an important condition of the government assuming long-term liability is that the operator remains liable for negligence. CO_2 storage developers will be liable if they fail to meet a duty of care. The statutory approach to long-term liability parallels the following approach adopted in the European Union (EU) as outlined in EU Directive 2009 that also provides for the transfer of responsibility to a competent authority for monitoring, corrective measures, and liability, but only under the following stringent conditions:

[90] Supra note 4, Appendix 2.

[91] Supra note 66, Sect. 121.

1. after a minimum of 20 years has passed since the closure of the long-term CO_2 injection and storage facility;
2. if all of the available evidence indicates that the stored CO_2 will be completely and permanently contained; and
3. after providing a financial contribution for 30 years of monitoring.[92]

In both the EU and Alberta, there are strict legal requirements that must be satisfied to complete a site transfer, including no negligence on the part of the facility developer and operator. The strict transfer conditions in the Alberta and EU approaches to manage long-term liability reflect the "Polluter pays" principle that is embedded into EPEA.[93] The CCS developer and operator must pay for the environmental damage that may arise from gas leaks and the damage that may arise from the industrial project, prior to the approved transfer by the Alberta Government.

4.2 *Potential Conflicts with Existing Rights Holders*

In 2010 the *Carbon Capture and Storage Statutes Amendment Act* was adopted to (i) clarify ownership for pore space, (ii) establish a pore space disposition scheme for CCS, (iii) enable the transfer of long-term liability for injected CO_2 to the province post-closure, and (iv) establish a stewardship fund for injected CO_2 monitoring and remediation costs.[94] Further important amendments were made to the MMA, including:

1. a declaration vesting title to pore space across all non-federal lands in the Crown right of Alberta (s 15.1);—de facto expropriation of the pore pace associated with freehold mineral title despite the accompanying clause to declare otherwise [s 15.1(4)]

[92] European Parliament, Directive 2009/31/EC on the Geological Storage of Carbon Dioxide, Article 18-20, 23 April 2009, online: https://eurlex.europa.eu/LexUriServ/LexUriServ.do?uri5OJ:L:2009:140:0114:0135:EN:PDF. (Accessed 20 October 2018).

[93] Supra note 37.

[94] Nick Ettinger, Renée Matthews & Rudiger Tscherning, "Alberta's Carbon Capture & Storage Land Grab And The Potential For Conflicts Of Subsurface Rights" (March 9, 2022), online: ABlawg, https://ablawg.ca/2022/03/09/albertas-carbon-capture-storage-land-grab-and-the-potential-for-conflicts-of-subsurface-rights/.

2. the Minister of Energy was granted discretionary authority to enter into "contracts" [s 9(a)] and "agreements" with third parties for the disposition of pore tenure rights for CCS (ss 16, 116); and
3. a pore space disposition scheme for carbon sequestration permit and lease agreements established (part 9).

These "agreements" with the Minister under Part 9 of the MMA could be in the form of a permit or a lease. As provided under Sect. 115(1) of the MMA an evaluation permit is required "to evaluate the geological or geophysical properties of a subsurface reservoir in a location to determine its suitability for use for the sequestration of captured carbon dioxide", and per Sect. 116(1) a lease "granting the right to inject captured carbon dioxide into a subsurface reservoir for sequestration". The *Carbon Sequestration Tenure Regulation*, Alta Reg 68/2011 ["CSTR"] sets out requirements for the Minister to grant a lease, including the following:

1. The Minister must be satisfied on the evidence submitted that the specific location in the application is suitable for sequestration of captured CO_2.
2. The Lesssee submits a monitoring measurement and verification (MMV) plan which accounts for the potential for operations under a CCS lease to interfere with mineral recovery, which include a wide variety of natural resources.
3. Pursuant to AER Directive 65 the accounting for third party mineral rights is to be submitted based on geological interpretations and calculations in its application under the OGCA.[95]

Among the requirements needed under the OGCA is the obligation of the applicant to obtain the consent of any existing lessees of Crown land whose subsurface tenure rights might intersect. This creates potential conflict, as it could give holders of Crown agreements a veto over the approval of CCS schemes that intersect their tenure. There is also a risk that future wells could pose a risk to the storage integrity of a reservoir, especially as freeholders will retain the right to drill in or through these reservoirs indefinitely.

[95] *Supra* note 91.

The OGCA sets out under Sect. 39(1) that "[n]o scheme for.. (d) the storage or disposal of any fluid or other substance to an underground formation through a well.. may be proceeded with unless the Regulator, by order, has approved the scheme on any terms and conditions that the regulator prescribes". Further, the AER will not approve of a CCS scheme unless the Regulator is satisfied that the injection of CO_2 will not interfere with (a) the recovery or conservation of oil or gas or (b) an existing use of the underground formation for the storage of oil or gas.[96] The AER will need to consider the implications of a CCS project for other proposed projects in the vicinity proposing to use the same formation. This has been the case in the AER decision regarding Pure Environmental Waste Management ("Pure") in the Hangingstone wastewater disposal project. Pure applied to dispose of wastewater into salt caverns in the Keg River formation, however Suncor was targeting the same caverns for disposal from its nearby in situ oilsands project. The AER initially rejected two of the proposed injection wells which were close to existing Suncor injection wells. Suncor and Pure later filled a joint submission that was accepted by the AER, however absent that Pure faced the risk that it would not obtain an approval for its injection wells.[97]

5 Recent CCS Tenure Changes and Agreements

On September 14, 2023, the Alberta Government released new carbon sequestration tenure provisions applicable to Small-Scale and Remote Carbon Sequestration projects. These provisions create a third form of sequestration rights.[98] The two types of sequestration rights that predate the 2023 tenure forms are (1) the sequestration permit and lease under Part 9 of the Alberta *Mines and Minerals Act*, RSA 2000, c M-17 ["MMA"] and the CSTR, and (2) Sect. 9 MMA Crown agreements for hub and spoke projects.

The 2023 tenure provisions incorporate projects not covered under proposed or existing hub, "small-scale" CO_2 projects, waste disposal

[96] Supra note 37 s 39(1.1).

[97] Ibid.

[98] The Government of Alberta, Small-Scale and Remote Carbon Sequestration Tenure, 13 September 2023, online: https://training.energy.gov.ab.ca/Guides/Small-Scale%20and%20Remote%20Carbon%20Sequestration%20Tenure%20-%20Application%20Guidelines.pdf.

projects, and experimental projects. These 2023 provisions differ from the two previous types of carbon sequestration tenures as it appears they contemplate that a CCS project proponent will proceed straight to a lease without the need for obtaining an evaluation permit.

Under the 2023 tenure changes, CCS project developers will navigate a complex application process that involves two agreements: (1) a pore space lease ("PSL") granting the right to sequester carbon dioxide into a subsurface reservoir (pore space) and (2) a pore space unit agreement ("PSUA") addressing various interests and activities within the location, including Crown interests. It would appear the complexity of these new agreements can lead to overlapping interests with other subsurface tenure rights holders, making negotiations challenging. Templates for these two agreements and SSR Application Guidelines (the "Guidelines") and several other associated documents were released by the Government of Alberta in connection with the SSR Carbon Sequestration Tenure announcement.

6 Pore Space Lease Form

The PSL appears to be a standard form agreement like Alberta Crown petroleum and natural gas leases for public lands. However, the same cannot be said for the combined granting clause and *habendum*. The PSL grants the exclusive rights to use pore space in subsurface formations to sequester injected substances such as CO_2, subject to AER approval. However, as Professor Nigel Bankes notes the language in the pore space lease suggests that there may be a potential conflict between the exclusive nature of the sequestration right and the substances authorized to be injected by the AER.[99] A literal reading of the provision indicates that the drafter conflated the two concepts in the sense that if the AER only provides approval for the injection of CO_2, the Crown could still authorize the use of the pore space for the injection of oilfield waste such as saline produced water.[100] In addition, the PSL does not appear to be the appropriate home for provisions related to rent, compensation, and liability management fees. Further, the *habendum* would appear to imply

[99] Nigel Bankes, "Crown Pore Space Lease and Pore Space Unit Agreement" (27 September 2023), online: https://ablawg.ca/2023/09/27/the-crown-pore-space-lease-and-pore-space-unit-agreement/.

[100] *Ibid.*

that a PSUA is necessary for the lease to be effective, giving rise to a potential PSUA negotiation challenge when a CCS proponent does not hold an existing lease.

7 Pore Space Unit Agreements

It would appear that the PSUA has been drawn from the proprietary form of hydrocarbon unitization agreements developed by the Petroleum Joint Venture Association and the Alberta Government (Crown) Gas Storage Unit Agreement. It incorporates preambular paragraphs, operative articles, and exhibits. A common theme throughout the operative articles is that an added degree of complexity arises when various interests are involved.[101] As Professor Bankes notes, while the operative articles of the pore space unit agreement appear logical when considering them under rudimentary examples, the addition of numerous parties with various interests creates significant complexity for practice under the current wording of the articles. Beyond this general observation, a few articles warrant further discussion.

Article I of the agreement incorporates guiding definitions and principles for interpretation of the contract. For example, "Working Interest" is defined as an ownership interest in the right to produce or sequester substances without limiting it to specific types of substances, making it applicable to both hydrocarbon production and sequestration activities. Further complexity arises under subclause (ii), where the definition may be difficult to understand and apply in circumstances where the relevant rights are more broadly framed and heterogeneous.

Article II outlines exhibits, including tract descriptions, well logs, and methodology for calculating payments to the Crown for hydrocarbons that the Energy Minister determines to be in the Unitized Zone.

[101] See Article III, that provides for lease continuation, even if there are no production activities in a unitized zone, which can be challenging to apply in a heterogeneous context. Although this provision's implications are relatively straightforward to apply in the context of a conventional oil production unitization, the implications are more far reaching in a unit that comprises both sequestration and production interests; Article IV grants authority to conduct operations across tract boundaries, but it is unclear how this works for diverse interests in the same area. It appears that the question of how to combine such heterogeneous working interests within a tract participation formula will be challenging and may stretch the concept of working interest to the breaking point; Article V, allows qualification for tract conclusion when 100% of the working interest is owned by a party. However, determining ownership can be complex when various interests are involved.

However, the methodology is left unspecified, indicating it is a work in progress.

Articles X and XI address the effective date and termination of the PSUA, including termination provisions on the cessation of operations. Given the ongoing monitoring, measurement, and verification ("MMV") activities and the fact that termination of the PSUA will automatically bring the pore space lease to an end, there are concerns in connection with the discretionary nature of the termination provisions.

Ultimately, the PSUA is more ambitious than traditional unitization agreements, as it seeks to combine ownership rights for various substances and activities. Beyond general clarification, the complexities that arise when combining diverse interests within a tract of participation formula call for more clarification.

8 SSR Application Guidelines and Tenure

Beyond the large and *prima facie* uncorrelated scope captured by the SSR Carbon Sequestration Tenure, as Professor Bankes notes a number of items found in the guidelines raise concern. For starters, including acid gas disposal ("AGD") projects in the SSR stream signifies a shift from the previous property authorization model, that was provided for under Sect. 54 of the MMA. This inclusion introduces complexity and administrative procedures for AGD operations. While this change may offer increased security of tenure, it also poses challenges in terms of governance and coordination. Another noteworthy development in SSR tenure is the permission for injection into depleted hydrocarbon reservoirs, expanding beyond the traditional focus on saline aquifers and pure CO_2 sequestration operations. This shift raises questions about the applicability of the Crown pore space vesting provisions.

Of notable concern is the absence of MMV requirements in the Guidelines, an essential requirement of CCS projects authorized under Part 9 of the MMA and the CSTR, and internationally. MMV is necessary for tracking and verifying the success of carbon sequestration projects and aligning with international CCS project standards. Although a rationale for its absence may result from this topic being the subject of a recent delegation of authority from the DEM to the AER, an outright failure to address MMV requirements would represent a significant departure from international practices and standards. Unlike other forms of sequestration tenure, the SSR stream is also silent on information regarding the proponent's financial capacity and benefits to Indigenous communities and information on sequestration risks and measures.

Finally, it is accepted practice that unitization is typically required where there are multiple tracts and owners within a single oil and gas pool, and the working interest owners in the various tracts decide to develop the pool cooperatively. However, in a pore space, the Crown is the only owner and unless it grants leases to multiple lessees, there should not be a need for unitization. Things become more complex in the case of a depleted reservoir where there may be pore space interests, petroleum and natural gas interests, and lithium brine interests. The PSUA requirement appears to deal with scenarios where multiple interests, such as pore space and mineral interests, co-exist. As Professor Bankes notes, considering the lack of compulsory unitization, it is reasonably foreseeable that such negotiations among these interests may be complex and time-consuming.

Alberta's SSR Carbon Sequestration Tenure represents a significant development in regulating underground pore space for carbon sequestration and related hydrocarbon and lithium brine activities. However, the process does not appear to be straightforward and there is uncertainty within the recent regulatory changes. As Professor Bankes notes the above discussion illustrates that the recent changes raise questions about transparency, clarity, and the need for effective risk management in this evolving regulatory space. As noted by Premier Smith in her keynote address at a CCUS Conference in the provincial capital city of Edmonton on September 12, 2023, industry members noted the importance of regulatory certainty to attract more investment in future CCS projects. Although desirable, whether the Government of Alberta's tenure agreements increase clarity in the CCS space remains unclear.

9 Conclusion

Governments in Canada with depleted oil and gas production reservoirs are using them to store CO_2 as part of a national GHG reduction strategy. Under Canada's constitution provincial governments have jurisdiction over energy resources and environmental matters on provincial lands, and therefore CCUS and CCS projects are for the most part provincially regulated. The province of Alberta, as the largest oil and gas producing province, relies on legislation initially adopted to regulate oil and gas well drilling, injection and withdrawal of natural gas, and the construction of pipelines to regulate oilfield carbon capture projects. Oil and gas legislation has been amended and new regulations have been created to avoid multiple use reservoir disputes, provide legal certainty and promote additional investment in more CCUS and CCS projects. The Alberta

Mines and Minerals Act provides that the provincial government owns the pore space for CO_2 sequestration schemes and assigns several technical decision-making powers concerning CCS to specified officials within the Alberta Energy Regulator (AER). The *Oil and Gas Conversation Act* and associated regulations provide approval criteria for CCS projects and increased requirements for monitoring and testing. In addition, the *Technology Innovation and Emissions Reduction (TIER) Regulation*, under *the Emissions Management and Climate Resilience Act*, that sets out Alberta's approach to reducing emissions from industrial emitters, was updated in 2023 to introduce a variety of new TIER tax credits for a range of industrial projects including CCS.

Due to the different types of lands and property rights holders in Alberta (i.e., provincial, federal, private and First Nation lands), the legal system in the province must protect, support, and balance competing commercial interests to provide fairness and security to encourage investment in new CCS projects. Project developers must obtain CO_2 injection and storage rights and surface rights. Under the *Alberta Surface Rights Act*, genuine efforts must be made by project proponents to negotiate reasonable compensation with surface rights owners for project surface access and use. Furthermore, projects must be in the public interest with social, economic, and environmental impacts considered for project approval. Environmental protection considerations must be reviewed and considered by the AER, and public consultation is an essential element of ensuring potential problems and solutions are addressed. A third issue discussed in this chapter is the potential short and long-term liability for environmental and property damage from CO_2 leaks and human injuries. To encourage private investment in CCS projects, the Alberta Government decided to assume long-term liability for damage arising from CO_2 storage projects under the *Carbon Capture and Storage Statutes Amendment Act*.

In Alberta, the MMA stipulates that the provincial government owns the pore space on provincial lands, and for CO_2 sequestration schemes, the Crown is accountable if there are CO_2 leaks from subsurface reservoirs in the long term. The Alberta Carbon Sequestration Tenure Regulation enables the Government of Alberta to issue evaluation permits, agreements, and leases for carbon sequestration projects. The OGCA allows the AER to approve CO_2 schemes if CCS injection activities do not interfere with the licensed recovery of oil, gas and lithium brine resources, or an existing right to use of a subsurface reservoir to store oil and gas. Surface rights issues for CO_2 sequestration and EOR projects are managed under the same administrative board (Law and Property Rights

Tribunal) created to resolve surface rights disputes for oil and gas wells and facilities under the *Surface Rights Act*.

The Alberta Minister of Energy and Minerals may enter into an agreement with a company to drill wells for the purpose of injecting and storing CO_2 into subsurface reservoirs. The agreement, requires the lessee to obtain a well license from the AER and submit a monitoring plan, comply with monitoring requirements, provide reports with respect to monitoring and complying with all work requirements. Under the MMA the agreement provides that leases cannot be transferred without the permission of the Minister. The lessee may apply to the Minister for a project closing certificate. The Minister may issue a closing certificate if the lessee has completed all of its end-of-life obligations under the OGCA and EPEA, all of the relevant regulations and the captured CO_2 is stable in the reservoir. Upon issuing a closing certificate, the Minister becomes the owner of the captured CO_2 and is liable for gas leaks, and all aspects relating to it. The AER can also designate CO_2 reservoirs as orphan facilities. This is similar to orphan oil and gas wells, the lessees who are a defaulting well license or working interest participant and therefore have not met their obligations under the OGCA and EPEA and the associated regulations.

Currently as part of the mineral rights agreement issued under the MMA, the Minister may issue an evaluation permit, providing the permit holder with the right to do geological testing for 5 years with respect to the suitability of the agreement area for carbon capture activities. The Minister may approve a monitoring plan for an evaluation permit that must contain an analysis of the likelihood that the operations or activities that may be conducted under the permit will not interfere with hydrocarbon recovery. The regulation also sets out the specific requirements to obtain a Carbon Sequestration Lease which include fees associated for a lease application, annual rent, evidence that is satisfactory to the Minister that the site is appropriate, and all monitoring and closure plans required under the regulations are adequate. The initial term of the lease is 5 years, which may be renewed with an additional application, monitoring plan, closure plan, and the lessee's approval from the AER that it is authorized to continue to inject the captured CO_2. The regulation also specifies that the Minister may approve a CCS project closure plan for the site in question, but that plan ceases to have effect three years from the date of the issuance of the closure plan approval or from the renewal date of the lease, if same is renewed.

CHAPTER 13

CCUS in Indonesia

Anton Latief and Randy Hendrika

1 Introduction

Indonesia is home to some of the largest industrial clusters in Southeast Asia, which unfortunately also means that it is home to some of the world's largest greenhouse gas emitters. In line with global decarbonization trends, Indonesia has announced plans to achieve carbon neutrality by 2060 while remaining highly dependent on natural resources—especially fossil fuels, such as oil and natural gas. Under these circumstances, Indonesia has seen a surge of interests on carbon capture and storage ("**CCS**") and carbon capture, utilization and storage ("**CCUS**")

Carbon Capture and Storage and Carbon Capture, Utilization and Storage Activities in Indonesia.

A. Latief (✉)
South Jakarta, Indonesia
e-mail: anton.latief@gmail.com

R. Hendrika
South Tangerang, Indonesia
e-mail: randy.hendrika@gmail.com

E. G. Pereira et al. (eds.), *Carbon Capture Utilization and Storage*, Sustainable Development Goals Series,
https://doi.org/10.1007/978-3-031-81272-9_13

technologies as some of the most promising means and strategies to decarbonize its economy.

In general, CCS and CCUS are understood as the processes related to the capture of CO_2 which may then either be stored by permanently injecting CO_2 into rock formations such as saline aquifers or depleted oil and gas reservoirs (i.e., CCS) or utilized in products or services (i.e., CCUS). In that regard, studies have shown that Indonesia has promising potential for CCS/CCUS activities. Carbon storage assessments conducted by LEMIGAS in 2009, 2012 and 2015 concluded that Indonesia has CO_2 storage capacity potential of 12.2 billion tons.[1] Such capacity is generally divided into two significant locations, i.e., 2.5 billion tons of CO_2 capacity located in depleted oil and gas reservoirs and 9.7 billion tons located in saline aquifers. The report also shows that CO_2 storage capacity in Indonesia is generally distributed regionally as follows: East Kalimantan region (140 million tons in depleted reservoir), Central-South Kalimantan (10 million tons in depleted reservoir), Central Sulawesi (19 million tons in depleted reservoir), Central Sumatra (229 million tons in depleted reservoir), West Java (402 million tons in depleted reservoir and 2029 million tons in saline aquifer), East Java (116 million tons in depleted reservoir), and West Papua (551 million tons in depleted reservoir). Nevertheless, the most significant capacity is located in the South Sumatra region (924 million tons in depleted oil and gas reservoir and 7.7 billion tons in saline aquifer).

Indonesia's oil and gas sector also has a very promising potential for the implementation of CCS/CCUS technologies. For example, one of the major challenges in the development of certain gas fields in Indonesia is the high CO_2 content of its natural gas, which has to be removed before the natural gas can be distributed and consumed. Pertamina's multi-trillion cubic feet East Natuna asset offshore Indonesia is Southeast Asia's largest unexploited gas resource (its gas has 71% CO_2 content), and the gas from Repsol's Kali Berau Dalam gas field development in Sumatra has 26% CO_2 content but Repsol has committed that all its greenfield projects will be net zero. The potential of such gas fields could be unlocked with CCS/CCUS technology, which when combined with enhanced oil recovery could rejuvenate the production of oil at mature oilfields in Indonesia.

[1] https://www.asiaccusnetwork-eria.org/articles/indonesia-tabled-ccs-and-ccus-in-net-zero-emission-road-map.

Against the above backdrop, it is no surprise that Indonesia was among the first countries in Southeast Asia to announce that CCS/CCUS would play an important role in decarbonizing their industrial sectors. Like many of its neighboring countries, it is planning to rely on CCS/CCUS technologies as one of its key strategies to meet its long-term climate goals of net zero by 2060 and Carbon Emissions reduction targets under its Nationally Determined Contribution goal of 31.89% by 2030 (43.2% with international support), while simultaneously boosting oil production to 1 billion barrels of oil and gas production to 12 billion cubic feet of gas per day by 2030.

2 Key CCUS Projects in Indonesia

Currently, there are about 15 CCUS projects reported in Indonesia and its overview updates are as follows[2]:

No	*Project*	*Expected CO_2 Storage*	*Expected on Stream*
1	ARUN CCS—Carbon Aceh and PEMA	Significant, under assessment	2028
2	Gemah CO_2 EOR (CCUS)—Petrochina International Jabung	Significant, under assessment	2028
3	Ramba CO_2 EOR (CCUS)—Pertamina	Significant, under assessment	2030
4	Central Sumatra Basin CCS-CCUS—Pertamina and Mitsui	Significant, under assessment	2028
5	Sakakemang CCS—Repsol Sakakamang	2 million tons per year and 30 million tons in total	2028
6	Jatibarang CO_2 EOR (CCUS)—ertamina and Jogmec	14.6 thousand tons/year	2031
7	Gundih CO_2 EGR (CCUS)—Pertamina, J Power, and Janus, ITB	3 million tons for 10 years	2027
8	Sukowati CO_2 EOR (CCUS)—Pertamina, Lemigas, and Japex	7 to 14 million tons for 15 years	2027 Pilot 2023 commercial

(continued)

[2] *Ibid*; https://ekonomi.bisnis.com/read/20231129/44/1719104/ri-jajaki-peluang-kerja-sama-teknologi-tangkap-karbon-dengan-korsel.

(continued)

No	*Project*	*Expected CO_2 Storage*	*Expected on Stream*
9	Sunda Asri Basin—Pertamina and ExxonMobil	2 gigatons in saline aquifer	2029
10	Kutai Basin CCS—CCUS Hub—Pertamina	Significant, under assessment	2028
11	CCU to Methanol, Pertamina Refinery Balikpapan—Pertamina and Air Liquide	Significant, under assessment	2028
12	East Kalimantan CCS/ CCUS—Kaltim Parna Industry	10 million tons for 10 years	2028
13	Blue Ammonia CCS—Panca Amara utama, JOGMEC, Mitsubishi, and Pertamina	19 million tons for 20 years	2028
14	Tangguh CO_2 EGR (CCUS)—BP Tangguh	25–33 million tons for 10–15 years	2026/2027
15	Abadi CCS/CCUS—Inpex Masela, Ltd	70 million tons by 2055	2029

Based on the above table, 8 (out of 15) projects will start their operation in the next decade and contribute to Indonesia's emission reduction target. In Indonesia CCUS is projected to reduce 2,5 million tons of CO_2 per year from 2026 onward and to gradually increase to 8,6 million tons per year in 2031, with no additional capacity expected from 2031 until 2035. CO_2 stored is projected to reach a cumulative of 68.2 million tons in 2035. A report produced in collaboration between IEA and the Ministry of Energy and Mineral Resources reported that carbon capture and injection capacity will reach 190 million tons in 2060.[3] In light of the above overview, several of the notable CCUS projects in Indonesia and their updates are as follows:

2.1 ARUN CCS—Carbon Aceh and PEMA

Arun is one of the largest oil and gas fields in the Asia Pacific with cumulative production of more than 16 trillion cubic feet of gas and 700 million barrels of condensate.[4] It is now one of the largest, if not

[3] *Ibid.*

[4] https://www.carbonaceh.com/stakeholder-updte.

the largest, depleted oil and gas fields in the region. There are undeveloped gas fields in the region potentially containing more than 500 million tons of CO_2.[5] The Arun field was developed with wells and infrastructure concentrated into four clusters, with more than 140 wells have been drilled across the field, most being equipped with 10,000 psi working pressure stainless-steel well heads to resist corrosion from high-pressure and high-temperature production fluids containing CO_2.[6] Gas was reinjected into the reservoir through 12 injection wells for pressure maintenance. Over the period 1976 to 1997, 5.2 trillion cubic feet of gas was reinjected into the reservoir providing valuable data on the behavior of the reservoir during injection.[7] The field had a very high initial reservoir pressure of more than 7,000 psi and a temperature of 1800°C at a depth of 3,000 m subsea.[8] By 2005 reservoir pressure had decreased to around 500 psi which is encouraging from a CO_2 injection perspective.[9]

In September 2022, Carbon Aceh Pte Ltd and PT Pembangunan Aceh (PEMA) entered into a Joint Venture Agreement (JVA) and in November 2022, created a joint venture company called PT PEMA Aceh Carbon (PT PAC), to evaluate and re-develop the depleted Arun gas field in the province of Aceh, Indonesia for CCS.[10] In addition to a detailed study of the Arun reservoir, the study will evaluate the potential for reusing the existing wells, pipelines, and all facilities and equipment considered necessary for the operation of a CCS project. As at March 2023, due to significant progress made in collating and curating the Arun data, the feasibility study is planned to commence.[11]

[5] Ibid.

[6] Ibid.

[7] Ibid.

[8] Ibid.

[9] Ibid.

[10] Ibid.

[11] Ibid.

2.2 Central Sumatra Basin CCS-CCUS—Pertamina and Mitsui

On 7 April 2022, Mitsui & Co announced that they have launched a joint feasibility study with PT Pertamina (Persero) on CCUS commercialization in Indonesia.[12] This study aims to evaluate the CO_2 subsurface storage capacity of the oil and gas fields whose production volume has been declining, among those in Rokan Block, which is currently operated by Pertamina in Central Sumatra and is one of the largest onshore oil and gas blocks in Indonesia containing Duri and Minas oil fields, and to examine potential commercialization toward the establishment of a CCUS value chain, including capture and transportation of CO_2 emitted from industrial plants, power generation plants, and other facilities.[13] This study also assesses the potential of receiving CO_2 from domestic and international suppliers, via ship transportation, aiming to create new low-carbon solution business in Indonesia.[14]

2.3 Gundih CO_2 EGR (CCUS)—Pertamina, J Power, Janus, CoE ITB

Pertamina entered into a Joint Study Agreement (JSA) with *Institut Teknologi Bandung* (ITB) and several Japanese partners, including JANUS, JGC Corporation, J Power, to explore the implementation of CCUS and Enhanced Gas Recovery (EGR) in the Gundih field project in Cepu, Central Java.[15] The joint study marks a significant initiative to reduce Carbon Emissions, which has the potential to reduce 3 million tons of CO_2 annually for ten years while increasing gas production.[16] The stored CO_2 will be considered carbon credits shared between Indonesian and Japanese governments.[17]

[12] https://www.mitsui.com/jp/en/topics/2022/1243212_13410.html; https://www.cnbcindonesia.com/news/20220407144542-4-329708/pertamina-gandeng-mitsui-kembangkan-teknologi-ccus.

[13] https://www.mitsui.com/jp/en/topics/2022/1243212_13410.html.

[14] https://www.mitsui.com/jp/en/topics/2022/1243212_13410.html.

[15] https://www.jpnn.com/news/pertamina-gandeng-perusahaan-jepang-itb-studi-ccus-di-lapangan-gundih.

[16] Ibid.

[17] Ibid.

Pertamina developed a Decarbonization Roadmap to support global climate control, with the CCUS initiative expected to operate by 2026 after feasibility studies period from June 2021 to February 2022, followed by the Front End Engineering Design (FEED) and Engineering Procurement, and Construction (EPC) in 2022 to 2024.[18] In June 2023, Pertamina, through its subsidiary Pertamina Hulu Energi (PHE), is seeking to obtain funding from the Just Energy Transition Partnership (JETP) to accelerate the implementation of CCUS technology for the upstream oil and gas sector.[19] In a news report Representative Director and President of JANUS, Kazuhiko Chikamoto, mentioned that the decarbonization project in Gundih field is believed to be the best decarbonization project in Asia considering the innovation that will be brought and implemented in the near future through this project.

2.4 Tangguh CO_2 EGR (CCUS)—BP Tangguh

In 2021, Tangguh is the largest gas producing field in the country with 1.4 billion cubic feet (Bcf) per day of gas production through two LNG trains and will reach 2.1 Bcf per day once its Train 3, currently under construction, is online.[20] Tangguh expansion project including the construction of its Train 3 has been declared as a National Strategic Project by the Government of Indonesia.[21]

The Tangguh LNG EGR (CCUS) project is conducted as a part of Project UCC which total investment is estimated to be worth US$2.6 billion in offshore and onshore facilities.[22] It comprises of Ubadari gas field development, enhanced gas recovery through carbon capture, utilization and storage (EGR/CCUS), and onshore compression.[23] Tangguh LNG EGR/CCUS itself will have 3 injection wells,

[18] https://biz.kompas.com/read/2021/06/19/141315628/komitmen-turunkan-emisi-co2-pertamina-gandeng-perusahaan-jepang-dan-itb-studi.

[19] https://indonesiabusinesspost.com/risks-opportunities/pertamina-hulu-energi-seeks-jetp-funding-to-accelerate-ccus-implementation-in-upstream-oil-and-gas-sector/.

[20] SKK Migas Setujui POD Lapangan Ubadari dan Vorwata CCUS BP Tangguh—Papua Kini.

[21] Ibid.

[22] Tangguh EGR/CCUS | Who we are | Home (bp.com).

[23] Ibid.

1 offshore injection platform, 1 offshore CO_2 pipeline, and onshore facilities for CO_2 removal, processing, and compression.[24]

Its Ubadari Field and Vorwata CCUS project are estimated to enable a potential additional recovery of 1.3 trillion cubic feet (Tcf) of gas.[25] The Vorwata CCUS development itself will see an approximately 25 million tons of CO_2 injected back into the Vorwata reservoir to reduce venting of the great majority of produced CO_2 and provide incremental gas production through enhanced gas recovery (EGR).[26] The CO_2 injection will remove up to 90% of the reservoir-associated CO_2 which is currently vented and represents nearly half of the Tangguh LNG emissions.[27] This would make Tangguh LNG plant to be one of the lowest greenhouse gas intensity LNG plants in the world.[28] The parties to the project are BP Berau Ltd, MI Berau B.V. (jointly established by INPEX and Mitsubishi Corporation), CNOOC, Nippon Oil Exploration (Berau), KG Berau, LNG Japan, and KG Wiriagar.[29]

The Government of Indonesia approved the Ubadari gas field development and Vorwata EGR/CCUS plan of development (POD) in August 2021 and the project is expected to be on stream in 2026/2027.[30]

3 Legal Framework, Policies, and Norms for CCUS Activities in Indonesia

As shown in previous sections, the Government of Indonesia has been made aware of the importance of CCS/CCUS and the role it may have in assisting Indonesia to reach its net Carbon Emission goals. Further, it has also identified depleted oil and gas reservoirs containing 8 gigatons of carbon dioxide (CO_2) that can be used for CCS and CCUS and an

[24] Ibid.

[25] Ibid.

[26] Tangguh EGR/CCUS | Who we are | Home (bp.com); SKK Migas Setujui POD Lapangan Ubadari dan Vorwata CCUS BP Tangguh—Papua Kini.

[27] SKK Migas Setujui POD Lapangan Ubadari dan Vorwata CCUS BP Tangguh—Papua Kini.

[28] Ibid.

[29] https://www.inpex.co.jp/english/business/asia.html.

[30] Ibid.

estimated 400 gigatons of CO_2 from saline aquifers.[31] In light of such potential, the Government of Indonesia along with many other players of the oil and gas industry emphasizes on the relevance of CCS/CCUS for Indonesia[32] and the importance of having CCS/CCUS regulated as soon as possible.[33] In light of the foregoing, the Government of Indonesia has prioritized the regulation of CCS/CCUS activities in the upstream oil and gas sector as the first step to realizing the technology in Indonesia.

3.1 MEMR Regulation 2/2023—CCS/CCUS Regulatory Framework in Upstream Oil and Gas Industry

In 2023, Indonesia has made a significant progress in developing the regulatory framework for CCS/CCUS activities with the issuance of the Minister of Energy and Mineral Resources ("**MEMR**") Regulation No. 2 of 2023 on Implementation of Carbon Capture and Storage and Carbon Capture, Utilization and Storage in the Upstream Oil and Natural Gas Business Activities ("**MEMR Regulation 2/2023**"), which came into effect on 3 March 2023. This regulation provides a welcome clarity on the CCS/CCUS regulatory environment in Indonesia and is intended to cover four principal areas of focus: technical, business, legal, and economic aspects.[34]

It is important to note that MEMR Regulation 2/2023 only regulates CCS/CCUS activities in the upstream oil and gas industry and does not provide for CCS/CCUS activities in other industries. In that regard, in the context of MEMR Regulation 2/2023, CCS/CCUS activities are conducted by production sharing contract Contractors ("**Contractors**"), which are business entities or permanent business entities that are determined to carry out exploration and exploitation activities in an oil and gas working area ("**Working Area**") based on a cooperation contract with the Special Task Force for Upstream Oil and Gas Business Activities

[31] https://migas.esdm.go.id/post/read/dorong-perdagangan-karbon-dan-penurunan-emisi-dalam-negeri-kementerian-esdm-kembangkan-ccs-di-indonesia.

[32] https://ekonomi.bisnis.com/read/20230814/44/1684866/gencarkan-ccus-pertamina-temukan-lokasi-penyimpanan-karbon-kapasitas-2-giga-ton.

[33] https://migas.esdm.go.id/post/read/kopi-senja-eps-2-kupas-tuntas-ccs-ccus.

[34] https://www-esdm-go-id.translate.goog/en/media-center/news-archives/aturan-ccs-ccus-diteken-upaya-indonesia-capai-rendah-emisi-dan-tingkatkan-produksi-migas?_x_tr_sl=id&_x_tr_tl=en&_x_tr_hl=en&_x_tr_pto=sc.

(*Satuan Kerja Khusus Pelaksana Kegiatan Usaha Hulu Minyak dan Gas Bumi*) ("**SKK Migas**") or the Aceh Oil and Gas Management Agency (*Badan Pengelola Migas Aceh*) ("**BPMA**") (as applicable). Please refer to the section "*CCS/CCUS Regulatory Framework in Other Industries*" further below for an update on the regulatory framework for CCS/CCUS activities in sectors other than the upstream oil and gas sector.

3.1.1 General Provisions on Capture, Transport, Utilization, and Injection

MEMR Regulation 2/2023 outlines the legal framework for the capture, transport, injection, and storage of CO_2 and other greenhouse gas that can be converted into CO_2 equivalents ("**Carbon Emissions**"). Carbon Emissions included in MEMR Regulation 2/2023 include those originating from upstream oil and gas exploration and production business activities (e.g., Carbon Emissions separated in oil and gas production facilities or resulted from combustion), other industries, or direct air capture, which are transported by pipeline, truck, shipping or other prescribed means in order to be injected and stored, or utilized.[35] Third parties' (defined as any Carbon Emission producer that is outside of a Working Area) Carbon Emissions that originated from other industries may be injected and stored by a Contractor in its oil and gas working area through a cooperation agreement between such Contractor and such third party.

Carbon Emissions may be transported by pipelines, truck, vessel, and/or other applicable transportation methods. Thereafter, Carbon Emissions to be injected and stored shall be injected into reservoirs in oil and gas fields, unconventional oil and gas reservoirs, salty aquifers, or coal seams for Coalbed Methane ("**CBM**") activities (collectively, "**Injection Target Zones**").[36] The injection of Carbon Emissions must be conducted by the relevant Contractor in accordance with good engineering principles based on the characteristics of each location and carried out during the production period.[37]

Apart from injection and storage, Carbon Emissions may be utilized to increase oil and gas production through enhanced oil, natural gas, or

[35] MEMR Regulation 2/2023, Articles 6 and 7.

[36] MEMR Regulation 2/2023, Article 8(2).

[37] MEMR Regulation 2/2023, Article 8(1) and (3).

CBM recovery.[38] CCS/CCUS projects will be able to rely on revenues from this increased production, as well as revenues from injection and storage of Carbon Emissions from outside the oil and gas working area. For more elaboration on this point, please see sub-section (d) (*Monetization of CCS/CCUS*) further below.

3.1.2 Implementation of CCS/CCUS Activities

MEMR Regulation 2/2023 provides that CCS/CCUS projects shall be conducted in two phases—(i) planning phase and (ii) implementation phase.

a. *Planning Phase*

During the planning phase, the Contractor must submit a plan to implement CCS/CCUS in its Working Area or the Working Area of another Contractor as part of its field development plan.[39] If the CCS/CCUS activity is part of the first field development plan, the plan should be submitted to MEMR.[40] However, if the CCS/CCUS activity is part of a subsequent field development plan, the proposal should be submitted to SKK Migas or BPMA (as applicable).[41]

Due to a decreased oil and gas production of depleted reservoirs, they may be utilized if there is potential for utilizing them as a storage location for Carbon Emissions from other Contractors. In such cases, the Contractor of the depleted reservoir may submit a proposal for the utilization of the depleted reservoir for CCS/CCUS activities by submitting a field development plan, or subject to requisite approval of SKK Migas or BPMA (as applicable), cooperate with other parties (and remain responsible) to manage the depleted reservoir for CCS/CCUS activities. Alternatively, MEMR may request the Contractor to partially relinquish part of the oil and gas working area where the depleted reservoir is located to be used by other Contractors determined by MEMR to carry out CCS activities.[42]

[38] MEMR Regulation 2/2023, Article 9.

[39] MEMR Regulation 2/2023, Articles 11 and 18.

[40] MEMR Regulation 2/2023, Articles 11(2) and 16(1).

[41] MEMR Regulation 2/2023, Articles 11(2) and 16(2).

[42] MEMR Regulation 2/2023, Articles 58(1) and 59(1).

MEMR Regulation 2/2023 provides that following the approval of MEMR, SKK Migas, or BPMA (as applicable) on the CCS/CCUS implementation plan of a Contractor's field development plan, such Contractor shall submit a proposal for the amendment of its cooperation contract to the MEMR. While such amendment to cooperation contract will be subject to further discussion and negotiation with SKK Migas prior to its approval by MEMR, it is likely that such amendment should cover among others the following key areas:

- the description of CCS/CCUS activities and its inclusion as petroleum operations and its costs as operating costs;
- the recognition for the potential use of third party funding for the CCS/CCUS activities;
- the implementation of the monetization from the CCS/CCUS activities including among others (i) the mechanisms to implement and handle carbon trading and/or cooperation with third party (for third party access); and (ii) allocation of revenues between the Government of Indonesia and Contractor from carbon trading and/or carbon injection and storage services (as applicable);
- the provision of any incentives (fiscal or otherwise);
- the creation and management of monitoring funds; and
- the right of the MEMR to request for the relinquishment of the portion of working area on which an unutilized depleted reservoir is located.

b. *Implementation Phase*

Subject to the prior-approval of the MEMR, SKK Migas, or BPMA (as applicable), the CCC/CCUS project then proceeds to the implementation phase, which consists of:

- preparation of the documents on mitigation and handling of environmental and social impacts and public involvement, while ensuring those documents are made in accordance with the applicable prevailing laws and regulations;
- engineering, procurement and construction processes;
- the commissioning and operation of CCS/CCUS activities;
- the implementation of operation safety management;
- the management of environmental aspects;

- the implementation of emergency response activities;
- the implementation of repair and maintenance activities;
- the implementation of monitoring and measurement, reporting and verification ("**MRV**")activities; and
- the closure of CCS/CCUS activities.[43]

The implementation plan is required to include technical, economic, operational and safety and environmental aspects, and closure of activities. Such plan shall be based on a study which assesses the feasibility of the proposed CCS/CCUS project according to relevant standards and good engineering principles. Such study shall cover geology, geophysics, reservoirs, transportation, storage and injection operations, economics, engineering, safety and environment, risk evaluation and mitigation, monitoring, and MRV aspects, as well as an estimate of the Carbon Emissions storage capacity which the Contractor may request to be independently certified.[44] The results of such study shall at least include the following:

- estimate of the Carbon Emissions storage capacity conducted through static and dynamic modeling in the Target Injection Zone;
- depth and thickness of Target Injection Zone;
- hydraulic conductivity of Target Injection Zone;
- Carbon Emission compositions and its impacts on the Target Injection Zone;
- Integrity of the supporting zone layer, impermeable zone layer, and geological trap that include at least: (i) the maximum injection pressure limit in well hole that does not exceed the fracture/gradient/minimum in situ stress), and (ii) geomechanics and geochemical studies;
- Well integrity of the injection well, production well, supervisory well, and/or abandoned well in the surrounding area which has the potential to be the source of Carbon Emission leakage;
- Injection flow rate and pressure;
- Injection period;

[43] MEMR Regulation 2/2023, Articles 19(1) and (2).

[44] MEMR Regulation 2/2023, Articles 12 and 13.

- Design and implementation plan for the drilling of injection well;
- Increase of pressure in the Target Injection Zone due to injection activity;
- Requirement and specifications of the surface facilities required for the injection activities;
- Integrity of the required surface facilities;
- Dynamic modeling of the Carbon Emission spread during and after a certain period following injection;
- Estimation on the increase of oil and gas production from CCUS activities;
- Estimation of reduction to Carbon Emission;
- Economics analysis;
- Risk assessment and mitigation for long-term storage as well as environmental and social impact and public involvement;
- Planned utilization of Carbon Emission storage capacity; and
- Monitoring and MRV plan.

3.1.3 Costs and Funding

Pursuant to Article 40 of the MEMR Regulation 2/2023, as long as the implementation of CCS/CCUS is within the framework of upstream oil and gas operations, all costs of such implementation can be treated as operating costs in accordance with the provisions of the relevant cooperation contract.[45] The implementation of CCS/CCUS with Carbon Emissions from upstream oil and gas operations are part of petroleum operations in the relevant oil and gas working area. For cases where the Carbon Emissions are sourced from operations other than the upstream oil and gas operations, the CCS/CCUS activities constitute petroleum operations from the point the Contractor[46] received the Carbon Emissions. MEMR Regulation 2/2023 contemplates that CCS/CCUS projects comprising of the following activities, can also be funded by third parties through grants, project financing, or other funding schemes permitted by the subsequent laws and regulations[47]:

[45] MEMR Regulation 2/2023, Article 40.

[46] MEMR Regulation 2/2023, Article 1(32) defines a "contractor" as a business entity or permanent business entity that is determined to carry out exploration and production in an oil and gas working area based on a cooperation contract with SKK Migas or BPMA.

[47] MEMR Regulation 2/2023, Article 41.

a. Feasibility study;
b. Construction of facilities for the CCS/CCUS activities; and/or
c. Certification to reduce greenhouse gas emission, including administration, validation, verification, and certification to reduce greenhouse gas emission.

The MEMR Regulation 2/2023 further stipulates that any goods and equipment used directly for the implementation of CCS/CCUS activities purchased by the Contractor is state property, and as such, needs to be managed in accordance with applicable laws and regulations.[48]

3.1.4 Monetization of CCS/CCUS

Presidential Regulation No. 98 of 2021 concerning the Implementation of Carbon Pricing to Achieve the Nationally Determined Contribution Target and Control Over Greenhouse Gas Emissions in National Development (**"PR 98/2021"**) is a regulation issued in October 2021 and had just taken effect in October 2022. It is issued as a climate financing instrument to achieve Indonesia's Nationally Determined Contribution (**"NDC"**) goal and empowers the Indonesian Government to introduce several carbon pricing mechanisms, including carbon trading. Up until the writing of this Indonesian chapter, PR 98/2021 is still waiting for its implementing regulations to be issued. MEMR Regulation 2/2023 complements the PR 98/2021 by paving the way for carbon pricing and trading to support the economics of CCS/CCUS projects.

Monetization of CCS/CCUS activities in connection with upstream oil and gas activities can be done based on MEMR Regulation 2/2023 at least through carbon trading and/or cost recovery from shared facilities. Meanwhile, monetization of CCS/CCUS activities in the context where the Carbon Emissions are not sourced from upstream oil and gas activities can be done through the provision of injection and storage services.[49] As long as the facilities meet technical, economic, and operational safety feasibility requirements, third parties are allowed to utilize the facilities operated by the Contractor for the implementation of CCS/CCUS activities.[50] With third parties allowed to utilize the facilities for

[48] MEMR Regulation 2/2023, Article 45.

[49] MEMR Regulation 2/2023, Articles 42(1) and (2).

[50] MEMR Regulation 2/2023, Article 21.

CCS/CCUS activities in an oil and gas working area, existing infrastructure and storage reservoirs can be used to create multi-user CCS hubs. MEMR had also encouraged oil and gas companies in Indonesia to receive carbon for injection from the oil and gas industry and other industries by installing carbon capture facilities in their working areas.[51] However, the MEMR Regulation 2/2023 does not stipulate on the regulatory documentation or framework required for cross-border transportation of CO_2 and other greenhouse gases and so bilateral agreements will likely need to be signed to facilitate the transfer of such gas from emitters to sequesters in Indonesia.

Tax incentives for CCS/CCUS activities are also offered to Contractors in accordance with laws and regulations regarding tax treatment of upstream oil and gas business activities.[52] Pursuant to Government Regulation No. 53 of 2017 on Tax Treatment for Upstream Oil and Gas with Gross Split Profit Sharing Contract and its implementing regulations, the tax incentives may include[53]:

a. exemption from import duties for goods used for petroleum operations,
b. exemption from value-added tax and/or luxury goods sales tax for the following goods and/or services used for petroleum operations:
   i. taxable objects and/or services;
   ii. taxable imported objects;
   iii. usage of taxable intangible objects from outside the Republic of Indonesia; and/or
   iv. usage of taxable services from outside the Republic of Indonesia.
c. exemption from income tax for imported goods that have received import duties exemption in paragraph (a) above; and
d. reduction of land and building tax by 100%.

[51] https://www.marketscreener.com/quote/stock/INPEX-CORPORATION-6498754/news/Indonesia-invites-oil-and-gas-operators-to-set-up-carbon-capture-facilities-43209385/.

[52] MEMR Regulation 2/2023, Article 43.

[53] Government Regulation No. 53 of 2017 on Tax Treatment for Upstream Oil and Gas with Gross Split Profit Sharing Contract, Article 25.

Separately, there are news reports which mention that SKK Migas is considering to provide 3 or 4 PSC Contractors with an increase of their production entitlements as an incentive to increase the CCS/CCUS project economics, especially for work areas that have high CO_2 contents. However, as at the date of this chapter is written, we have yet to see an implementation of this incentive in practice.[54]

Apart from injection and storage of Carbon Emission, oil and gas production can also be increased through enhanced oil, natural gas or CBM recovery.[55] CCS/CCUS projects will be able to rely on revenues from this increased production, as well as revenues from injection and storage of Carbon Emissions from outside the oil and gas working area. By injecting CO_2, oil production can be boosted while Carbon Emissions are simultaneously reduced through permanently storing CO_2 when utilized for enhanced oil recovery. The Asian Development Bank conducted a study on the assessment of CO_2-enhanced oil recovery in Indonesia. The study found that much of the CO_2 injected to rejuvenate the production of oil at mature oilfields serves as a climate mitigation measure as it remains trapped in the field. A favorable fiscal and legal framework are among the number of factors to which enhanced oil recovery success depends on aside from the availability of inexpensive sources of CO_2, which could be in the form of CO_2 extracted from natural gas processing plants located close to oil fields with declining production in Indonesia. With regard to financial incentives to implement enhanced oil recovery, there are currently no regulation that provide much guidance, although the current soaring oil prices are likely to support the economic potential of the technology for implementation at mature oil fields.

[54] https://katadata.co.id/happyfajrian/berita/643e3cb4a9fec/pemerintah-siapkan-insentif-bagi-kkks-migas-yang-pasang-teknologi-ccus).

[55] MEMR Regulation 2/2023, Article 9.

3.1.5 Monitoring, Measurement, Reporting, and Verification

In order to ensure worker safety, installation and equipment safety, environmental safety and public safety, monitoring activities are to be carried out in accordance with the monitoring plan in the approved CCS/CCUS implementation plan.[56] The monitoring plan is essential to identify any potential risks due to the injection of Carbon Emissions, such as potential risks of leakage, contamination of ground water, the integrity of the buffer zone layers, impermeable zone layers, and geological traps.[57]

MEMR Regulation 2/2023 requires monitoring activities to be carried out **up to 10 years after the completion of the closing of CCS/CCUS activities** based on the approved CCS/CCUS implementation plan (more on this in the "Closure of CCS/CCUS Activities" section below).[58] Costs for monitoring activities are to be reserved by the Contractor of the relevant oil and gas working area in a joint account on behalf of the Contractor and SKK Migas or BPMA (as applicable) and the reserved costs shall be included as operating costs.[59] The cooperation contract shall have the monitoring activities to be a part of the petroleum operations, while the amount and installments of monitoring costs reserves shall be proposed to and approved by SKK Migas or BPMA (as applicable) during the annual Work Program & Budget approval process.[60] Since the MEMR Regulation 2/2023 does not provide much explanation on the subject of terms and scope of the agreements to establish a structure to fund such joint accounts for monitoring costs, including the method of calculating such costs, the elements comprising such costs (i.e., among others, whether such monitoring costs would include an amount reserved to conduct any remediation activities during the monitoring period), and the handling of any excess or insufficiency of the reserved costs at the end of the monitoring period, further clarity from the Indonesian regulators would be required on these matters. Results from the monitoring are to be reported every six months to the Director General of Oil and Gas of the Ministry of the Energy and Mineral Resources ("**Director General**") through SKK Migas or BPMA (as applicable) and should the monitoring

[56] MEMR Regulation 2/2023, Article 27(1).

[57] MEMR Regulation 2/2023, Article 28.

[58] MEMR Regulation 2/2023, Article 27(3).

[59] MEMR Regulation 2/2023, Articles 27(5), 27(6) and (8).

[60] MEMR Regulation 2/2023, Article 27(7).

results shows any leak, contamination of ground water or unplanned shift of Carbon Emissions, they are to be reported to the Director General within the prescribed time frames.[61] Additionally, emergency response systems must be put in place by the Contractor, to ensure dangerous conditions with the potential to threaten workers safety, installation and equipment safety, environmental safety, and/or public safety can be dealt with quickly and accurately.[62]

Aside from monitoring activities, MRV activities such as the following shall be performed by the Contractor conducting CCS/CCUS activities[63]:

- *Measurement—A measurement program shall be carried out* at least once a year by Contractor conducting CCS/CCUS activities. The Contractor shall prepare a measurement program in accordance to the specifications of the CCS/CCUS location, which shall contain at least (a) the inventory of Carbon Emissions during the activities and (b) the CCS/CCUS operating parameters. The measurement program can use either the direct method or the indirect method.[64]
- *Reporting*—The CCS/CCUS and CEV (i.e., Carbon Economic Value, being the value of each unit of greenhouse gas emissions resulting from human activities and economic activities) reporting shall include general data and technical data (such as the method of calculating the reduction in Carbon Emissions).[65] General data comprises of (a) the identity of the Contractor as the party conducting and is liable for the CCS/CCUS activities, (b) title and type of activity, (c) the mechanism of CCS/CCUS activities as well as the opted CEV; and (d) transfer of technology, increase in capacity, and financing. Meanwhile, technical data comprises of (a) calculation of the baseline greenhouse gas emissions, (b) choosing the period of reference to determine the baseline greenhouse gas emissions, (c) calculation method to achieve Carbon Emissions reduction, (d)

[61] MEMR Regulation 2/2023, Articles 30(1) and (2).

[62] MEMR Regulation 2/2023, Article 46.

[63] MEMR Regulation 2/2023, Article 32(1).

[64] MEMR Regulation 2/2023, Articles 34(1), (2), and (3).

[65] MEMR Regulation 2/2023, Articles 35(1) and (3).

monitoring result on activities data, including measurements, location, and period when CCS/CCUS is conducted, (e) the achieved target for reduction and/or absorption of Carbon Emission; and (f) description of the managerial system, including the name of the person in charge, and the system constructed to supervise and collect activity data in relation to the performance of the CCS/CCUS and the CEV.

- *Verification*—The control and quality assurance of measurement results and monitoring of implementation of CCS/CCUS activities shall be carried out through verification in accordance with applicable laws and regulations.[66]

3.1.6 Closure of CCS/CCUS Activities

Closure of CCS/CCUS activities shall be conducted in the event:

a. the storage capacity of the Injection Target Zone is full;
b. there are no more Carbon Emissions to be injected;
c. the cooperation contract expires and is not extended;
d. an unsafe condition arises; or
e. a force majeure event (as defined in the cooperation contract) occurs causing closure of the CCS/CCUS activities to be the best option to be pursued.[67]

The Contractor shall submit the plan to close the CCS/CCUS activities based on either reason (a), (b), or (c) above to the MEMR through the Head of SKK Migas or the Head of BPMA. The plan must contain at least the following descriptions[68]:

a. information on the reservoir, equipments, installations, facility, and wells closed due to the closure of CCS/CCUS activities;
b. Total Carbon Emission reduced;
c. a cost estimate;
d. timeline of the closure; and

[66] MEMR Regulation 2/2023, Article 38.

[67] MEMR Regulation 2/2023, Article 22.

[68] MEMR Regulation 2/2023, Article 23(2).

e. a prevention plan against potential danger to the environment, humans, resources and equipment, installations and facilities due to the closure of the CCS/CCUS activities.

The closure of the CCS/CCUS activities can be performed based on the approved closure plan upon obtaining the MEMR's approval of such closure plan.[69] The Director General, or any independent verifier it appointed, shall verify the closure of CCS/CCUS activities in accordance with the approved closure plan, and that the completion of the closure met the referenced standards and good engineering principles.[70] The Contractor shall bear the cost of this verification and such cost shall form part of the operating costs.[71]

If an unsafe condition arises or a force majeure event occurs, the Contractor shall close the CCS/CCUS activities and provide reports on the closure to the Director General within the following time frames[72]:

a. 1 × 24 hours, if submitted through electronic messaging;
b. 2 × 24 hours, if submitted in writing; and
c. Once every 24 hours to report on the progress in closing the CCS/CCUS activities.

The report must contain at least the following details[73]:

a. The chronology of the unsafe condition or force majeure;
b. The cause for the closure of the CCS/CCUS activities;
c. Impact that arises if the closure of the CCS/CCUS activities are not conducted;
d. Procedure to close the CCS/CCUS activities;
e. The safety actions performed; and
f. Information on the reservoir, equipments, installations, facility, and wells closed due to the closure of CCS/CCUS activities.

[69] MEMR Regulation 2/2023, Article 23(6).

[70] MEMR Regulation 2/2023, Article 25.

[71] MEMR Regulation 2/2023, Article 26.

[72] MEMR Regulation 2/2023, Articles 24(1) and (2).

[73] MEMR Regulation 2/2023, Article 24(3).

3.2 CCS/CCUS Regulatory Framework in Other Industries

As at the writing of this Indonesian chapter, there is no regulation which specifically and expressly regulates CCS/CCUS activities in other industries than the oil and gas industry in Indonesia. Nearing the end of 2023, there were news reports that mentioned that the Government of Indonesia is currently preparing a draft presidential regulation to implement CCS/CCUS in other sectors, allow for Indonesia to develop CCS/CCUS Hub, conduct cross-border CCS and other applications of CCS/CCUS. The Government of Indonesia recognizes that a robust legal basis is required to support the development of a safe and effective CCS/CCUS activities as well as to provide legal certainty for investors.[74]

4 Concerns and Challenges for CCUS Projects Development in Indonesia

CCUS technology is still relatively nascent and requires upfront capital-intensive investments. The costs will also vary greatly depending on the concentration of the carbon source and the technology required to conduct CCS/CCUS activities. The carbon capture costs alone are estimated to be around USD 16 per metric ton[75] in oil and gas industry and on average around USD 45 per metric ton[76] in other sectors (but may easily exceed USD 200 per metric ton depending on the capture method and technology required)[77]. In addition, in Indonesia carbon transportation costs may also become a considerable factor considering that there is a significant geological difference between where the emissions are generated (i.e., mostly in Java) and where they can be sequestered (i.e., mostly saline aquifer basins in Sumatra and Kalimantan). Such geological differences will result in high transportation costs. On top of such high carbon capture and transportation costs, the environmental risks of leaks are also

[74] https://indonesiabusinesspost.com/risks-opportunities/indonesia-aims-to-lead-southeast-asia-in-ccs-ccus-technology/#:~:text=According%20to%20Tutuka%2C%20implementing%20CCS,clear%20impact%20on%20environmental%20risks.

[75] *Economic viability of full-chain CCUS-EOR in Indonesia,* Aldilla Noor Rakhiemah, Yuan Xu, 2022.

[76] https://indonesiabusinesspost.com/risks-opportunities/indonesia-aims-to-lead-southeast-asia-in-ccs-ccus-technology/.

[77] https://iea.blob.core.windows.net/assets/78633715-15c0-44e1-81df-41123c556d57/DirectAirCapture_Akeytechnologyfornetzero.pdf.

generally still unknown and may also result in high financial exposures. Given the foregoing, currently CCS/CCUS activities are beginning to be realized mainly in oil and gas sector while other industries are still lagging behind. Although there is potential for certain sectors other than the oil and gas sector (such as cement, iron, steel, chemical industries) to be candidates for CCS/CCUS applications, these sectors are rather price sensitive and will affect the commodity markets' prices and competitiveness. Even in the oil and gas sector, the developments are still going at a slow pace as the incentives for implementing CCS/CCUS are still not clear. MEMR Regulation 2/2023 provides that the Contractors will be provided with incentives. There are news reports which mention that SKK Migas is considering to provide higher production split for oil and gas Contractors, but as at the date this chapter is written neither SKK Migas nor MEMR has issued any regulation stipulating clear incentives for CCS/CCUS applications.

In stark contrast to the high investment requirements and lack of incentives, monetization of CCS/CCUS in Indonesia is still not as lucrative as they are in other more developed countries. One of the main ways to monetize is through the carbon market. However, Indonesian carbon market was established only on 26 September 2023 and until January of 2024 there are only around 40 participants in the market. With average carbon price of around USD 4 per metric ton and highest price reported of only IDR 77,000 (around USD 4.9) per metric ton, the Indonesian carbon price is also very low compared to the price offered for similar products in more established markets such as the European Union (which record high is EUR 100.34 or around USD 109 per metric ton)[78].

5 Lessons Learned

A few of the takeaways from the development of CCS/CCUS in Indonesia are as follows:

- regulatory support is required to allow for investment in and development of CCS/CCUS. Indonesia has taken a major step when MEMR released the MEMR Regulation 2/2023. The clarity that

[78] https://www.statista.com/statistics/1322214/carbon-prices-european-union-emission-trading-scheme/.

MEMR Regulation 2/2023 brought to the implementation of CCS/CCUS in the oil and gas sector was very welcomed by the industry, and there is palpable confidence in the oil and gas sector toward the implementation since. Nevertheless, other than the oil and gas industry, CCS/CCUS applications are still not expressly and specifically regulated. There are news reports that a presidential regulation is being drafted for the application of CCS/CCUS in other industries. However, until such presidential regulation has been enacted, the lack of legal certainty and structure is hampering investors from investing in CCS/CCUS in other sectors. Among others, it is also for this reason that Global CCS Institute estimates that Indonesia's CCUS readiness score is 30 (out of 100) as at January 2024[79];

- development of CCS/CCUS in clusters may be required to optimize costs. Given the geological differences between the locations of where most of the emissions are generated and where most of the carbon may be injected in Indonesian archipelago, clusters in which emitters that are situated together should develop a joint CCS/CCUS infrastructures to present significant costs optimization opportunities; and
- considering that the technology is relatively nascent and still requires capital-intensive investments, clear incentives from the government would be required in order to allow for adaptation in other sectors (especially those which markets are price sensitive) than the oil and gas sector.

6 Conclusion

With a potential of around 12.2 billion tons for CO_2 storage capacity, Indonesia has a tremendous potential to become CCS/CCUS hub in southeast Asia. The Government of Indonesia has also taken the first step by enacting MEMR Regulation 2/2023 for the implementation of CCS/CCUS in the upstream oil and gas sector. With a presidential regulation being drafted for implementation of CCS/CCUS in other sectors, it is clear that Indonesia is preparing for wider application of CCS/CCUS in other sectors than the upstream oil and gas sector. Although widespread

[79] https://co2re.co/ccsreadiness.

implementation is not yet in sight for the near or medium term, with sufficient regulatory support and provision of strategic incentives, Indonesia is poised to implement CCS/CCUS as part of its key strategies to reach its medium-term and long-term Carbon Emission reduction goals.

CHAPTER 14

CCUS in China: Legal and Regulatory Perspectives

Lan WU

1 Introduction

1.1 Background on Carbon Capture Utilisation and Storage (CCUS) Development in China

China is the largest carbon emitter worldwide, accounting for 35% of the global total in 2023.[1] China's large carbon footprint contributes significantly to global greenhouse gas emissions, exacerbating climate crisis. Therefore, China plays a critical and influential role in the global efforts to combat climate change. In this context, developing the Carbon Capture Utilisation and Storage (CCUS) technology is highly relevant for China.

[1] International Energy Agency, *CO2 Emissions in 2023* (2024) https://www.iea.org/reports/co2-emissions-in-2023/the-changing-landscape-of-global-emissions accessed 8 January 2025.

L. WU (✉)
Ocean University of China, Qingdao, China
e-mail: wulan@ouc.edu.cn

E. G. Pereira et al. (eds.), *Carbon Capture Utilization and Storage*, Sustainable Development Goals Series,
https://doi.org/10.1007/978-3-031-81272-9_14

The development of CCUS in China is driven by multiple considerations, including climate change mitigation, environmental protection, energy security, technological innovation, sustainable development, and international competitiveness.[2] The CCUS technology has the potential to capture a substantial portion of its CO2 emissions from industrial processes, power generation, and other sources, thereby mitigating its impact on the climate.[3] China is under the pressure of achieving the 2030 carbon peak and 2060 carbon neutrality pledges, which are ambitious climate targets require significant improvements in emission reduction. CCUS technology can play a role in reaching these goals by capturing CO2 emissions from various sectors and offsetting those that are difficult to decarbonise, such as heavy industry and transportation. Furthermore, China's rapid economic growth has led to increased energy demand,[4] with coal remaining the primary source of energy, accounting around 55.3% in total energy consumption in 2023.[5] Transitioning to cleaner energy sources is essential for China's energy security and reducing reliance on imported fuels. CCUS can help enhancing the energy security by gradually facilitate the transition from coal to cleaner energy sources while minimising carbon emissions. Moreover, investing in CCUS research and development fosters technological innovation and creates opportunities for Chinese companies to become leaders in clean energy technologies. This can enhance China's competitiveness in the global market and contribute to economic growth and job creation. As the world's largest emitter, China's actions are crucial in global efforts to combat climate change, and CCUS development is an important part of the climate actions.

[2] Gu Y, *Carbon Capture and Storage Policy in China* (Centre for Climate Change Law White Paper, 2013).

[3] Bhavsar A and others, 'The Current Scope and Stand of Carbon Capture Storage and Unitlization: a Comprehensive Review' (2023) 8 Case Studies in Chemical and Environmental Engineering 100368 https://www.sciencedirect.com/science/article/pii/S2666016423000737 accessed 8 January 2025.

[4] Peng X, Adams PD and Liu J, 'China's New Growth Pattern and its Effect on Energy Demand and Greenhouse Gas Emissions' (2018) 1(4) Global Energy Interconnection 428 https://www.sciencedirect.com/science/article/pii/S209651171830063X accessed 8 January 2025.

[5] National Bureau of Statistics, Statistical Bulletin on National Economic and Social Development of the People's Republic of China 2023 https://www.stats.gov.cn/sj/zxfb/202402/t20240228_1947915.html accessed 8 January 2025.

Nowadays, with the expansion of application scenarios, CCUS technology has become an important part of China's carbon neutral technology system, there has been analysis of the role of CCUS in various scenario, which demonstrate the role of CCUS in reducing emissions to meet China's 2060 carbon neutrality target.[6] And CCUS is considered as a feasible technological solution for the deep decarbonisation of iron, steel, cement and other hard-to-reduce emission industries, and the main technological means to support the recycling of carbon in the future. There has been a comprehensive analysis of the application of CCUS technology in the whole industry and its future demand for emission reduction, which revealed that, it is predicted that China's demand for CCUS emission reduction in pursuit of carbon neutrality pledges will be around 14–31 million tonnes per year in 2025 and will grow to 58–147 million tonnes/year in 2030.[7] By industry, the thermal power industry will be the focus of CCUS application because of China's current installed capacity of power generation and the hard constraints on energy security; and it is estimated that one Gigatonne CO2 abatement per year through CCUS can be expected by 2060 in this industry.[8]

The CCUS technology had emerged from the concept of carbon capture and storage (CCS). CCUS takes the concept of CCS a step further by adding the aspect of 'utilisation', which means that captured CO2 can be used for various purposes instead of just being stored.[9] Most of the CCUS projects at the early stage in China primarily focused on geological utilisation for enhanced energy resource recovery such as CO2 Enhanced Oil Recovery (CO2-EOR) and CO2 Enhanced Coal

[6] Wang P and others, 'CCUS Development in China and Forecast its Contribution to Emission Reduction' (2023) 13(1) Scientific Reports 17811 https://www.nature.com/articles/s41598-023-44893-y accessed 8 January 2025.

[7] Zhang Xian and others, *CCUS Progress in China—a Status Report 2023* (The Administrative Center for China's Agenda 21, Global CCS Institute, and Tsinghua University, 2023), at 5.

[8] Zhang Xian and others, *CCUS Progress in China—a status report 2023.* (The Administrative Center for China's Agenda 21, Global CCS Institute, and Tsinghua University, 2023), at 7.

[9] Bhavsar A and others, 'The Current Scope and Stand of Carbon Capture Storage and Utilization: A Comprehensive Review' (2023) 8 Case Studies in Chemical and Environmental Engineering 100368 https://www.sciencedirect.com/science/article/pii/S2666016423000737 accessed 8 January 2025.

Bed Methane Recovery (CO2-ECBM). However, chemical and biological utilisation projects are steadily increasing each year nowadays.[10]

China's CCUS technology has made remarkable progress in various segments in recent years. For example, China has acquired the capability to design a large-scale CO_2 capture, pipeline transport, utilisation, and storage system, and the foundation for large-scale application has been established for the near future.[11] By demonstrating commitment to CCUS development and deployment, China can lead by example and encourage other nations to adopt similar measures, fostering international cooperation and collective action. Lastly, implementing CCUS aligns with China's broader agenda of promoting sustainable development and transitioning to a low-carbon economy. By reducing emissions while supporting economic growth, CCUS contributes to a cleaner, more sustainable future for China and the world.

1.2 *Significance of CCUS Law and Regulation in Addressing Climate Change*

The improvement of Chinese CCUS-related laws and regulations has gain momentum since the announcement of the dual carbon pledges in 2020.[12] The significance of CCUS policies, laws, and regulations in addressing climate change is multi-faceted. These legal frameworks play a crucial role in promoting the adoption and development of CCUS technology in several ways. Firstly, CCUS laws and regulations can offer financial incentives, tax breaks, or subsidies to encourage industries to invest in CCUS projects. These incentives help offset the high costs associated with implementing CCUS technology and provide a strong economic motivation for companies to reduce their carbon emissions.

[10] Zhang Xian and others, *CCUS Progress in China—a Status Report 2023* (The Administrative Center for China's Agenda 21,Global CCS Institute, and Tsinghua University, 2023), at 14.

[11] Zhang Xian and others, *CCUS Progress in China—a Status Report 2023* (The Administrative Center for China's Agenda 21,Global CCS Institute, and Tsinghua University, 2023), at 10.

[12] President Xi Jinping's remarks at the general debate of the 75th session of the United Nations General Assembly, announcing that China aims to have CO2 emissions peak before 2030 and achieve carbon neutrality before 2060. Xinhuanet, 2020 September 23, available at http://www.xinhuanet.com/english/2020-09/23/c_139388764.htm accessed 8 January 2025.

By making CCUS economically viable, these laws spur innovation and drive the deployment of CCUS at a larger scale. Moreover, CCUS laws and regulations define the rights, responsibilities, and benefits of stakeholders involved in CCUS projects. They establish clear guidelines for the capture, transportation, and storage of CO2, ensuring the safe and responsible operation of CCUS facilities. These regulations also address liability issues, environmental protection measures, and the equitable distribution of benefits among project participants, fostering transparency and accountability within the CCUS sector. Furthermore, CCUS laws and regulations are expected to include provisions to protect the environment during the process of capture, utilisation, and storage of CO2. They establish criteria for selecting suitable storage sites, monitoring and verification requirements, and protocols for addressing potential leakage or environmental impacts. By imposing stringent environmental standards, these regulations mitigate the risks associated with CCUS and ensure that carbon storage is conducted safely and sustainably.

In summary, CCUS laws and regulations are significant in addressing climate change by providing incentives, regulating rights and responsibilities, and ensuring environmental safeguards. These legal frameworks create a conducive environment for the development and deployment of CCUS technology, contributing to the reduction of greenhouse gas emissions and the mitigation of climate change.

This chapter focuses on the analysis of the legal framework of CCUS in China. Besides the newly passed Energy Law,[13] the governance of CCUS in China also relies heavily on policies and regulatory documents. Therefore, the legal analysis of this chapter adopts a broad view of 'law' and engages these policies and regulatory documents. This chapter continues with an overview of climate change policy in China, which set the scene for the development of the CCUS. Then it dives into an analysis of the legal framework of CCUS in China. Additionally, this chapter engages a case study of the CCUS projects in China, which is followed by providing policy recommendations for strengthening the CCUS legal governance in China. The last section concludes.

[13] Energy Law of the PRC, adopted at the 12th Session of the Standing Committee of the 14th National People's Congress on 2024 November 8, available at https://www.nea.gov.cn/2024-11/09/c_1310787187.htm accessed 8 January 2025.

2 Overview of Climate Change Policy in China

2.1 China's Commitment to Reducing Greenhouse Gas Emissions and the 1 + N Policy Framework

President Xi Jinping has made pledges of having CO2 emissions peak before 2030 and achieving carbon neutrality before 2060.[14] The top-level design for the carbon pledges gradually takes the shape of a '1 + N' system—the '1' refers to the top-level design guidance that plays a leading role in the overall policy system[15]; and the 'N' refers to policies and regulatory measures of different economic sectors implementing the guidelines of the top-level policies, consisting of regulatory documents in various industries providing detailed measures for implementation to fulfil ten key tasks including green and low-carbon energy transition; energy conservation and efficiency improvement; carbon peaking in the industry sector; carbon peaking in the urban–rural areas; low-carbon transportation; circular economy; low-carbon technology innovation; carbon credit enhancing; and low-carbon society.[16]

[14] Announcement by President Xi Jinping at the general debate of the 75th session of the United Nations General Assembly (2020 September 22) https://news.un.org/en/story/2020/09/1073052 accessed 8 January 2025.

[15] Central Committee of the CPC and State Council, 'Working Guidance for Carbon Dioxide Peaking and Carbon Neutrality in Full and Faithful Implementation of the New Development Philosophy' 见*Xinhua News* (2021 September 22) https://www.gov.cn/zhengce/2021-10/24/content_5644613.htm> accessed 8 January 2025; State Council, Notice of the Action Plan for Carbon Dioxide Peaking Before 2030, Guo Fa [2021] No. 23, 2021 October 24 https://www.gov.cn/zhengce/content/2021-10/26/content_5644984.htm accessed 8 January 2025.

[16] State Council, Notice of the Action Plan for Carbon Dioxide Peaking Before 2030, Guo Fa [2021] No. 23, 2021 October 24, Sect.3 https://www.gov.cn/zhengce/content/2021-10/26/content_5644984.htm accessed 8 January 2025.

2.2 *National Policies and Targets Related to Climate Change Mitigation*

In the subsequent three years following the announcement of dual carbon goals in 2020, Chinese governments and related government departments have issued more than 90 policy documents as of the end of 2023,[17] forming the bedrock of the low-carbon transition policy architecture under the '1 + N' policy system.

China has been developing a comprehensive policy and regulatory framework to support CCUS deployment. This includes the inclusion of CCUS in national plans on climate change, the establishment of emission trading systems, and the introduction of administrative measures specific to CCUS projects.

Besides the 1 + N policy system, national plans relating to fundamental aspects of economic development activities are also important to climate change mitigation. The five-year plans, which are the economic and social development plans implemented by the Chinese governments since 1953. These plans set out specific targets, policies, and strategies to guide the country's overall development over a five-year period. The current five-year plan is the fourteenth five-year plan (2021–2025), which aims to promote high-quality development with a focus on innovation, green development, opening up the economy, improving social welfare, and achieving technological self-reliance.[18] The Chapter 37, Chapter 38, and Chapter 39 of the fourteenth five-year plan present roadmaps for the enhancing ecosystem quality and stability, continuous improvement of environmental quality, and accelerating the green transformation of development patterns in China during the 2021–2025. These aspects of green development are closely related to the climate change mitigation efforts in China. The fourteenth five-year plan in China includes targets for reducing CO2 intensity by 18% and energy intensity by 13.5% from 2021 to 2025. This marks the first time that China's longer-term climate goals are mentioned in a Five-Year Plan, and the concept of a

[17] Caixin Think Tank, 'China's Path to the "3060" Target: An Analysis of the "1 + N" Policy Framework for Carbon Neutrality' (2024 March 1) https://index.caixin.com/2024-03-01/102170557.html accessed 8 January 2025.

[18] Outline of the 14th Five-Year Plan for National Economic and Social Development of the People's Republic of China and the Long-Range Objectives Through the Year 2035 , *Xinhuanet* (2021 March 13) https://www.gov.cn/xinwen/2021-03/13/content_5592681.htm accessed 8 January 2025.

CO2 emissions cap is introduced, although no specific cap is set.[19] The 2024 annual plenary sessions of the National People's Congress and of the Chinese People's Political Consultative Conference (referring to as the Two Sessions) redefined the energy intensity goals to further promote the development of renewable energy, which represent a progress in the transition from the energy consumption dual control (controlling the total amount of energy consumption and intensity of energy consumption) mechanism to carbon emission dual control (controlling the total amount of carbon emission and intensity of carbon emission) mechanism.[20]

In addition, China's national emission trading scheme (Chinese ETS) has been launched in 2021, with progress made in regulatory efforts represented by the issuance of the *Interim Regulations on the Administration of Carbon Emission Trading (Trial)*.[21] The Chinese ETS is expected to provide market-based incentives for regulated entities (mostly industrial carbon emitters such as power, steel, chemistry, among the others) to reduce carbon emissions. The CCUS technology can help regulated entities to conform with the emission reduction targets under the Chinese ETS. Regulated entities can reduce carbon emissions by adopting CCUS technologies to reduce the cost of purchasing additional emission allowances in the ETS system. ETS can incentivise regulated entities to invest in Research & Development (R&D) and adopt more efficient CCUS technologies, thereby reducing abatement costs and improving their market competitiveness, and such technological innovation can further promote the development and adoption of CCUS.

[19] Carbon Brief, 'Q&A: What does China's 14th 'Five Year Plan Mean for Climate Change?' (2021 March 12) https://www.carbonbrief.org/qa-what-does-chinas-14th-five-year-plan-mean-for-climate-change/ accessed 8 January 2025.

[20] Research institution of Carbon Neutrality, 'An Analysis of the Policy Implications and Impacts of the Transition from Energy Consumption Dual Control to Carbon Emission Dual Control in China' (2023 December 5) https://ricn.sjtu.edu.cn/Web/Show/1011 accessed 8 January 2025.

[21] Opening a New Chapter in the Legal Framework for Carbon Emissions Trading, *Guangming News* (2024 February 27) https://www.gov.cn/zhengce/202402/content_6934526.htm accessed 8 January 2025.

2.3 *Importance of CCUS for Implementing China's Climate Change Strategy*

Recognising the importance of CCUS in addressing climate change, the Chinese government has provided policy support and financial incentives to promote the development and deployment of CCUS technologies, which is evidenced by the establishment of large-scale demonstration projects in different sectors, such as the Shenhua Group's CCS project in the coal-to-liquids industry.

CCUS technology in China has been rapidly developing, forming a complete process system. China's journey with CCUS technology has spanned several years, marked by notable developments and milestones. China's initial ventures into CCUS began in the tenth Five-Year Plan (2005–2010) with a focus on research and development.[22] The related plans mentioned that the government would implement fiscal and tax policies that stimulate low-carbon technological innovation, implement financial policies that promote low-carbon innovation and entrepreneurship, improve the scientific and cultural quality of the entire nation, and create a social environment conducive to scientific and technological innovation.[23] The *Notice on the Release of China's Scientific and Technological Specialized Actions to Address Climate Change* issued in 2007 designated CCUS as a key technology area for focused support, concentrated research and development efforts, and demonstration projects.[24] After this, multiple ministries and departments, including the State Council and the Ministry of Science and Technology, issued relevant policy documents further promote the development of CCUS.

Adopting a learning-by-doing approach, China initiated pilot projects to explore the feasibility of capturing and storing CO2 emissions from

[22] Asian Development Bank, 'Study on the Roadmap for Demonstration and Promotion of Carbon Capture and Storage in China' (2015 November) https://www.adb.org/sites/default/files/publication/175347/roadmap-ccs-prc-zh_0.pdf accessed 8 January 2025.

[23] State Council, Outline of the National Medium- and Long-Term Scientific and Technological Development Plan (2006–2020), Guo Fa [2006] No. 6,2006 February 7, https://www.gov.cn/gongbao/content/2006/content_240244.htm accessed 8 January 2025.

[24] Ministry of Science and Technology, National Development and Reform Commission and other departments, Notice on the Release of China's Scientific and Technological Specialised Actions to Address Climate Change, Guo Ke Fa She Zi [2007] No. 407, 2007 June 13, available at https://www.mee.gov.cn/gkml/hbb/gwy/200910/t20091030_180716.htm.

various industries, including coal-fired power plants, oil refineries, and cement production facilities. Recognising the importance of CCUS in addressing greenhouse gas emission issues, especially for those hard-to-be-reduced emissions, China gradually expanded its CCUS efforts, with a focus on scaling up the technology and increasing its application in various industries. This expansion included the construction of more demonstration projects and the integration of CCUS into large-scale industrial complexes, such as the integration of CCUS with coal chemical plants in the Ordos region, located in the Inner Mongolia Autonomous Region of China. Moreover, China has invested significantly in R&D activities related to CCUS. Academic institutions, research organisations, and industry players have conducted extensive R&D to enhance capture technologies, develop new storage sites, and explore the utilisation of captured CO2 for enhanced oil recovery and other applications. CO2 capture technologies for emissions from different concentration sources have been established, and a large number of pilots have been carried out in industries such as coal power, petrochemicals, cement, and steel.[25] However, challenges remain for the CCUS industry in China to scaling up in order to achieve the 2030 carbon peak and 2060 carbon neutrality goals, which will be analysed in the next section.

3 CCUS Projects Overview and Its Legal Framework in China

3.1 Brief Overview of Ongoing CCUS Projects in China

The deployment of full-chain CCUS projects in China (including capture, transportation, and storage) is primarily driven by the need to enhance oil recovery (EOR) in aging oilfields located in northern China.[26] While China has conducted lab research on the potential of saline formations

[25] Li Yang, Wang Rui, Zhao Qingmin, and Fang Xin 'Status and Prospects of Carbon Capture, Utilisation and Storage Technology Application in China', 2023 August, Petroleum Science Bulletin, 8(4) 391–397, at 393.

[26] Toby Lockwood, 'Reducing China's Coal Power Emissions with CCUS Retrofits', IEA Clean Coal Centre, 2018 November, https://iea.blob.core.windows.net/assets/3765e352-61ef-4926-8371-1ea3a5ec94cd/CCUS_Retrofit_CIAB_report_IEACCC.pdf accessed 8 January 2025, at 32.

for storage, full-chain CCUS dedicated storage projects remain limited to date.[27]

Despite of this, China has become a global leader in patents related to the CCUS in recent years. Meanwhile, the construction of CCUS demonstration projects in China has been developing rapidly, with significant increases in number and scale, and the application of CCUS technology in more industries and fields.[28] China currently has approximately 100 CCUS demonstration projects of different sizes that are either operational or in the planning stage as of November 2022.[29] As of 2023 February, there are 49 projects in operation or under construction.[30] China has successfully implemented CCUS technology in its key industrial sectors, showcasing its effectiveness. The power sector and manufacture industry has become the key areas of CCUS regulation.[31] The demonstration projects majorly targeting areas such as petroleum, coal chemical industry, and power industry, but the industrial chain of large-scale CCUS full-process demonstration projects in China is not yet complete.[32] In terms of cost, China's CCUS demonstration projects are relatively affordable

[27] China Energy Technology and Economics Research Institute, 'Achieving Carbon Emission Reduction in China's Coal-fired Power Sector through CCUS Technology Retrofitting', (2019) https://iea.blob.core.windows.net/assets/dc3450c2-4694-4c8c-9db3-9843214ead8a/ChinaCCSretrofitreportCCUS.pdf accessed 8 January 2025, at 43.

[28] Zhang Xian and others, *CCUS Progress in China—a Status Report 2023* (The Administrative Center for China's Agenda 21, Global CCS Institute, and Tsinghua University, 2023), at 12.

[29] Zhang Xian and others, *CCUS Progress in China—a Status Report 2023* (The Administrative Center for China's Agenda 21, Global CCS Institute, and Tsinghua University, 2023), at 12.

[30] Zhao Zhenyu, Yao Shun, Yang Shuopeng, and Wang Xiaolong, 'Under the "Dual Carbon" Goal: Current Status, Problems and Suggestions for CCUS Development in China', Environmental Science 2024 (44) 1128, at 1132.

[31] Zeng Lixuan, Xing Hongfei, 'A Sample Review of Carbon Capture, Utilisation and Storage Policies in China under a "Dual Carbon" Governance Perspective', Journal of Hohai University (Philosophy and Social Science Edition) 2024 (2) 147, at 155.

[32] Zeng Lixuan, Xing Hongfei, 'A Sample Review of Carbon Capture, Utilisation and Storage Policies in China under a "Dual Carbon" Governance Perspective' , Journal of Hohai University (Philosophy and Social Science Edition) 2024 (2) 147, at 155.

compared to global standards, with costs ranging from medium to low levels.[33]

Since the carbon neutrality target was proposed, the scale of CCUS demonstration projects that have been put into operation and are under planning and construction in China has significantly expanded. For example, the Qilu Petrochemical-Shengli Oilfield Million-tonne CCUS Project[34] is the largest CCUS industry chain demonstration base and benchmark project currently in China, marking that China's CCUS industry has entered the stage of mature commercial operation, which is representative to a certain extent.[35]

3.2 *The CCUS Regulations in Promoting Project Development: The Interaction Between Law and Technology*

Just like other types of technology, the development and deployment of CCUS projects require a solid regulatory framework that effectively addresses the complex interactions between law and technology.

CCUS policies play a crucial role in promoting project development by fostering technological innovation, ensuring safety and environmental protection, and facilitating project scaling and deployment. These regulations can incentivise research and development, protect intellectual property rights, and encourage collaboration between regulators, academia, and industry to drive technological advancements in CCUS. They are also necessary to also establish risk assessment and mitigation measures, permit and licensing requirements, and monitoring and verification protocols to ensure the safety and environmental integrity of CCUS projects. For example, the CCUS laws and policies issued after the 18th National Congress of the Communist Party of China (CPC) in 2012, which boosted the R&D momentum of CCUS technologies, ushering

[33] Zhang Xian and others, *CCUS Progress in China—a Status Report 2023* (The Administrative Center for China's Agenda 21, Global CCS Institute, and Tsinghua University, 2023), at 13–17.

[34] National Energy Administration, 'China's First Million-Tonne CCUS Project Fully Completed and Put into Operation', *Xinhuanet* (2022 September 2) http://www.nea.gov.cn/2022-09/02/c_1310658658.htm accessed 8 January 2025.

[35] Qiao Zhen and Xiong Yueting, 'Study on the Current Situation, Problems and Countermeasures of Carbon Capture, Utilisation and Storage Technology in Shandong Province', Science and Technology Plaza, 2023 (2) 55, at 57.

in the first peak period of development.[36] Based on this, the Chinese CCUS policies started to explore the development of CCUS fiscal subsidy provisions and diversified fiscal incentives in line with China's national conditions.[37]

Since the introduction of China's '1 + N' policy system for emission peaking and carbon neutrality, climate actions had gained momentum, and more CCUS-related policies have been released. A series of policy documents has been issued by the Central Committee of the Communist Party of China (CPC) and the State Council at central level involves the R&D, application and promotion, and project construction of CCUS.[38] Provincial governments has also been active in issuing local regulatory documents providing more details for the localised development of CCUS. These policies include plans, standards, roadmaps, and technology catalogues accumulatively, and CCUS has been included for the first time in China's national Five-Year Plan (2021–2025).[39]

3.3 *Challenges and Barriers Faced by Regulating CCUS Projects in China*

Although the rapid development of Chinese CCUS in recent years, challenges remain at this stage. These challenges are at least fourfold: the key technologies for CCUS in the large-scale full chain system need to be established; the cost for CCUS remains high; and the existing construction projects are mainly at small scale with incomplete full chain deployment; and the policies for CCUS is insufficiently supporting further development as these policy documents are mainly of a guiding nature,

[36] Zeng Lixuan, Xing Hongfei, 'A Sample Review of Carbon Capture, Utilisation and Storage Policies in China under a "Dual Carbon" Governance Perspective', Journal of Hohai University (Philosophy and Social Science Edition) 2024 (2) 147, at 152.

[37] Zeng Lixuan, Xing Hongfei, 'A Sample Review of Carbon Capture, Utilisation and Storage Policies in China under a "Dual Carbon" Governance Perspective', Journal of Hohai University (Philosophy and Social Science Edition) 2024 (2) 147, at 152.

[38] Zhang Xian and others, *CCUS Progress in China—a Status Report 2023* (The Administrative Center for China's Agenda 21, Global CCS Institute, and Tsinghua University, 2023), attachment 2, at 33.

[39] Norton Rose Fulbright, Global Carbon Capture and Storage Regulations: A Driver or Barrier to CCS Project Development? (2023 September) https://www.nortonrosefulbright.com/en-sg/knowledge/publications/3f25baee/global-carbon-capture-and-stora accessed 8 January 2025.

supporting major R&D institutions or enterprises to carry out technology R&D and engineering demonstration, and are not binding on enterprises.[40]

The main legal and compliance issues that may need to be focused on in the development process of CCUS involve several aspects of legal issues in practice.[41] Firstly, the legal definition of CO2 is a key issue to be addressed by CCUS legislation, including the classification and technical standards for CO2, which are determining the ownership and rights of CO2 throughout the CCUS chain. Secondly, laws and regulations regarding environmental impact assessment is a critical consideration for CCUS projects development, especially concerning the project approval and authorisation issues, as well as the public engagement and consultation. Additionally, regulations regarding the safety of storage activities needs to address the issues of storage site inspections and infrastructure, defining the ownership or right to use the pore space, operational liabilities and financial security, operational procedure, site closure process, post-site closure and also long-term storage liability.[42] These aspects entail significant complexities of the design and implementation of a legal framework of CCUS, which play a significant role in facilitating CCUS project financing, adopting different financing instruments such as green bonds, thereby enabling the scaling-up and deployment of CCUS technologies.

From an institutional perspective, the State Council is the highest administrative body in China's government institutional system, responsible for formulating the general plan in the Five-Year Plans, which outline the country's key economic and social development goals, priorities, and policy measures for a 5-year period. The State Council is also responsible for overseeing the formulation of relevant sub-plans and regulations by its various ministries. Previously, issues related to climate change and carbon emissions reduction were mainly handled by the National Development and Reform Commission (NDRC), which was responsible for

[40] Li Yang, Wang Rui, Zhao Qingmin, and Fang Xin, 'The Current Status and Prospects of Carbon Capture, Utilisation, and Storage Technology Application in China', Petroleum Science Bulletin, 8(4) 391–397, at 393.

[41] International Energy Agency, CCUS Legal and Regulatory Database, https://www.iea.org/data-and-statistics/data-tools/ccus-legal-and-regulatory-database accessed 8 January 2025.

[42] Gao Yan, Li Zhuohui, 'Introduction to the International Legal Framework of CCUS and its Significance for China', (*Zhonglun Law Firm Research Articles*, 29 January 2024) https://www.zhonglun.com/research/articles/52590.html accessed 8 January 2025.

managing overall macroeconomic planning, including policymaking in the areas of energy and climate change. In 2018, the Ministry of Ecology and Environment (MEE) was established, replacing the previous Ministry of Environmental Protection (MEP). In addition to continuing MEP's responsibilities in the areas of air and water pollution, the MEE has also taken over responsibilities related to climate change and CCUS from the NDRC. The National Energy Administration (NEA), on the other hand, is responsible for coordinating other energy policies and regulations among relevant ministries. The further development of CCUS depends on the deployment of demonstration pilots, which means that the MEE requires higher level national support. Even though provincial governments have the political will to deploy CCUS projects, the necessary coordination among different governmental departments has always been a challenge for policymaking, as the CCUS management responsibilities were transferred from NDRC to MEE, the policymaking issue became more complex.[43] The establishment of the coordination mechanism is conducive for providing new motivation for CCUS deployment, but this also depends on the integration of the important progress into Five Year Plans.[44]

In addition to the issue of governmental department coordination, there is a lack of a clearly defined legal and regulatory framework for CO2 geological storage in China, which also adds uncertainty to the further development of the CCUS projects.[45] Currently most of the regulatory documents of CCUS are local governments working document at low level, with insufficient credibility and mandatory degree of the policy.[46]

[43] Toby Lockwood, 'Reducing China's Coal Power Emissions with CCUS Retrofits', IEA Clean Coal Centre, 2018 November, https://iea.blob.core.windows.net/assets/3765e352-61ef-4926-8371-1ea3a5ec94cd/CCUS_Retrofit_CIAB_report_IEACCC.pdf accessed 8 January 2025, at 30.

[44] Toby Lockwood, 'Reducing China's Coal Power Emissions with CCUS Retrofits', IEA Clean Coal Centre, 2018 November, https://iea.blob.core.windows.net/assets/3765e352-61ef-4926-8371-1ea3a5ec94cd/CCUS_Retrofit_CIAB_report_IEACCC.pdf accessed 8 January 2025, at 30.

[45] Yang Xiaoliang, Li Xiaochun, 'A Gap Analysis of China's Regulatory Framework for CO2 Geological Storage', Global CCS Institute, 2024 April, https://www.globalccsinstitute.com/wp-content/uploads/2024/04/CCS-in-China-15-April.pdf accessed 8 January 2025.

[46] Zeng Lixuan, Xing Hongfei, 'A Sample Review of Carbon Capture, Utilisation and Storage Policies in China under a "Dual Carbon" Governance Perspective', Journal of Hohai University (Philosophy and Social Science Edition) 2024 (2) 147, at 156.

Most of the existing policy documents are guiding and encouraging, with no detailed working rules or guidelines, and only local governments such as those of Beijing, Shanghai, Jilin, Hubei, Shanxi, and Inner Mongolia have made some specific arrangements for some of the pilot demonstration projects or the construction of research centres in their respective areas.[47]

Moreover, the absence of a robust market mechanism and strong policy incentives continues to pose obstacles to the advancement of CCUS in China.[48] The current CCUS projects are majorly led by public investment that has the issues with low competitiveness and poor overall economics of product, coupled with difficulties in financing the technology and high energy consumption, making it difficult for enterprises to sustain them alone.[49] Market incentive policies such as financial guarantees, tax breaks, green bonds, and low-carbon funds have not been given sufficient attention, and more efforts are needed for China to form an incentive environment suitable for the commercialisation of CCUS technologies.[50]

Although the above-mentioned cost reduction, the technology costs of CCUS remains relatively high and the technological advancement is not obvious, which put the power, steel and cement industry under pressure.[51] Challenges also include the lack of effective business models, insufficient incentives and regulatory measures, and difficulties in sources-sinks matching.[52] Most of the market incentives issued for CCUS in

[47] Wu Helai, Li Wangfan, and Ding Xian, 'Analysis of Carbon Capture, Utilisation and Sequestration Policies in China under the "Double Carbon" Target and Recommendations', Electric Power Construction 2022 (43) 28, at 33.

[48] Norton Rose Fulbright, 'Global Carbon Capture and Storage Regulations: A Driver or Barrier to CCS Project Development?', (2023 September) https://www.nortonrosefulbright.com/en-sg/knowledge/publications/3f25baee/global-carbon-capture-and-stora accessed 8 January 2025.

[49] Zhao Zhenyu, Yao Shun, Yang Shuopeng, and Wang Xiaolong, 'Under the "Dual Carbon" Goal: Current Status, Problems and Suggestions for CCUS Development in China', Environmental Science 2024 (44) 1128, at 1133.

[50] Zeng Lixuan, Xing Hongfei, 'A Sample Review of Carbon Capture, Utilisation and Storage Policies in China under a "Dual Carbon" Governance Perspective', Journal of Hohai University (Philosophy and Social Science Edition) 2024 (2) 147, at 153.

[51] Zhang Xian and others, *CCUS Progress in China—a Status Report 2023* (The Administrative Center for China's Agenda 21, Global CCS Institute, and Tsinghua University, 2023), at 20.

[52] , Zhang Xian and others, *CCUS Progress in China—a Status Report 2023* (The Administrative Center for China's Agenda 21, Global CCS Institute, and Tsinghua University, 2023), at 20.

China are guidance and encouragement, and there is a lack of financial subsidies and fiscal incentives for key areas and specific segments.[53] Besides, the overall cross-sectoral cooperation and linkages remain weak among different government departments, with a lack of basic consensus and willingness to work together on governance.[54]

In addition, challenges remain for integrating CCUS into carbon markets as there remains gaps between the CCUS project rules and carbon market rules.[55] The Chinese ETS has not become mature enough to operate nationally with meaningful carbon prices due to the insufficient degree of financialisation and market participation, the small scale and few trading varieties, the inadequate policy support system and the lack of specific industry-wide guiding policies.[56] Consequently, the Chinese ETS has been slow to promote in other industries across the country, which in effect restricts the promotion of the commercialisation of CCUS in China.

4 Strengthening CCUS Legal Framework in China

4.1 Identification of Gaps and Shortcomings in the Current Legal Framework

The development of CCUS-related legal framework in China is based on basic laws include *the Environmental Protection Law of the PRC, The Land Administration Law of the PRC, the Mineral Resources Law of the PRC, the Coal Law of the PRC, the Power Electric Law of the PRC, and*

[53] Zeng Lixuan, Xing Hongfei, 'A Sample Review of Carbon Capture, Utilisation and Storage Policies in China under a "Dual Carbon" Governance Perspective', Journal of Hohai University (Philosophy and Social Science Edition) 2024 (2) 147, at 154.

[54] Zeng Lixuan, Xing Hongfei, 'A Sample Review of Carbon Capture, Utilisation and Storage Policies in China under a "Dual Carbon" Governance Perspective' , Journal of Hohai University (Philosophy and Social Science Edition) 2024 (2) 147, at 156.

[55] Zhai Mingyang, Zhou Changbo, Zhang Yongbbo, Wang Xingzhi, Wang Xiaomeng, Liu Xiaoyu, Li Mingkui, 'Status Quo, Challenges and Countermeasures for Carbon Dioxide Capture, Utilisation And Storage Projects to be Integrated in the Carbon Market', China Environmental Management, 2023 (6) 87, at 88.

[56] Zhao Zhenyu, Yao Shun, Yang Shuopeng, and Wang Xiaolong, 'Under the "Dual Carbon" Goal: Current Status, Problems and Suggestions for CCUS Development in China', Environmental Science 2024 (44) 1128, at 1134.

the Clean Production Promotion Law of the PRC, and the Air Pollution Prevention and Control Law, among others.

Besides these basic laws, there are regulations, implementation measures, management measures and standards that specify the guiding principles in the basic laws to implement the specific projects. These regulations are issued by governments and different governmental departments. The policymaking and project approval of CCUS in China involves many government organisations such as the State Council, the NDRC, and the provincial Development and Reform Committees. Local governments are allowed to mandate stricter environmental standards or regulations than the national ones tailored to the local circumstances, according to the *the Environmental Protection Law of the PRC* article 15 and article 16. Therefore, the development of CCUS project has to comply with national laws and seek approval from the central government, and then further obtain permits from local authorities, and meanwhile must obey the special environmental rules made by local governments.

As for regulatory documents supporting technology development, China has listed the CCUS as one of the important technologies for fostering and developing energy-saving and environmental protection strategic emerging industries in the Twelfth National Five-Year Plan (2011–2015) for Science and Technology Development.[57] To further implement this, a special plan for the CCUS was issued, which provides guarantee measures including (a) Improving the coordination mechanism of science and technology (b) Increasing investment in science and technology (c) Strengthening international cooperation in science and technology (d) Establishing strategic alliances for industrial technological innovation (e) Strengthening the construction of innovation teams and talents.[58] Further, the issuance of the *Study on China's CCUS Technology Development Roadmap* in 2011 clarified the development goals, research

[57] Ministry of Science and Technology, Notice on the issuance of the National Twelfth Five-Year Plan for Science and Technology Development, Guo Ke Fa Ji [2011] No. 270, 2011 July 4, https://www.most.gov.cn/xxgk/xinxifenlei/fdzdgknr/qtwj/qtwj2011/201107/t20110713_88228.html accessed 8 January 2025.

[58] Ministry of Science and Technology, Specialised Plan for the Twelfth Five-Year Plan for National Science and Technology Development of Carbon Capture, Utilisation and Storage, Guo Ke Fa She [2013] No.142,16 February 2013 https://www.most.gov.cn/xxgk/xinxifenlei/fdzdgknr/fgzc/gfxwj/gfxwj2013/201303/t20130315_100225.html accessed 8 January 2025.

priorities and demonstration deployment strategies.[59] The implementation of the demonstration projects was further facilitated by the NDRC by issuing a regulatory document to clarify the implementation measures and tasks of CCUS demonstration projects in 2013.[60] And the NDRC issued the *Notice on Promoting Carbon Capture Utilisation, and Storage Demonstration Pilot* in 2013, aiming at pushing forward the research and deployment of CCUS technologies, as well as establishing related industry regulations and environmental standards for CCUS.[61] The national emission trading scheme (ETS) together with the target of setting an average carbon emission intensity of no more than 550 grams of carbon dioxide per kilowatt-hour for large state-owned power enterprises, provided positive impetus for CCUS technology application.[62] From an institutional perspective, the transfer of CCUS technology management responsibilities in to the newly established of MEE in 2018 also facilitated the development of more incentive mechanisms for CCUS demonstration deployment.[63]

Despite of the progress made in developing the legal framework for CCUS in China, the current legal framework surrounding CCUS in China exhibits certain gaps and shortcomings that need to be addressed. Firstly, there is a lack of specific and comprehensive regulations that are exclusively dedicated to CCUS. While China has made progress in integrating CCUS within existing environmental and energy-related laws and

[59] The Study on China's CCUS Technology Development Roadmap was updated in 2019, and the 2024 version is in the drafting process as of the writing of this chapter.

[60] NDRC, Notice of the NDRC on Promoting Experimental Demonstration of CCUS, Fa Gai Qi Hou [2013] No. 849, 2013 April 27, https://www.gov.cn/zwgk/2013-05/09/content_2398995.htm accessed 8 January 2025.

[61] NDRC, Notice of the NDRC on Promoting Experimental Demonstration of CCUS, Fa Gai Qi Hou [2013] No. 849, 2013 April 27, https://www.gov.cn/zwgk/2013-05/09/content_2398995.htm accessed 8 January 2025.

[62] Toby Lockwood, 'Reducing China's Coal Power Emissions with CCUS Retrofits', IEA Clean Coal Centre, 2018 November, https://iea.blob.core.windows.net/assets/3765e352-61ef-4926-8371-1ea3a5ec94cd/CCUS_Retrofit_CIAB_report_IEACCC.pdf accessed 8 January 2025, at 10.

[63] Toby Lockwood, 'Reducing China's Coal Power Emissions with CCUS Retrofits', IEA Clean Coal Centre, 2018 November, https://iea.blob.core.windows.net/assets/3765e352-61ef-4926-8371-1ea3a5ec94cd/CCUS_Retrofit_CIAB_report_IEACCC.pdf accessed 8 January 2025, at 10.

policies, there is a need for a dedicated legal framework that comprehensively addresses the unique challenges and opportunities associated with CCUS technologies. Secondly, there is a need for clearer guidelines and standards regarding liability and long-term stewardship of CO2 storage sites in China. The absence of specific provisions addressing these aspects creates uncertainty for project developers and may hinder private sector investment in CCUS. Additionally, the current legal framework may not adequately incentivise research and development of CCUS technologies or provide sufficient financial incentives for project deployment. Stricter regulations and financial mechanisms that promote innovation, investment, and collaboration within the CCUS sector are necessary to accelerate its development in China. Lastly, there is a need for enhanced monitoring, reporting, and verification protocols to ensure the safe and effective implementation of CCUS projects and to address potential environmental risks associated with CO2 storage. Addressing these gaps and shortcomings in the Chinese legal framework will be crucial to foster a supportive and conducive environment for CCUS development, deployment, and utilisation in the country.

4.2 Proposed Recommendations for CCUS Legal Framework

Key considerations for the further development of the CCUS development should include the construction of a carbon neutral technology system, R&D of key technologies, improvement of regulations and standards, targeted incentive mechanisms, and international cooperation and sharing.[64] To strengthen the existing laws and regulations concerning CCUS in China, the following recommendations are proposed.

Firstly, there is a need to establish a dedicated and comprehensive legal framework specifically tailored to CCUS in China, outlining clear guidelines, standards, and procedures for project development, operation, and long-term liability in legal instruments at higher level such as the State Council regulatory document and departmental regulatory documents issued by central government departments. This framework should address key aspects such as storage site selection, monitoring, and closure, as well as liability and financial mechanisms to incentivise

[64] Zhang Xian and others, *CCUS Progress in China—a Status Report 2023* (The Administrative Center for China's Agenda 21, Global CCS Institute, and Tsinghua University, 2023), at 20.

private sector investment and private enterprises engagement. Secondly, it is crucial to enhance collaboration and coordination between relevant government agencies, research institutions, and industry stakeholders to foster innovation, knowledge sharing, and technology transfer. This can be achieved through the establishment of interagency task forces or consortia that facilitate information exchange and joint research initiatives. Additionally, the introduction of financial incentives, such as tax credits or subsidies, and streamlined permitting processes can encourage increased investment in CCUS projects. Moreover, strengthening monitoring, reporting, and verification protocols will ensure the safe and effective implementation of CCUS projects, mitigating potential environmental risks. Lastly, fostering international cooperation and aligning Chinese CCUS regulations with international standards and best practices will facilitate cross-border collaboration, knowledge dissemination, and technology diffusion. By implementing these recommendations, China can develop a robust legal framework that supports the widespread adoption and deployment of CCUS technologies, contributing to the country's efforts in mitigating greenhouse gas emissions and combating climate change.

4.3 Suggestions for Enhanced Supportive Policies

In addition to addressing the above-mentioned gaps in the legal framework of CCUS development in China, it is also important to enhance the supportive policies, which are policy documents supplementing the implementation of the legal framework. Firstly, it is suggested to establish a dedicated regulatory body or task force specifically focused on CCUS project approvals can help streamline the process and provide expertise in assessing project feasibility, environmental impact, and safety considerations. This regulatory body should have clear guidelines and timelines for project evaluation, ensuring efficiency and transparency. This requires comprehensive top-level design. Although it would be unrealistic to establish a legal framework in a short term given that it is now still at the early stage of CCUS development, using a framework law can be desirable for establishing regulatory principles for each stage of CCUS from a systemic perspective, laying the foundation for subsequent development of specific regulations.[65] Temporary policies would be desirable for specific project construction needs, targeting pilot regions or projects. Given the limited regulatory experience with CCUS projects at present and the vast difference between CCUS projects across different locations, it may be difficult to establish uniform regulations. Therefore, temporary regulations or policies can be initially implemented for specific projects, tailoring them to individual project needs.[66] Following a learning-by-ding approach, after accumulating regulatory experience from different projects and drawing lessons, universal regulations and policies can be formulated and optimised.[67] Meanwhile, policies and regulations are needed to address the overlapping issues of underground space use rights and mineral rights of CCUS projects.

Secondly, implementing a one-stop-shop approach to permit and licensing procedures can simplify the administrative burden for project developers. This approach would involve coordinating with multiple government agencies involved in the approval process to consolidate and streamline the necessary permits, licences, and environmental assessments. The permit and licensing mechanisms need to be coordinated with the legal provisions of the property right defined in the new Civil Code in China to allocate the right and responsibility of storage space of CCUS projects.[68]

Additionally, providing financial incentives such as grants, subsidies, or tax incentives for CCUS projects can encourage investment and expedite project development. These incentives can be tailored to different stages of project development, including research and development, demonstration, and large-scale deployment. This need to take into account of the long-term responsibility allocation between the local government and

[65] Gao Yan, Li Zhuohui, 'Introduction to the International Legal Framework of CCUS and its Significance for China', (*Zhonglun Law Firm Research Articles*, 29 January 2024) https://www.zhonglun.com/research/articles/52590.html accessed 8 January 2025.

[66] Gao Yan, Li Zhuohui, 'Introduction to the International Legal Framework of CCUS and its Significance for China', (*Zhonglun Law Firm Research Articles,* 29 January 2024) https://www.zhonglun.com/research/articles/52590.html accessed 8 January 2025.

[67] Gao Yan, Li Zhuohui, 'Introduction to the International Legal Framework of CCUS and its Significance for China', (*Zhonglun Law Firm Research Articles*, 29 January 2024) https://www.zhonglun.com/research/articles/52590.html accessed 8 January 2025.

[68] Gao Yan, Li Zhuohui, 'Introduction to the International Legal Framework of CCUS and its Significance for China', (*Zhonglun Law Firm Research Articles*, 29 January 2024) https://www.zhonglun.com/research/articles/52590.html accessed 8 January 2025.

project operator so that a balance can be achieved between the project investors' willingness to invest, government's willingness to take over, and the long-lasting impacts of the project on local public safety and environment. This is important for the sustainable development of a CCUS project in a long term.[69]

Moreover, fostering collaborations between industry stakeholders and research institutions can accelerate knowledge sharing, technology transfer, and capacity building. Finally, establishing a supportive policy framework that includes long-term price signals or carbon pricing mechanisms can create a stable investment environment for CCUS projects by ensuring a revenue stream for captured CO2. By implementing these suggestions, China can streamline approval processes, reduce administrative barriers, and create supportive policies that facilitate the widespread deployment of CCUS projects and contribute to the country's climate goals.

4.4 Importance of Public Engagement and Stakeholder Involvement in CCUS Decision-Making

Public engagement and stakeholder involvement are critical components of decision-making processes related to CCUS in China. The successful implementation of CCUS projects requires more than just technological advancements and policy frameworks. It necessitates active engagement with the public and stakeholders to build trust, ensure transparency, address concerns, and foster inclusive decision-making. Firstly, public engagement serves as a vital tool for building trust and ensuring transparency in CCUS decision-making. By providing opportunities for open dialogue, information sharing, and accountability, China can establish a foundation of trust between policymakers, project developers, and the public. Transparent decision-making processes increase credibility and demonstrate a commitment to involving diverse perspectives, fostering public confidence in the CCUS initiatives.

Secondly, public engagement allows for the identification and addressing of concerns and misconceptions surrounding CCUS technologies. By actively involving the public and stakeholders, China can gain a

[69] Gao Yan, Li Zhuohui, 'Introduction to the International Legal Framework of CCUS and its Significance for China', (*Zhonglun Law Firm Research Articles*, 29 January 2024) https://www.zhonglun.com/research/articles/52590.html accessed 8 January 2025.

better understanding of the worries, fears, and misunderstandings held by different parties. For example, businesses, as the main implementers of a CCUS project, especially large state-owned energy enterprises, can take the initiative to invest in the initial promotion phase, accumulate technical, commercial and management experience through valuable demonstration projects, and actively explore the establishment of a platform for cooperation and exchange of experiences and achievements, such as technology and techniques, emission reduction effects, environmental impacts, economic benefits, operation processes, and the effectiveness of risk monitoring and control, so as to serve as a reference for other relevant enterprises, and then as a basis for the country to formulate relevant policies.[70] This knowledge sharing enables targeted communication, education, and clarification efforts, helping to dispel myths, provide accurate information, and alleviate concerns about the safety, environmental impact, and long-term viability of CCUS projects.

Moreover, public engagement efforts play a crucial role in educating and raising awareness among the public and stakeholders about the importance of CCUS in mitigating climate change. Through information dissemination and outreach programmes, China can ensure that accurate knowledge reaches the public. This helps individuals grasp the potential of CCUS technologies, their role in reducing greenhouse gas emissions, and their alignment with national and global climate goals. Informed discussions and decision-making are fostered when the public has a solid understanding of the benefits and challenges associated with CCUS.

Stakeholder involvement is essential for achieving inclusive decision-making in CCUS initiatives. By engaging a diverse range of stakeholders, including local communities, NGOs, academia, industry representatives, and affected parties, China can ensure that different perspectives and interests are considered. Inclusive decision-making promotes equity, social acceptance, and shared responsibility, as it allows for the integration of various viewpoints into the policy and project development processes. By involving stakeholders, China can generate more comprehensive, balanced, and socially just outcomes.

[70] Zhao Zhenyu, Yao Shun, Yang Shuopeng, and Wang Xiaolong, 'Under the "Dual Carbon" Goal: Current Status, Problems and Suggestions for CCUS Development in China', Environmental Science 2024 (44) 1128, at 1134.

Lastly, engaging the public and stakeholders enhances the effectiveness of CCUS policies and regulations. By actively involving them in decision-making, China can tap into the collective wisdom, expertise, and local knowledge of various stakeholders. Participation of different stakeholders is an effective means to mobilise all parties to actively participate in the CCUS industry chain. The government hopes to actively mobilise the public, enterprises, experts and other subjects to actively participate in the research and operation of the CCUS project, encourage the free operation of business models in line with the market mechanism, and promote the development of the CCUS industry in the direction of clustering, so as to reduce the cost through the economies of scale.[71] China has been using policies including Talent Cultivation and Recognition Reward system to support CCUS technology innovation.[72] Public input and stakeholder perspectives contribute to more comprehensive and informed decision-making, leading to policies and measures that better reflect the needs and aspirations of the society. This increases the likelihood of successful CCUS project implementation and public acceptance, ensuring that policies are not only technically sound but also socially and politically feasible.[73]

[71] Zeng Lixuan, Xing Hongfei, 'A Sample Review of Carbon Capture, Utilisation and Storage Policies in China under a "Dual Carbon" Governance Perspective', Journal of Hohai University (Philosophy and Social Science Edition) 2024 (2) 147, at 154.

[72] Zeng Lixuan, Xing Hongfei, 'A Sample Review of Carbon Capture, Utilisation and Storage Policies in China under a "Dual Carbon" Governance Perspective', Journal of Hohai University (Philosophy and Social Science Edition) 2024 (2) 147, at 154.

[73] Zeng Lixuan, Xing Hongfei, 'A Sample Review of Carbon Capture, Utilisation and Storage Policies in China under a "Dual Carbon" Governance Perspective', Journal of Hohai University (Philosophy and Social Science Edition) 2024 (2) 147, at 154.

5 Conclusion

China has set ambitious targets to reduce greenhouse gas emissions and achieve carbon neutrality. CCUS plays a crucial role in these goals, and China's commitment to developing and deploying CCUS technologies is expected to continue and strengthen in the coming years.

In this chapter, we explored the legal framework of CCUS in China. We highlighted the significance of CCUS in addressing climate change and the need for a robust legal framework to support its development. Through an overview of China's climate change policies, we established the importance of CCUS in the country's mitigation strategy. We examined the current CCUS projects in China and identified the challenges and barriers they face. Furthermore, we emphasised the role of regulations in promoting CCUS project development and the interplay between law and technology. We also identified gaps and shortcomings in the existing legal framework and proposed recommendations for improvement, including streamlined approval processes and enhanced supportive policies. Lastly, we emphasised the importance of public engagement and stakeholder involvement in CCUS decision-making.

The case study on CCUS law and regulation in China has several implications for the development of an effective legal framework. It highlights the need for comprehensive and tailored regulations that address the unique challenges and opportunities of CCUS projects. The findings call for increased coordination and collaboration among government agencies, industry stakeholders, and research institutions to streamline approval processes and provide consistent guidance. Moreover, the case study underscores the importance of incorporating public engagement and stakeholder involvement in decision-making processes to ensure transparency, build trust, and address concerns. These insights can inform policymakers and regulators in China as they refine and enhance the legal framework for CCUS development.

Looking ahead, the future prospects for CCUS development in China are promising. The country's strong commitment to reducing greenhouse gas emissions and its recognition of the role of CCUS in achieving climate goals create a conducive environment for project advancement. The insights gained from this chapter can guide policymakers and regulators in developing targeted measures to overcome challenges and accelerate CCUS deployment. This includes strengthening the legal framework, incentivising research and development, fostering international collaboration, and providing financial support. Furthermore, continued public engagement and stakeholder involvement will be crucial to ensure the social acceptance and successful implementation of CCUS projects in China. By leveraging these opportunities and addressing the identified gaps, China can position itself as a leader in CCUS technology and contribute significantly to global efforts in combating climate change.

CHAPTER 15

CCUS in Nigeria

Opeyemi Omotuyi

1 Introduction

Climate change remains a major challenge facing the world in the twenty-first century. Its complex and dynamic nature requires multi-dimensional and multi-sectoral mitigation and adaptation approaches enabled by a dynamic policy framework. One of such approaches is the adoption of carbon management technologies such as carbon capture, utilization and storage (CCUS). CCUS consists of methods and processes of capturing carbon from large point sources, such as industrial facilities that use fossil fuels, with the goal of mitigating or preventing greenhouse gases (GHG) emissions. Such captured carbon is either stored permanently in various geological storages or used in other industrial applications.[1] CCUS involves three major processes; first is the capturing of carbon

[1] International Energy Agency, "About CCUS", www.iea.org/reports/about-ccus [accessed 19 November 2023]; International Energy Forum, "What is CCUS and Why is it Necessary?" www.ief.org/news/what-is-ccus-and-why-is-it-necessary [accessed 19 November 2023].

O. Omotuyi (✉)
Federal University Oye-Ekiti, Oye-Ekiti, Nigeria
e-mail: opeyemi.omotuyi@fuoye.edu.ng

E. G. Pereira et al. (eds.), *Carbon Capture Utilization and Storage*, Sustainable Development Goals Series,
https://doi.org/10.1007/978-3-031-81272-9_15

from the point of emission or directly from the atmosphere; second is the compression and transportation of the captured carbon to storage sites either through pipelines, rail or trucks; and third is the injection of captured carbon into the ground for storage in deep geological formations which trap the carbon permanently under the ground.[2] Captured carbon may also be stored temporarily in depleted oil reservoirs for utilization in a range of industrial applications. Summarily therefore, CCUS is primarily characterized by processes, methods or techniques, of preventing carbon and other GHG from escaping into and remaining in the atmosphere. This primary feature of CCUS makes it a significant aspect of the climate actions needed to combat global warming and climate change globally. At the COP 27 held in 2022, the opportunities provided by CCUS for countries to develop and industrialize in a manner involving reduced carbon footprint were emphasized. The Conference highlighted the importance of CCUS in reducing carbon emissions of large-scale industries, and as a business opportunity capable of creating new industry for decarbonizing hard-to-abate industrial sectors.[3] Thus, countries around the world are increasingly embracing CCUS methods and techniques as they strive to meet up with their climate commitments and targets.

Nigeria is not exempted from the impacts and grave consequences of climate change as a result of the persistence of gas flaring and GHG emissions from her various industrial sectors. This posits the country as a major actor with enormous responsibilities in the global climate change mitigation efforts. Major mile stones in Nigeria's response to the challenges posed by climate change include the signing of the Paris Agreement in 2015, ratification of same in 2017 and submission of an updated nationally determined contribution (NDC) to the UNFCCC Secretariat in 2021. By these actions, Nigeria indicated her commitment to transition to a low-carbon economy as set out by the Paris Agreement. Through the updated NDC, Nigeria specifically commits to 20% unconditional and

[2] Ibid.

[3] International Institute for Sustainable Development, "Potential of CCS in Africa Highlighted During COP 27" 16 November 2022, https://sdg.iisd.org/news/potential-of-ccs-in-africa-highlighted-during-cop-27/ [accessed 21 November 2023].

47% conditional emissions reduction targets by 2030.[4] Furthermore, at the COP 26 held in November 2021, Nigeria made a commitment to achieve net-zero emissions by 2060.[5] This implies that Nigeria commits to taking necessary actions to curb GHG emissions from all sources within her jurisdiction and achieve carbon neutrality by 2060.

In furtherance of these climate commitments, the Nigerian government has taken several actions including the deployment of climate policies, programs and frameworks, for the regulation of climate change in the country. First is the development of a National Climate Change Policy (NCCP) consisting of a holistic framework which prescribes sectoral and cross-sectoral strategic policies and actions relevant for the management of climate change in the country.[6] Other actions include implementation of a long-term emissions reduction plan aimed at cutting down emissions by 50% by 2050 (LTV-2050),[7] as well as the enactment of a Climate Change Act,[8] and Petroleum Industry Act in 2021.[9] In addition, the Nigerian government launched an 'Energy Transition Plan' (ETP) in 2022 as a path to achieving carbon neutrality by 2060.[10] In line with international trend, the Nigerian government through these various policies and plans identifies CCUS as a key technology to help meet up with the country's climate commitments and targets. Nigeria has also indicated her commitment to the development of CCUS by subscribing to the

[4] Federal Ministry of Environment, "Nigeria's First Nationally Determined Contribution- 2021 Update", https://climatechange.gov.ng/wp-content/uploads/2021/08/NDC_File-Amended-_11222.pdf [accessed 21 November 2023].

[5] https://statehouse.gov.ng/news/at-cop26-president-buhari-pledges-net-zero-emissions-by-2060-says-nigeria-will-maintain-gas-based-energy-transition/ [accessed 20 November 2023].

[6] Federal Ministry of Environment, "National Climate Change Policy for Nigeria 2021–2030", https://climatechange.gov.ng/wp-content/uploads/2021/08/NCCP_NIGERIA_REVISED_2-JUNE-2021.pdf [accessed 21 November 2023].

[7] 2050 Long-Term Vision for Nigeria (LTV-2050) 2021, https://unfccc.int/sites/default/files/resource/Nigeria_LTS1.pdf [accessed 20 November 2023].

[8] Climate Change Act 2021, https://faolex.fao.org/docs/pdf/NIG208055.pdf [accessed 20 November 2023].

[9] Petroleum Industry Act, http://www.petroleumindustrybill.com/wp-content/uploads/2021/09/Official-Gazette-of-the-Petroleum-Industry-Act-2021.pdf [accessed 20 November 2023].

[10] https://energytransition.gov.ng/ [accessed 20 November 2023].

Clean Energy Ministerial CCUS Initiative, an inter-governmental initiative that aims to accelerate CCUS as a viable carbon mitigation option, facilitate diffusion of knowledge on technologies, regulations and policies and lead to strategic partnerships to accelerate both near and longer-term investment in CCUS.[11]

Subsequent sections of this chapter explore some of the key CCUS projects in Nigeria; the legal framework, policies and norms relevant to CCUS in Nigeria; as well as the gaps and challenges relating to CCUS development in Nigeria. The chapter concludes with some reflections on CCUS development in Nigeria and the lessons learnt.

2 Key CCUS Projects in Nigeria

The Nigerian government recognizes the long-term role that CCUS is capable of playing in meeting up with the national ETP, as well as economic development and energy security, through natural gas development in the country. Thus, various CCUS projects have been deployed in the petroleum sector under the 'Clean Development Mechanism' (CDM) of the Kyoto Protocol. This project-based mechanism is a form of emissions trading scheme which involves developed countries offsetting their emissions by financing the implementation of emissions reduction projects in developing countries and obtaining investment incentives in form of certified emission reduction (CER) credits.[12] The CDM have been utilized for capturing and processing associated natural gas in the Nigerian Petroleum Sector thereby, reducing the volume of gas flared. Such captured associated natural gas is utilized in various gas-to-power projects in Nigeria including, the 'Recovery of Associated Gas at Kwale Oil–Gas Processing Plant', 'Ovade Ogharefe Gas Capture and Processing', 'Recovery and marketing of gas that would otherwise be flared at the Asuokpu/Umutu Marginal Field, Nigeria', 'Recovery and Utilization of Associated Gas from the Obodugwa and neighbouring oil fields in Nigeria'. These projects involve the capturing of associated gas from

[11] https://www.cleanenergyministerial.org/initiatives-campaigns/carbon-capture-utilization-and-storage/.

[12] United Nations Framework Convention on Climate Change, "The Clean Development Mechanism", https://unfccc.int/process-and-meetings/the-kyoto-protocol/mechanisms-under-the-kyoto-protocol/the-clean-development-mechanism [accessed 30 November 2023].

various oil fields in Nigeria, the compression and transportation of the captured gas through pipeline networks to domestic gas market to be marketed for use by end users of gas. Any captured gas remaining unsold for domestic consumption is utilized by the projects proponents in an independent gas-fired power plant for electricity generation. These projects have contributed to energy security in Nigeria through natural gas development as a source of clean energy production.

The first CCUS project outside the petroleum sector is the landfill gas (LFG) project which was registered under the CDM in 2012. The project is targeted at building, operating and maintaining a landfill gas collection and utilization system on three landfills in Nigeria including Olusosun, Abule Egba and Solous landfills. The first phase of the project involves an initial LFG flaring system consisting of a gas collection network and an extraction and flaring system. The second phase consists of the installation of an electricity generation plant interconnected to the local grid at the Olusosun landfill site. The project is undergoing implementation by the Lagos Waste Management Authority and Ably Carbon, a corporate organization in France.[13] The project has been successful based on some predetermined progress indicators such as a substantial volume of GHG emissions reduction and a better environment.[14]

However, in recognition of the need for large-scale development and deployment of CCUS within the industrial sector generally, the Nigerian government is in partnership with international organizations to chart a path for a comprehensive CCUS framework in Nigeria. For instance, the Nigerian government in conjunction with the International Energy Agency (IEA) organized a joint workshop in 2021 with the objective of building CCUS capacity and identifying short-term needs and opportunities for CCUS development and deployment in the country.[15] The workshop highlighted a national CCUS work program outlining the

[13] See the LFG Project in Nigeria, https://cdm.unfccc.int/UserManagement/FileStorage/YI6CGF0E7VM31JASRZXKQDWULB8OHT [accessed 30 November 2023].

[14] Nigeria, "Second Biennial Update Report (BUR2) to the United Nations Framework Convention on Climate Change (2021) 61, https://unfccc.int/sites/default/files/resource/NIGERIA%20BUR%202%20-%20Second%20Biennial%20Update%20Report%20%28BUR2%29.pdf [accessed 30 November 2023].

[15] International Energy Agency and the Office of the Vice President of Nigeria, "Facilitating Nigeria's Energy Transition through CCUS Development" 10 September 2021.

requisite actions relevant for the adoption of CCUS in Nigeria. First, the workshop identified the need for technical assessments including assessment of potential storage resources, mapping stationary carbon sources and emissions clusters and a pre-feasibility assessment through pilot CCUS projects as important aspects of CCUS development in Nigeria. Other actions outlined in the work program include development of legal and regulatory frameworks for CCUS; strengthening of national support for CCUS through robust stakeholder engagement on CCUS strategy development; and building national and institutional capacity including technical capacity for CCUS development and deployment in Nigeria.

Similarly, the Nigerian government has collaborated with the International Finance Corporation (IFC) and the World Bank to carry out a study termed 'the Industrial Carbon Dioxide Utilization and Storage in Nigeria: Diagnostics and Scoping' (ICCUS D&S) aimed at providing information on the effectiveness of industrial CCUS in Nigeria. The objectives of the ICCUS D&S include the development of a domestic market for CCUS for industrial emissions, and mapping out sources of carbon emissions and potential sites for underground carbon storage in Nigeria.[16] Under the initial phase of the ICCUS D&S, the IFC partners the Nigerian government to undertake national carbon source and an initial national storage assessment, as well as identify the most viable sectors and companies that can pilot new CCUS technologies in Nigeria. The World Bank partners the government to outline a legal framework that can accelerate the uptake of CCUS technologies and help domestic CCUS industry meet international standards.[17]

The initial phase of the ICCUS D&S concluded that commercial-scale development and deployment of industrial CCUS is viable in Nigeria. The initial results from the ICCUS D&S disseminated to key Nigerian stakeholders in October 2023 included the first carbon storage assessment maps of Nigeria, with major carbon sources mapped out. The maps indicated that major carbon sources are concentrated in the South and South-Western part of Nigeria, with many close to the Niger Delta region. Hence, because of the abundance of large oil fields in this region, the ICCUS D&S focused on depleted oil and gas fields in the Niger

[16] International Finance Corporation, "IFC and World Bank to help Nigeria Pave the Way for Domestic Carbon Storage" 10 February 2022, https://pressroom.ifc.org/all/pages/PressDetail.aspx?ID=26819 [accessed 21 November 2023].

[17] Ibid.

Delta as potential carbon storage resources in Nigeria as against saline formations. On the basis of some key features such as feasibility, impact, scale, etc., fifteen potential CCUS projects were identified and ranked, with the top two being a fertilizer plant and a cement plant, both of the scale of about 2 to 2.5 metric tons of carbon emissions per annum. Both projects are estimated to have between 100 to 250 km pipelines from the respective plants to offshore storage in the Niger Delta. Pilot versions of the projects were suggested at a scale of 20kt per annum, and 10 km pipelines to local storage.[18] The results from the initial phase of the ICCUS D&S facilitated a proposed new CCUS R&D Centre of Excellence by the Nigerian Geological Survey Agency. This is notwithstanding the Africa Centre of Excellence for Carbon Management Technology and Innovation launched by the Nigerian government in partnership with the United States Department of Energy in 2022. The Centre was launched with the objective of serving as a platform for knowledge dissemination and a reference center for policymakers and industries on issues related to carbon management and the development of various carbon technologies in Nigeria and Africa as a whole.[19]

The phase two of the ICCUS D&S is intended to finalize and publish a national strategy on CCUS and a national storage atlas, commence pre-feasibility work on the identified potential pilot projects, develop appropriate regulations and implement capacity building and stakeholder awareness on CCUS.[20]

[18] Tim Dixon, "IFC Nigeria Dissemination Workshop on Industrial CCUS and a New Centre of Excellence" 23 October 2023, https://ieaghg.org/ccs-resources/blog/ifc-nigeria-dissemination-workshop-on-industrial-ccus-and-a-new-centre-of-excellence [accessed 22 November 2023].

[19] See Africa Centre of Excellence for Carbon Management Technology and Innovation, https://www.energy.gov/fecm/articles/africa-centre-excellence-carbon-management-technology-and-innovation-cop-27 [accessed 22 November 2023].

[20] Tim Dixon, op cit.

3 Legal Framework, Policies and Norms for CCUS in Nigeria

Currently, there is an absence of a specific and comprehensive legal framework on CCUS in Nigeria. There are however existing laws, regulations and policies that supports and enables the development and deployment of CCUS in the country. Pursuant to her climate responsibilities under the Paris Agreement, Nigeria is committed to certain emissions reduction targets including 20% unconditional and 47% conditional emissions reduction by 2030, 50% emissions reduction by 2050 and net-zero emissions by 2060. In furtherance of these climate commitments, the Nigerian government had implemented some climate policies, programs and laws to address the challenges of climate change in the country. Through these climate change framework and similar laws, Nigeria identifies with the opportunities presented by the development and deployment of CCUS in the country. Some of these climate-related policies and laws are highlighted below;

i. Climate Policies

First is the NCCP, which has as one of its strategic objectives the promotion of technology and innovations to address the challenges of climate change. The NCCP also affirms the importance of policy measures that promote climate change mitigation such as those that support low-cost and technically feasible solutions to reduce carbon emissions in petroleum operations, including recovery and use of escaping gas.[21] Likewise, the ETP incorporates a national technology roadmap including CCUS in the pathway toward carbon neutrality. For instance, the ETP stipulates the application of Bioenergy with Carbon Capture and Storage (BECCS) in cement production as a decarbonization strategy for the industrial sector.[22] The ETP also recognizes the application of CCUS as a decarbonization strategy for emissions from petroleum operations.[23] The LTV-2050 emphasizes the role of innovation in the achievement of low

[21] National Climate Change Policy, op cit.

[22] Nigeria Energy Transition Plan, 'Industry', https://www.energytransition.gov.ng/industry-2-2-2/ [accessed 26 November 2023].

[23] Nigeria Energy Transition Plan, 'Oil and Gas', https://www.energytransition.gov.ng/oil-and-gas-2-2/ [accessed 26 November 2023].

emission development in Nigeria. Particularly, the plan advances innovation in technology as a crucial element for climate change mitigation. Hence, it adopts a climate technology led approach, including technological innovations in CCUS, as one of the strategies for the achievement of its target.[24]

ii. Climate Change Act

The Climate Change Act provides a legislative framework for the achievement of Nigeria's climate goals on the one hand, while promoting inclusive growth and sustainable economic development on the other hand.[25] The Act was enacted to facilitate the coordination of climate actions necessary for achieving Nigeria's climate commitments,[26] including implementation of mitigating measures that promote low-carbon economy in the country.[27] In so doing, the Act provides a legal basis for the application of a carbon tax and emissions trading system as one of the strategies for climate change mitigation and adaptation in Nigeria.[28] Likewise, the Act makes provision for an economic regulatory system through the stipulation of fines and charges as disincentives for flouting climate change mitigation and adaptation obligations in the country.[29] Furthermore, the Act provides a system of carbon budgeting whereby periodic carbon budgets are set in line with Nigeria's NDCs, and in a manner that enables compliance with her climate commitments and targets.[30]

In addition, the Act establishes a Climate Change Fund and prescribes the application of the fund toward financing innovative climate change mitigation and adaptation projects, and incentivizing corporate entities for their efforts toward transiting to clean energy and sustaining reduced GHG emissions, among others.[31] Although the Act does not refer to

[24] 2050 Long-Term Vision for Nigeria, op cit., p31.

[25] The Climate Change Act, op cit., Sect. 1.

[26] Ibid, Sect. 1(b).

[27] Ibid, Sect. 1(h).

[28] Ibid, Sect. 15(1)(e).

[29] Ibid, Sect. 15(1)(d).

[30] Ibid, Sect. 19(1).

[31] Ibid, Sect. 15(1) & (2).

any specific climate technology, the nature of CCUS makes it a viable mitigating measure for the achievement of Nigeria's climate goals. Thus, CCUS could constitute one of several innovative climate change mitigation projects envisaged by the Climate Change Act. Moreover, the stipulation of a system of carbon tax, emissions trading, carbon budgeting and penalties for defaulting, provides an indirect legal basis for the development and deployment of climate technologies such as CCUS. Hence, it is safe to conclude that the Climate Change Act indirectly provides a legal basis for the development and deployment of CCUS technologies in Nigeria.

It is noteworthy that the Climate Change Act establishes the National Council on Climate Change (the Council) and empowers it to make policies and decisions on all matters concerning climate change in Nigeria.[32] At its inaugural meeting held early 2023, the Council proposed relevant mechanisms necessary for achieving Nigeria's climate goals in line with the provisions of the Climate Change Act. Such mitigating mechanisms proposed include target emissions limit for the public and private sectors of the economy, penalties for exceeding the prescribed limits and a carbon tax system.[33] As stated earlier, these climate mechanisms are indirect approaches to encouraging the adoption of clean technologies such as CCUS by the public and private sectors in Nigeria. The adoption of such technologies allows entities comply with any prescribed emissions limits thereby reducing liability for the stipulated penalties and carbon taxes. In other words, whenever these proposed climate mechanisms become formally implemented by the Council, it will provide an indirect legal basis for the development and deployment of carbon reduction technologies such as CCUS in order to escape the potential liability for carbon taxes and other defaulting penalities.

iii. The Petroleum Industry Act (PIA)

Having envisaged the need for decarbonization in Nigeria, the PIA just like the Climate Change Act indirectly provides a legal basis for the development and deployment of CCUS in the country. The PIA

[32] Ibid, Sect. 3(1).

[33] Abdulkareem Mojeed, "Climate Change: FG to Unveil Carbon Tax System for Nigeria" Premium Times (Nigeria) 13 February 2023.

requires every petroleum concessionaire to prepare an environmental management plan stating the measures intended for the mitigation of the negative environmental impacts of its operations.[34] Arguably, this requirement envisages the deployment of environmental management measures including carbon reduction technologies such as CCUS by petroleum operators as part of their environmental obligations under the Act.

In furtherance of the objectives of the NCCP which include the adoption of policy measures that support the reduction of carbon emissions in petroleum operations including the recovery and use of escapiong gas, the PIA prohibits the unlawful flaring of associated gas in Nigeria.[35] In so doing, the Nigerian Upstream Petroleum Regulatory Commission (NUPRC) is vested with the right to take gas associated with petroleum production at the emission sites.[36] These provisions envisage the need for decarbonization of petroleum operations through the deployment of technologies that capture associated natural gas from production sites rather than flaring.

Furthermore, pursuant to its regulatory functions under the PIA, the NUPRC expressly identifies with the development and deployment of CCS in Nigeria. Pursuant to a draft regulation proposed to be implemented upon the requisite approval, the NUPRC expects petroleum mining lessees to provide CCS services with the reservoirs on their lease areas.[37] Upon the approval and issuance of this regulation, petroleum mining lessees will be expected to utilize the dry wells or abandoned reservoirs on their lease areas for carbon storage.

iv. Associated Gas Re-Injection Act & Flare Gas (Prevention of Waste and Pollution) Regulations

These enactments were made with the objective of minimizing gas flaring in Nigeria. To achieve this, all oil and gas producers in Nigeria are mandated to implement programs and plans for the re-injection of all

[34] Pteroleum Industry Act, op cit., Sect. 79(2)(h) & 102.

[35] Ibid, Sect. 104.

[36] Ibid, Sect. 105(2).

[37] Draft Acreage Management (Drilling and Production) Regulations 2022, regulation 36(3), https://www.nuprc.gov.ng/wp-content/uploads/2022/08/Nigeria-Acreage-Management-Drilling-and-Productions-Regulations.pdf [accessed 27 November 2023].

associated gas or put in place schemes for the viable utilization of all associated gas in industrial projects.[38] The federal government is vested with the right to take associated gas and conduct competitive bid processes whereby qualified applicants are granted a permit to access and take associated gas from any specified site on behalf of the government.[39] Also, any licensee may apply to the Minister of Petroleum Resources for a permit to utilize associated gas for commercialization and upon the grant of such permit, the licensee may utilize associated gas for its own purposes.[40]

These regulatory provisions highlight the viability of capturing and utilization and promote CCUS as a form of emissions control even though with respect to associated natural gas from petroleum operations.

4 CCUS Development in Nigeria: Gaps and Challenges

As a result of the novelty of CCUS in Nigeria, major barriers and challenges may inhibit the industrial implementation and up-scaling of CCUS applications in the country. These barriers and challenges ranges from absence of a holistic legal framework, to lack of expertise in CCUS technology deployment and high capital expenditure for CCUS infrastructure.[41] The most significant of these barriers is the absence of a comprehensive legal framework on CCUS. As put by the IEA, the successful development and deployment of CCUS relies on the establishment of legal and regulatory frameworks to ensure the effective stewardship of CCUS activities and the safe and secured storage of carbon.[42] According to the United Nations (UN) sustainable development agenda, the prevention of climate change requires the integration

[38] Associated Gas Re-Injection Act 1979, Sect. 1 & 2.

[39] Flare Gas (Prevention of Waste and Pollution) Regulations 2018, regulation 2(1) & 3(1).

[40] Ibid, regulation 3(2) & (3).

[41] Adedola Betiku and Bassey O. Bassey, "Exploring the Barriers to Implementation of Carbon Capture, Utilisaton and Storage in Nigeria" A Paper Presented at the International Petroleum Technology Conference, Riyadh, Saudi Arabia, 21–23 February 2022, https://onepetro.org/IPTCONF/proceedings-abstract/22IPTC/1-22IPTC/479969 [accessed 30 November 2023].

[42] International Energy Agency, *Legal and Regulatory Frameworks for CCUS: An IEA CCUS Handbook* (IEA Publications, 2022) 3.

of preventive and mitigating measures, in this case CCUS, into national policies, strategies and planning.[43] This is because a well implemented CCUS framework will typically address other barriers and enablers of CCUS development and deployment. A CCUS legal framework may take the form of government regulation through the command and control approach. This may range from setting a legislative framework that prescribes decarbonization thus making business models such as CCUS viable, to setting clear framework on the risks and potential costs of CCUS projects, as well as the responsibilities of CCUS operators. It may also take the form of a market-based regulation or economic approach consisting of policies providing specific support to cover for the high costs of CCUS operations or prescribing disincentives for GHG emissions.[44]

Currently, there are legislative provisions in the PIA and Climate Change Act that indirectly addresses the adoption of CCUS technologies in Nigeria. Similarly, Nigeria encourages the development and deployment of CCUS technologies through its various climate policies which are all principles-based. There is however an absence of specific legislations or regulations that set out a clear framework or specific requirements on rights, responsibilities, liabilities, etc., with regard to CCUS. Also, no specific economic incentives or disincentives on CCUS have been fully implemented and enforced. These forms of requirements and provisions are drivers and enablers of CCUS development and where they are lacking, the development of CCUS is hindered. This is because regulatory compliance is one of the major drivers of corporate responsibility and where there are no specific regulations to be complied with, there is no motivation for industries to do more than the barest minimum. In this instance, there is no motivation for industries to invest in such capital-intensive carbon technologies as CCUS.

Second, it is trite that the realization of an effective CCUS regime hinges on infrastructural development including expansion of carbon storage facilities, as well as enhancement of carbon transportation networks, among others. This is plagued by the issue of high costs required for the development of such infrastructure and facilities. There is

[43] United Nations, "Sustainable Development Goals", goal 13.

[44] Elisabeth Dutschke and Vicki Duscha, "Is there a Future for CCUS in Europe? An Analysis of the Policy Framework and Societal Support" 16th International Conference on Greenhouse Gas Control Technologies (GHGT-16), 23–27 October 2022, Lyon, France.

currently a lack of policy incentives addressing such capital challenge thus inhibiting the industrial scaling of CCUS development in Nigeria.

To ensure a widespread adoption and public acceptance of CCUS, as well as enhance its effectiveness as a carbon reduction measure, it is essential to ensure the safety and security of the geological areas in which captured carbon is injected for storage.[45] According to the initial results from the ICCUS D & S, depleted oil and gas fields in the Niger Delta region are potentially carbon storage resources in the country especially because of the concentration of major carbon sources in the region. This may pose significant geological challenges especially with regard to ensuring the safety and security of stored carbon in a manner that prevents leakage into the environment. First, addressing this requires high technical skill, as well as availability of research and development support.[46] Meanwhile, lack of technical expertise for CCUS is a major challenge to the development of CCUS in Nigeria.[47] Second, this may give rise to the issue of the long-term liability for carbon storage sites. Presently, Nigeria lacks any framework that addresses the long-term liability for carbon storage sites.

In addition, stakeholders' acceptance of CCUS will depend on the economic viability of CCUS projects which is governed by several factors including the availability of a viable carbon market and policy. Similarly, the development of CCUS may be plagued by the challenge regarding the availability of a viable business model that enables investors to recover the costs incurred on a CCUS project. In other words, the financial returns on CCUS projects must be reasonable to attract investments. There is currently a lack of a robust carbon market and policy that may enhance a viable business model for CCUS in Nigeria.

[45] International Energy Agency, *Legal and Regulatory Frameworks for CCUS: An IEA CCUS Handbook*, op cit., 54.

[46] Mohammed Dahiru Aminu, "Potential for CCS in Nigeria" (2015) *Carbon Capture Journal* 15.

[47] Adeola Betiku and Bassey O. Bassey, "Exploring the Barriers to Implementation of Carbon Capture, Utilisation and Storage in Nigeria", op cit.

5 Lessons Learnt

The achievement of carbon reduction and negative emissions through the adoption of CCUS technologies calls for the implementation of appropriate regulations and policies that support the development and deployment of such technology. Principles-based regulatory approach such as is currently obtainable in Nigeria may be inefficient without complementing it with other regulatory approaches such as the command-and-control and economic or market-based approaches. This portends that the quest for the industrial development and deployment of CCUS in Nigeria must be backed by the implementation of a comprehensive legal and regulatory framework on the subject-matter. To ensure an effective CCUS framework in Nigeria, relevant lessons may be observed from other jurisdictions that have scaled up CCUS development particularly, Europe and North America.

First, the possibility of a large-scale deployment of CCUS may significantly hinge on the availability of incentives and deterrents such as incentives for the adoption of clean technologies and disincentives for GHG emissions. Hence, an incentive system adopted by the Inflation Reduction Act (IRA) which provides enhanced tax credits to companies for every metric ton of carbon captured and utilized occasioned the scaling of CCUS technologies in the United States. The IRA ensures this development by the provision of certain tax credits to owners and developers of CCUS projects. Similarly, the Canadian government encourages the development and deployment of CCUS technologies through adequate budgetary allocations to CCUS projects. Particularly, the Canadian government budgets enhanced investment tax credit for CCUS projects periodically ranging from investments in direct air capture and other CCUS equipment, to investment in equipment for carbon transportation, storage and use.[48] This encourages investments in the development of CCUS technologies, as well as a rapid scale-up of CCUS deployment in Canada. Norway adopted the deterrent system by the

[48] See for instance the Government of Canada Budget 2022, https://www.budget.canada.ca/2022/report-rapport/toc-tdm-en.html [accessed 30 November 2023].

implementation of a carbon tax system in 1991 whereby emissions generated by offshore oil and gas activities are subjected to tax. This led to the first commercial offshore CCS project in the North Sea.[49]

Second, such framework must address the issue of the long-term liability for carbon storage sites for example, through provisions on liability transfer whereby government accepts liability for stored carbon. For instance, the government of Alberta assumes the long-term liability of licensed carbon sequestration projects in Alberta. Upon assumption of liability, the government becomes the owner of any captured carbon and assumes all obligations of the lessee. The government also indemnifies a lessee against liability for damages attributable to such licensed carbon sequestration project.[50] This has fostered the wide acceptance and deployment of CCUS technologies by the relevant stakeholders in Alberta.

Third, a viable CCUS framework must provide for policy incentives that address the high costs and capital challenge associated with CCUS including provisions on capital grants and subsidies for eligible investors in CCUS technologies, loan guarantees to owners and developers of CCUS projects, as well as innovation funding and investment in CCUS research and development by governments.[51] Aside the provision on tax credits, the United States for instance is committed to the provision of funding across its CCUS value chain. The funding support is aimed at the development of CCUS in the US through mechanisms including research and development funding, loan grants and permitting support.[52] Likewise, addressing the large upfront capital expenditure required for the development of CCUS infrastructure and facilities, the UK government announced the Carbon Capture and Storage Infrastructure Fund (CIF) in 2020 with substantial budgetary allocations for investments needed to realize the development and deployment of CCUS facilities in the

[49] International Finance Corporation, "Carbon Capture and Storage: A Critical Tool for Fighting Climate Change" 26 April 2023.

[50] Mines and Minerals Act, Revised Statutes of Alberta 2000, Chapter M-17, Sect. 121.

[51] International Energy Agency, "20 Years of Carbon Capture and Storage: Accelerating Future Deployment" (2016).

[52] See generally the Infrastructure Investment and Jobs Act, H.R.3684, 117th Cong. (2021–2022), Title III—Fuels and Technology Infrastructure Investments, https://www.congress.gov/bill/117th-congress/house-bill/3684/text [accessed 30 November 2023].

UK.[53] The CIF provides a policy incentive and revenue support for the UK industrial sector to adopt CCUS solutions thereby resulting in the proliferation of investments in the UK CCUS industry.

Furthermore, to enhance the industrial scaling of CCUS in Nigeria, there must be implemented a policy framework that ensures a viable business model for CCUS investments. For instance, in order to encourage private investments in the UK CCUS industry, a proposed commercial framework for business models that enable and support stable investment in CCUS projects was published in 2018.[54] Alongside the business models, the UK government ensures an improved regulatory stability and transparency by establishing an independent economic regulation for carbon transport and storage networks thereby, enabling investors to earn a fair return on their investments under a predictable and stable regulatory framework.[55]

Aside the regulatory lessons that may be learnt from other jurisdictions in order to ensure the implementation of an effective CCUS framework in Nigeria, some lessons may also be observed on CCUS within the Nigerian context. First, while the large-scale deployment of CCUS is significant for the achievement of Nigeria's climate goals, it is also relevant for driving economic development, technological innovation and higher employment opportunities in the country. CCUS presents an opportunity for Nigeria to continue attracting investments in her petroleum sector thereby achieving economic prosperity from her vast petroleum resources while maintaining her climate commitments. In other words, CCUS presents an opportunity for responsible and sustainable petroleum exploration and production. This is in addition to the potential role of CCUS in the resolution of the dilemma faced by the hydrocarbon industry. This is more so with regard to the seemingly conflict posed by the need to meet the market demand for energy, fuel and other by-products on the

[53] "The Carbon Capture and Storage Infrastructure Fund: An Update on its Design" 2021, https://www.gov.uk/government/publications/design-of-the-carbon-capture-and-storage-ccs-infrastructure-fund/the-carbon-capture-and-storage-infrastructure-fund-an-update-on-its-design-accessible-webpage [accessed 29 November 2023].

[54] "Industrial Carbon Capture Business Models" 2018, https://assets.publishing.service.gov.uk/government/uploads/system/uploads/attachment_data/file/759286/BEIS_CCS_business_models.pdf [accessed 29 November 2023].

[55] See the UK Department for Business & Trade, "Carbon Capture, Usage and Storage", https://www.great.gov.uk/international/content/investment/sectors/carbon-capture-usage-and-storage/ [accessed 29 November 2023].

one hand, and the climate goals on the other hand. CCUS can positively impact the achievement of a just energy transition and ensure energy justice and security by improving energy efficiency and promoting access to advanced and cleaner fossil-fuel technology in Nigeria.

According to the UN sustainable development agenda, global partnership and collaboration is a veritable means of implementing any of the sustainable development goals. The agenda emphasizes the significance of collaboration and international cooperation on issues such as science, technology and innovation, among others, as well as knowledge sharing on mutually agreed terms, as a means for the achievement of sustainable development globally.[56] The Nigerian experience indicates that international collaboration, particularly in the area of funding and technological support, enhances the development of CCUS especially in developing countries. With the support of international organizations such as the IEA, IFC and the World Bank, the move toward the scaling of CCUS in Nigeria has received the much needed attention. Likewise, the feasibility analysis for potential gas recovery and utilization opportunities from the Olusosun, Abule Egba and Solous landfills in Nigeria was made possible by funding support from the United States Environmental Protection Agency under the Global Methane Initiative.[57]

Naturally, technology is dynamic and evolving rather than static hence, the need to keep up to date with new technologies in the carbon industry. This calls for a subsisting system of research and development support for CCUS. The establishment and funding of research institutions or agencies, such as the CCUS R & D Centre of Excellence proposed by the Nigerian Geological Survey Agency, as well as the Africa Centre of Excellence for Carbon Management Technology and Innovation, serve as potential solution to addressing this and other challenges relating to research and development support for CCUS. In addition, such institutions or agencies are viable instruments for capacity building in the deployment of innovative CCUS technologies.

Further, the development and deployment of CCUS in Nigeria may be scaled up through the CDM. The CDM had proven effective in some projects involving the capturing and processing of associated gas

[56] United Nations, "Sustainable Development Goals", goal 17.

[57] Global Methane Initiative, "MSW Project Opportunity: Olusosun, Abule Egba and Solous Lagos, Lagos State, Nigeria", https://www.globalmethane.org/expo-docs/posters/MSW/MSW_NI_Project_Lagos-LFG_FINAL.pdf [accessed 30 November 2023].

in Nigeria, such as the 'Kwale Oil–Gas Processing Plant project' and the 'Ovade Ogharefe Gas Capture and Processing' project. A critical evaluation of the features of CDM identifies it as a viable instrument for overcoming the financial and technological barriers inhibiting the large-scale development of CCUS in Nigeria. A study on the achievement of the CDM in Nigeria within the context of investment flow reveals an investment flow of over four billion dollars into the country within a seven-year period.[58] The potential benefits of CDM to the development of CCUS will therefore include; increased flow of capital investments in CCUS projects and infrastructural development, stimulation of CCUS expertise and technological innovation and creation of viable carbon markets, among others. Moreover, with an uneven distribution of CDM projects across countries around the world, Africa had benefitted very little from the CDM.[59] Therefore, investments in industrial CCUS projects present viable opportunities for better access to CDM benefits in Africa generally and Nigeria particularly. Furthermore, stakeholders' acceptance of CCUS will depend significantly on the economic viability of CCUS projects which is primarily determined by several factors including the availability of a viable carbon market and policies. For instance, an analysis of a viable implementation of a landfill gas-to-energy project in Nigeria indicate flaring as the most economically viable option for dealing with captured landfill gas as against utilization.[60] Implementation of such projects under the CDM however addresses the economic barriers and allows for the up-scaling of landfill gas-to-power projects in Nigeria.

Also, lessons from the existing gas-to-power projects in Nigeria indicates that the development and deployment of CCUS poses positive impacts on some aspects of sustainable development in the country

[58] Akpofure Rim-Rukeh, "Cleaner Development Mechanism (CDM) in Nigeria: Assessing Progress and Looking Forward" Nigerian Environmental Society, FCT Chapter, 23 September 2020, https://nesfct.org/cleaner-development-mechanism-cdm-in-nigeria-assessing-progress-and-looking-forward/ [accessed 30 November 2023].

[59] Surendran Pillay, "An Assessment of Clean Development Mechanism Project Contribution to Sustainable Development in Nigeria" (2016) 15(6) *International Business & Economics Research Journal* 315–328; Philip Ejoor Agbonifo, "Reforming theClean Development Mechanism (CDM) to Tackle the Environmental Policy Gap in the Nigerian Oil and Gas Industry" (2015) 17(2) *Journal of Sustainable Development in Africa* 65–76.

[60] Aboyade A.O., Rabiu A., and Amigun B., "The Potential for Climate Change Mitigation in Solid Waste Disposal: A Case Study of Lagos Landfills" (2012) 2(4) *The Online Journal of Science and Technology* 84–89.

particularly the environment, economy, health, employment and technology transfer.[61] Large-scale industrial development and deployment of CCUS promises multifaceted positive impacts on sustainable development including economic development through a viable carbon market, emissions reduction leading to improved air quality, as well as positive social impact through creation of green jobs, accessibility to clean energy, technology transfer and technical skills acquisition, among others.

The viability of a robust business model for carbon may be enhanced in Nigeria through the instrumentality of a feed-in-tariff policy (FIT). A FIT is an energy policy that supports the development of renewable energy by providing developers of renewable energy with a purchase guarantee on a long-term basis. Although a FIT is a widely accepted strategy for developing the scale of renewable energy, it is capable of adjustment to align with the features of the energy market, infrastructure and technological conditions of individual countries.[62] The features of a FIT policy including long-term investment security and profitability guarantee, among others, make it a suitable policy strategy for scaling the development of CCUS in Nigeria. Although Nigeria already has in place FIT regulations for renewable electricity, the same may be applied for the development of carbon markets in the country whereby a premium price or tariff is set for carbon products so that profitability is ensured for CCUS investor and developers. The Nigerian government may also commit to taking all unutilized captured carbon from qualifying CCUS facilities on the basis of a premium price set by a FIT policy thereby providing a purchase guarantee to operators of CCUS facilities.

In addition, the industrial development and deployment of CCUS technologies in Nigeria may enhance business competitiveness and corporate sustainability especially in the petroleum sector where there is high incidence of conflicts between oil companies and the host communities as a result of the adverse environmental impacts of the sector.

[61] Surendran Pillay, "An Assessment of Clean Development Mechanism Project Contribution to Sustainable Development in Nigeria" op cit.

[62] "Low Carbon Green Growth Roadmap for Asia and the Pacific: Fact Sheet—Feed-In Tariff", https://www.unescap.org/sites/default/files/26.%20FS-Feed-In-Tariff.pdf [accessed 30 November 2023].

6 Conclusion

The various oil-to-gas projects ongoing in Nigeria indicate the potential success of large-scale development and deployment of CCUS in the country. Meanwhile, the Nigerian government, through the various climate policies and targets, has indicated a strong commitment to climate mitigation mechanisms and strategies including climate technologies such as CCUS. To drive this commitment, the Nigerian government is undergoing partnership with various international organizations with the goal of scaling the development and deployment of CCUS in Nigeria. The various partnerships have been successful in placing Nigeria on the path toward the industrial scaling of CCUS. A major barrier that may however hinder the scaling of CCUS in the country is the absence of a specific and comprehensive legal framework to drive such development. This calls for the implementation of a specific and comprehensive CCUS framework in Nigeria.

To ensure a viable and an effective CCUS framework in the country however, significant lessons may be observed from other jurisdictions that have successfully scaled the development of CCUS. First, such framework must set clear and specific requirements with regard to rights, responsibilities and liabilities, etc., on CCUS. It must also address major drivers of and barriers to the development of CCUS including economic and other forms of incentives and disincentives.

It is noteworthy that most of the gaps and challenges with CCUS are already noted by the various preliminary studies conducted and/or ongoing in Nigeria including the IEA-Nigeria joint CCUS workshop and the ICCUS D& S. It is therefore a possibility that these challenges will be addressed upon the completion of the proposed national CCUS work program and phase two of the ICCUS D & S. Notwithstanding, relevant lessons from other jurisdictions that may enhance the industrial scaling of CCUS in Nigeria are proffered.

Meanwhile, there are other strategic policy measures already in place in Nigeria that may be adopted to scale the industrial development and deployment of CCUS in the country including the CDM and the FIT regulations.

CHAPTER 16

CCUS in the Netherlands

Martha M. Roggenkamp

1 Introduction

The possibility to capture CO_2 for industrial use or underground storage has been considered since the turn of the century. While the possibility to employ captured CO_2 for industrial uses was part of an energy saving policy, the underground storage of CO_2 is increasingly part of international, EU and national climate policies. The Netherlands has made several attempts to store CO_2 in depleted gas fields. The first two attempts—on land and at sea—failed. However, currently several CCS projects are being developed offshore. These projects involve the transport by pipeline or ship of CO_2 to underground reservoirs on the Dutch continental shelf and to the Norwegian continental shelf.

This chapter will first discuss (Sect. 2) the policy and practices with regard to carbon capture and use. Thereafter, the focus will be on carbon

M. M. Roggenkamp (✉)
University of Groningen, Groningen, The Netherlands
e-mail: M.M.Roggenkamp@rug.nl

University of Pau, Pau, France

E. G. Pereira et al. (eds.), *Carbon Capture Utilization and Storage*, Sustainable Development Goals Series,
https://doi.org/10.1007/978-3-031-81272-9_16

capture and underground storage. Section 3 will discuss the legal framework regarding underground storage of CO2 and how it has developed since turn of the century. This will be followed by a brief analysis of the rules governing the transport of CO_2, i.e. the transport of CO_2 as part of the CCS chain. The next section will present the CO_2 storage projects, which have been initiated in the previous two decades as well as the current projects that will start in 2026. Section 5 will present some findings and recommendations based on the Dutch experiences.

2 CO_2 Capture and Use in the Netherlands

CO_2 capture, transport and use is not a new phenomenon. In the Netherlands, CO_2 has been captured for many decades in order to be used for various industrial purposes such as the soft drink industry and the greenhouse sector. Yara in Sluiskil, for example, aims at capturing 1,4 Mtonne CO_2 per year, and part of this is reused as raw material or auxiliary material for, among other things, the soft drink industry.[1] CO_2 for the (soft drink) industry is largely transported by road and inland waterways. Such transport is subject to general transport law such as the Road Transport of Goods Act, the Transport of Hazardous Substances Decree and the External Safety of Transport Routes Decree.[2]

Shell Refinery in Rotterdam is also capturing CO_2 but this is primarily used for the greenhouse horticulture sector in the area near Rotterdam (Europoort).[3] Greenhouse growers need CO_2 as a nutrient to stimulate crop growth.[4] Reuse of captured CO_2 also means significant energy savings because greenhouse growers no longer have to burn natural gas

[1] Yara Sluiskil B.V., 'Kijk op Yara—Samen klaar voor de toekomst', p. 4. See https://www.yara.nl/contentassets/6ca92ab7997f4f3faa50fd9ada6d6855/2210781-duurzaamheidsverslagsite.pdf/.

[2] Legislation can be found at https://wetten.overheid.nl/ No English translation is available.

[3] Shell, 'Zes vragen over het afvangen en gebruiken van CO_2' en 'Shell Nederland', 31 maart 2021, https://www.shell.nl/over-ons/nieuws/nieuwsberichten-2021/zes-vragen-over-het-afvangen-en-gebruiken-van-co2-en-shell-nederland.html.

[4] Since the discovery of the Groningen field, Dutch greenhouse growers have for many centuries been supplied with natural gas at a very low price. At the end of the 1990s, the European Court of Justice had to assess whether these low gas rates should be classified as state aid. See, among others, MM Roggenkamp et al. (ed), *Energy Law in Europe*, Oxford University Press, 2001, p. 254.

to produce CO_2. By replacing natural gas with captured CO_2, growers can save significant amounts of natural gas per year, which also reduces CO_2 emissions per year. While CO_2 for the (soft drink) industry is largely transported by road and inland waterways, greenhouse growers receive CO_2 via a network of pipelines. Since 2005 this pipeline system transports 'organic CO_2 for assimilation by plants' (abbreviated to OCAP). It has a total length of approximately 350 kilometres to which some 580 greenhouse growers are connected.[5] These greenhouse growers are supplied with 500.000 tonnes CO_2 per year, and for 2020 it envisaged some 280 million cubic metres gas savings.[6] Currently, there is no sector-specific legislation in place governing the transport of CO_2 via pipelines. However, since 2014 the Decree and the Regulation on External Pipeline Safety applies with regard to those pipelines with at least an external diameter of 70 mm (or an internal diameter of 50 mm) and a pressure of 16 bar or more.[7] On this basis, the operator of the CO_2 pipeline meeting these basic requirements must, among others, take into account zoning obligations between the pipeline and certain buildings.

Although industrial reuse of CO_2 can provide significant energy savings and/or cost savings, it does not lead to a relevant (long term) reduction in CO_2 emissions as crops do not permanently store CO_2. Therefore, the Netherlands considered the possibility for CO_2 emission reduction via (permanent) underground storage. Such storage is based on sector-specific legal arrangements, which are discussed below.

3 The Regulation of CO_2 Storage and Transport

3.1 Underground Storage of CO_2

Underground storage of CO_2 has been part of Dutch climate policy for about 25 years. The 1998 climate policy plan already referred to the possibility of underground storage of CO_2.[8] At that time, however, the goal was not so much to store CO_2 permanently, but mainly to provide for

[5] OCAP pipeline system consists of almost 100 km transportation pipelines and 250 km distribution pipelines.

[6] https://www.ocap.nl/nl/images/OCAP_Factsheet_Nederlands_tcm978-561157.pdf.

[7] Art 1 sub 2 and art. 2 sub e Regeling externe veiligheid buisleidingen.

[8] TK 1998–1999, 26,603, no. 2.

temporary storage to create a potential buffer of CO_2 for industrial users, like the greenhouse growers.[9]

The possibility to develop underground storage was facilitated by the Mining Act of 2003, which offered the possibility to obtain a permit for storing 'substances'.[10] Although the term 'substance' was not defined in the act, parliamentary documents show that the storage of CO_2 was considered, and that CO_2 thus can be considered as a 'substance'.[11] Based on such a permit, CO_2 can be stored underground and, if necessary, retrieved for industrial use. However, this option has never been used.

Instead, policy increasingly focused on the development of the so-called clean fossil, i.e. the capture and permanent underground storage of CO_2. The idea was even that the Netherlands could play a leading role in CO_2 storage in Europe due to the availability of many onshore and offshore (soon to be depleted) oil and gas fields, aquifers and coal seams where potentially CO_2 could be stored.[12]

3.2 Permanent Storage of CO_2

3.2.1 EU Directive on Geological Storage of CO_2

The need to focus on the *permanent* storage of CO2 follows the introduction of the EU Directive 2009/31/EC on the geological storage of CO_2.[13] Although the directive considers the permanent underground storage of CO_2 as one of the options for reducing CO_2 emissions, it also explicitly states that capture and permanent storage of CO_2 (or CCS) is not a goal itself but should be seen as a transitional technique that may assist in contributing to climate change mitigation.[14]

The main focus of Directive 2009/31/EC is on geological *storage* of CO_2. It hardly provides any provisions on CO_2 capture and transport. The EU legislator was of the opinion that it was sufficient to amend

[9] TK 1998–1999, 26,603, no. 2, p. 42, p. 57, p. 72.

[10] Art. 25 Mining Act.

[11] TK 1998–1999, 26,219, nr. 7, pp. 19–20.

[12] https://edepot.wur.nl/177047.

[13] Directive 2009/31/EC of 23 April 2009 on the geological storage of carbon dioxide and amending Directive 85/337/EEC, Directives 2000/60/EC, Directive 2001/80/EC, Directive 2004/35/EC, Directive 2006/12/EC, Directive 2008/1/EC and Regulation (EC) No 1013/2006, OJ (2009) L 140/114.

[14] Recital 4 Directive 2009/31/EC.

others EU laws relevant for these activities. Therefore, the directive provides for the amendment of, *inter alia*, the Directive on Integrated Pollution Prevention and Control (IPPC Directive), the Environmental Impact Assessment Directive, the Environmental Liability Directive, the EU Emissions Trading Scheme (EU ETS) Directive and the Waste Directive.[15] As a result, the following main principles apply: (i) captured CO_2 is not considered as 'waste' within the framework of the directive, and (ii) CO_2 that is captured and permanently stored is considered as not emitted so that companies holding an emissions permit—and capture and permanently store CO_2—can trade unused emission allowances deriving therefrom.

Given the above, the directive primarily provides a regime that ensures that captured CO_2 can be stored safely and permanently in suitable geological storage locations on land or at sea, i.e. in the exclusive economic zone ('EEZ'). Moreover, when designing this regime the EU legislator has applied provisions that are in line with the principles of the EU Hydrocarbons Directive (Directive 94/22/EEC). In other words, the award of an exploration and/or storage permit has to be based on transparent and non-discriminatory conditions. Member States were required to transpose the directive into national law before 25 June 2011[16] and submit reports to the Commission every four years assessing the national implementation of the directive.[17] Nevertheless, the application of CCS for climate mitigation has been slow. Firstly, many Member States have been late in transposing the directive in national law.[18] Secondly, several Member States were opposed to applying CCS and made use of their right not to allow for any storage in their territory.[19] Thirdly, the incentives to apply CCS were limited and the costs often too high, especially in relation to the price of CO_2 emission allowances. So when

[15] See Art. 31—Art. 37 Directive 2009/31/EC.

[16] Art. 39 Directive 2009/31/EC.

[17] Art. 27 Directive 2009/31/EC. Since 2014 the Commission has published four implementation reports of which the most recent report was published in October 2023 and covers the period May 2019-April 2023.

[18] Whereas in July 2011 some 26 Member States had not yet implemented the directive, all Member States had notified transposition measures in October 2013. See COM(2014).

[19] Art. 4 Directive 2009/312/EC.

the directive had to be assessed in 2015, there was a lack of practical experience.[20] The evaluation report, nevertheless, concluded that some issues needed attention, such as the need to regular update the guidance documents. These guidance documents not only provide Member States with a methodological approach for implementing some key provisions of the directive but also have a harmonising effect so that Member States are less inclined to take very different views when applying these measures. So far, four guidance documents have been published and they focus on, inter alia, financial security and mechanisms (art. 19 and art. 20) and the transfer of responsibility (art. 18).[21]

In 2011, the Dutch legislator transposed Directive 2009/31/EC into the Mining Act.[22] When doing so the text of the directive was almost transposed into the act literally, but since then the act has been expanded and supplemented.

3.2.2 Permitting

The Directive provides for two different permits: an exploration and a storage permit. An exploration permit can be awarded in order to assess whether a geological formation is suitable for underground storage of CO_2.[23] Also in the Netherlands it is prohibited to search for CO_2 storage facilities without a permit.[24] However, there is an exception to this rule if sufficient knowledge and information is available regarding the suitability of the subsoil for CO_2 storage. This will often be the case in the Netherlands as so far parties tend to make use of (almost) depleted gas fields. In such case, an exploration permit will not be necessary as sufficient geological information is available about the reservoir and thus the possibility to apply immediately for a storage permit.

In general, the licensing regime applying to CO_2 storage is similar to the regime applying to the exploration and production of hydrocarbons. This means that, inter alia, a storage permit is exclusive and cannot be

[20] Evaluation report was published under contract number No. 340201/2014/679421/SER/CLIMA.C1.

[21] See https://climate.ec.europa.eu/eu-action/carbon-capture-use-and-storage/implementation-ccs-directive_en#documentation.

[22] See Chapter 3 of the Mining Act.

[23] Art 4 and Art 5 Directive 2009/31/EC.

[24] Art. 25 para 1 Mining Act.

granted for an area for which a storage or extraction permit has already been granted.[25] If the holder of an exploration permit has demonstrated the suitability of a location, this person will in principle also be granted a storage permit.[26] However, if a storage permit is applied for without a prior exploration permit, the storage permit has to be awarded in competition instead. In that case, the application for the storage permit has to be published in the Official Journal of the EU and the Dutch Official Gazette so that others also can apply for a storage permit for the same area.[27] When assessing applications, consideration is given to, among others, the technical and financial capabilities of the applicant, the manner in which storage will be carried out and previous experiences of the applicant, the safety of local residents and nearby infrastructure and the environment. In the event of multiple applications, the efficient and expeditious storage is of decisive importance.[28] The question can be asked whether such a situation will arise if a storage permit is applied for in relation to an (almost) depleted gas field by a party that previously held the production permit. After all, this party will have an advantage in terms of geological and technical knowledge of the storage complex.

A storage permit can be granted to one or more persons.[29] It will be subject to several conditions, such as the period of CO_2 injection, the composition of the CO_2 stream, the total amount of CO_2 to be stored, the maximum pressure of the reservoir and requirements regarding monitoring.[30] In addition, the permit holder is required to provide third parties access to the storage facility on the basis of conditions that are reasonable, transparent and non-discriminatory, provided that sufficient capacity is available and the CO_2 stream of third parties corresponds to the technical specifications.[31] In addition, the permit holder will be required to provide financial security to guarantee that all obligations under the storage permit will be met and that costs and damages related

[25] Art. 26 Mining Act.

[26] Art. 26a Mining Act.

[27] Art. 26b Mining Act. In case of a prior exploration permit, this competitive procedure would apply on this stage of the process.

[28] Art. 27 Mining Act.

[29] Art. 22 Mining Act.

[30] Art. 31b Mining Act.

[31] Art. 31 b Mining Act.

to storage can be recovered from the permit holder. The injection and storage of CO_2 takes place on the basis of an approved monitoring plan in order to check whether the injected CO_2 behaves as expected. Last but not least, the permit holder must hold an emissions permit just in case CO_2 would escape from the reservoir.

Unlike a hydrocarbon permit, the application for and the draft storage permit must be submitted to the European Commission who may issue a non-binding opinion on this draft.[32] The condition was included in the Directive[33] in order to provide the Commission insight into the development of CCS.[34]

3.2.3 Closure, Post-Closure and Transfer of Responsibility

There may be several reasons to close a storage site. These vary from a situation that the conditions of the permit are not being met and the competent authority decides to withdraw the permit, to the situation that the storage site has been completely filled with CO_2. In any case, such closure needs to be based on a closure plan that has been approved by the competent authority.[35] In the Netherlands, it is the Ministry of Climate and Green Growth who is entitled to approve of a closure plan on the basis of an advice by the State Supervision of Mines. After the site has been closed on the basis of the approved closure plan, the permit holder remains responsible for the closed site for another 20 years.[36] Only if all available information shows that the stored CO_2 will remain completely and *perpetually* contained, the responsibility of the site can be transferred to the competent authority (= the State) after a period of 30 years after closure. Once responsibility has been transferred, costs can no longer be recovered from the permit holder unless there is fault or negligence on its part before the transfer of responsibility. In addition to the above-mentioned financial security, the permit holder must cover the costs relating to the storage site, such as monitoring costs, during this period of 20 years after closure.[37]

[32] Art. 31c and Art. 31d Mining Act.

[33] Art. 10 Directive 2009/31/EC.

[34] Art. 29 a.f. Mining Act.

[35] Art. 31i Mining Act.

[36] Art. 31 j (c) Mining Act.

[37] Art. 31j (d) Mining Act.

3.3 *Transport of CO_2*

As mentioned above, there are many different ways to transport CO_2. However, when considering the CCS value chain, it seems that transportation by pipeline and ship is the most important option. Both of them will be discussed below.

3.3.1 Transport by Pipeline

Although directive 2009/31/EC does not prescribe any particular mode of transportation, it is clear that the European legislator had piped transportation in mind as the provisions governing transportation in the directive seem to apply to piped transportation in particular.

The provisions on transport in the directive relate to the protection of the environment and climate change. It requires that an environmental impact assessment will be carried out for the purposes of the construction and operation of a CO_2 pipeline, which have a certain minimum length and diameter.[38] Additionally, it requires that the operator of a CO_2 pipeline holds an emissions permit.[39] Like the holder of the storage permit, the pipeline operator must be committed to purchase carbon emission allowances if an unexpected leakage would occur.

The directive also regulates, to some extent, the use of CO_2 pipelines. Like the regime in the Gas Directive governing upstream gas pipelines, Directive 2009/31/EC provides for a simple regime governing third-party access, i.e. negotiations between a third party and a pipeline operator with a limited ability to refuse access. The grounds for refusal include (a) insufficient capacity, (b) incompatibility of technical specifications and (c) reasonable needs of the owner/operator.[40] Because CO_2 pipelines like upstream gas pipelines are considered as natural monopolies or can be regarded as an essential facility, it is generally assumed that third parties can rely on competition law if a refusal to provide third

[38] Recital 17 and art. 31 Directive 2009/31/EC.

[39] Art. 4 EU ETS Directive. See also Martha M. Roggenkamp, 'Transportation of CO_2 in the EU', in: I. Havercroft, R. Macrory and R. Stewart, *Carbon Capture and Storage—Emerging Legal and Regulatory Issues,* Hart Publishing, 2018, p. 251.

[40] Art. 21 Directive 2009/31/EC. See for a more detailed analysis of the TPA regime Martha M. Roggenkamp, 'Transportation of CO_2 in the EU', in: I. Havercroft, R. Macrory and R. Stewart, *Carbon Capture and Storage—Emerging Legal and Regulatory Issues,* Hart Publishing, 2018, p. 245–p. 266.

parties access could be considered as an abuse of the pipeline companies' dominant position.[41]

These principles are also included in Article 32 of the Mining Act. If requested by a third party, the operator of a CO_2 pipeline is obliged to transport CO_2 on the basis of reasonable, transparent and non-discriminatory conditions and can only refuse such request on the basis of the aforementioned grounds. An appeal based on a lack of capacity or connection opportunities will be honoured if the pipeline operator can remedy the refusal by carrying out work that increases pipeline capacity and/or size and the number of pipeline connections, provided this work is economically justified, the applicant is prepared to pay for the work and the work does not have a negative impact on the environment and/or safety of the transport.

It is important to note that the above provisions only apply to CO_2 pipelines that are part of the CCS value chain[42] and thus not to other CO_2 pipelines such as the OCAP pipeline. In other words, neither an emissions permit nor an environmental impact assessment is required for pipelines such as the OCAP pipeline. There is neither any sector-specific access regulation for such pipelines, but here too, third parties can in principle rely on competition law.

3.3.2 *Transport by Ship*

The transport of CO_2 by ship falls outside the scope of Directive 2009/31/EC, but is becoming more and more important. Transport by ship takes place under an entirely different legal framework. Several international treaties regulate the safety and liability of CO_2 transport by ship. The most important treaty governing the safety of merchant ships is the SOLAS Convention, which requires ships sailing under their jurisdiction (flag-state jurisdiction) to comply with some specific safety standards. The International Gas Carrier Code, for example, governs the construction of gas tankers and the safety of ships transporting various gases, including CO_2. Liability for any CO_2 leakage or outflow falls under the International 'Convention on Liability and Compensation for Damage

[41] Art. 17 of the Dutch Gas Act provides that the Competition Act applies to the transport of gas via upstream pipelines situated on the continental shelf.

[42] See Art. 2 of the Dutch Emissions Trading Decree and the Dutch Environmental Impact Assessment Decree (part C 8.1) explicitly refer to the transport of greenhouse gases via pipelines for geological storage.

in Connection with the Carriage by Sea of Dangerous and Noxious Substances' of 1996 (HNS Convention).[43] The latter treaty is expected to enter into force in 2024.[44]

Another issue concerns the emission of CO_2 from ship engines during the propulsion of the ship and the possible escape of CO_2 due to accidents or damage to the ship. It is of special importance that as of 2024 larger seagoing vessels (vessels carrying more than 5,000 tonnes) will be subject to the EU ETS and in particular the monitoring and verification requirements of Regulation (EU) 2015/757.[45] Consequently, ship owners and/or operators will be obliged to purchase and to submit sufficient emissions allowances to compensate for the CO_2 emissions registered each year.[46] Not yet clear and unresolved is the question whether the same regime will apply to ships transporting CO_2 and potentially the outflow of CO_2 from these ships due to a leak or other damage.[47]

[43] See also Martha M. Roggenkamp, 'Transportation of CO_2 in the EU', in: I. Havercroft, R. Macrory and R. Stewart, *Carbon Capture and Storage—Emerging Legal and Regulatory Issues*, Hart Publishing, 2018, p. 245–p. 266.

[44] Entry into force of the treaty requires twelve signatories. By the end of 2023, eight countries had ratified the treaty. The Netherlands will ratify jointly with Belgium and Germany. More information about the status of the entry into force can be found via https://www.hnsconvention.org/status/ The HNS Convention provides for an insurance obligation, a maximum amount in the event of damage and the establishment of a fund for victims if their damage is not fully compensated by the ship owner.

[45] See Directive 2023/959 of 10 May 2023 amending Directive 2003/87/EC establishing a system for greenhouse gas emission allowance trading within the Union and Decision (EU) 2015/1814 on the establishment and operation of a market stability reserve for the Union greenhouse gas emission trading scheme. The application of the scheme depends on whether the ports of departure and/or call are located inside the EU or outside the EU. The European Commission will assess in 2026 whether the obligation will also apply to smaller sea going vessels (between 400 and 5000 tonnes). See also https://climate.ec.europa.eu/eu-action/transport/reducing-emissions-shipping-sector/faq-maritime-transport-eu-emissions-trading-system-ets_en.

[46] The CO_2 allowances will be submitted to the national competent authority. The obligation for sea going vessels will be introduced gradually, starting with 40% in 2024 and 70% in 2025 to 100% in 2026.

[47] See also 'Communication from the Commission to the European Parliament, the Council, the European Economic and Social Committee of the Regions—Towards an ambitious Industrial Carbon Management for the EU', 6 February 2023, COM(2024) 62 final, p. 11.

3.3.3 *Cross-border Transport of CO_2*

An important obstacle for any cross-border transport and storage of CO_2 was the London Protocol.[48] Article 6 of the London Protocol prohibits the export of wastes or other substances by Contracting Parties for the purpose of dumping, i.e. the storage of wastes or other substances in the seabed and its subsoil by means of ships and man-made structures at sea (=pipelines).[49]

As this provision also hampers cross-border transport and storage of CO_2, a number of Contracting Parties submitted a proposal to adapt the provision in order to facilitate cross-border transport of CO_2 for the purpose of CCS. As amendments to the London Protocol require a 2/3 majority,[50] a proposal was adopted in 2009 to amend this provision accordingly. Given the lack of the required 2/3 majority, the Netherlands and Norway submitted a resolution in 2019 to allow countries that have ratified the amendment, and pending the adoption of the amendment, to enable cross-border transport and storage of CO_2. Such provisional application only applies to those countries that have ratified this amendment, i.e. Norway, UK, the Netherlands, Finland, Estonia, Sweden, Belgium and Denmark.[51] This means that currently cross-border transport and storage of CO_2 in the North Sea area is possible.

3.4 *Financing CCS*

As previously mentioned, the rationale behind Directive 2009/31/EC is that parties holding a greenhouse gas emissions permit and subsequently capture and permanently store specific quantities if CO_2 will be able to sell 'unused' allowances and thus generate an income from trading these emission allowances, which again could incentivise investments in the CCS value chain. Although recently the price of emission allowances

[48] 1996 Protocol to the Convention on the Prevention of Marine Pollution by Dumping of Wastes and Other Matters, 1972. See https://wwwcdn.imo.org/localresources/en/OurWork/Environment/Documents/PROTOCOLAmended2006.pdf.

[49] Art. 1 sub 4 London Protocol.

[50] Art. 21 London Protocol.

[51] France and Germany are in the process of ratification. In addition, the Islamic Republic of Iran and the Republic of Korea have also ratified the amendment. See further EU-London_Protocol_Analysis_paper_final0930.pdf (europa.eu).

has increased considerably,[52] the market price of these allowances has in general been too low to cover the costs of CCS.

Therefore, additional measures have proven necessary to facilitate the development of CCS. At EU level, various possibilities for further financing have been developed. These include financial instruments to promote socio-economic developments (such as the Recovery and Resilience Facility[53] and Just Transition Fund),[54] innovative techniques (Innovation Fund)[55] or specifically support the development of cross-border CO_2 transport networks (Connecting Europe Facility or CEF).[56] The latter fund can also be used for the development of Trans-European Energy Networks (TEN-E). As part of the TEN-E policy, five CCS projects have been designated in 2021 as Projects of Common Interest (PCI) and the Commission decided in December 2023 to use a significant part of the CEF funds for the development of four CO_2 storage and transport projects, including projects in the port of Rotterdam.[57]

At national level, in 2020 the subsidy scheme for promoting sustainable energy production (SDE) was amended in order to financially support climate mitigation projects. Hence, the new SDE++ subsidy is no longer limited to financing sustainable energy production methods, but is also used to support innovative techniques reducing CO_2 emissions, including CCS.[58] In the meantime, several Dutch projects have made use of the possibility to apply for an SDE++ subsidy to capture and/or store CO_2. These are discussed below.

[52] See https://www.statista.com/statistics/1322214/carbon-prices-european-union-emission-trading-scheme/.

[53] Regulation (EU) 2021/241 of 12 February 2021 establishing the Recovery and Resilience Facility.

[54] Regulation (EU) 2021/1056 of 24 June 2021 establishing the Just Transition Fund.

[55] Regulation (EU) 2023/2537 of 15 September 2023 amending Delegated Regulation (EU) 2019/856 supplementing Directive 2003/87/EC of the European Parliament and of the Council with regard to the operation of the Innovation Fund.

[56] Regulation (EU) 2021/1153 of 7 July 2021 establishing the Connecting Europe Facility and repealing Regulations (EU) 1316/2013 and (EU) 283/2014.

[57] https://energy.ec.europa.eu/news/connecting-europe-facility-nearly-eu600-million-energy-infrastructure-contributing-decarbonisation-2023-12-08_en.

[58] See the 2022 Decree on the Promotion of Sustainable Energy Production and Climate Change. Article 1(y) of the Decree explicitly refers to the reduction of greenhouse gas emissions through CCS.

4 CCS Projects in the Netherlands

For almost two decades, attempts have been made to develop CCS projects in the Netherlands. The first attempts have not been successful, but have shaped the process and ideas on CCS. Therefore, this section will first present these initial attempts and discuss why they initially proved unsuccessful. This will be followed by an analysis of the most recent projects.

4.1 First Attempts

4.1.1 Storage on Land

The first attempt to permanently store CO_2 involved a 'pilot' by Shell in Barendrecht (a town close to Rotterdam). The intention was for Shell to capture CO_2 at its refinery in Pernis and then store the captured CO_2 nearby in an (depleted) gas field in Barendrecht. Shell also held via NAM[59] the production licence (e.g. concession) of this relatively small gas field, which could store up to 0.8 megatons CO_2. This project was intended as a pilot and, depending on its outcome, another storage permit could be awarded to a second field near Barendrecht.[60]

This pilot started in 2007 and was based on the assumption that a storage permit would be awarded on the basis of the (then) Mining Act in 2003 (see 3.1 above). However, the project quickly encountered considerable resistance from local residents who were concerned about the safety of the stored CO_2. Although at the time Directive 2009/31/EC had not yet been adopted and had therefore had not yet been transposed into the Mining Act, the Minister agreed to take into account the more rigorous principles of the directive when granting the storage permit.[61] Still, the Minister of Economic Affairs decided in 2010 not to award the permit due to the very large public opposition.[62] An important consequence of this failed CCS project was the decision not to store CO_2 on land and to focus on CO_2 storage offshore.

[59] NAM or *Nederlandse Aardoliemaatschappij* is a joint venture of Shell (= operator) and Exxon that only operated in the Netherlands.

[60] This field could potentially store about 9 megatons of CO_2.

[61] TK 2008–2008, 31,510, no. 36.

[62] See also https://www.globalccsinstitute.com/archive/hub/publications/8172/barendrecht-ccs-project-case-study.pdf.

4.1.2 Storage Offshore

Shortly afterwards, Industrial parties started to develop projects aiming at CO_2 storage offshore. This involved the 'Green Hydrogen' project and the 'Road' project.

The 'Green Hydrogen' project was intended to start in 2016 and was designed as a cross-border CCS project in which capture would take place at Air Liquide near Rotterdam (and possibly later from the Antwerp facility). The captured CO_2 would then be shipped to the Danish continental shelf where it would be injected into an almost depleted oil field operated by Maersk and used for enhanced oil recovery. This project failed for two reasons. Firstly, there was a financing problem because the committed NER300 funds were withdrawn by the EU at the end of 2012 and the price of the emissions allowances was too low to cover operational costs. Secondly, this project faced the prohibition of Article 6 of the London Protocol (see above para. 3.3.3) involving cross-border transport of CO_2 (e.g. prohibition of 'dumping').

During the same period, the ROAD project was initiated by a joint venture[63] of the following electricity producers and oil companies: Uniper (formerly E.ON Benelux), GDF SUEZ Energie Nederland and TAQA Energy. The intention was to capture CO_2 at Uniper's MPP3 power plant on the Maasvlakte and then transport it via a 25 km (20 km offshore) pipeline to some depleted gas fields of TAQA Energy in block P15 and P18 situated about 20–40 km from Rotterdam. Storage was supposed to start in 2015. Although all permits had been granted for both the capture of CO_2 and the permanent storage of CO_2 (including the assessment by the European Commission) as well as the transport of CO_2 (construction and use of the pipeline), the project nevertheless could not go ahead. Like the 'Green Hydrogen' project, the ROAD project faced financing problems due to the low price of the emission allowances. However, the main reason for terminating the project was the Dutch government's policy involving early closure of the coal-fired power plants.[64] Consequently, there was no longer a long-term perspective for dedicated CCS projects involving coal-fired electricity generators, i.e. the ROAD project. This development also illustrates that future CCS projects will no longer

[63] The companies operate via Maasvlakte CCS Project CV.

[64] L.S. Braaksma, K.J. de Graaf, M.M. Roggenkamp, "Afscheid van kolen(centrales): waar een wil is, is een wet?", *Ars Aequi*, 2019, p/ 449–p. 458.

involve dedicated point-to-point projects involving electricity producers as generators will increasingly replace fossil fuels with renewable energy sources. However, CCS will be and remain relevant for other industries, which cannot make such a transition.

4.2 Recent CO_2 Storage Projects

Various CCS projects are currently being developed. This includes a cross-border CCS project by Yara (Cactus) and two 'national' CCS projects, namely Porthos and Aramis. In contrast to the Cactus and Porthos projects, no investment decision has yet been taken regarding Aramis.

4.2.1 Cactus: Dutch-Norwegian Cross-border CCS Project

In 2023, Yara Sluiskil concluded an agreement with the Norwegian CO_2 transport and storage company 'Northern Lights' (see Chapter 16 below) to permanently store volumes of CO_2 captured by Yara, in the Norwegian seabed. While part of the captured CO_2 still will be used for industrial purposes in the Netherlands (see paragraph 2 above), the remaining part will be liquefied and shipped to Norway as of 2026. As a result, Yara can reduce its CO_2 emissions by 800,000 tonnes per year from 2026 and some 1.5 million tonnes by 2030.[65]

This is the second attempt by Yara to store CO_2 outside the Netherlands (see Sect. 4.1.2 above). However, this time the project is facilitated by the provisional application of the amendment to Article 6 London Protocol (see paragraph 3.3.3 above). Another potential obstacle for developing this project was the potential lack of financial support as the SDE++ scheme only applies to CCS projects in the Netherlands (see above 3.4).In order to enable this project and to achieve the Dutch CO_2 emission reduction goals for 2030 and 2050, the Ministry of Economic Affairs and Climate Policy decided on 30 October 2023 to grant Yara a subsidy of Euro 30 million for this specific CCS project, but on the condition that the project will be carried out between 1 November 2023 and 1 July 2026.[66]

[65] https://www.yara.com/corporate-releases/yara-invests-in-ccs-in-sluiskil-and-signs-binding-co2-transport-and-storage-agreement-with-northern-lights-the-worlds-first-cross-border-ccs-agreement-in-operation2/.

[66] Letter (reference number DGBI/38212616) of the Ministry of Economic Affairs and Climate of 30 October 2023. Subsidy is granted pursuant to Article 2, paragraph 1

4.2.2 Porthos

The next CCS project that will take of is Porthos, which is short for 'Port of Rotterdam CO_2 Transport Hub & Offshore Storage'. This project will be developed by a joint venture[67] of three government-owned entities: the Rotterdam Port Authority, the Dutch gas transport company Gasunie and the public energy company participating on behalf of the Dutch State in upstream energy activities 'Energiebeheer Nederland' (EBN).[68] It involves the transport of CO_2 via a pipeline between the harbour of Rotterdam and a storage site in the Dutch part of the North Sea. The tasks have been divided in line with their expertise, i.e. the Rotterdam Port Authority is charged with the developments onshore (e.g. in the port itself), Gasunie will be involved in the transport of CO_2[69] and EBN in CO_2 storage.[70] Unlike the above-mentioned ROAD project, CO_2 emitters have no direct participating interest in the project. Porthos is designed as an open access infrastructure project and therefore exclusively provides transport and storage services to third parties that aim at permanently storing CO_2 in order to meet their greenhouse gas emission reduction goals.

The total infrastructure consists of approximately 30 km of gathering pipelines in the Rotterdam port area. These gathering pipelines will be connected to a compressor station where the pressure will be increased from 35 to 130 bar. A 22-km-long pipeline runs from the compressor station to the existing gas production installation P18-A, which will be reused for CO_2 storage, and from where CO_2 will be injected in two 'depleted' gas fields located in block P18-2 and block P18-4. These reservoirs can accommodate approximately 37 Mtonne CO_2 over the next 15 years. For the purpose of such storage, TAQA and EBN have

(a) and (h) of the Framework Act on Subsidies of the Ministry of Economic Affairs and Climate and the Ministry of Agriculture.

[67] The joint venture company is the 'Porthos CO_2 Transport and Storage C.V'.

[68] As carbon transport and storage falls outside the scope of regular state participation, EBN was given special consent pursuant to Art. 82 para 4 of the |Mining Act by the Ministry of Economic Affairs and Climate to participate in Porthos. See also letter to Parliament of 3 October 2023 on CCS market developments, TK 2023–2024, 32,813, nr. 1298, p. 4.

[69] Art. 10 para 2 (d) Gas Act.

[70] Following art. 82 para 4 Mining Act, the Minister of Economic Affairs and Climate has permitted EBN to be involved in a commercial capacity.

been granted a CO_2 storage permit for block P18-2. In addition, TAQA Offshore BV has been awarded an amended storage permit for block P18-4.[71] As a CO_2 storage permit had already been awarded in 2013 for this reservoir as part of the ROAD project (see above 4.1.2), a change to the existing permit was considered sufficient.[72] The European Commission has received and assessed both draft permits and subsequently approved them.[73]

In connection with the development of Porthos, it is relevant that the European Commission designated Porthos as a project of common interest (PCI) in February 2022. Subsequently, Porthos was also selected as one of the projects eligible for financial support under the Connecting Europe Facility (see above in Sect. 3.4). Even though the TEN-E Regulation provides for accelerated permitting procedures, delays may still occur in practice due to appeal procedures. In the case of Porthos, the dune crossing of the landfall pipeline met with considerable resistance. The Administrative Court of the Council of State ruled on 16 August 2023 that it nevertheless agreed on the physical planning and environmental permits for the Porthos project and that the pipeline could be constructed. The final investment decision was made in October 2023 and construction has started in summer 2024.[74] The system is expected to be operational in 2026.

As mentioned above, Porthos is designed as an 'open access' project and consequently all interested third parties will be able to access the system on a non-discriminatory basis. This non-discrimination principle is included in Directive 2009/31/EC and the Mining Act (see above in 3.3.1.), but still offers considerable scope to interpretation. Porthos has applied the principle of non-discrimination by holding an 'open season' in which everyone was offered the opportunity to express interest in transporting and storing CO_2 permanently. Based on the interest shown, Porthos has concluded joint development agreements with Air Liquide,

[71] Please note that EBN participates in both gas production licences.

[72] The application for a storage permit for P18-6 was submitted on 25 July 2023 (see www.nlog.nl and *Staatscourant* no. 20 984).

[73] See https://www.rvo.nl/sites/default/files/2021/12/Advies-RVO-op-CO2-opslag vergunning-reservoir-P18-4-Porthos-fase-2.pdf. The change also relates to the financial security that all permit holders are required to.

[74] Volkskrant, 4 September 2024, p. 21.

Air Products, ExxonMobil and Shell.[75] According to these agreements, these four parties are obliged to start building the necessary capture facilities and obtaining all necessary permits relating thereto. In addition, they have agreed to make every effort to conclude the required transport and storage agreements, for which Porthos has developed a set of standard conditions.[76] Based on the principles of these standard agreements and the decision in 2021 to award a SDE++ subsidy to Air Liquide, Air Products, ExxonMobil and Shell,[77] these four parties and Porthos concluded the final transport and storage contracts in December 2021. By doing so, they contracted all available storage capacity.

Although Porthos currently has no more storage capacity available, it cannot be ruled out that additional storage capacity could be added to the system in the future so that other (Dutch and foreign) parties could also permanently store CO_2 offshore or, potentially, supply CO_2 to industrial users. One of the options raised is to connect Porthos with the OCAP system.

4.2.3 Aramis

The Aramis project is currently being developed by a partnership of Gasunie (25%), EBN (25%),[78] TotalEnergies (25%) and Shell (25%). The corporate legal design and exact participation in the company operating the Aramis project is still to be decided and depends on the award of the permits for constructing the necessary infrastructure. Based on an environmental impact assessment of 2022,[79] a final decision on the

[75] The first cooperation agreement was concluded in the autumn of 2019 and was given a follow-up one year later (autumn 2020). The common goal of both agreements was to conclude a transport and storage contract.

[76] The 'Standard CO2 Transport and Storage Conditions in respect of the Porthos System' can be found at https://www.porthosco2.nl/wp-content/uploads/2022/03/Porthos-standard-CO2-Transport-and-Storage-Conditions.pdf.

[77] This subsidy covers the difference in costs between CO_2 emission rights and the costs for CCS. See letter to Parliament, TK 2020–2021, 31,239, no. 332.

[78] Similar to Porthos, EBN has been granted permission from the Minister to participate in Aramis. See fn 71.

[79] The assessment took into account storage in two fields operated by TotalEnergies (L4-A and K6-C) and another field operated by of Shell (K14-FA). See Netherlands commission for environmental impact assessment, Environmental Impact Assessment Aramis—CO2 transport (CCS Project)—*Advies over reikwijdte en detailniveau van het milieueffectrapport*, 18 August 2022, projectnumber: 3630, p. 6.

preferred route was made in February 2024[80] and the developers could start applying for the necessary permits. In the same period, the European Commission has also classified Aramis as a project of common interest (PCI project) and, like Porthos, Aramis would thus be able to claim financial support under CEF. In the Netherlands, Aramis has also been designated as a project of national importance in the context of the Multi-Year Infrastructure Energy and Climate Program (MIEK).[81] A final investment decision is expected in 2025/2026, after which the project could be completed and ready for use in 2028/2029.[82]

Despite several similarities, the Aramis project differs from the Porthos project. Instead of developing one integrated transport and storage project, the developers of Aramis need to focus on three different components of the CCS chain, i.e. a CO_2 receiving terminal plus compressor station, a subsea pipeline for the transport of CO_2 and several geological CO_2 storage locations on the Dutch continental shelf, including the gas fields permitted to Total Energies and Shell (as identified in the environmental impact assessment) but also other fields operated by, e.g. Neptune. Apart from these differences, Aramis also foresees several means to cooperate with Porthos.

Aramis intends to be an open access project. For this purpose, it will need a receiving terminal for liquid CO_2 that is brought in by ship or otherwise, temporarily stored and prepared for transport offshore via the Aramis subsea pipeline. This receiving terminal will be constructed and operated by CO_2next—a joint venture of the gas infrastructure company Gasunie and the (oil) storage company Vopak—on the site of Vopak's Petroleum Terminal at the Maasvlakte. CO_2 next operates this terminal on the basis of the 'open access' principle and has therefore made use of an 'open season' to assess the interest of third parties.[83] If CO_2 is

[80] Ministry of Economic Affairs and Climate, *Voorkeursalternatief (VKA) van het Aramis initiatief (Transport van CO_2 (kooldioxide)*, see www.rvo.nl/sites/default/files/2024-03/Voorkeursalternatief-Aramis.pdf.

[81] www.aramis-ccs.nl. See also Letter to Parliament of 3 October 2023 on the CCS market, TK 2023–2024, 32,813, no. 1298, p. 5.

[82] www.aramis-ccs.nl.

[83] The 'open season' was organised between December 2021 and January 2022. The storage capacity is 48,000m3 and the annual throughput capacity is about 5.4 Mt. See co2next.nl/status/www.aramis.nl.

supplied by pipeline, use can be made of the Porthos compressor station, which also will be connected to the Aramis pipeline.[84]

The next part of the CCS chain is located offshore. The offshore infrastructure will involve a pipeline of approximately 200 km connecting the receiving terminal at the Maasvlakte with an offshore distribution platform that can serve as a hub for several offshore fields where CO_2 can be stored. This pipeline will be built and operated by Aramis and capable to transport 22 Megatons of CO_2 per year. As regards CO_2 storage, the initial focus was on some of the fields operated by TotalEnergies and Shell, but since then also Neptune (via Neptune L10 Carbon Storage) and Wintershall Dea (via Winterhall Carbon Management Solutions) offer geological storage. TotalEnergies and Shell have committed to offer 5Mt per year of storage capacity in the initial phase,[85] but Neptune and Wintershall and possibly other parties will be able to offer storage capacity afterwards. Early 2024, the necessary CO_2 storage permits had not yet been applied for and consequently an opinion by the EU Commission is still pending.

In principle, third parties (e.g. CO_2 emitters) can apply for a SDE++ subsidy to cover part of the extra costs. They also need to enter into an agreement with a storage operator to store certain volumes of CO_2. Third parties could thus be faced with two agreements: a transport agreement with Aramis and an agreement with a storage operator. This would require some close cooperation between all parties involved and could lead to coordination problems with regard to the commissioning of the pipeline and the storage locations or the amount of available capacity. What happens if the pipeline is ready for use, but the storage locations are not ready yet and vice versa?[86] Aramis seems to foresee such problem by opting for a situation where the operator of the CO_2 storage facility also will take care of the transport of CO_2 and thus offers a combined transport and storage service to all users. In other words, Aramis will then provide the necessary CO_2 transport and compression services to the operator of the reservoir, i.e. holder of the CO_2 storage permit.

[84] See www.rvo.nl/onderwerpen/bureau-energieprojecten/lopende-projecten/aramis.

[85] This will at least cover a period of 15 years.

[86] This situation would be similar to the connection of several offshore wind parks to one offshore converter station and subsea cable in Germany and the Netherlands.

A major point of concern has involved Aramis' access and tariff structure. As required by Directive 2000/31/EC, Aramis needs to offer third parties (e.g. emitters) access to the system. Refusal is limited to conflicting technical specifications and a lack of capacity. Aramis has addressed the first ground of refusal by publishing an overview of the CO_2 specifications that users can take into account when applying for access.[87] With regard to the latter, there is currently sufficient capacity available. Nevertheless, agreements need to be made about the commercial terms and the tariffs. However, third parties have voiced a concern that some of the participating parties in the Aramis project could behave as monopolists. Parties like TotalEnergies and Shell will have their own (climate) interests and will thus be in need of storage capacity. Their interests will thus differ from those of the state-owned companies Gasunie and EBN.[88] If parties are of the opinion that Aramis (or those directly involved in Aramis) would abuse its dominant position, they need to turn to the relevant provisions in competition law. However, competition law always applies *ex post* and can be time consuming. In order to facilitate the development of a CCS market, one of the options would be to introduce a regime of regulated access instead of the current regime based on negotiations until the CCS market has become a mature market.[89] This suggested approach is the opposite of the idea included in Directive 2009/31/EC that negotiated access is more favourable in developing markets as it limits administrative costs, which are often related to a regime of tariff regulation. So far, the existing regime has not (yet) been amended.

[87] www.aramis-ccs.com/nl/nieuws/co2-specifications-for-aramis-transport-infrastructure.

[88] B. Wientjes, *Analyse van de mogelijke governance modellen voor het CCS-project Aramis*, 14 juli 2022, p. 4. This report is published at www.rijksoverheid.nl/documenten/rapporten/2022/06/14/bernard-wientjes-analyse-van-de-mogelijke-governancemodellen-voor-het-ccs-project-aramis-rapport-14-6-2022.

[89] See report by prof. M. Mulder, *Verkenning van de marktordening voor Carbon Capture and Storage (CCS)—Een onderzoek op verzoek van het Ministerie van Economische Zaken en Klimaat*, March 2024. This report can be found at https://open.overheid.nl/documenten/b4aba868-154c-427b-a81a-e733e20456c3/file.

5 Conclusion

The Netherlands opted about 25 years ago for carbon use and storage. This chapter illustrates that ambitions are not enough to develop permanent geological storage of CO2. Public opposition and market changes have resulted into several failed projects. However, a new era is now emerging. The EU and Dutch ambition to become carbon neutral in 2050 has revitalised the need to apply CCS. On EU level, changes have been made to the EU ETS as a result of which the price of emission allowances is increasing and CCS is also included in the Taxonomy Regulation and the draft regulation for a net-zero industry (Net-Zero Industry Act). It may be that the Commission also will propose some changes to the legal framework following the publication of the Revised Guidance to the CCS Directive in July 2023.[90] Irrespective of any changes on EU level, however, Member States are entitled to amend their national laws on CCS as long as the minimum norms are adhered to.

Although the Dutch legislator has indicated that there is no specific reason to suggest any changes, unless the market is not developing as expected. At the moment, it seems that the main concerns involve the transport and access regime. The current directive contains a lack of guidance regarding infrastructure development as well as the third-party access regime. More guidance seems to be required with regard to tariff design and the extent to which transport and storage operators may abuse their position and the involvement of competition authorities. More coordination also seems desirable with regard to the alignment of various transport modalities, such as transport by ship and via pipelines. Finally, attention should be paid to the difference between the legal framework applying to piped transport for permanent storage on the one hand, and for reuse of CO_2 on the other hand. There are no obvious reasons for distinguishing between the governance of both systems, i.e. why one system is subject to an environmental impact assessment and an emissions permit and the other is not. Connecting both systems could benefit all parties involved as it would facilitate greenhouse growers in receiving sufficient volumes of CO_2 and emitters to have sufficient transport routes to the available storage sites.

[90] See https://climate.ec.europa.eu/news-your-voice/news/european-commission-publishes-revised-guidance-documents-ccs-directive-2024-07-23_en#:~:text=The%20update%20of%20the%20non-binding%20CCS%20Directive%20GDs%20in%202024.

Although important steps have been made to develop CCS, practical experience still has to be gained and these may shape future projects and most likely also the applicable legal regime.

CHAPTER 17

CCUS in Norway

Rune Omdahl and Rune Tjomsås Andersen

1 Introduction

On the first day of 2007, the then prime minister of Norway, Jens Stoltenberg, proudly announced that Norway should take a pioneering role on carbon emission control through capturing the CO_2 from the gas-fired power plant at Mongstad. This was the infamous «moonlanding» where a test plant for capturing approximately 100 000 tonnes of CO_2 annually should be built. The plan was that Norway should take the number one spot in the world on handling of CO_2 emission. The

The authors would like to greatly acknowledge the valuable input made by our esteemed colleague Anders Rødland, without this chapter would not have seen the crack of dawn.

R. Omdahl (✉) · R. T. Andersen (✉)
Oslo, Norway
e-mail: rom@adeb.no

R. T. Andersen
e-mail: rta@adeb.no

E. G. Pereira et al. (eds.), *Carbon Capture Utilization and Storage*, Sustainable Development Goals Series,
https://doi.org/10.1007/978-3-031-81272-9_17

«moonlanding» at Mongstad did not go as planned[1] but Norway has continued the work to build a sustainable chain for handling of CO_2 emission. Norway is now committed through the Langskip project[2] to develop and construct a full-scale carbon capture, transport and storage[3] (CCS)[4] value chain.

The current regulatory regime in Norway only allows for CCS offshore Norway on the Norwegian Continental Shelf ("**NCS**"). Onshore storage is not yet allowed. Carbon "storage" is not a novel issue on the NCS. CO_2 has since 1996 been reinjected into underground offshore reservoirs as part of the production of gas from the NCS. Approximately one million tonnes annually of CO_2 are injected into the Utsira Formation from the gas production at the Sleipner Vest field and the Utgard field. Since 2008 approximately 700 000 tonnes of CO_2 have annually been separated from the LNG production at the Melkøya LNG plant and reinjected into the Snøhvit formations on the NCS.[5] The handling of CO_2 at Sleipner and Snøhvit is regulated by the Norwegian Petroleum Activities Act 1996 ("**PAA**") and associated regulation (the "**Petroleum Regulation**").[6]

CCS activities on a commercial basis (stand-alone of the petroleum activities) are regulated by the act relating to Scientific Research and Exploration for and Utilisation of Subsea Natural Resources other than Petroleum Resources[7] (the "**Subsea Natural Resources Act**"), and more specifically the Regulation relating to the Utilisation of Subsea Reservoirs on the Continental Shelf for Storage of CO_2 and Transport of CO_2 on the Continental Shelf (the "**CCS Regulation**").[8]

[1] However, the project paved way for the TCM—Technology Centre Mongstad. TCM is considered the leading centre in the world when it comes to testing of CCS technology. TCM is owned by the Norwegian State, Equinor, Shell and Total Energies.

[2] See White Paper no. 33 (2019–2020) on the project (Meld. St. 33 (2019–2020)).

[3] The transportation and storage part of the project is named Northern Lights and is done in cooperation between Equinor, Shell and Total.

[4] "CTS" (Capture, transport and storage) would probably be a better acronym for these activities, but CCS has been used in the business for a long time and we have elected to keep that.

[5] https://www.norskpetroleum.no/miljo-og-teknologi/fangst-transport-og-lagring-av-co2/.

[6] See in particular Chapter 4a of the Petroleum Regulation.

[7] Act No 12 of 21 June 1963, (Nw: *«Lov om andre undersjøiske naturforekomster»*).

[8] Regulation 2014 No 1517. The CCS Regulation is based on the EU CCS Directive.

The legal framework for petroleum activities has been developed in Norway since the late 1960s and establishes what is commonly known as a concession system for the regulation of offshore petroleum activities. The system for licencing in accordance with the CCS Regulation is built on this framework, and there are therefore many provisions similar to those found in the petroleum system.

The CCS Regulation stipulates that the Norwegian State has the proprietary right to subsea reservoirs for exploitation for storage of CO_2 and an exclusive right to management of such reservoirs.[9] Only the State may thus carry out carbon storage and transport activities, unless a private entity is awarded licenses, approvals and consents pursuant to the CCS Regulation.

In addition to the CCS Regulation, the Ministry of Labour and Social Inclusion (the "**MLSI**") has enacted a separate CO_2 Safety Regulation[10] on the basis of the Subsea Natural Resources Act and the Working Environment Act,[11] inter alia establishing the Norwegian Petroleum Safety Authority (the "**PSA**") as the administrative body responsible for ensuring compliance with the applicable safety regulation.

Additionally, the Ministry of Climate and Environment (the "**MCE**") has amended the Pollution Control Regulation[12] on the basis of the Pollution Act[13] and the EU CCS Directive, to include a Chapter 35 on storage of CO_2 in geological formations. Chapter 35 inter alia stipulates a separate requirement for an injection and storage permit and authorises the Norwegian Environment Agency ("**NEA**") to consider applications for permits. Chapter 35 also contains regulation on requirements for financial security. The CO_2 Safety Regulation and the Pollution Control Regulation will not be reviewed separately in this chapter, but will be referred to in passing when otherwise relevant.

There has been an increasing commercial interest for CCS activities on the NCS. At the time of writing, nine concessions have been granted by the MoE. Notwithstanding this, the CCS industry is still in its inception phase, and much is to be done, and much is to be learned.

[9] See sec. 1-2 of the CCS Regulation.

[10] Regulation 2020 No 186.

[11] Act of 17 June 2005 No 62, (Nw: *«arbeidsmiljøloven»*).

[12] Regulation 2004 No 931.

[13] Act of 13 March 1981 No 6, (Nw: *"forurensningsloven"*).

The topic of this chapter will be the regulation of the CCS activities on the NCS on a commercial basis and in accordance with the CCS Regulation. This chapter is a short introduction to the rules and regulations for CCS in Norway and does not pretend to be exhaustive.

A party interested in applying for an exploration license (and an exploitation license) related to utilisation of subsurface reservoirs for storage of CO_2 will have to meet certain criteria[14] in order to be eligible for the possible granting of a license. If not previously qualified, a party interested in a license (whether exploration or exploitation, by application or acquisition) must qualify either as a licensee or as an operator. The main criteria for qualification are financial strength, technical (and geological) competence and reliability.

All applicants do not have to possess the same qualities. Complementary qualities are seen as a strength for an application. However, the operator, who will be in charge of the day-to-day management and operation of the partnership, is expected to be experienced in the drilling of subsea wells, geological evaluations and reservoir technical matters (i.e. qualities that a typical oil company in Norway would have). We will touch upon the qualifying criteria in item 2 below.

The CCS concession system provides the framework for the regulation of the CCS business but delegates a number of decisions on concrete matters to the Norwegian authorities and the Ministry of Energy (the "**MoE**") in particular. A brief outline of the CCS concession system will be provided in item 3 below.

The authorities will enjoy considerable discretionary powers when making their decisions. So far relatively few decisions related to the CCS Regulation and pertaining documents have been taken so there is not much governmental practice to draw upon when evaluating how the authorities will use their discretionary powers. However, considering the background for the regulation, the similarities with the petroleum concession system and that many of the same administrative bodies are involved, it cannot be ruled out that previous decisions under the Petroleum Regulation may provide indications as to how the authorities will use their powers pursuant to the CCS Regulation.

[14] See third paragraph of sec. 3-1 and third paragraph of sec. 4-1 of the CCS Regulation.

If eligible to be a participant, an interested party may apply for a licence and subsequently be offered a participating share in such a licence. A brief outline of the application process is set out in item 4 below.

If accepting the offer from the MoE, the prospective participant will have to formally accept the offer and accede to the concession documents, mainly the work program and the model Agreement for Storage Activities (the "**Agreement**"). The Agreement consists of three parts: the Special Conditions ("**SC**"), the Joint Operating Agreement ("**JOA**") and the Accounting Agreement ("**AA**"). The main part of the Agreement is the JOA. It is this document that contains the provisions that mainly regulate the relationship between the parties.

The licence will (usually) be offered to a group of companies. The participants in the licence will form a general partnership in accordance with the Act no 114 of 21 June 1985 on general and limited partnerships (the "**Companies Act**").[15] The Companies Act and the JOA are not fully compatible so the corporate agreement between the parties will usually contain deviations from the provisions in the Companies Act in order to align the law with the concession documents. The relationship between the JOA and the Companies Act will be addressed in item 5 below.

Depending on the provisions of the awarded licence and the agreement between the parties the partnership will as licensee have the right to store CO_2, and the expectation at the outset is that the operator as part of its authority pursuant to the JOA may represent the partnership in marketing and selling storage capacity to customers. Some operational issues will be addressed in item 6 below.

An eligible part may in addition to or as an alternative to applying for a license acquire on a commercial basis a participating share from any licence holder on the NCS; we address this in short in item 7.

Once the storage reservoir has been fully utilised, the licensees can apply to the authorities for the consent to shut down operations. The licensees' will in the period after the shutdown is approved remain responsible for monitoring the reservoir, etc., and liable for leakages, but may after a minimum of (at the outset) 20 years apply for the transfer of responsibilities to the Norwegian State. Such transfer may be accepted on certain conditions, one of which is a financial contribution

[15] See second paragraph of sec. 3-1 of the CCS Regulation.

by the operator. The final chapter relates to the rules for withdrawal and relinquishment.

2 Eligibility—Qualifying as a Licensee

Exploration licences (and exploitation licenses) shall be awarded on "*objective, published and non-discriminatory criteria*"[16] among eligible interested parties. The CCS Regulation stipulates certain qualities that the potential entrants should possess.

The main criteria are financial strength, the technical (and geological) competence[17] and reliability deemed necessary to carry out the relevant activities under the regulations, and that the quality requirements on the licensee and its organisation have been met. All participants must satisfy the conditions, but this does not mean that all companies must have the same qualities. The intention of the Ministry of Energy is to ensure that the group has complementary qualities that will ensure the proper operation of the activities to be undertaken. However, at least one of the licensees, usually the one that also will be appointed operator, must have geological, reservoir technical, and drilling competence, i.e. qualities that a typical oil company in Norway would have. In general, the qualification requirements are very similar to the requirements that have been applied to companies qualifying as licensees and operators in the oil and gas business on the NCS.[18]

The licensee shall have an organisation that enables it to make informed decisions on the activities to be performed under the licence.[19] Consequently, any participant must have in-house capabilities reflecting the necessary core competencies. This does not necessarily mean that the participant immediately must build a separate organisation or have to specifically hire employees in Norway. Depending on the role in the partnership, the participant can use staff from group companies or contracted

[16] See CCS Regulation sec. 3-1 and 4-1, respectively, and sec. 11-6.

[17] Geological competence is only listed in relation to the exploitation license. Possibly reflecting that type of competency is more critical in this phase that in the developing and operating phase.

[18] Presumably, the Ministry of Energy will lean on the experience obtained under the Petroleum Regulation, but considering that the main activities will be CO_2 reservoir exploration, transportation, injection, and storage it is not given.

[19] See sec. 11-2 of the CCS Regulation.

personnel (independent consultants) for a considerable period of time.[20] However, as the activities grow, we assume that a core organisation must be built in Norway and be employed by the Norwegian entity.[21] Companies wishing to qualify as an operator are subject to more extensive requirements than is the case for other licensees.

It must also be considered that the partnership is organised as a company, a General Partnership, and this will be a separate entity registered in the Norwegian Companies Register.[22] This company may have a separate organisation, see in item 6 below.

The applicant must document its ability to service financial obligations expected to be incurred in the CCS business. This includes both procurement obligations in the drilling, development and operation of the storage (and transportation) facilities, and latent liabilities that may be incurred as a consequence of the activities, such as pollution. The authorities have shown flexibility in its evaluation of how the financial ability is documented.

Notwithstanding the applicant's financial strength, it must be expected that the authorities will require that security for the potential liabilities and obligations under the applicable regulation[23] is posted. The ability to provide such security shall be documented in the application, and the security provided shall at the latest be effective as of commencement of injection in the reservoir. The form of such security can be discussed, and the MoE has, in so far as it relates to the petroleum industry, required parent company guarantees, and it seems that they uphold the same principle for the CCS industry.

It must be noted that the MoE is not the sole authority under the CCS Regulation. As for security there is also a requirement for security under the Pollution Control Regulation. The relation between the two

[20] In the petroleum industry, it is required that a core organisation is in place before granting of a license with a minimum number of employed personnel of about 9 people, with technical and HSE experience. Currently, such requirement has not been put upon the participants in the CCS licenses.

[21] Inter alia to comply with the required emergency preparedness, see sec. 10-2 of the CCS Regulation.

[22] Norwegian Central Coordinating Register for Legal Entities (*Norw.* Enhetsregisteret) and Norwegian Register of Business Enterprises (*Norw.* Foretaksregisteret).

[23] See inter alia the CCS Regulation sec. 5-9, 5-10 and 1-9, as well as the Pollution Control Regulation sec. 35-15.

security arrangements is not yet clarified. It remains to be seen how the discussion between the ministries turns out, not only concerning form of security required, but also the amount for which security shall be required and for how long the security shall be in place.

By law, the licensees are obliged to have sufficient insurance cover, as a minimum cover for liabilities connected to damage on facilities, pollution and other third-party responsibilities, removal of wreck and clean-ups due to accidents, as well as insurance covering the employees of the licensee.[24]

3 The CCS Concession System

The CCS concession system is structured around a stepwise approach following the expected lifespan of a CCS project on the NCS.

Interested parties will have to search for and explore relevant subsea reservoirs suitable for CO_2 storage. This will be done on the basis of a reconnaissance license.[25] When potential suitable reservoirs are identified, more detailed research must be performed. Such work is done in accordance with an exploration licence.[26] Then, if the parties wish to develop and operate a reservoir site, they will apply for an exploitation licence[27] and approve a Plan for Development and Operation (“**PDO**”). When the development phase is completed, the parties will enter into the commercial phase of the project where the various services will be provided to customers. This phase will end when the reservoir is full. At some time, the reservoir site may be abandoned.

The MoE may grant a reconnaissance license to parties that are interested in surveying the NCS for subsurface reservoirs that may be suitable as future sites for storage of CO_2. The reconnaissance license grants a non-exclusive right to conduct such reconnaissance activities.[28] The license is usually granted by the MoE for a period of three years.[29] This license has a rather limited scope, due to which the requirements for being

[24] See the CCS Regulation sec. 11-14.

[25] “undersøkelsestillatelse” in Norwegian.

[26] “letelisens” in Norwegian.

[27] See sec. 2-3 of the CCS Regulation.

[28] See second paragraph of sec. 2-1 and sec. 2-4 of the CCS Regulation sec. 2-1 second paragraph, sec. 2-4.

[29] See third paragraph of sec. 2-1 of the CCS Regulation.

awarded a reconnaissance license are less strenuous than with the exploration and exploitation licenses. Granting of a reconnaissance license is not a prerequisite for an exploration license.

The next step is the exploration license. At this point, there is a specific area of the NCS that has been preidentified as containing a reservoir that has the potential for storage of CO_2. This license is an exclusive license to explore a particular subsea reservoir site within a specific delineated area in order to further assess the potential reservoir for CO_2 storage purposes.[30]

The exploration license is awarded by the Norwegian King in Council (the Government) and may (usually) be awarded two or more licensees incorporated pursuant to Norwegian law and registered in the Norwegian Company Register.[31] The approached parties will be required to enter into a Partnership Agreement and a JOA with each other and an operator will be appointed.[32]

An exploration license will normally be awarded on the condition that a work program stipulated in the license document will be completed.[33] The work program, for the time being, is structured around a milestone approach, whereby certain activities shall be carried out within defined deadlines. If these deadlines are not met, without any extension being granted by the MoE, the license will be surrendered to the authorities.

Anyone who meets the qualifications criteria may apply for an exploitation license[34] for the exclusive use of a specific subsea reservoir for injection and storage of CO_2.[35] The authority to award an exploitation license rests, as with the exploration license, with the Norwegian King in Council (the Government). In general, the same conditions as for an exploration license apply: it may be awarded to one or more participants, if the license is awarded to a group of licensees, these licensees shall be joint and several liable,[36] the licensees being offered a license will

[30] See sec. 3-1 of the CCS Regulation.

[31] See note 23.

[32] See sec. 2-1 of the CCS Regulation.

[33] See sec. 3-3 of the CCS Regulation.

[34] "Utnyttelseslisens" in Norwegian.

[35] The area initially defined in the license may subsequently be changed if the CO_2 in the reservoir migrates into new areas.

[36] See the CCS Regulation sec. 4-1 second paragraph.

be required to enter into a Partnership Agreement and a JOA with each other, etc.

Although anyone may apply for the license, the CCS Regulation specifically states that applicant(s) who already are licensees under an exploration license, who applies for an exploitation license within the exploration license term, and who have completed the work program and fulfilled the other conditions of the exploration license shall have a preferential right to an exploitation license that covers the relevant reservoir.[37] Based on the experience from the petroleum regulation regime, it is expected that the exploitation license will be awarded to the licensees of the exploration license. At the time of writing, only one exploitation license has been awarded.

We would also expect the authorities to be willing to announce the award of the exploitation license before the licensees under an exploration license finally commit to a PDO and the investments required according to the plan. We believe the authorities in this respect recognise the companies' and investors' need for a stable, legal framework within which investments may be made and are willing to encourage such investments by awarding an exploitation license as a basis for such investment commitments.

According to the CCS Regulation, the duration of the exploitation license will be decided by the authorities upon its award.[38] The duration of the one exploitation license awarded so far was set to 25 years after the completion of a work program. Based on this experience and experience from the petroleum regulation regime, we would expect that the duration is set for a time period that is long enough for the licensees to complete the development of the reservoir as storage facilities and to make full use of the facilities once they have become operational. The initial duration set in the license may according to the CCS Regulation be extended by application from the licensees in special circumstances.

Although the exploitation license and approved PDO grants the licensees the overall right to carry out injection and storage activities, separate consents from the MoE and the MLS are required to commence injection of CO_2 for storage.[39] This consent shall be "compatible" with

[37] See the CCS Regulation sec. 4-1 fourth paragraph.

[38] See the CCS Regulation sec. 4-4.

[39] See the CCS Regulation sec. 5-2 first paragraph.

the storage permit to be awarded by the NEA pursuant to Chapter 35 of the Pollution Control Regulation[40] (the "**Storage Permit**"). It is expected, considering the rights already granted the partnership, that the primary objective of these latter consents and permits is to ensure compliance with regulations and plans that have already been approved.

The MoE may according to the CCS Regulation award a separate license for installation and operation of facilities, typically for transportation of CO_2 in pipelines (a "**Transportation License**"), when such facilities are not encompassed by an approved PDO. An application for such license shall contain a plan for installation and operation of the facilities ("**PIO**"), which would be approved upon the award of the license. Supply and utility vessels and vessels for the transportation of CO_2 in bulk are not covered by the term "facilities".

A Transportation License may, subject to the discretion of the MoE, be awarded to the same licensee or group of licensees that have been awarded an exploitation license, or to a completely or partly different group of licensees. The latter would typically be relevant in the event a transportation pipeline from a terminal on land transports CO_2 to a cluster of storage reservoirs in the same vicinity offshore.

4 Applying for a License

The MoE currently applies an "open door" policy enabling interested parties to apply for a license when and if interested in carrying out CCS activities. Apart from each licensee fulfilling the minimum requirements for qualification reviewed in 2 above, the application will be evaluated upon its own merits. The NPD has published guidelines for "applications for licenses" and "for the application letter and company information".

The main content of the application seems to be "geological and technological" evaluations of the area and potential reservoir concerned, but also information on the development solution envisaged and an outline of the surveillance of the reservoir shall be part of the application. The applicant(s) shall also propose a work program, and the applicant(s) willingness to commit to various activities within the relevant area is traditionally regarded as an important factor in the authorities' evaluation of the application and in any competition with other applicants.

[40] See the Pollution Control Regulation sec. 35-4 first paragraph, and the CCS Regulation sec. 5-2 second paragraph.

With respect to the expected duration of the application procedure, the authorities do not make pre-commitments, and the period may vary depending on several factors, such as the area for which the application has been made, the relevant reservoir in question, the identity of the applicants and whether or not they have applied previously. Some guidance may, however, be found in the time spent on previous applications. Looking at the time between announcement and awards in attachment 1, we see that the authorities have spent approximately 3 to 6 months on making these decisions. This is also comparable to the typical time spent on the similar awards of petroleum production licenses.

Upon receipt of an application, the Ministry of Energy will normally publicly announce the possibility for other interested parties in applying for licenses in the same area within a deadline, thus ensuring compliance with the conditions of the EU CCS Directive.

5 The General Partnership

One or more legal entities can apply for an exploration or exploitation licenses on the NCS.[41] If two or more applicants are offered a licence and accepts it, they will be considered as a general partnership[42]; ANS for short in Norwegian (an abbreviation for Ansvarlig Selskap), upon the award and acceptance of the license. To our knowledge, it is a standard requirement from MoE in all JOAs that the CCS business shall be conducted through a general partnership, cf. also the CCS Regulation § 3-1 (2).

The general partnership will be subject to the Companies Act.[43] The Companies Act requires that the members of the company enter into a Partnership Agreement.[44] The Partnership Agreement is the main constitutional document of the general partnership and can to a certain extent deviate from the solutions set out in the Companies Act.

The Partnership Agreement shall, as a minimum, contain information on the partnership's name, name and address of all the participants, the

[41] See second paragraph of sec. 3-1 and 4-1 of the CCS Regulation respectively.

[42] See paragraph 1b) of sec. 1-2 of the Companies Act.

[43] See sec. 1-1 of the Companies Act.

[44] Se sec. 2-3 of the Companies Act.

partnership's purpose, the municipality of the partnership's headquarters and whether the participants shall make capital contributions.[45]

There is no requirement for capital contributions when starting up a general partnership, but whether or not contributions are made at incorporation must be reflected in the Partnership Agreement. A contribution may be made by other means than capital, but if so, the Partnership Agreement must also reflect the value of the invested, non-monetary assets.

The participants in the partnership will have joint and several liability for the obligations and liabilities incurred by the partnership.[46] However, the general partnership is a separate legal entity. Once the partnership has been incorporated, each participant becomes, pursuant to the JOA, an owner of an ideal share of the partnership's capital assets, such share equalling the participant's participating interest in the partnership. The participating interests of the participants will be determined in the exploration / exploitation license. This entity, when incorporated, will be the holder of the license.[47] Thus, it will also be obligated under the concession documents to follow the provision inter alia set out in the Agreement. The JOA is not fully aligned with the regulation in the Companies Act[48] and contains several provisions that must be harmonised and thus addressed in the Partnership Agreement. Usually, the Partnership Agreement contains a general clause stipulating that the provisions of the JOA shall for all purposes be considered implemented in the Partnership Agreement to the extent the law permits.

The partnership meeting is the supreme governing body of the general partnership. Pursuant to the JOA, however, the partnership shall upon incorporation establish a management committee (the "MC") consisting of a representative of each participant, and, thus, often a Partnership

[45] Sec. 2-3 (2) of the Companies Act.

[46] See sec. 2-4 of the Companies Act and second paragraph of sec. 3-1 and 4-1 of the CCS Regulation.

[47] This symbiosis between the Companies Act and the concession documents is that the term "licensee" will refer to both the partnership as such and the individual participants in the partnership.

[48] The JOA is modelled after the agreement used in the Petroleum Industry. However, joint ventures established in accordance with the Petroleum Act and the licensing documents are specifically exempted by law from the regulations in the Companies Act.

Agreement contains a clause stating that MC shall be considered as the supreme body of the partnership under the Companies Act.

The default voting rule of the Companies Act is unanimity between the representatives of the participants, but this is usually derogated by the voting rules stipulated in the Partnership Agreement. The structure of these voting rules follows the structure applied in a petroleum joint venture, where a decision as a main rule shall be considered adopted if at least x members representing at least y% of the participating interests have voted in favour of a proposal.

Exceptions to this "majority" voting rule are set out in various provisions in the Partnership Agreement and the JOA, most importantly perhaps in connection with the decision to submit a PDO to the MoE for approval. The decision itself is subject to the majority vote, but each participant shall within a three-month period after the decision notify the MoE on an individual basis as to whether or not it adheres to the plan. Unless all participants adhere to the PDO within the deadline, the approval process is put on hold by MoE.

The JOA does not specifically regulate the handling and allocation of income received through the operation of the partnership.

The differences in regulation in the Companies Act and the JOA entails that the provisions in the Companies Act on inter alia organisation,[49] distribution of income among the participants,[50] transfer of participating share,[51] and withdrawal and dissolution[52] are replaced with inter alia the operator as the representative of the partnership and the Management Committee as the supreme governing body.

A partnership is considered a separate legal entity but is not an entity subject to tax.[53] The Norwegian partnerships are taxed at the level of the participants, but each participant's share of the taxable income is computed at the level of the partnership as if the latter was a taxable entity. The net result is then allocated to each participant pro rata to their participating interest and taxed together with such participant's other income. If the partnership's net result is a loss, the participants' share of the loss will

[49] See sec. 2-20 and 2-21 of the Companies Act.

[50] Se sec. 2-25 and 2-26 of the Companies Act.

[51] Se sec. 2-28 of the Companies Act.

[52] See sec. 2-32 through 2-37 of the Companies Act.

[53] See second paragraph of sec. 2-2 of the Tax Act.

reduce other income liable to tax at the hands of the individual participant the year in question.[54]

A partnership share will be comprised as a qualifying investment under the Norwegian participation exemption method for the participants (Nw: *"fritaksmetoden"*). Note however that 3% of any distribution from the general partnership of dividends received by the partnership as well as capital gains from qualifying shares owned and realised by the partnership shall be deemed taxable income for the partners.[55]

6 Operational Issues

Once a company has entered the CCS business as either a licensee or an operator, certain questions arise with respect to how the business pursuant to the relevant exploration or exploitation license shall be conducted.

The day-to-day operations are conducted by the operator.[56] However, it is expected that the licensees participate in the activities, and the licensees and the operator are imposed with a fiduciary duty to "see-to-it".[57] This duty goes in two directions: internally and externally:

- Internally, the licensee is obligated to ensure that its own organisation complies with applicable laws, regulations and individual administrative decisions laid down pursuant to the CCS Regulation;
- Externally, the licensee is obligated to ensure that those contracted to perform work, either personally, through employees or through contractors or subcontractors, comply with applicable requirements.

Furthermore, and perhaps most important, the «see-to-it» duty entails a specific obligation to have in place quality assurance and management systems that enables the licensees to inter alia follow up and control that the operator's performance, contract execution, quality assurance, etc., are carried out in compliance with the applicable regulation.

The operator's fiduciary duty is more extensive than the licensees' obligation. The operator must inter alia ensure, prior to entering into

[54] See sec. 10-41 of the Tax Act.

[55] See second paragraph of sec. 10-41 of the Tax Act.

[56] See Art. 3-1 of the JOA.

[57] See the CCS Regulation sec. 11-8.

contracts, that its contractual partners are competent and qualified and able to perform the work. During the continuation of the work, the operator must monitor its contractual partners and check that the facilities and equipment being used, and the work being done, meet acceptable standards and complies with applicable legislation, especially the requirements specified in the HSE regulations.

The licensees must implement systematic measures and systems (management systems) that are adequate to fulfil the fiduciary duty and to, from time to time, audit the operator's management system ensuring the correct discharge of the operator's fiduciary duty.

The operator is responsible pursuant to the JOA for the day-to-day management of the partnership and shall represent and act on behalf of the participants in the affairs of the partnership. In this respect, issues arise as to how the management of storage rights shall be carried out, the interface with transportation and terminal operations and the monitoring of the storage location and subsequent shut down and transfer of responsibility to the authorities. The main asset of the partnership in the storage phase of the operations (after the exploration and development phases) is the right pursuant to the exploitation license to inject and store CO_2 in the storage reservoir. This right usually belongs to the partnership as such, and this right is a prominent part of the partnership's joint capital assets. The operator's authority under the JOA includes a right to enter into agreements on behalf of the partnership and to manage such agreements.[58] Consequently, if not otherwise agreed, it is the operator that disposes over the partnership's joint capital assets on a day-to-day basis and may, on the authority of entering into agreements on behalf of the partnership, at the outset probably also be authorised to sell capacity in the reservoir. As part of the sales activities, one would expect the operator also to carry out marketing.

However, a joint marketing and disposal of storage capacity raises questions in relation to the competition regulation, and whether a cooperation in that respect by companies in a partnership is legal. We have not carried out a competition law evaluation in this respect but note that some industry representatives have argued that the incorporation of a general partnership such as the "*ANS*" is beneficial in relation to

[58] See Art. 3.2 of the JOA.

the competition regulation. The argument is that the partnership structure allows for greater cooperation, also in relation to the joint marketing and sale of storage capacity, whereas the participants in a joint venture exempted from the Partnership Act to a greater extent would be required to market and sell the capacity individually. We do, however, expect that the licensees will have a certain amount of leeway in this respect, and that alternative arrangements may be agreed in the Partnership Agreement or otherwise.

The MoE has so far not disclosed any proposal for a model joint operating agreement applicable to the exploitation phase of a CCS project. We would expect such model agreement to be similar to the JOA, but the marketing and sale of storage capacity is a subject matter that only becomes relevant in the exploitation phase. The core nature of the subject matter makes us believe that this is an area where we may well see further development in the regulation of the matter in the JOA and perhaps elsewhere. At the least, we would expect there to be some freedom of choice within the various partnerships to decide whether the storage capacity should be marketed and sold jointly, or whether this should be carried out by the individual participants, provided of course that the preferred solution is in compliance with applicable competition regulation.

Third parties that do not participate in the injection and storage partnership have certain third-party access rights to the storage facilities pursuant to the CCS Regulation.[59] Considering that the activities according to the CCS Regulation principally are of a commercial nature and that sales to third persons probably are the main driver of the project, the issue of third-party rights raises questions that the format of this chapter does not allow to elaborate on, so this issue will not by further analysed herein.

In the event the storage reservoir is in a geographical area on its own, we would expect the transportation and terminal solution, provided such solution is part of the overall technical concept, to be part of a PDO.

If the reservoir on the other hand is in a cluster of reservoirs being developed at the same time, and all require pipeline transportation, we would expect that the transportation part of the project(s) may well be developed as a separate project under a PIO. There could also be other scenarios where separate developments and PIOs are appropriate.

[59] See sec. 5-12 of the CCS Regulation.

In any case, the pipeline transportation and terminal projects under the CCS Regulation are formally organised as independent partnerships and thus legal entities separate from the storage partnership. In the first scenario, the partnerships will probably be owned by the same participants. But in the second scenario, the ownership of the transportation partnership may be different to the ownership of the various storage partnerships. In the latter case especially, transportation agreements would be required between the partnerships before operation starts.

A particular issue with the onshore terminals which is currently unresolved is that the CCS Regulation only applies offshore the Norwegian mainland. The MoE therefore does not have the formal authority required to allow for onshore activity in connection with a CCS project. These are one of the issues that the MoE plans to correct in upcoming proposals for amendments to the regulation.

An important aspect of the partnership's responsibilities during the operational phase of the storage facilities is the duty pursuant to both the CCS Regulation and the Pollution Control Regulation to monitor the injection facilities and the storage location. This monitoring shall include the possible dispersion of CO_2 from the initial storage reservoir. The monitoring shall be carried out in accordance with a monitoring plan prepared by the operator and approved by the MoE and the NEA. The plan shall be revised as required and in any event no later than every fifth year. Revisions are also subject to the approvals by the authorities.

A storage facility shall be shut down when the conditions stipulated in the storage permit have been satisfied and the Government gives its consent to such shutdown.[60] The operator shall, however, remain responsible for the monitoring, reporting and implementation of corrective measures after the consent to a shutdown is given and until the Norwegian State takes over the responsibilities for the storage facilities. These monitoring activities shall be carried out in accordance with a post-operation plan which is also subject to the approval by the authorities. According to the Pollution Control Regulation, the operator shall also remain liable for the transfer of quotas for leakage of CO_2 from the storage facilities, if otherwise required pursuant to the Norwegian Greenhouse Gas Emission Trading Act.

[60] See CTS Regulation sec. 5-7 and Pollution Control Regulation sec. 35-13.

Subsequent to an approved shutdown of the storage facilities, both the partnership and the State may take the initiative to have the responsibilities of the facilities transferred to the State, provided certain conditions stipulated in the CCS Regulation and the Pollution Control Regulation are satisfied. These conditions include, but are not limited to, the substantiated indication that the stored CO_2 will remain permanently enclosed, that a minimum period of no less than 20 years has expired (exceptions to the latter may be granted under certain circumstances), and that the storage location has been prudently abandoned and the injection facilities removed.

As an additional regulatory condition, the operator shall have made a financial contribution available to the State as further instructed by the MoE. This contribution or security shall as a minimum cover anticipated monitoring expenses for a period of 30 years.[61] The State may use this contribution or security to cover its expenses after the responsibility for the storage facilities has been transferred to the State.

7 Transactions

An eligible party may acquire a participating share from an existing licensee on the NCS. Such an interest can be acquired directly, or indirectly through, e.g. the purchase of shares in a company holding a participating interest in a license. Both forms of acquisition are subject to the approval by the MoE.[62]

A direct transfer is also regulated in the JOA and the Companies Act, although we expect the JOA regulation to derogate the regulation in the Companies Act going forward. According to the JOA, a licensee may transfer its participating interest in an exploration license to a buyer on the condition that the buyer becomes bound by the JOA and the license. Such transfer is, however, subject to the consent of the MC if the work program in the license has not been completed yet. With the current structure of the work program and the milestone approach, by which the decision on whether or not to submit a PDO typically is the last milestone decision of the program, this means in practice that an exploration license may not be transferred to a third party without the consent of the MC.

[61] See sec. 5-10 of the CCS Regulation.

[62] See the CCS Regulation sec. 11-13.

The licensees do not, however, have any right of first refusal or the like in connection with a transfer.

Neither the JOA nor the Companies Act contain any change of control provision, meaning that indirect transfers may be carried out without the requirement of a consent by the MC of the partnership or other licensees.

The seller in both a direct and indirect transfer may remain liable for certain obligations after the completion of the transfer, such as certain abandonment obligations, but this would have to be stipulated as a condition in the exploitation license as it does not directly follow from the CCS Regulation.[63] The abandonment obligations will mainly consist of the plugging of injection wells and removal of any facilities installed to operate the injection and storage operations, such as a platform used to carry injection equipment.

If the interested buyer is not an existing licensee to an exploration or exploitation license, the company must qualify as a new licensee in the relevant license. Such qualification is carried out either at the time of application for a license or at the time of notification to the MoE. Presently, the MoE is not offering a pre-qualification process like the one that is offered in the oil and gas business.

If a potential buyer is considering assuming the operatorship in the target licence, it must also qualify as an operator. The threshold for qualifying as an operator is higher than the threshold for a licensee.

8 Withdrawal and Relinquishment

8.1 Withdrawal[64]

A party is entitled to withdraw from the exploration licence once the work obligation has been completed.[65] However, the right to withdraw may not be a real option considering how the authorities have elected to structure the work program (the milestone approach). Presently, the work

[63] This is contrary to the oil and gas business, where the secondary abandonment liability in direct transfers is stipulated in the Petroleum Act. The omittance of this provision in the CCS Regulation may indicate that the authorities do not wish to apply the same policy on secondary liability in relation to the CCS business.

[64] A party withdrawing from the licence must also withdraw from the partnership. The provision for this is set out in Sects. 2-32 to 3-25 of the Companies Act. However, the Partnership Agreement will replace these provisions with the provision in the JOA.

[65] See model JOA article 20.

program issues seem to conclude with the PDO decision. Consequently, for all practical purposes, the option of withdrawal in accordance with the JOA may not be available to the parties in an exploration licence.

A participant that has acceded to a PDO may only withdraw from the agreement once the MoE has determined that the plan has been completed.[66]

In the event a participant is able to withdraw from exploration licence, the JOA stipulates a 4 months' notice period before the withdrawal may be effectuated. The withdrawing participant is liable for costs incurred by the partnership up to the date of notice of withdrawal.

8.2 Relinquishment

If a milestone set out in the work program is not met, and no extension has been applied for or granted, or the necessary milestone decision has not got beyond the passmark required for a valid decision, the rights under the exploration license is forfeited and the license has to be relinquished.[67]

Pursuant to the CCS Regulation, the partnership may relinquish the whole or parts of the area of the exploration license on 3 months' notice.[68] However, if there are obligations that remain pursuant to the exploration license that has not yet been completed (typically uncompleted work program), this relinquishment is subject to the MoE's approval.

Minimum two years prior to the expiry of an exploitation and/or transportation license or the use of the facilities finally ceases, the licensees shall according to the CCS Regulation submit a cessation plan to the MoE for approval. The cessation plan shall contain a proposal for future use of the storage facilities, or the discontinuation of such use. The MoE

[66] We expect that participants to an exploration licence will, to the extent possible, apply for an exploitation licence and not accede to the PDO prior to receiving such license from the Ministry of Energy. Consequently, this provision will most likely not be applicable for an exploration license. A participant under an exploration license is not obliged to apply for an exploitation license.

[67] There seems to be no provisions on sole-risk in the applicable concession documents, so minority participants that wish to continue will have to rely on approval from the Ministry of Energy to continue the activities.

[68] See sec. 3-5 of the CCS Regulation.

shall decide on the disposal of the facilities and stipulate a deadline for implementing the decision. The licensees (and owners of the facilities) are obliged to ensure that this decision is implemented. In the event an interest in the licenses is transferred to a third party, both licensee and owner are jointly liable for the implementation of the decision. The Norwegian State also has a right under the CCS Regulation to take over a licensees' fixed facilities when the license expires, or the use of such facilities finally ceases.

The MoE may stipulate more detailed conditions in connection with the disposal decision, inter alia regarding continued monitoring of the reservoir and responsibility in relation to permanent storage of the injected CO_2.

9 Closing Remarks

The CCS Regulation was enacted in 2014 and is closing in on its tenth anniversary. Notwithstanding this, carbon storage and transportation on a commercial basis is still an emerging business in Norway. The MoE continuously work with the industry on the development of the system and although the concession documents are now in place, adjustments and amendments to these model documents are expected as further industry experience is obtained. The above presentation of the current regulation on the subject matter should be read with the caveat that the regulation is developing and changing as further experience with the business is gathered.

CHAPTER 18

CCUS in Qatar

Eduardo G. Pereira and Elena I. Athwal

1 Introduction

Carbon is a necessary component of all living things; it can be found in the air (atmosphere), water (hydrosphere), and earth (geosphere). The carbon cycle refers to the recycling of carbon between these three domains. Carbon is combined with oxygen in the atmosphere to form CO_2. This molecule was steady in the atmosphere until 200 years ago, when it began to increase due to increased human activity, which currently injects about 33 billion metric tonnes of CO_2 into the atmosphere every year, causing serious adverse effects on climate change and global warming.[1] One country that has been greatly affected by the increased

[1] Mohammad A T Alsheyab, 'Qatar's Effort for the Deployment of Carbon Capture and Storage (2017) 19 Global NEST Journal 3, 453–457, https://journal.gnest.org/publication/gnest_02269, accessed 24 March 2024. See also: Shatha M Obeidat, Anas A

E. G. Pereira (✉)
Institute of Energy and Environment, University of São Paulo (IEE-USP), São Paulo, Brazil
e-mail: egp.portugal@gmail.com

E. I. Athwal
Harvard Law School, Cambridge, MA, USA

E. G. Pereira et al. (eds.), *Carbon Capture Utilization and Storage*, Sustainable Development Goals Series,
https://doi.org/10.1007/978-3-031-81272-9_18

carbon emissions is Qatar. As a major oil and gas producer, Qatar's economy heavily relies on the extraction and export of fossil fuels. This has led to a significant increase in carbon emissions, making Qatar one of the highest per capita emitters of CO_2 in the world. Recognizing the urgent need to address this issue, Qatar has been actively working towards reducing its carbon footprint through various initiatives and investments in renewable energy sources.

According to Grantham Research Institute on Climate Change and the Environment, regarding carbon capture utilization and storage (CCUS), 'CCUS refers to a suite of technologies that enable the mitigation of carbon dioxide (CO_2) emissions from large point sources such as power plants, refineries, and other industrial facilities, or the removal of existing CO_2 from the atmosphere'.[2] While carbon capture is the process of capturing carbon dioxide emissions from industrial processes and power generation facilities, storage involves storing the captured carbon dioxide underground, usually in depleted oil and gas reservoirs or deep saline aquifers and utilization refers to using the captured carbon dioxide as a feedstock for the production of valuable products such as chemicals, fuels, or building materials. And because fossil fuels are currently the dominant source of primary global energy consumption and are expected to stay so in the foreseeable future, the goal of carbon capture and storage (CCS) is to allow the continued use of fossil fuels while lowering CO_2 emissions and thereby lowering the climate effect of fossil fuel use. Because fossil fuels now supply more than 85% of all primary energy globally, achieving the 2 °C warming target will necessitate CCS making a significant contribution to decreasing GHG emissions. With this in view, CO_2 capture, storage, and utilization technologies have gained significant attention in Qatar. As one of the world's largest producers of natural gas,

Al Bakri and Said ElBanna, 'Leveraging "Green" Human Resource Practices to Enable Environmental and Organizational Performance: Evidence from the Qatari Oil and Gas Industry' (03 December 2020) 164 Journal of Business Ethics, 371–388, https://link.springer.com/article/10.1007/s10551-018-4075-z, accessed 24 March 2024.

[2] The London School of Economics and Political Science (LSE), 'What Is Carbon Capture, Usage and Storage (CCUS) and What Role Can It Play in Tackling Climate Change?' Grantham Research Institute on Climate Change and the Environment (11 December 2023), https://www.lse.ac.uk/granthaminstitute/explainers/what-is-carbon-capture-and-storage-and-what-role-can-it-play-in-tackling-climate-change/, accessed 24 March 2024.

Qatar is committed to reducing its carbon footprint and mitigating the environmental impact of its fossil fuel industry.

CO_2 is one of the greenhouse gases that retains sunlight's energy and keeps the world at a tolerable temperature. However, recent increases in CO_2 emissions are having a negative impact on global warming and climate instabilities, which are changing weather patterns. This sparked public concern, leading to the founding of the United Nations Framework Convention on Climate Change in 1992, with the goal of stabilizing greenhouse gases at a level that limits human involvement with the climate.[3] According to the International Energy Agency assessments, CO_2 emissions in 2050 should be half of those in 2005.[4] This view reflects the increasing urgency to mitigate climate change and reduce greenhouse gas emissions, not just in Qatar but in the global space.

There are numerous mitigation techniques available to lower CO_2 emissions into the atmosphere. These could be divided into two categories: (1) the relatively inexpensive technologies, such as increasing the supply of energy and its efficiency at the point of use, moving away from coal and oil towards gas, using renewable energy sources like forests; and (2) the more expensive mitigation technologies, such as nuclear power, carbon capture and storage, and large-scale renewable energy production. These techniques are crucial for countries like Qatar, which heavily rely on fossil fuels for their energy needs.

While the key disadvantages of the first group are limited impact and short-term applicability, the disadvantages of the second category are safety, public acceptance, and the high cost of renewable energy.[5] In addition, another disadvantage of the second category is the lack of infrastructure for CCUS. Meanwhile, Qatar has been actively exploring CCUS as a potential solution to mitigate its carbon emissions.

[3] Zoya Banan and Abbas Maleki, 'Carbon Capture & Storage Deployment in Iran' (2013) 37 Energy Procedia, 7492–7501.

[4] International Energy Agency (IEA), 'Carbon Capture and Storage—Legal and Regulatory Review', International Energy Agency (Paris, France, May 2011), https://origin.iea.org/reports/carbon-capture-and-storage-legal-and-regulatory-review-2011, accessed 24 March 2024.

[5] United States Department of Energy, CO_2 Capture, Reuse, and Storage Technologies for Mitigating Global Climate Change White Paper, DE-AF22-96PC01257 (1997).

Carbon capture and storage has been identified as a viable technique for CO_2 mitigation in the medium to long term.[6] This is based on the availability of suitable geological formations for CO_2 storage in Qatar, such as depleted oil and gas reservoirs. Irrespective of the potential benefits of carbon capture and storage, Qatar is also exploring the option of carbon utilization. By converting captured CO_2 into valuable products, such as fuels, chemicals, and building materials, Qatar aims to create a circular economy and reduce its overall carbon footprint. This approach not only helps in reducing greenhouse gas emissions but also provides economic opportunities and diversification for the country's resource-dependent economy.

2 Highlights of New Initiatives

Qatar has several new initiatives to mitigate its reliance on fossil fuels. Qatar happens to have two of the six CCUS projects in the MENA region—the other projects are located in Saudi Arabia, the UAE, and Iraq.[7] One important example is its new partnership with General Electric (GE), in which Qatar aims to develop a carbon capture roadmap and to establish the largest carbon dioxide recovery and sequestration facility in the Middle East and North African (MENA) region.[8] QatarEnergy and GE entered into a memorandum of understanding, in order to create a world-scale carbon hub in Qatar, underscoring the commitments of both parties to reduce carbon emissions and to enhance sustainability in the energy sector.[9] Qatar, as the leading exporter of liquified natural gas (LNG), has built the most extensive carbon capture recovery and sequestration facility in the MENA region, planning to capture over 5 million

[6] United States Energy Information Agency (EIA), Country Analysis Briefs: Qatar (2013).

[7] Oil and Gas News, 'Kingdom Drives CCUS Growth to Reach Net-Zero by 2060' (01 April 2024), https://ognnews.com/Article/46685/Kingdom_drives_CCUS_growth_to_reach_net-zero_by_2060, accessed 2 April 2024.

[8] QNA 'Qatar CCUS Forum' (n 1).

[9] GE Vernova, 'QatarEnergy and GE to Develop Carbon Capture Roadmap and Low Carbon Solutions for Qatar's Energy Sector' Press Release (21 September 2022), https://www.gevernova.com/news/press-releases/qatarenergy-and-ge-to-develop-carbon-capture-roadmap-and-low-carbon-solutions-for, accessed 24 March 2024.

tonnes of CO_2 annually by 2025; this demonstrates Qatar's firm commitment to environmental sustainability, while also supporting global energy security.[10]

Based on the above highlights, this paper aims to evaluate the potential of CCUS in Qatar. Qatar is one of the largest producers of natural gas in the world, leading to significant carbon emissions. Implementing CCUS technologies can help mitigate these emissions by capturing and storing carbon dioxide underground or utilizing it in various industrial processes. This paper will analyse the feasibility, economic viability, and environmental impact of implementing CCUS in Qatar, providing valuable insights for policymakers and stakeholders in the region.

2.1 Carbon Dioxide Emissions Status in Qatar

Carbon dioxide is the primary greenhouse gas emitted in Qatar, largely due to its heavy reliance on fossil fuels for energy production and industrial processes. Qatar has one of the highest per capita carbon dioxide emissions in the world, primarily driven by its oil and gas industry, at 37.6 metric tonnes per capita in 2022. This high level of emissions is primarily due to the burning of fossil fuels for energy production and industrial processes.[11] Qatar now has an energy-based economy because it is the world's third largest dry natural gas producer, after Russia and Iran, as shown in Table 1. Qatar is the world's largest LNG exporter as well as the location of the world's largest gas-to-liquid (GTL) production plant. This contributes significantly to government revenue. In 2013, the daily production of liquid fluids was 1.6 million barrels per day, with crude oil accounting for approximately 45% and non-crude oil accounting for 55%.[12] As of 2023, a decade later, Qatar's daily production of liquid fluids remains significant, with current figures reflecting a production rate

[10] OilPrice.com, 'Qatar Builds the Biggest Carbon Capture Plant in the Middle East' (08 October 2019), https://oilprice.com/Latest-Energy-News/World-News/Qatar-Builds-The-Biggest-Carbon-Capture-Plant-In-The-Middle-East.html, accessed 24 March 2024.

[11] Global Carbon Budget, 'Annual CO_2 emissions (per capita)' [dataset] (2023), https://ourworldindata.org/co2/country/qatar, accessed 15 July 2024.

[12] UNDP, 'Human Development Report 2013: The Rise of the South: Human Progress in a Diverse World' (1 January 2013), https://hdr.undp.org/content/human-development-report-2013, accessed 15 July 2024.

of around 2 million barrels per day, highlighting the continued importance of the oil and gas sector in the nation's economy.[13] The abundance of natural gas resources in Qatar has presented the country with an opportunity to explore CCUS technologies. CCUS technologies have the potential to capture and store carbon dioxide emissions from industrial processes, helping Qatar reduce its carbon footprint.

Table 1 World's largest dry natural gas producers (tcf)

Country	*Trillion Cubic Feet*
Russia	1688
Iran	1200
Qatar	843
United States	473
Turkmenistan	400
Saudi Arabia	333
UAE	215
Nigeria	203
Venezuela	200
Algeria	159

Source US Energy Information Administration (2024). The latest data available is from 2021[14]

According to the latest data, Qatar's carbon dioxide emissions were estimated at 101.34 million tonnes per year in 2022, with a per capita emission of approximately 37.6 MMt/year of CO_2, reflecting ongoing efforts and challenges in managing emissions.[15] Although Qatar emits

[13] Our World in Data, 'Oil Production', https://ourworldindata.org/grapher/oil-production-by-country, accessed 15 July 2024. See also, Energy Institute, 'Statistical Review of World Energy' (2024), https://www.energyinst.org/statistical-review#regional-overview, accessed 23 July 2024.

[14] US Energy Information Administration, 'International Natural Gas Reserves' (2024), https://www.eia.gov/international/data/world/natural-gas/dry-natural-gas-reserves?pd=3002&p=00000000000000000000004&u=0&f=A&v=mapbubble&a=-&i=none&vo=value&t=C&g=00000000000000000000000000000000000000000000000000000000000001&l=249-ruvvvvvfvtvnvv1vrvvvvfvvvvvvfvvvou20evvvvvvvvvvvvvvs&s=315532800000&e=1609459200000&, accessed 15 July 2024.

[15] Global Carbon Budget, 'Annual CO_2 Emissions (per capita)' [dataset] (2023), https://ourworldindata.org/co2/country/qatar, accessed 15 July 2024. See also, Oxford

less CO_2 overall than other GCC nations (KSA = 662.55 MMt/year, UAE = 243.9 MMt/year, Kuwait = 109.19 MMt/year), the country's estimated CO_2 emissions per person are significantly higher than those of these nations (18.2 tonnes in KSA, 25.8 tonnes in the UAE, and 25.6 tonnes in Kuwait) (see Figs. 1 and 2).[16] This discrepancy can be attributed to Qatar's small population size compared to its GDP and energy consumption. Qatar has the highest GDP per capita in the world at US$87,480.40[17] and relies heavily on its oil and gas industry, which contributes to its high energy consumption.

Recent studies indicate that approximately 70% of Qatar's CO_2 emissions are attributed to fuel burned for energy production, with gas flaring accounting for 10% and the petrochemical, water, and electricity sectors contributing 20% of the total emissions, while household and business users are responsible for the remaining 33% of the total emissions.[18]

One important note regarding per capita emissions is that the indicator is deceptive, because it employs production-based emissions accounting, which links Qatar to significant exports of fossil fuels rather than the nations that actually import and utilize the fuel. This method appears to show Qatar as the world's top emitter of CO_2 per person, when in fact, if based on consumption, the figure would be significantly altered, from 101.34 MMt/year to 71.79 MMt/year of CO_2 annually, and from 37.6 MMt per person to 26.71 MMt per person (see Figs. 3 and 4).[19]

Business Group, 'Qatar's Increasing Renewable Energy Generation Capacity' (2024), https://oxfordbusinessgroup.com/reports/qatar/2024-report/energy-utilities/renewing-interest-increasing-generation-capacity-in-renewable-energy-solutions-aligns-with-the-countrys-long-term-sustainability-drive-overview/#:~:text=Qatar%20aims%20to%20have%20more,from%20renewable%20energy%20by%202030, accessed 15 July 2024.

[16] Global Carbon Budget, 'CO_2 and Greenhouse Gas Emissions' [dataset] (2023), https://ourworldindata.org/grapher/annual-co2-emissions-per-country?country=QAT~SAU~KWT~ARE, accessed 15 July 2024.

[17] World Bank Group, 'GDP per capita (current US$)—Qatar 2022' (2024), https://data.worldbank.org/indicator/NY.GDP.PCAP.CD?locations=QA, accessed 15 July 2024.

[18] Gulf Times, 'Qatar Eyes Sustainable Future with Eco Initiatives' (19 June 2024), https://www.gulf-times.com/article/685079/qatar/qatar-eyes-sustainable-future-with-eco-initiatives, accessed 15 July 2024.

[19] Global Carbon Budget, 'How do CO_2 Emissions Compare When We Adjust for Trade? Which Countries are Net Importers and Exports of Emissions? How Much CO_2 is 'Offshored'? [dataset] (2023), https://ourworldindata.org/grapher/annual-co2-emissions-per-country?country=QAT~SAU~KWT~ARE, accessed 15 July 2024.

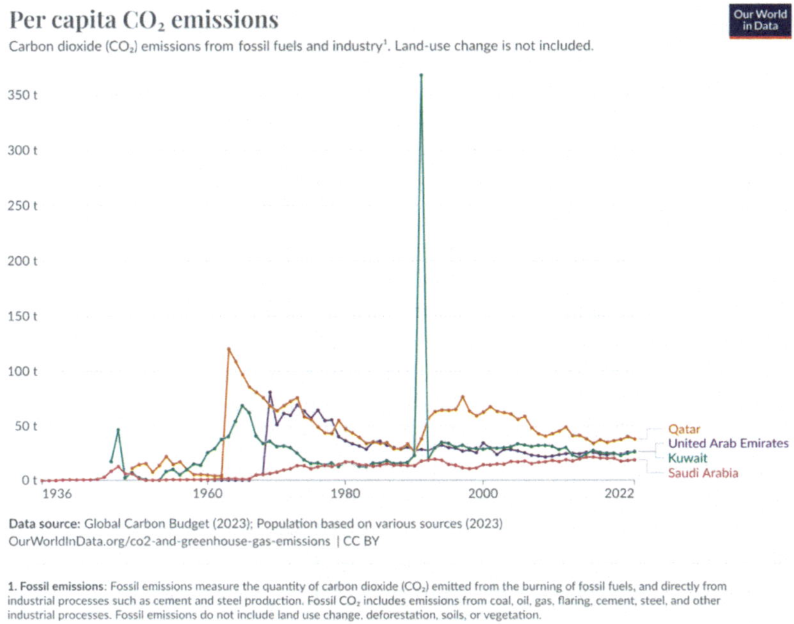

Fig. 1 Per capita emissions of the Gulf countries

Qatar ratified both the Kyoto Protocol in 2005[20] and the United Nations Framework Convention on Climate Change (UNFCCC) in 1996. In recent years, Qatar's CO_2 emissions have continued to rise, but there has been a notable decrease in per capita emissions, with respect to consumption (see Figs. 3 and 4). This trend indicates that Qatar has been making significant efforts to reduce its carbon footprint and improve its environmental performance. One of the key strategies adopted by the country is CCUS.[21] As discussed in Sect. 1, this technology enables the capture of CO_2 emissions from industrial processes and either stores or utilizes them, preventing their release into the

[20] UNFCCC, Ratification of the Paris Agreement and Kyoto Protocol by Qatar, https://unfccc.int/node/61146, accessed 15 July 2024.

[21] QatarEnergy, 'Your Energy Transition Partner: Sustainability Report 2023', https://www.qatarenergy.qa/en/MediaCenter/Publications/QatarEnergy%202023%20Sustainability%20Report.pdf, accessed 15 July 2024.

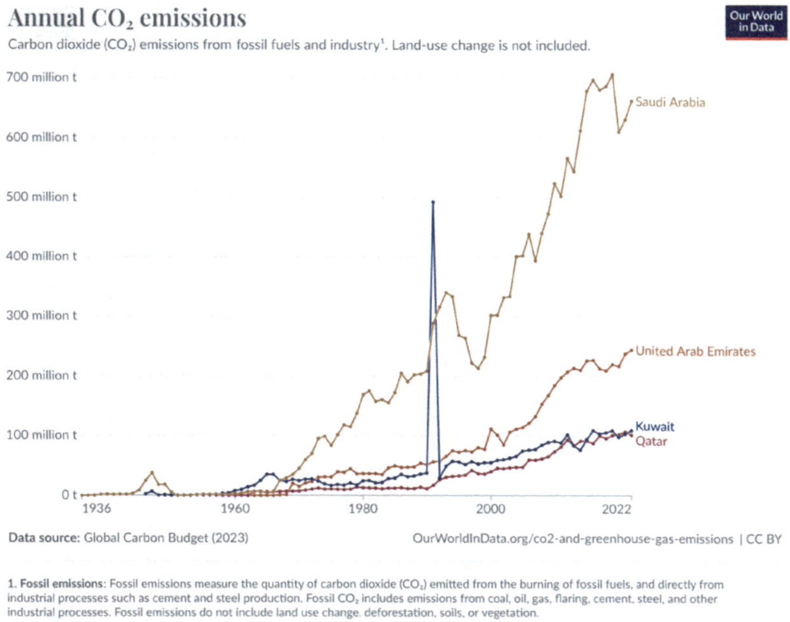

Fig. 2 Total annual CO_2 emissions of the Gulf countries

atmosphere and contributing to global warming. Additionally, Qatar is investing heavily in renewable energy sources, such as solar power, and implementing energy efficiency programmes to further mitigate its environmental impact. These initiatives reflect Qatar's commitment to sustainability and its proactive approach to addressing climate change.

2.2 *Carbon Capture, Storage, and Utilization (CCUS) Status in Qatar*

Qatar, which has the world's third largest natural gas reserves and the 13th largest oil reserves, is fully reliant on oil and natural gas for its energy needs. In 2022, Qatar maintained its position as a leading LNG exporter with 79 million tonnes (Mt) of LNG exported. Qatar, along with Australia, the United States, and Russia, collectively contributed

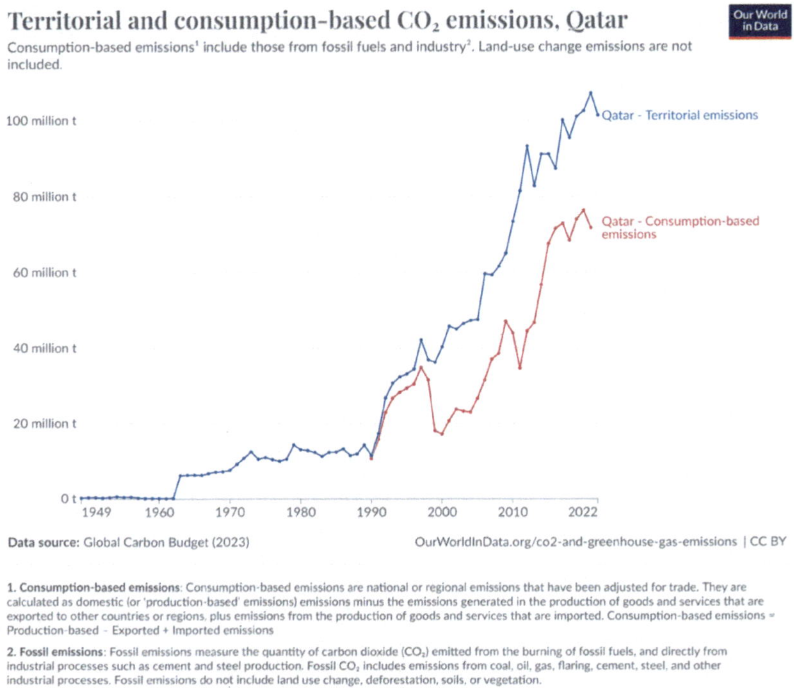

Fig. 3 Territorial and consumption–based CO_2 emissions—Qatar

68% of the global LNG supply. These contributions underscore Qatar's significant role in the global LNG market.[22]

The Qatar CCUS Forum in 2023, was able to foster global collaboration in order to mitigate carbon emissions and to highlight Qatar's role in leading the development of the first international carbon credit programme.[23] Qatar is actively investing in eco-friendly measures to reduce its carbon footprint.[24] These initiatives include a strong emphasis

[22] Gas Exporting Countries Forum (GECF), 'Global Gas Outlook 2050' (8th ed, March 2024) 114–115, https://www.gecf.org/_resources/files/pages/global-gas-outlook-2050/gecf-global-gas-outlook-20231.pdf, accessed 22 July 2024.

[23] QNA, 'Qatar CCUS Forum' (n 1).

[24] QatarEnergy, 'Climate Change Mitigation' (2023), https://www.qatarenergy.qa/en/Sustainability/Pages/ClimateChangeMitigation.aspx, accessed 22 July 2024.

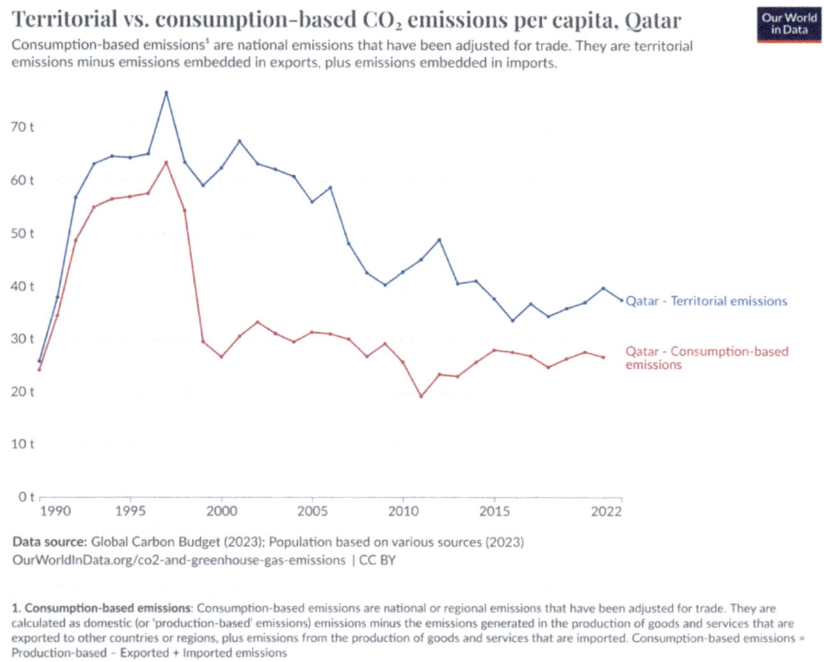

Fig. 4 Territorial and consumption-based CO_2 emissions per capita—Qatar

on CCS technologies and targeted efforts to reduce methane emissions. By focusing on these areas, Qatar aims to mitigate the environmental impact of its extensive oil and gas operations and promote sustainable development.[25]

Carbon capture and storage, also known as sequestration, as mentioned in Sect. 1, is a method that aims to reduce the CO_2 content of exhaust emissions from power plants and industrial units by capturing and safely storing it underground to prevent it from entering the atmosphere. This technology is particularly beneficial in Qatar, where it can be applied across vital sectors such as cement, steel, fertilizers, power generation, and natural gas processing, as well as in the production of clean hydrogen. CCS includes capturing CO_2 from large emission sources (point-source

[25] GECF, 'Global Gas Outlook' (n 28) 107.

capture) and directly from the atmosphere through direct air capture and storage (DACCS) or bioenergy with capture and storage (BECCS), thus playing a crucial role in mitigating climate change.[26]

The usage of CCS in Qatar offers numerous benefits. Qatar is actively developing CCUS technologies to mitigate CO_2 emissions from fossil fuel consumption. Efforts include leveraging subsurface geological formations, such as saline aquifers and depleted oil and gas reservoirs, for CO_2 sequestration.[27] The global interest in CCUS is growing, with implementations across various industrial sectors and potential economic benefits from integrating CCUS with enhanced oil recovery (EOR).[28]

QatarEnergy has successfully deployed 2.2 million metric tonnes per annum (MMTPA) of CCUS capacity and aims to increase this to 7–9 MMTPA by 2030 and over 11 MMTPA by 2035. This current capacity captures CO_2 from the feed gas to LNG trains and sales gas assets, which is vital for producing lower-carbon intensity LNG for export. Future plans include integrating CCUS with existing LNG trains, capturing CO_2 in ammonia production, from new natural gas processing facilities, post-combustion carbon from gas-fired turbines, and building CO_2 transport pipeline infrastructure. As of 2024, Qatar has two of the six CCUS projects in the MENA region.[29]

However, the financial risks and environmental effects of CCS remain unknown, and it has not yet shown to be a viable commercial solution.

[26] Global CCS Institute, 'CCS Explained: The Basics. What is Carbon Capture and Storage and How Does it Work?', https://www.globalccsinstitute.com/resources/ccs-101-the-basics/, accessed 22 July 2024.

[27] Mohammad Jafari, Shuang Cindy Cao and Jongwon Jung, 'Geological CO_2 Sequestration in Saline Aquifers: Implication on Potential Solutions of China's Power Sector' (2017) 121 Resources, Conservation and Recycling, 137–155, https://www.sciencedirect.com/science/article/abs/pii/S0921344916301240, accessed 22 July 2024. See also: John Wilkinson and others, 'Subsurface Design Considerations for Carbon Dioxide Storage' (2009) 1 Energy Procedia 1, 3047–3054, https://www.sciencedirect.com/science/article/pii/S1876610209007267, accessed 22 July 2024.

[28] Manal Al-Shafi and others, 'A Review on Underground Gas Storage Systems: Natural Gas, Hydrogen and Carbon Sequestration' (2023) 9 Energy Reports, 6251–6266, https://www.sciencedirect.com/science/article/pii/S2352484723009794, accessed 22 July 2024. See also: Osama Massarweh and Ahmad S Abushaikha, 'CO2 Sequestration in Subsurface Geological Formations: A Review of Trapping Mechanisms and Monitoring Techniques' (2024) 253 Earth-Science Reviews, 104793, https://www.sciencedirect.com/science/article/pii/S001282522400120X#bb0050, accessed 22 July 2024.

[29] QatarEnergy, 'Sustainability Report 2023' (n 28).

Since natural gas generation is 50% less carbon intensive than coal, the economic return of commercializing CCS in Qatar is uncertain.[30] The reduced volume of CO_2 emissions from natural gas plants means that the total amount of CO_2 available for capture and storage is lower, leading to diminished financial returns. For example, since natural gas generation is 50% less carbon intensive than coal, it emits significantly less CO_2 per unit of energy produced. This means that the absolute amount of CO_2 available for capture and storage in CCS projects is reduced when dealing with natural gas plants.[31] Economically, the return on CCS investments is highly dependent on the volume of CO_2 captured and stored. Lower emissions from natural gas plants lead to reduced financial returns because the cost per tonne of CO_2 captured remains relatively constant, but the total volume of CO_2 captured is lower. Financial incentives, such as carbon credits or taxes, are typically based on the amount of CO_2 reduced, so with less CO_2 to capture, these incentives or credits will be lower, reducing the overall financial benefit. Additionally, the high costs of CCS infrastructure, which include capture equipment, transportation pipelines, and storage facilities, do not decrease proportionally with the reduced CO_2 output, leading to a higher cost per unit of CO_2 captured and stored. Maintenance costs for CCS systems also remain high regardless of the CO_2 output, further impacting economic viability. Environmentally, the reduced amount of CO_2 captured from natural gas plants means that the overall impact on reducing greenhouse gas emissions is lower compared to CCS applied to coal plants. This makes the justification for high CCS investment less compelling, as the environmental benefits, such as mitigation of climate change, are diminished. Consequently, the economic return of commercializing CCS in Qatar, where natural gas is the predominant energy source, is uncertain due to these factors.

Despite challenges related to the transition to renewable energy and the high costs associated with CCUS infrastructure, these technologies offer a viable pathway for Qatar to reduce its carbon footprint while

[30] Ibid.

[31] Ibid.

maintaining energy security.[32] The potential application of CCS in industrial operations is extensive, particularly given that Qatar is home to the world's largest GTL plant and the world's largest LNG producer.[33]

2.3 Carbon Capture, Utilization, and Storage Initiatives in Qatar

Carbon Capture, Utilization, and Storage Initiatives in Qatar are an important step towards reducing greenhouse gas emissions and mitigating climate change. Qatar, being a major producer of natural gas, has recognized the importance of carbon capture and storage technologies in reducing its carbon footprint. The country has invested in several projects aimed at capturing carbon dioxide emissions from its industries and storing them underground or utilizing them for enhanced oil recovery. One such initiative is the Ras Laffan CCS facility, the largest in the MENA region, aiming to capture and sequester over 5 million tonnes of CO_2 annually by 2025, reflecting Qatar's commitment to both global energy security and environmental challenges. These initiatives not only promote sustainable development but also position Qatar as a leader in the global effort to combat climate change.

2.3.1 History

Qatar is heavily investing in carbon capture and storage. The Qatar US and Carbon Storage Research Centre, which includes approximately 40 academic staff, postdoctoral researchers, and PhD students, was founded to strengthen Qatar's engineering talent and knowledge, as well as to expand research capacity in carbon capture and storage (CCS) and cleaner fossil fuels. It is a 10-year, US$70 million programme based at Imperial College London.[34]

[32] Laxman Prasad Ghimire and Yeonbae Kim, 'An Analysis on Barriers to Renewable Energy Development in the Context of Nepal using AHP' (2018) 129 Renewable Energy, Part A, 446–456, https://www.sciencedirect.com/science/article/abs/pii/S0960148118306414, accessed 22 July 2024. See also: Kasereka Ndwali, Jackson G Njiri and Evan M Wanjiru, 'Multi-objective Optimal Sizing of Grid Connected Photovoltaic Batteryless System Minimizing the Total Life Cycle Cost and the Grid Energy' (2020) 148 Renewable Energy, 1256–1265, https://www.sciencedirect.com/science/article/abs/pii/S0960148119315563, accessed 22 July 2024.

[33] QatarEnergy, 'Sustainability Report 2023' (n 28).

[34] Sadah Mohammed and others, 'A Systematic Review: The Role of Emerging Carbon Capture and Conversion Technologies for Energy Transition to Clean Hydrogen' (2024)

In addition, from 2007 to 2010, the Qatar Foundation financed 130 grants totalling $118 million in environment and energy research, $9.3 million of which was allocated to carbon capture and storage.[35] These grants were used to support various initiatives aimed at developing and implementing carbon capture and storage technologies in Qatar. One notable project funded by the Qatar Foundation was the Qatar Carbonates and Carbon Storage Research Centre (QCCSRC), collaboration between Qatar Petroleum, Shell, and Imperial College London, focusing on the feasibility of storing captured carbon dioxide in Qatar's abundant carbonate reservoirs. The largest blue ammonia project by Qatar Fertilizer Company (QAFCO) is expected to capture 11 million tonnes per annum of CO_2.[36] The North Field East (NFE) project, with an investment of $28.75 billion, includes CCS technology as a key environmental aspect.[37]

The Dukhan oilfield, which saw a production decline from 350,000 barrels per day in 2003 to 250,000 barrels per day at the end of 2008, is receiving CO_2 injections from Qatar's Pearl GTL plant,[38] as part of Qatar's CCS utilization initiatives, aimed at reducing greenhouse gas emissions through enhanced oil recovery (EOR). In 2013, Qatar University's Gas Processing Centre (GPC) announced the introduction of a carbon capture and management Road Map,[39] aiming to capture and store 5 million tonnes of carbon dioxide per year by 2025.

447 Journal of Cleaner Production, 141506, https://www.sciencedirect.com/science/article/pii/S0959652624009545, accessed 22 July 2024.

[35] Ibid.

[36] MEES, 'Qatar Launches 'World's Largest' Blue Ammonia Project' (2 September 2022), https://www.mees.com/2022/9/2/refining-petrochemicals/qatar-launches-worlds-largest-blue-ammonia-project/dcc199b0-2ab0-11ed-a8a5-913ffd3e63d0, accessed 22 July 2024.

[37] S&P Global, 'QP to Spend $200 Million on Emissions Reduction Technology for LNG Expansion Project' (30 June 2021), https://www.spglobal.com/commodityinsights/en/market-insights/latest-news/natural-gas/063021-qp-to-spend-200-million-on-emissions-reduction-technology-for-lng-expansion-project, accessed 22 July 2024.

[38] A Benamor and A Aroussi, 'Towards a Technology Roadmap for Carbon Capture and Management for Qatar' (2013) 2 International Journal of Biological, Ecological and Environmental Sciences (IJBEES) (3).

[39] International Energy Agency, (2013), Technology Roadmap: Carbon Capture and Storage.

Additionally, in 2012, the Qatar Fuel Additives Company (QAFAC) ordered a large-scale CO_2 recovery unit from Mitsubishi Heavy Industries to be deployed in its methanol manufacturing plant.[40]

2.3.2 Recent Developments

In 2019, Qatar, the leading exporter of liquified natural gas, or LNG, inaugurated the most extensive carbon capture recovery and sequestration facility in the MENA region, in Ras Laffan. This facility, as stated by Qatar's Minister of State for Energy Affairs, Saad Sherida Al Kaabi, highlights Qatar's commitment to both 1) support global energy security of supply and 2) tackle environmental challenges such as the sequestration of GHGs.[41]

As mentioned above, QatarEnergy has successfully deployed 2.2 MMTPA of CCUS capacity and aims to increase this to 7–9 MMTPA by 2030 and over 11 MMTPA by 2035. The CO_2 Export Project is progressing on schedule, with around 94% completion, and will export captured CO_2 from QatarEnergy LNG South facilities to Dukhan for enhanced oil recovery. The Front-End Engineering Design (FEED) project to capture CO_2 from seven LNG North trains and three LNG South trains, with over 50% progress by year-end, aims to capture over 4 MMTPA of CO_2. A memorandum of understanding with an original equipment manufacturer (OEM) to develop a CCS roadmap includes a feasibility study to capture around 2.5 MMTPA of CO_2 from power plants, expected to be completed in 2024, with plans for a similar study for refrigeration gas turbines at LNG facilities.[42]

As of 2024, Qatar has two of the six CCUS projects in the MENA region: the other projects are located in Saudi Arabia, the UAE, and Iraq.[43] Qatar has also given a minimal regulatory framework to facilitate CCS work. This is covered by one of the regulations that states, 'Counteract the effects of pollution in its various forms' and prevents damage as well as the immediate and long-term environmental repercussions of building, industrial, agricultural, and economic operations.[44]

[40] Ibid., 19.

[41] OilPrice.com, 'Qatar' (n 10).

[42] QatarEnergy, 'Sustainability Report 2023' (n 28).

[43] Oil and Gas News, 'Kingdom Drives CCUS Growth' (n 8).

[44] OilPrice.com, 'Qatar' (n 10).

These regulations demonstrate Qatar's commitment to address environmental concerns and promote sustainable development. By providing a regulatory framework for carbon capture and storage utilization initiatives, Qatar aims to mitigate the negative impact of pollution and ensure the preservation of the environment for future generations. This proactive approach showcases Qatar's dedication to finding innovative solutions and becoming a leader in the field of carbon capture and storage technology.

QatarEnergy plans to dramatically scale up CCS capabilities, aiming to capture over 11 million tonnes per annum of CO_2 by 2035.[45] This substantial increase in capacity is part of a broader strategy to enhance the sustainability of Qatar's major industries, including its liquefied natural gas facilities. The firm's roadmap for carbon capture, developed in collaboration with a major US company, emphasizes this commitment to reducing carbon emissions and transitioning to cleaner energy sources.

These initiatives demonstrate Qatar's proactive approach to integrating advanced CCS technologies within its energy sector, positioning the country as a leader in the global effort to combat climate change.

3 Challenges and Opportunities of Carbon Capture, Storage, and Utilization in Qatar

In Qatar, CCUS present both challenges and opportunities for sustainable development. As a major producer of natural gas and oil, Qatar has a significant carbon footprint that needs to be addressed. However, the country also has the resources and technological capabilities to implement CCUS projects effectively.

3.1 Challenges

As mentioned above, with the energy sector accounting for 96.5% of emissions, Qatar is among the countries with the highest emissions per person in the world. Gas-fired electrical generators provide about half of Qatar's energy-related emissions of heat and electricity, with industrial operations accounting for the second-highest share of GHG emissions. Qatar possesses the third largest natural gas reserves globally. The latest data indicates that Qatar's gas reserves stand at over 2,000 trillion cubic

[45] QatarEnergy, 'Sustainability Report 2023' (n 28).

feet, following a recent discovery that added about 240 trillion cubic feet to its reserves.[46] Qatar is the 14th largest oil producer in the world, with its proven oil reserves remaining substantial.[47] Thus, Qatar remains heavily reliant on fossil fuels for its energy needs. As of 2024, nearly all of Qatar's energy demands are met by oil and natural gas, with renewable energy sources contributing a very small fraction. Significant investments are being made to increase the contribution of renewable energy, notably solar power, which is expected to expand its capacity to over 5 gigawatts in the next decade.[48]

One major challenge is the high cost associated with implementing CCUS technology. The cost of capturing CO_2 can range from US$15–25 per metric tonne for processes with high CO_2 concentrations (like ethanol production) to US$40–120 per metric tonne for more dilute sources such as power generation and cement production.[49] The infrastructure required for capturing, transporting, and storing carbon dioxide is expensive and requires significant investment. The costs associated with CCUS infrastructure, including capture equipment, transportation pipelines, and storage facilities, are substantial and do not decrease proportionally with reduced CO_2 output.

The economic return of commercializing CCUS in Qatar is uncertain due to the lower volume of CO_2 emissions from natural gas plants compared to coal plants. Lower emissions from natural gas lead to reduced financial returns because the cost per tonne of CO_2 captured remains relatively constant, but the total volume of CO_2 captured is lower.[50]

[46] The Peninsula, Qatar Business, 'Qatar Announces Significant Increase in North Field Gas Reserves: Raising LNG Production Capacity to 142 MTPA before the End of 2030' (25 February 2024), https://thepeninsulaqatar.com/article/25/02/2024/qatar-announces-significant-increase-in-north-field-gas-reserves, accessed 22 July 2024.

[47] The Peninsula, 'Qatar Leverages LNG Demand to Attain Clean Energy Transition' (25 June 2024), https://thepeninsulaqatar.com/article/25/06/2024/qatar-leverages-lng-demand-to-attain-clean-energy-transition, accessed 23 July 2024.

[48] Ibid.

[49] Congressional Budget Office, 'Carbon Capture and Storage in the United States' (December 2023), https://www.cbo.gov/system/files/2023-12/59345-carbon-capture-storage.pdf, accessed 23 July 2024.

[50] Reuters, 'Explainer: Why Carbon Capture is No Easy Solution to Climate Change' (27 November 2023), https://www.reuters.com/business/environment/why-carbon-capture-is-no-easy-solution-climate-change-2023-11-22/, accessed 23 July 2024.

Also, ensuring that stored carbon dioxide does not leak back into the atmosphere over the long term requires careful assessment of storage sites. The environmental impact of storing captured carbon underground is not fully understood and raises concerns about potential leaks or accidents that could harm the ecosystem.[51]

3.2 Opportunities

Qatar's strong financial position and substantial natural resources provide significant opportunities for the development and implementation of CCUS technologies. One of the primary opportunities lies in leveraging CCUS to EOR. By injecting captured CO_2 into mature oil fields, Qatar can boost oil production while simultaneously sequestering carbon dioxide. This dual benefit can make CCUS projects economically viable, providing a substantial return on investment. The Ras Laffan CCUS facility, for instance, is already contributing to enhanced oil recovery efforts, which demonstrates the potential economic benefits of integrating CCUS with EOR techniques.

Additionally, Qatar's commitment to expanding its CCUS infrastructure presents numerous opportunities for sustainable development. The country aims to scale up its CCUS capabilities significantly, targeting over 11 million tonnes of CO_2 captured annually by 2035. This ambitious goal includes integrating CCUS systems with existing LNG trains, developing new natural gas processing facilities, and establishing CO_2 transport pipeline infrastructure. These initiatives not only support Qatar's goal of reducing its carbon footprint but also position the country as a leader in the adoption of advanced environmental technologies in the region.

Qatar's investment in research and development is another critical opportunity. The Qatar Carbonates and Carbon Storage Research Centre (QCCSRC), established through a collaboration between Qatar Petroleum, Shell, and Imperial College London, underscores Qatar's dedication to advancing CCUS technologies. This centre focuses on the

[51] British Geological Survey, 'Understanding Carbon Capture and Storage: Discovering Geology—Climate Change' (2024), https://www.bgs.ac.uk/discovering-geology/climate-change/carbon-capture-and-storage/, accessed 23 July 2024.

feasibility of storing captured CO_2 in Qatar's abundant carbonate reservoirs and exploring innovative solutions for carbon capture and utilization. By fostering a robust research environment, Qatar can drive technological advancements that enhance the efficiency and cost-effectiveness of CCUS, creating a model that can be replicated globally.

The global push for sustainability and the transition to a low-carbon economy further enhance Qatar's opportunities in the CCUS sector. As countries worldwide strive to meet their climate goals, there is an increasing demand for technologies that can mitigate carbon emissions. Qatar's proactive approach to developing and deploying CCUS positions it as a key player in the global market for carbon management solutions. This can open up new avenues for international collaborations, technology exports, and the creation of a knowledge hub for CCUS in the Middle East.

Moreover, Qatar's initiatives to incorporate CCUS in the production of lower-carbon ammonia and hydrogen present significant economic opportunities. By capturing CO_2 emissions from industrial processes and utilizing them in the production of clean energy carriers, Qatar can diversify its economy and reduce its reliance on traditional fossil fuels. This strategic shift towards sustainable energy sources aligns with global trends and enhances Qatar's competitiveness in the evolving energy market.

In conclusion, the opportunities for CCUS in Qatar are vast and multifaceted. From enhancing oil recovery and expanding infrastructure to fostering research and tapping into global sustainability trends, Qatar is well positioned to leverage CCUS technologies for economic and environmental benefits. These efforts not only contribute to Qatar's climate goals but also reinforce its status as a leader in sustainable development and innovation in the energy sector.

4 Future Outlook and Conclusion

In conclusion, the future outlook for CCUS in Qatar looks promising. The country has already made significant progress in implementing CCUS technologies, with various initiatives and projects underway. Qatar's commitment to reducing carbon dioxide emissions and its investments in research and development indicate a strong commitment to the future of these technologies. While the country's vast reserves of natural gas provide ample opportunities for carbon capture and utilization, it further enhances its potential as a leader in this field. With continued investment and collaboration, Qatar is well positioned to address the challenges and capitalize on the opportunities presented by CCUS.

Qatar's commitment to the future of these technologies is evident in its efforts to develop advanced infrastructure and research facilities dedicated to carbon capture and utilization. Through partnerships with international organizations and the establishment of research centres, Qatar is actively promoting knowledge exchange and innovation in the field. By leveraging its natural gas reserves and investing in cutting-edge technologies, Qatar can not only reduce its own carbon emissions but also become a global leader in promoting sustainable and environmentally friendly practices.

CHAPTER 19

CCUS in Trinidad and Tobago

Saelese Haynes

1 Introduction

1.1 History and Background

As highlighted, CCUS (also CCS)—Carbon Capture Utilisation and Storage, or in some circles, Carbon Capture Use and Sequestration—describes processes and systems for the capture and usage of carbon generated in various industrial processes including in upstream oil and gas production or in downstream activity and operations, such as at petrochemical plants.

The Republic of Trinidad and Tobago has been in the oil and gas game longer than most. A not very well-known fact is that Trinidad and Tobago's 1857 Oil Well drilled by the Merrimac Oil Company from the USA[1] predates the Drake Well in Titusville, Pennsylvania, by two (2) years although the latter is better recognised as the first modern, successful

[1] Mulchansingh, Vernon C. (1971), "The Oil Industry in the Economy of Trinidad" Institute of Caribbean Studies, Vol. 11, No. 1, pp. 73–100 at p. 74.

S. Haynes (✉)
Toco, Trinidad and Tobago
e-mail: saelesehaynes@gmail.com

E. G. Pereira et al. (eds.), *Carbon Capture Utilization and Storage*, Sustainable Development Goals Series,
https://doi.org/10.1007/978-3-031-81272-9_19

commercial oil well in the Western hemisphere and has a rich and well-documented history. In contrast, the exact site of Merrimac oil well has never been verified, but it was said to have been dug/ drilled to a depth of sixty-one (61) feet in La Brea in the close vicinity of Trinidad's Pitch Lake.[2]

The peak of Trinidad and Tobago oil production was in 1981 when there was approximately 240,000 barrels of oil production[3] per day. Just a few years prior, the Government of the Republic of Trinidad and Tobago realised and reported that there was a high volume of associated natural gas that was essentially being wasted. Around that time, over 180 MMscf of natural gas was being flared on a daily basis in the Teak and Poui marine fields situated off Trinidad's southeast coast.[4]

The commercial development of natural gas markets has predated oil markets—in England, natural gas was commercially produced since 1785[5] however, Trinidad and Tobago had no ready market for its associated natural gas—hence the high levels of flaring in the industry. The World Bank, in assuming a "typical" associated gas composition, a flare combustion efficiency of 98% and a Global Warming Potential for methane of 25, found that each cubic metre of associated gas flared results in about 2.8 kilograms of CO_2-equivalent emissions (CO_2e), resulting in over 350 million tons of CO_2-equivalent emissions annually. In addition, the methane emissions resulting from the inefficiency of the flare combustion contribute significantly to global warming.[6]

In the Administration Report 1960 produced by the Ministry of Energy and Energy Industry's predecessor, it was highlighted that "the under utilisation of natural gas is a problem that has long plagued both the oil industry and the territorial economy. In any associated gas

[2] Ministry of Energy and Energy Industries "Historical Facts on the Petroleum Industry of Trinidad and Tobago," https://www.energy.gov.tt/historical-facts-petroleum/ accessed April 14, 2024.

[3] The Energy Chamber of Trinidad and Tobago (2019), "Need to Increase Oil Production," https://energynow.tt/blog/need-to-increase-oil-production?rq=need%20to%20increase%20oil%20production accessed April 5, 2024.

[4] National Gas Company of Trinidad and Tobago "The Early Years", https://ngc.co.tt/about/history/the-early-years/ accessed April 6, 2024.

[5] American Public Gas Association "History of Natural Gas", https://www.apga.org/apgamainsite/aboutus/facts/history-of-natural-gas accessed April 5, 2024.

[6] World Bank Group "Gas Flaring Explained," https://www.worldbank.org/en/programs/gasflaringreduction/gas-flaring-explained accessed April 5, 2024.

economy such as ours, a market must exist for the gas or else it goes to waste".[7] Alarmingly, it was further highlighted that sixty-one per cent (61%) of gas produced was being flared. The report continued by indicating that "the Petroleum Division has been actively concerned with this problem from a conservation viewpoint".

Addressing the wastage,[8] and also possibly the harmful environmental impacts of flaring of associated natural gas, becomes one of the drivers for the establishment, in 1975, of the National Gas Company of Trinidad and Tobago ("NGC")—a state-owned and run enterprise with a mandate to create and execute a blue print for the Government of the Republic of Trinidad and Tobago ("GORTT")'s gas development. Prior to this, the country's first Prime Minister, the Right Honourable Dr. Eric Williams had announced in his budget speech that "Government or a designated agency would become the sole seller of gas which would be a 'trigger' for industrialisation".[9]

Even in its early years of operations, NGC had fully entrenched itself as an integrated multi-energy company, as a merchant of natural gas, purchasing, transporting, supplying, and selling natural gas to four heavy consumers of natural gas and 18 light industrials/manufacturers located mainly in the east–west corridor. By 1978, total gas sales amounted to 150 MMscf/d.[10] By contrast, as indicated above, over 180 MMscf of natural gas was being flared and lost daily in the Teak and Poui fields.

In July 1979, in an effort to stem this problem and to prevent wastage of this valuable resource, NGC was mandated to implement the "*Flare Gas Conservation Project*", which comprised the design, construction, and installation of two (2) offshore compression facilities through which the

[7] Ministry Responsible for Petroleum "Administration Report 1960", records now held by the Ministry of Energy and Energy Industries.

[8] Central Bank of Trinidad and Tobago 19th Dr. Eric Williams Memorial Lecture, (2005), "Eric Williams and the Emergence of the National Energy Sector," https://www.central-bank.org.tt/sites/default/files/lectures/19th%20Dr.%20Eric%20Williams%20Memorial%20Lecture.pdf accessed April 19, 2024.

[9] National Gas Company of Trinidad and Tobago "NGC is Born," https://ngc.co.tt/about/history/ngc-is-born/ accessed April 6, 2024.

[10] National Gas Company of Trinidad and Tobago "The Early Years," https://ngc.co.tt/about/history/the-early-years/ accessed April 6, 2024.

majority of the gas could be saved and compressed, prior to being transferred to land for commercial use in areas such as power generation. This project came on stream in 1982.

Such an approach to natural gas development has been recognised internationally as a unique model of development called the "Trinidad Model". Even in more recent times, countries such as Ghana with later discoveries of commercial quantities of hydrocarbon resources have approached Trinidad and Tobago for its support and advice in basing development on this "Trinidad Model".

From the 1970s, Trinidad and Tobago has actively encouraged and/or participated in utilisation of gas in the upstream chain as an alternative to flaring the gas through enhanced oil recovery (EOR) activity, where associated natural gas and/or other fluids were reinjected into performing wells to enhance production. While "enhanced oil recovery" is now a trending concept, significant CO_2-EOR projects in Trinidad and Tobago date back to as early as 1972 and T&T was one of the earliest nations, certainly in the region, to adopt such technology on a significant scale.[11]

Trinidad tested the first CO_2-enhanced oil recovery (CO2-EOR) project in 1972 by operator Texaco in the areas of Guayaguayare and Brighton which then ramped up by 1974.[12] In 1973, when Four (4) CO_2 pilot floods were implemented in what became the Petroleum Company of Trinidad and Tobago Limited (Petrotrin)'s reservoirs at its Forest Reserve and Oropouche fields, both in the southern basin of Trinidad.[13] The pilots were deemed to be successful in that significant amounts of oil were recovered as a direct result of the CO_2 injection. In particular, over the course of almost one hundred years, the Forest Reserve field

[11] University of Trinidad and Tobago (2018), "Postgraduate Research Seminar: Carbon Dioxide Enhanced Oil Recovery in Trinidad and Tobago," https://utt.edu.tt/index.php?articles=1&article_key=6943&wk=11&show_archive=1#:~:text=CO2%2DEOR%20projects%20in,pipelines%20were%20compromised%20by%20corrosion accessed April 14, 2024.

[12] Ministry of Energy and Energy Industries "Carbon Capture Utilization and Storage (CCUS)," https://www.energy.gov.tt/our-business/carbon-capture-utilization-and-storage-ccus/ accessed April 5, 2024.

[13] Global CCS Institute (2012), "Carbon Capture and Storage Regulatory Review for Trinidad And Tobago" citing Mohammed-Singh, L.J. Singhal, A.K, 2005, Lessons from Trinidad's CO2 Immiscible Pilot Projects, SPE Reservoir Evaluation & Engineering, pp. 397–403 https://www.globalccsinstitute.com/archive/hub/publications/54126/ccs-regulatory-review-trinidad-tobago.pdf accessed April 14, 2024.

has produced more than 500 million barrels of crude, said to be helped significantly by enhanced oil recovery methods.[14]

Around that time, the Ministry of Petroleum and Mines highlighted the El Blanco "410" water flood in Guayaguayare, and an "experimental" water flood project in Herrera Sands in Trinity as "proven to be very successful".[15]

In particular, it is noteworthy that the Ministry of Petroleum and Mines (later the Ministry of Energy and Energy Industries/"MEEI"), in its Annual Administrative Report for 1973, highlights that between 1969 and 1973, local injection operations included gas, water, and thermal projects which contributed between 7.9% and 15.5% of annual total oil production.[16]

1.2 Reflection on the Need for Change

As a mature participant in the oil and gas industry as well as an island nation/a "small island developing state", Trinidad and Tobago is well poised to make considered plans to address its roles and responsibilities, as well as its contributions and its continuing legacy.

Carbon dioxide can stay in the atmosphere for centuries and it is likely that CO_2 resulting from gas flared over the years in Trinidad's oil production is still adversely affecting our environment today and will likely continue to have deleterious impacts in the world that our descendants inherit.

While fairly negligible on a global scale, Trinidad and Tobago's per capita carbon emissions were the second highest in the world in 2018 as ranked by the World Bank's Global Carbon Project 2019. Per capita emissions experienced a fairly consistent upsurge and peaked in 2010.[17] Even though there was a decline after 2010, emission levels remained elevated, and T&T continued to unenviably rank in the top ten (10) countries in

[14] The Geographical Society of Trinidad and Tobago "The History of Petroleum Exploration in Trinidad and Tobago," https://thegstt.org/history/petro accessed April 14, 2024.

[15] Ministry of Petroleum and Mines "Annual Report—1973" (1973), records held by the Ministry of Energy and Energy Industries.

[16] *Ibid.*

[17] World Bank data, "World Bank Global Carbon Project Report 2019."

the world with the highest emissions per capita as at the end of 2022.[18] In addition, another layer in the Trinidad and Tobago emissions conversation is that the high level of emissions is led in no small part by aged and inefficient plant and processes, which leads to another area of concern in that there is an element of "extra-unnecessary" pollution.

On the other hand, the GoRTT is an active signatory to the Paris Agreement which seeks to restrict global temperature rise to 2 degrees Celsius above pre-industrial levels and with aspirations to further restrict it to no more than 1.5 degrees Celsius.

GORTT's commitment to reduction is in line with the commitments of other similar countries, rather than addressing its disproportionate contributions to the global pollution problem, given the ratio of its small population and moderate Gross Domestic Product, and arguably possibly ignoring the urgent need for reform in its policies addressing its finding or re-establishment of its place in the global energy transition. Nevertheless, there is a recognised thrust and Trinidad and Tobago has continued to re-affirm its intentions to de-carbonise its energy chain.

Circa May 2017, a Carbon Dioxide-Enhanced Oil Recovery Road Map, later CO_2 Emissions Reduction Mobilisation ("CERM"), initiative started their meetings as a collaborative project between academic institutions: the University of the West Indies (UWI) and the University of Trinidad and Tobago (UTT) as well as government energy players—the Ministry of Energy and Energy Industries (MEEI), PETROTRIN, and the National Gas Company (NGC). CERM aimed to make representations for sustainable development of known oil reserves using carbon dioxide-enhanced oil recovery (CO_2-EOR). In particular, CERM reiterated that CO_2-EOR is in alignment with the climate change policy of Trinidad and Tobago with the potential to reduce CO_2 emissions. In particular, it was suggested that the use of CO_2 for enhanced oil recovery has the additional benefit of the possibility of sequestering the CO_2 after the project is complete.

It was intended that the CO_2-EOR Pilot Project would have been implemented by September 2018 through PETROTRIN or its selected Joint Venture partner or Lease Operator. However, in 2018 PETROTRIN's operations were terminated and the company was closed with its former operations being handed over to Trinidad

[18] https://www.statista.com/statistics/270508/co2-emissions-per-capita-by-country/ accessed 14 June 2024.

Petroleum Holdings Limited and Four (4) subsidiaries: Heritage Petroleum Company Limited (Heritage); Paria Fuel Trading Company Limited (Paria); Guaracara Refinery Limited (Guaracara); and Petroleum Company of Trinidad and Tobago (Petrotrin).

1.3 A Peek into the Future

Anthropogenic environmental pollution from carbon remains a concern for GoRTT as is its ability to meet the targets that it has set for itself whether internally or by alignment and co-signing with various international obligations.

Trinidad and Tobago's activities in industry will continue to translate to significant CO_2emissions, and therefore, policy measures and laws around curbing and/or netting off CO_2 emissions (with the latter being the seemingly preferred approach) will continue to be relevant and crucial.

According to the MEEI, "As a signatory to the United Nations Framework Convention on Climate Change, Kyoto Protocol and Paris Agreement steps must be taken to reduce carbon emissions and to meet the target of cutting emissions by 15% by 2030 (UNDP, 2018). CCUS provides an avenue for Trinidad and Tobago to do so while simultaneously **boosting** our oil production"[19] (own emphasis).

2 Describe the Key CCUS Projects in Your Jurisdiction

At present, Trinidad and Tobago does not have any large-scale[20] CCUS projects tracked by international agencies. However, identified groundwork and/or possible projects for development include the following major projects, the majority of which are government-led or owned, and

[19] Ministry of Energy and Energy Industries "Carbon Capture Utilization and Storage (CCUS)," https://www.energy.gov.tt/our-business/carbon-capture-utilization-and-storage-ccus/ accessed April 5, 2024.

[20] "Large scale" is used as defined by the International Energy Agency as covering all CO_2 capture, transport, storage and utilisation projects worldwide that have been commissioned since the 1970s and have an announced capacity of more than 100 000 t per year (or 1 000 t per year for direct air capture facilities)—https://www.iea.org/data-and-statistics/data-tools/ccus-projects-explorer accessed April 10, 2024.

for the purposes of this research identified projects are those initiated within the last fifteen (15) years.

Carbon capture has been proposed as a way for the oil and gas industry to continue production at current or increased levels without adversely impacting the environment and been one of the largest drivers of climate change.

The status of some of the significant CCUS projects is provided below and those projects will be grouped into those doing preliminary groundwork as compared to those that identify key projects that are/may be ready for implementation.

1. *Carbon Reduction Strategy Task Force (CRSTF) [2010]*

In April 2010, the Carbon Reduction Strategy Task Force was initialised and empowered to:

1. develop a national "Carbon Reduction Strategy";
2. develop a regulatory and policy environment for carbon capture utilisation and storage and credit trading; and
3. recommend suitable projects and incentives to attract investment from companies and research bodies interested in exploring and pursuing carbon reduction initiatives.[21]

The CRSTF presented its findings for suitable projects including identifying sources of concentrated CO_2, viable transportation options, and new business opportunities.[22]

2. *Carbon Capture and Storage Regulatory Review for Trinidad and Tobago [2012]*

[21] Global CCUS Institute https://www.globalccsinstitute.com/archive/hub/publications/54126/ccs-regulatory-review-trinidad-tobago.pdf *op. cit.*

[22] Ministry of Energy and Energy Industries "Carbon Capture Utilization and Storage (CCUS)," https://www.energy.gov.tt/our-business/carbon-capture-utilization-and-storage-ccus/ accessed April 11, 2024.

This review resulted from a workshop intended to help the government and other stakeholders understand the potential role Carbon Capture and Storage ('CCS') could play in Trinidad and Tobago.

The dossier identified possible legislative amendments to permit, promote, and regulate CCS in Trinidad and Tobago, including[23];

- Amendment of the Petroleum Act[24] to create a licensing system for CO_2 capture, transport, and storage;
- Amendment of the Certificate of Environmental Clearance (Designated Activities) Order 2001 to cover CO_2 capture, transport, and storage; and
- Amendment of Pipelines Act to include references to licences for CO_2 pipelines.

This legal and regulatory permitting matrix outlined a way forward for legislative amendment to facilitate and encourage local CCS projects.

3. *Carbon Dioxide-Enhanced Oil Recovery Road Map*

The overarching goal of the CERM Project at its inception was the implementation of CO_2-EOR on a (relatively) large scale to increase Trinidad and Tobago's oil revenue by circa TT$ 200–500 million per year.[25]

The CERM Project's 2019 Report identified and evaluated some of the benefits of a CCUS project for Trinidad and Tobago, including: overall reduction in carbon footprint, building local expertise with a new generation of talent and increasing oil production by 15–20%.[26]

[23] Global CCUS Institute https://www.globalccsinstitute.com/archive/hub/publications/54126/ccs-regulatory-review-trinidad-tobago.pdf *op. cit.*

[24] Chap. 62:01.

[25] Ministry of Energy and Energy Industries "Carbon Capture Utilization and Storage (CCUS)," https://www.energy.gov.tt/our-business/carbon-capture-utilization-and-storage-ccus/ accessed April 11, 2024.

[26] The CERM Project CO2 Emission Reduction Mobilisation Project (2019), "Developing a National CCS Program in Trinidad and Tobago," http://www.thecermproject.com/uploads/1/1/9/7/119795923/cerm_symposium_summary_report.pdf accessed June 19, 2024.

This report also evaluated the options for funding CCUS projects including the availability of international funding through the Green Climate Fund and the World Bank, as well as access for technical assistance through the Climate Technology Centre and Network.

4. *A Techno-economic Analysis of Carbon Management in Trinidad and Tobago through coupled Enhanced Oil Recovery and Geological Storage [2018]*

This 2018 Carbon Management Study was funded by GoRTT and resulted in a proprietary screening tool for identifying suitable CCUS projects based on known properties of Trinidad's oil fields as well as models for reservoir simulation with assumptions and economic evaluation.[27]

5. *MEEI Guidelines [2020]*

Circa 2020, approved guidelines made under the existing Petroleum Act and Regulations were issued by sector regulator, the Ministry of Energy and Energy Industries (MEEI) for operators wishing to engage in CO_2 flooding.

Those guidelines establish that sector regulator, the MEEI, is in favour of assisting and encouraging operators to carry out EOR projects on the condition that it is duly kept abreast with the actions of the operators to ensure that the projects are successful and beneficial.

Under the guidelines, the operator is required to submit a CO_2 Injection Study Proposal for approval which must include, *inter alia*, "proof that the environment would not be affected in a negative way while undertaking CO_2 injection operations. ... Precautionary measures should be met with relevant supporting documents to ensure that the operation

[27] The University of Trinidad and Tobago (2018), "A Techno-economic Analysis of Carbon Management in Trinidad and Tobago through coupled Enhanced Oil Recovery and Geological Storage," https://www.energy.gov.tt/wp-content/uploads/2023/05/Carbon-Management-in-Trinidad-and-Tobago-Final-Report-June-21st.pdf accessed June 10, 2024.

of the proposed CO_2 injection would be taken in a safe manner with minimal harm to the environment upon approval".[28]

Post-approval, the operator is also required to submit implementation reports on agreed schedules and engage with the MEEI for monitoring visits and discussions.

6. *Steering Committee for Carbon Capture and Carbon Dioxide-Enhanced Oil Recovery (CO_2-EOR) [2021]*

On 18 February 2021, the Cabinet appointed a Steering Committee for Carbon Capture and Carbon Dioxide-Enhanced Oil Recovery to oversee the implementation of a Large-Scale CO_2-EOR Project.[29]

This initiative aims to boost Trinidad and Tobago's oil revenue while addressing carbon dioxide emissions reduction, carbon capture, and carbon sequestration.

Chaired by the Permanent Secretary of the MEEI, the committee focuses on CO_2-EOR in the oil acreages managed by Heritage Petroleum Company (formerly PETROTRIN) in Point Fortin Central, Guapo, Gran (*sic*) Ravine, and Forest Reserve.

Heritage Petroleum leads and is responsible for implementing the project, while MEEI funds the technical studies conducted by the University of the West Indies (UWI) and/or the University of Trinidad and Tobago (UTT), utilising research and development funds contributed under the country's Production Sharing Contracts.

The committee remains actively engaged in its exploratory and research roles, and in September 2023, it was agreed that the term of the Carbon Capture and Carbon Dioxide (CO_2)-Enhanced Oil Recovery Steering Committee be extended for two (2) years until 30 September 2025.

[28] Ministry of Energy and Energy Industries (2020), "Guidelines To Operators for the Approval of a CO_2 Injection Project by the Resource Management Division," https://www.energy.gov.tt/wp-content/uploads/2020/09/MEEI-CO2-Flood-Guidelines-.pdf accessed June 10, 2024.

[29] The Energy Chamber of Trinidad and Tobago (2021), "Carbon Capture and Carbon Dioxide (CO_2) Enhanced Oil Recovery Steering Committee Established," https://energynow.tt/blog/meei-establishes-carbon-capture-and-carbon-dioxide-co2-enhanced-oil-recovery-steering-committee accessed June 1, 2024.

As at 2023, from reports provided by Heritage Petroleum, a subsurface study and opportunity framing workshop are well underway with an engineering study on track for completion in 2024.[30]

7. *National Storage (Atlas) Project [2021]*

The National Storage Atlas project, scheduled over a one-year period beginning in December 2021, was approved by the MEEI for funding, with an estimated cost of US$146,941.00 or TT$1,998,400.00. In January 2022, multinational British Petroleum, a key partner in the initiative, pledged an additional US$50,000.00 or TT$340,000.00 towards the project.[31] The Atlas is a key component of the programme that is intended to develop full-scale Carbon Capture and CO_2- Enhanced Oil Recovery deployment.

The primary objective of the project is to construct a CO_2 storage atlas, with the following sub-objectives:

1. Gathering data on relevant onshore and offshore basins in Trinidad and Tobago for geological screening to identify plausible and implausible storage sites;
2. Surveying literature on properties necessary for reservoir storage calculations, describing laboratory testing methods, and conducting subsurface surveys essential for appraising storage formations;
3. Locating selected drill core boreholes to perform geotechnical tests, validating the literature values used for aquifer storage-capacity calculation;
4. Estimating the CO_2 storage capacities of prospective formations; and
5. Conducting an analysis of technical, logistical, legislative, and regulatory gaps in the Carbon Capture and Storage (CCS) chain to develop an appropriate training programme in collaboration with international partners.

[30] Heritage Petroleum Company Limited "Heritage ESG Report" 2022–2023, https://heritage.co.tt/wp-content/uploads/2024/05/HPCL_ESGReport2022-3_Spread_FINAL2.pdf accessed June 18, 2024.

[31] Ministry of Energy and Energy Industries "Carbon Capture Utilization and Storage (CCUS)," https://www.energy.gov.tt/our-business/carbon-capture-utilization-and-storage-ccus/2/ accessed April 15, 2024.

8. Fiscal Incentives [2021]

In the 2022 National Budget (approved in October 2021),[32] the Minister of Finance introduced allowances in determining the chargeable profits of a petroleum company for that year of income.

This provision allows for an allowance of 30% per cent of the actual expenditure, up to a maximum of $500,000.00 (US$73,529.00), for companies investing in Carbon Capture and Storage and Enhanced Oil Recovery.

The measures became law in December 2021 through the Finance Act No. 16 of 2021, and were effective from 1 January 2022.

It is clear that Trinidad and Tobago needs to transition to low-carbon status to meet its targets and commitments and that advanced work is being done to cover the assessment for building frameworks for the deployment of CCUS, although none of those initiatives have led to active deployment of the technology at this time.

In addition, at the Energy Conference 2024 held in Trinidad and Tobago, Chairman of the Steering Committee for Carbon Capture and Carbon Dioxide-Enhanced Oil Recovery (CO_2-EOR) and Permanent Secretary, MEEI admitted: "CCUS has not taken off in the manner that was expected. On review it has been determined that the form and the size of the incentive may have been barriers to the undertaking of CO_2 projects in Trinidad and Tobago".[33]

While the development of CCUS industry guidelines has been facilitated through the work of the committees identified above and GoRTT has developed and legislated the grant of fiscal incentives to encourage CCUS and even as most of our industry majors have publicly committed to their own environmental net-zero carbon goals, uptake for projects has been slow, perhaps in the absence of legislation requiring that companies undertake and implement CCUS where applicable and appropriate.

[32] Ministry of Finance (2021), "Budget Statement 2022: Resilience in the Face of a Global Pandemic," https://www.finance.gov.tt/wp-content/uploads/2021/10/Budget-Statement-2022.pdf accessed April 4, 2024.

[33] Latin American and Caribbean Petroleum Engineering Conference (2023), "Engineered Opportunities: Technologies Shaping the Energy Landscape," https://www.energy.gov.tt/wp-content/uploads/2023/06/SPE-Latin-American-and-Caribbean-Petroleum-Engineering-Conference-2023.pdf accessed April 20, 2024.

As at the time of writing, Trinidad and Tobago has recently been awarded funding from the Green Climate Fund (GCF) to support the implementation of climate action in line with national policy documents such as those referenced herein.[34]

3 Describe and Analyse the Legal Framework, Policies, and Norms in Relation to CCUS in Your Jurisdiction

An important foundational work that was completed for T&T in its legal planning for a CCUS matrix was the 2012 Carbon Capture and Storage (CCS) Regulatory Review previously referenced. This research work resulted in a comprehensive report on the country's regulatory framework with the emphasis on identifying and proposing solutions for key regulatory gaps in creating an enabling environment for CCUS.

The existing regulatory framework was deemed suitable for accommodating a CCUS/ CCS project, especially one involving the capture of CO_2 from an industrial facility, such as an ammonia plant, and its transport either offshore or onshore to an existing oil or gas reservoir for enhanced oil recovery (EOR) purposes. As indicated, that approach has been successfully implemented in the past locally with Trinidad and Tobago having over forty (40) years' experience—at the time of the report—in CO_2-EOR.

However, the report also identified several regulatory gaps, overlaps, policy issues, and areas for improvement that the Trinidad and Tobago government needs to address for comprehensive and effective regulation of CCS/CCUS.

Some of these significant gaps included:

Competition for Acreage: There is a potential for storage of CO_2 to compete with other subsurface uses, such as hydrocarbon production, especially in a jurisdiction like Trinidad and Tobago where land space is limited. It was noted that the current legislative framework (Petroleum Act) does not provide guidance on competing rights between

[34] Press Release May, 2024—https://ieaghg.org/news/green-climate-fund-supports-its-first-ccs-activity-developments-with-a-global-impact-from-trinidad-tobago/ accessed June 19, 2024.

different users but considered that this could be addressed in the existing application/ screening stages for petroleum activity.

Regulation of Injection into Non-hydrocarbon Reservoirs: It was also considered that the regulation of injecting CO_2 into saline aquifers not located within petroleum acreages is unclear.

The report posited that the MEEI (at that time known as the Ministry of Energy and Energy Affairs / MEEA) might still be the most appropriate regulatory authority for such activities, but clarification on the issue was needed. However, given that this activity was not immediately likely within the proposed initial CCUS projects for T&T, it was not considered particularly high risk at the time of completion of the report.

Environmental Management Act[35] Designated Activities (DAs): While it was considered that the current DAs under the Certificate of Environmental Clearance (Designated Activities) Order 2001 issued under the Environmental Management Act were broad enough to cover most CCS/ CCUS activities, there were scenarios, such as capturing CO_2 from a naturally occurring source, that are not explicitly covered.

Recommendations were made to amend the Certificate of Environmental Clearance (Designated Activities) Order 2001 to specifically cover CO_2 capture, transport, and storage.

Transportation of CO_2 (Pipelines): The Pipelines Act 1933[36] has no specific provisions to refer to pipelines transporting CO_2 pipelines. The recommendation made was to amend the Act to include CO_2 pipelines to ensure effective accommodation within the existing regulatory framework.

Short- and Long-Term Liability: The existing legislative framework does not explicitly prescribe liability for stored CO_2 in the event of a leak. This issue could be addressed through a Certificate of Environmental Clearance (CEC) obligation issued by the Environmental Management Authority (EMA) or through dedicated legislation and regulation.

In particular, it was considered that T&T also needed to consider how to manage long-term liability for CO_2 when stored long term. There are different approaches to be considered as some jurisdictions that have developed CCS frameworks (e.g. the European Union) propose transferring liability after a period of time and after certain conditions have

[35] Chap. 35:05.

[36] Chap. 35:51.

been meet from the project operator to the government. In other jurisdictions, such as in Wyoming, USA, the liability remains with the operator in perpetuity.

Generally in Trinidad and Tobago, the long-term liability for activity is thought to remain in the hands of thee operator/licensee, although exceptions to this provision are also encapsulated in law.

Applicable parts of the Petroleum Regulations made under the Petroleum Act provide that:

> (q) … before the expiry or determination of a licence, the licensee shall, unless the Minister otherwise determines, plug all wells drilled by him. Such plugging shall be done in accordance with specifications from time to time approved by the Minister;
> (r) the Minister may require in every case where the licensee intends to abandon a borehole or well from which potable water can be produced or which may be used for any public purpose that such borehole or well be transferred in ownership to the Water and Sewerage Authority without the payment of any compensation. The licensee shall be relieved of all liability in respect of such borehole or well and the supply thereof as from the date of such transfer, without prejudice to any rights or obligations which may have arisen before such date (own emphasis).

Trinidad and Tobago's model Production Sharing Contract provides, in relation to abandonment and decommissioning that:

> From the date of the first Commercial Discovery, the Minister and Contractor shall agree to establish an interest bearing escrow account in the name of the Minister at a financial institution determined by the Minister to accumulate cash reserves for use to fund against possible pollution and eventual abandonment of wells and decommissioning of facilities related to Petroleum Operations in the Contract Area.
> …
> (a) Upon determination of the Contract, where Contractor fulfils all obligations in respect of environmental remediation, abandonment of wells and decommissioning of facilities to the satisfaction of the Minister, all existing funds in the escrow account shall remain with the Minister.
> (b) If the escrow amount is insufficient to complete the approved programme or environmental remediation, Contractor shall pay all such additional required costs. (c) In the event the Minister elects to have the facility delivered up to him, the Minister, shall assume all responsibility for the facility and its abandonment and hold Contractor harmless against any

liability with respect thereto accruing after the date of such transfer to the Minister.

As such, it is clear that "long-term" liability (at least certainly for up to the relevant provisions of the country's Limitation Act) remains with the operator and it is proposed that liability for CCUS projects maintains a similar model.

Financial Incentives: Some time after the publication of the Carbon Capture and Storage (CCS) Regulatory Review, GoRTT implemented a tax incentive regime for operators involved in CCUS projects (effective from January 2022). However, to date the MEEI has recognised that CCUS has not taken off in the manner that it expected.

The review has determined that it was likely that the form and the size of the incentive might have been barriers to the undertaking of CO_2 projects in Trinidad and Tobago and a commitment was made by the MEEI to continue working closely with the Ministry of Finance with a view to implementing a revised regime.

3.1 Comparing Legislative Regimes (Canadian Perspective)

Trinidad and Tobago has been actively monitoring legislative incentives and regulatory regimes for operationalising CCUS projects. Canada is one of the other countries legislating for the inclusion of taxes and incentives around the regulation of carbon production and usage. The *Fall Economic Statement Implementation Act, 2023* (also known as Bill C-59), includes a variety of taxation measures introduced in the Canadian Parliament in November 2023 legislating a federal approach to carbon projects.

The Bill includes provisions for amending Canada's *Income Tax Act* to implement clean energy tax measures such as:

- A Carbon Capture, Utilization and Storage Investment Tax Credit (CCUS ITC) and
- A regime for wages and apprenticeship requirements applicable to CCUS ITC programmes.

Passage and implementation of the Act, however, has not been without difficulty. The Bill is currently stalled at the House of Representatives while meetings with key stakeholders continue.[37] In particular, the Bill has been challenged, mainly on the basis that it is said to provide new additional incentives to operators who are seen as engaging in activity that is harmful to the environment.

Trinidad and Tobago would do well to continue to monitor all activity associated with this legislative undertaking in considering its position, in particular in relation to the finalisation of incentives for operators.

In conclusion, while Trinidad and Tobago's existing regulatory framework is well placed to accommodate CCS/ CCUS projects, there is a need for targeted amendments and improvements to address specific regulatory gaps and ensure effective regulation of those activities.

4 What are the Main Concerns and Challenges to Develop CCUS Projects in Your Jurisdiction?

Trinidad and Tobago has established a strong and continuing commitment to support and engage in the development of necessary policies and regulations to support the development and deployment of a CCUS programme.

It is clear that locally there is an immediate opportunity for CCUS. Specific opportunities identified consider that the country has already had success with the use of CO_2 in enhanced oil recovery (e.g. in Forest Reserve) as well as the fact that when an oil or gas field does not offer the opportunity for EOR, it can be a good location for CO_2 storage. Further, the use of mature oil or gas fields makes sense as the geological properties of the field would have been probed prior to production. The use of such a field will minimise the level of analysis required to confirm its suiting for CO_2 storage.

The main concerns and challenges to developing Carbon Capture Utilisation and Storage (CCUS) projects in Trinidad and Tobago include:

1. **Regulatory Reform**: While the existing regulatory framework is generally recognised as well placed to accommodate CCUS projects, there are significant gaps and overlaps to be addressed.

[37] https://www.parl.ca/LegisInfo/en/bill/44-1/c-59 accessed April 19, 2024.

These include issues related to land use planning, liability, injection into non-hydrocarbon reservoirs, and the need for specific regulations for CO_2 pipelines.

Implementing CCUS projects requires advanced technology for CO_2 capture, transport, and storage. Additionally apart from the infrastructure for transporting and storing CO_2 needing to be in place, which may require significant investment, regulatory positions on the sharing of resources to maximise opportunities should be identified.

2. **Policy and Coordination**: There is a need for a streamlined and coordinated approach to the approval and licensing process for CCUS projects.

   In particular, this complaint transcends the entire oil and gas industry locally; Stakeholders have identified a need for greater coordination among regulatory bodies to curb the long lead time for projects.

3. **CCUS Incentives in Industry**: Apart from an enabling regulatory framework, there may be some hesitation in industry to complete non-mandated CCUS projects without Financial Support Mechanisms and/or Tax Credits.

   Developing CCUS projects can be expensive, and financial incentives, such as tax allowances and subsidies, may be necessary to encourage investment. The availability of funding and the cost-effectiveness of projects are key considerations. Preliminary work and research completed in Trinidad and Tobago have outlined various elements of the value chain from sources of CO_2 for a large-scale project, transportation and management initiatives and business development opportunities.

   Operators may be unlikely to adopt CCS without incentives and regulatory mechanisms, and which is perhaps reflected in the low/disappointing take up of current incentives introduced in Trinidad and Tobago, such as those prescribed in the Finance Act, 2022.

4. **Public Perception and Stakeholder Engagement**: Public perception of CCUS projects and stakeholder engagement is crucial for project success, and as shown in the Canadian example used, may need to be carefully managed.

   While, historically, T&T has not had as active environmental warriors and champions as may pertain in jurisdictions such as

> Canada, education of the population has led to more sophistication and demand for accountability in projects that involve the use of the country's natural resources and that have possible impacts on the environment. Ensuring transparency, addressing concerns about safety and environmental impact, and engaging with local communities are important considerations.

Addressing these concerns and challenges will require a comprehensive approach that involves collaboration among government agencies, industry stakeholders, and the public. Developing a clear regulatory framework and policy, providing financial incentives, and fostering innovation will be key to advancing CCUS projects in Trinidad and Tobago.

While globally CCUS is recognised as a "necessary and relatively low-risk piece of the decarbonisation puzzle", the technology is not moving fast enough to achieve the environmental commitments set by governments.[38] It is clear that this is also the position for Trinidad and Tobago. The International Energy Agency concludes that thus far, the history of carbon capture "has largely been one of unmet expectations". Significant work is needed if the impact required is to be met in the timeframes expected. Globally (and in Trinidad and Tobago), we are already lagging behind and in danger of breaking our environmental promises.

5 Lessons Learned

In Trinidad and Tobago, GoRTT has committed to continuing to explore the place and space for CCUS projects as well as identifying and promoting specific CCUS projects with promise as appropriate for Trinidad and Tobago.

Around the industrial world, Carbon Capture projects run the gamut from upstream to downstream activity. In addition, identified utilisation and/or storage of excess carbon produced includes options such as trucking/ piping of carbon (1) for storage underground or (2) for use in industry—such as for injection/ compression activities.

[38] "Scaling the CCUS Industry to Achieve Net-Zero Emissions" (2022), https://www.mckinsey.com/industries/oil-and-gas/our-insights/scaling-the-ccus-industry-to-achieve-net-zero-emissions.

Trinidad and Tobago has identified that the best space or niche for it within the realm of CCUS projects will be the use of CO_2 in EOR and/or possible underground storage in matured and non-producing oil fields.

Overall, the Trinidad and Tobago government, like many others, is likely to have learned that implementing CCUS requires a comprehensive approach that addresses technological, legal and regulatory, financial, and social aspects. Flexibility, adaptability, and continuous learning are key to successful CCUS implementation.

It seems that the "build it and they will come" approach originally adopted by GoRTT has not borne much fruit. GoRTT has expressed its willingness to actively support CCUS projects and has done significant groundwork in establishing a permitting and fostering framework for CCUS projects (including legislation offering incentives to operators). In addition, our local industry includes multinationals with their own ambitious and targeted objectives for low carbon production (such as in country multinationals).

However, it seems that those positive factors have not yet converged into actual committed projects. As senior GoRTT officials (the Honourable Minister of Energy and Energy Industries and the Permanent Secretary, MEEI) have recognised, there is still significant work to be done to bring the horses to the water troughs and make them drink, as it were, in providing for CCUS projects in Trinidad and Tobago.

Some of the steps taken in getting to that welcoming and incentivised regulatory environment include:

1. **Assessing Technological Challenges:** Implementing CCUS requires advanced technologies for capturing, transporting, and storing carbon dioxide. GoRTT has done significant research in understanding the local industry's typologies as well as methods for overcoming technological challenges to improve the efficiency and cost-effectiveness of CCUS systems.
2. **Regulatory Framework:** Establishing a supportive regulatory framework is crucial for the successful implementation of CCUS projects. Likewise, GoRTT has completed background research and modelling to highlight the importance of developing clear regulations and policies that address issues such as liability, storage rights, and permitting requirements.

3. **Public Acceptance:** Where public acceptance and buy in are present to support CCUS projects. This can lead to more robust accountability and can accelerate factors in the success of CCUS projects. GoRTT should evaluate the importance of communicating the benefits of CCUS, addressing concerns about safety and environmental impacts, and involving stakeholders in the decision-making process through open public consultations and other methods of engagement.

   Public engagement can also be a driver for increasing collaboration and partnerships between governments, industry, civil society and research institutions to share knowledge, expertise, and resources in fostering CCUS friendly policies and practices.
4. **Financial Incentives:** Providing financial incentives, such as tax credits or subsidies, can help incentivise the deployment of CCUS technologies. However, as GoRTT has recently recognised, incentives alone may not be the answer. It is important to design effective financial mechanisms to support CCUS projects and attract private sector investment.
5. **Monitoring and Verification:** Monitoring, reporting, and verifying the storage of carbon dioxide are critical for ensuring the long-term effectiveness of CCUS projects. GoRTT has committed to ensuring that CCUS projects are robustly monitored with verification systems to track the use and/or storage of carbon dioxide and ensure compliance with regulations.

There is room and scope for the "Trinidad model" should now be reviewed and, where appropriate, modified to include CCS as a valuable new dimension. "Today's vented CO_2 from natural gas based operations can be monetized in much the same way as was done in the case of natural gas, with the main differences being the nature and composition of the gas used and the direction of flow".[39]

Legislation and legislative reform will continue to be needed for CCUS for at least two (2) reasons. First of all, the activity is not without risks. Protection of people, the environment, and also investment is

[39] Sobers, Lorraine & Lashley, Selwyn (2012) "Win–Win: Enhanced Oil Recovery and Carbon Storage in Trinidad & Tobago" cited via https://www.energy.gov.tt/wp-content/uploads/2023/05/Final-Report-Feasibility-of-CCS-Projects-in-Trinidad-Tobago-20May13._.pdf.

key. Secondly, in the short-term review, it was found that voluntary involvement in CCUS projects by operators was not immediate.

6 Conclusion

Our world today is tinged and toned with a marked shift from positions held fifteen to twenty years ago. It is no longer the prevailing line of thinking that less fossil fuel usage is the answer. Instead, solutions are being sought in *safe and responsible* fossil fuel usage.

In a June 2023 address on "Engineered Opportunities: Technologies Shaping the Energy Landscape"[40] Minister of Energy and Energy Industries, the Honourable Stuart Young recognised:

> The oil and gas industry is at a crossroad. In the face of climate change alternative forms of energy in particular renewables are being seen as the answer to the development of clean and sustainable energy. It is this scenario that is facing the industry today. To its credit the oil and gas industry has been cleaning up its act. There has been a concerted effort by the industry to reduce greenhouse gas emissions, the flaring of gas and the elimination of methane leaks through the review of processes in oil and gas exploration and development.
>
> However, the industry may have an ally in the form of carbon capture and sequestration that has the potential to contain or permanently store greenhouse gases. It is a policy that is taking root not only due to the potential to remove greenhouse gases but to the recognition that oil and gas are essential to the long-term energy requirement of a growing global population.[41]

After all, "Oil and gas in the ground is akin to leaving money on the table".[42] The prevailing line of thinking is that on a macro-level fossil fuels will continue to play a major part in providing energy security and meeting the energy requirements of the global economy.

[40] *Op. Cit.*

[41] Ministry of Energy and Energy Industries, 2023 "Engineered Opportunities: Technologies Shaping the Energy Landscape," https://www.energy.gov.tt/wp-content/uploads/2023/06/SPE-Latin-American-and-Caribbean-Petroleum-Engineering-Conference-2023.pdf, accessed on April 12, 2024.

[42] Ibid.

In his opening address to the 2024 Trinidad & Tobago Energy Conference, Minister Young summarised the work of the Carbon Capture and CO_2-Enhanced Oil Recovery (CO_2-EOR) Steering Committee. He shared that the work to map the use of Carbon Capture as a method to reduce the carbon footprint of fossil fuels would continue to dominate the energy market in the future. In addition, he indicated that the committee was actively working to characterise potential reservoirs, both onshore and offshore, for CO_2 storage and to select a pilot reservoir for CO_2 injection and increased oil recovery.

Is Trinidad and Tobago ahead of the curve? Not many producing countries around the world have legislation that directly addresses CCUS as part of a state strategy to regulate projects. However, what is also clear is that collectively, even those countries that are warm to CCUS (including Trinidad and Tobago) are still woefully behind on projects needed to meet established environmental targets. Trinidad and Tobago is recognised as the CCUS "centre of expertise" for the Caribbean region and through the groundwork done to date, is well poised to lead this transition.

CHAPTER 20

CCUS in the UAE

Felipe Villasuso Lago

1 Introduction

The United Arab Emirates ("UAE") is a major oil and gas producer and exporter and one of the highest per capita emitters of CO_2 in the world. The UAE has recognized the need to diversify its energy mix and reduce its carbon footprint and has taken steps to promote carbon capture, utilization, and storage ("CCUS") as part of its national strategy. This article provides an overview of the existing and planned CCUS projects in the UAE (based on publicly available sources), as well as the potential challenges and opportunities associated with such deployment.

The UAE is in a region particularly vulnerable to the impact of climate change. The nation faces extreme temperatures that have already increased more than the global average. Challenges such as water scarcity,

This chapter reflects the personal research and opinion of its author and not necessarily the opinion of its employees.

F. Villasuso Lago (✉)
London, UK
e-mail: lagof@lsbu.ac.uk

E. G. Pereira et al. (eds.), *Carbon Capture Utilization and Storage*, Sustainable Development Goals Series,
https://doi.org/10.1007/978-3-031-81272-9_20

high solar radiation, aridity, high sea water salinity, and sea level rise—all of which are projected to become more significant with climate change—are already part of the day-to-day reality in the UAE.

The UAE therefore recognizes the need to urgently act on climate change. For this reason, the nation has been increasing its climate action ambitions steadily throughout the years. It was the first nation among the Gulf Cooperation Council ("GCC") countries to ratify the Paris Agreement in 2016. Since then, the UAE has advanced four increasingly ambitious Nationally Determined Contributions ("NDCs"), each with improved targets more ambitious than the last. This level of ambition is embedded in each strategy, project, and plan put forward to tackle climate change. The UAE's First Long-Term Strategy for Demonstrating Commitment to Net Zero by 2050 ("LTS"), presented in this article, is intended to act as an enabler of the UAE's ambitions set out in the UAE Net Zero by 2050 Strategic Initiative.[1]

As part of these initiatives, the UAE hosted the 28th edition of the Conference of the Parties ("COP") to the UN Framework Convention on Climate Change ("UNFCCC") ("COP28") with the presence of world leaders to work together on solutions to tackle climate change. There are 198 members (197 countries and the European Union) of the Convention, constituting near universal membership.

COP28 provided a milestone opportunity for the world to come together, course correct, and drive progress to keep 1.5C within reach, in order to meet the goals and ambitions of the Paris Agreement.

The event included two zones:

1. The Blue Zone was a UNFCCC-managed site, open to accredited parties (country negotiators) and observer delegates (NGOs, IGOs, UN Agencies), media, and world leaders. The Blue Zone hosted formal negotiations throughout the duration of the conference, including hundreds of official side events, panel discussions, speaking and cultural events.
2. The Green Zone was a space managed and delivered by the COP28 UAE Presidency. It offered a platform to non-accredited delegates,

[1] UAE Ministry of Climate Change and Environment. (2023). The United Arab Emirates' First Long-Term Strategy (LTS) Demonstrating Commitment to Net Zero by 2050. Retrieved from https://unfccc.int/sites/default/files/resource/UAE_LTLEDS.pdf.

including youth groups, civil society, the private sector, and indigenous groups to have their voices heard, promoting dialogue and awareness about climate action. Through the Green Zone, the COP28 UAE Presidency offered the private sector the opportunity to showcase their contributions and solutions to the global climate challenge. The Green Zone also hosted technical conferences, panels, and activations with a focus on existing and future solutions to climate change mitigation.

The following extracts from the [Chairman's] summary at COP28 illustrated the importance of CCUS for the UAE's carbon ambitions[2]:

1) "Renewable capacity must be trebled by 2030 to increasingly substitute demand for fossil fuels and enhance energy access whilst ensuring continued energy security:
   (...)
   a. Permitting and grid-connections must be streamlined, and liability regimes reformed, to accelerate deployment of low-carbon technologies.
2) In parallel, fossil fuels must phase-down significantly this decade to keep 1.5°C within reach:
   a. Carbon capture, utilisation and storage (CCUS) technologies have potential to support the energy transition, but should be focused where most needed, e.g., in heavy emitting industrial sectors, and to decarbonise existing infrastructure in developing countries that is not planned to be decommissioned in the near term.
   b. National policies, planning and investments must be coherently aligned to a 1.5°C pathway, and implementation plans with timeframes are required to enable this."

Equally, the decarbonization charter published during COP28 contained similar principles:

Accelerated decarbonization: Implementation of the above ambitions will need increasing alignment with industry best practices to accelerate the decarbonization of the industry's operations. Reducing methane emissions and stopping routine flaring will provide the largest reductions in emissions from oil and gas operations by 2030. Increasing alignment with broader

[2] Summary of Global Climate Action at COP 28. Retrieved from https://unfccc.int/documents/636485?gad_source=1&gclid=EAIaIQobChMI2NKJ9vHThQMV-0NBAh29VACcEAAYASAAEgIYNfD_BwE.

> industry best practices would mean, for example, additional reductions from electrification of upstream operations, deployment of CCUS, and use of low-carbon hydrogen.

Similarly, the Summary of Global Climate Action at COP 28 further highlighted these principles:

> The High-Level Champions have also contributed to the expansion of the Breakthrough Agenda launched at COP 26. The Cement and Concrete Breakthrough, launched by Canada and the United Arab Emirates, aims to focus the industry on speeding up decarbonization by sharing 3 best practices, working on policy and standards, and supporting new innovation in emerging areas like carbon capture and storage or utilization, or circular economy.

Moreover, as far back as 2013 His Excellency, Dr. Sultan Al Jaber, (future President of COP28), outlined the importance of CCUS in the UAE[3]:

> From investments in renewable energy and energy efficiency to the development of new natural gas infrastructure and peaceful nuclear power capacity, the UAE is making progress in addressing the nation's growing energy demands. But that progress is just a start. Infrastructure projects alone can't provide a long-term solution. It is now time for countries to develop clear national energy strategies that coherently address national goals in a regional and global context. For the UAE, we express these goals as follows:
>
> 1. Provide enough energy to power economic growth and serve growing populations, so that energy sufficiency does not impede development.
> 2. Ensure the long-term viability of the region as a major exporter of hydrocarbons to the world economy.
> 3. Minimize the overall energy costs to government and society by clarifying pricing signals to reduce consumption.
> 4. Maximize inward investment and job creation for new energy sectors.
> 5. Minimize greenhouse gas emissions and the environmental damage they can create.

[3] Al Jaber, S. MENA Energy Transition Strategy: A Call for Leadership in Energy Innovation. Retrieved from https://www.sciencedirect.com/journal/energy-strategy-reviews/vol/2/issue/1.

6. Identify and capture the benefits of scale and synergistic planning that would result from coherent national policies and work plans. (...)

> In the UAE, we also firmly believe that oil and gas are a bridge to future energy. As such, we are investing heavily in large-scale carbon 6 Energy vision / Energy Strategy Reviews 2 (2013) 5e7 capture, utilization and storage (CCUS) projects that capture carbon emissions from power generation and industry, and use them for enhanced oil recovery in Abu Dhabi's oil reservoirs, rather than re-injecting natural gas. The program will both mitigate carbon emissions and help sustain our supplies and production of hydrocarbons for domestic and international markets.

Undoubtedly, the UAE views the deployment of technological innovations as a conducive means of achieving its objectives, which includes a focus on CCUS, which has long been considered an under-used technology, like offshore wind.[4]

By way of regional comparison, the three major carbon capture projects in the GCC are located in the Kingdom of Saudi Arabia, Qatar, and the UAE, which make up approx. 10% of the 40 million tons per annum ("Mt p.a.") of CO_2 captured globally. The GCC lags behind the USA, which is the global leader in CCUS, and accounts for approximate 60% of global CO_2 capture.[5]

However, there is scope to change this. Gaffney Cline, (2022) provides a helpful overview of the potential for CCUS activities in the GCC, stating that member countries have a prevalence of high purity CO_2 emitting industries such as natural gas processing, fertilizer, and petrochemical sectors which could provide significant advantages for leveraging CCUS as a decarbonization tool. The study identified the following key opportunities for deploying CCUS in the GCC:

- CCUS has promising applications across multiple industrial activities and at a lower cost compared to other regions.

[4] Ashurst. (2024). Powering Change Technologies Fueling the Future, page 30, Retrieved from https://www.ashurst.com/en/powering-change-technologies-fuelling-the-future/

[5] Russel, D., Ridge, S., Ahmad, I. COP28 Ahead: Carbon Capture, (Utilisation) and Storage. Retrieved from https://cms-lawnow.com/en/ealerts/2023/11/cop28-ahead-carbon-capture-utilisation-and-storage.

- There is significant storage potential close to emissions clusters. This could also support cost-effective direct air capture.
- The region has potential for many promising large-scale CCUS hubs, with cross-border cooperation providing additional options.

The study concluded that the GCC could use CCUS to become a global leader in low-carbon hydrogen production and the export of low-carbon goods, which could translate into significant upsides by 2050. Additionally, the deployment of CCUS could support the creation of highly skilled jobs in various economic sectors.

2 Key CCUS Projects in the UAE

The UAE has been a pioneer in CCUS in the GCC and implemented the first large-scale CCUS project in the region. Al Reyadah project, which started operation in 2016, captures up to 800,000 tons of CO_2 per year from the Emirates Steel Arkan plant in Abu Dhabi, and transports it via pipeline to the Rumaitha and Bab oil fields, where it is injected for enhanced oil recovery ("EOR").

The UAE has also announced several other CCUS projects that are under development or feasibility study. These include:

- The "Dubai CCUS project," which is a collaboration between the Dubai Electricity and Water Authority (DEWA), the Dubai Carbon Centre of Excellence (DCCE), and the UK-based Carbon Clean Solutions Limited (CCSL). The project plans to capture CO_2 from a 600 MW natural gas-fired power plant in Dubai, and use it for various applications, such as concrete production, algae cultivation, and methanol synthesis. The project is expected to capture up to 360,000 tons of CO_2 per year.[6]
- The "Fujairah CCUS project," which is a partnership between the Fujairah Cement Industries (FCI), the Abu Dhabi Future Energy Company (Masdar), and the US-based CO_2Concrete. The project aims to capture CO_2 from the FCI plant in Fujairah and use it to

[6] DEWA. (2019). DEWA Signs MoU with DCCE and CCSL to Implement Pilot Project to Use Carbon Capture Technology. Retrieved from https://www.dewa.gov.ae/en/about-dewa/news-and-media/press-and-news/latest-news/2019/10/dewa-signs-mou-with-dcce-and-ccsl-to-implement-pilot-project-to-use-carbon-capture-technology.

produce low-carbon concrete. The project is expected to capture up to 100,000 tons of CO_2 per year.[7]

- In 2021, an agreement with TotalEnergies to explore joint research, development, and deployment partnership opportunities in the areas of CO_2 emission reductions and CCUS.[8]
- American oil company Occidental, a leading proponent, agreed to evaluate direct air capture in the UAE and the US. And in July 2023, another UAE player, Sharjah National Oil Company, announced a partnership with Japan's Sumitomo to develop carbon capture around Sharjah.[9]
- Research shows the following projects: (1) In September 2023, the Habshan CCUS project was approved which may have the capacity to store 1.5 Mt p.a. of CO_2; (2) Eni, from Italy, has been exploring CCUS opportunities concerning innovative geo-mechanical and geo-chemical CCUS programs, (2) In July 2023, the Sharjah National Oil Corporation signed a deal with Sumitomo Corporation for work on a CCS project.[10]
- In 2023, Shell's Cansolv CO_2 Capture System technology has been selected for the Habshan carbon capture, utilization, and storage (CCUS) project with the capacity to capture and permanently store 1.5 million tons of carbon dioxide (CO_2). It is one of the largest post-combustion carbon capture projects today and among few which have taken final investment decision globally.[11]

[7] Masdar. (2020). Masdar and FCI to Develop First Industrial-Scale Carbon Capture Project in MENA. Retrieved from https://masdar.ae/en/news-and-events/news/2020/01/13/masdar-and-fci-to-develop-first-industrialscale-carbon-capture-project-in-mena.

[8] TotalEnergies. (2020). Total and ADNOC Sign Strategic Framework Agreement on CO_2 Emissions Reduction and CCUS retrieved from https://totalenergies.com/media/news/communiques-presse/total-and-adnoc-sign-strategic-framework-agreement.

[9] Mills, R. the National News, 2023, https://www.thenationalnews.com/business/comment/2023/09/11/how-the-uae-is-advancing-efforts-to-strengthen-its-carbon-capture-commitment/

[10] Russel, D., Ridge, S., Ahmad, I. COP28 Ahead: Carbon Capture, (Utilisation) and Storage. Retrieved from https://cms-lawnow.com/en/ealerts/2023/11/cop28-ahead-carbon-capture-utilisation-and-storage.

[11] Shell. (2023). ADNOC Gas Awards $615m Contract for One of MENA's Largest Integrated Carbon Capture Projects. Retrieved from https://www.shell.com/business-customers/catalysts-technologies/resources-library/adnoc-shell-cansolv-technology-large-scale-carbon-capture.html.

- In 2024, an investment of 10.1% stake was made in UK-based CCUS player Storegga, represented an international equity investment in carbon management and supports its long-term strategy from the side of the UAE. The stake in Storegga will provide new CCUS partnerships with Eni (via Neptune), Shell, and Talos and build upon existing agreements with the likes of TotalEnergies, BP, Santos, and Oxy. Such deal benefits the UAE in order to build upon its existing UK position via the BP-operated H2 Teeside project and provide entry into Norway and the US. It will take net capacity up to 5.2 Mtpa and a step closer toward meeting its 10 Mtpa by 2030 target.[12]
- Furthermore, in 2024 the UAE has considered a pilot project with a British-Omani sustainability company 44.01 to permanently convert carbon dioxide from the air into a mineral within rock formations in Fujairah. The project is being carried out in partnership with the Fujairah Natural Resources Corporation and Masdar.[13]
- Finally, it has been noted the allocation of USD 23bn for deployment in its low-carbon solutions projects, more strategic moves are expected.[14]

3 Applicable Policies, Legal Framework, and Norms with respect to CCUS in the UAE

As outlined below, the UAE has adopted a few policies and strategies that indirectly support the deployment of CCUS (albeit not having a dedicated CCUS regulation) as part of its broader vision and commitment to sustainable development, energy diversification, and climate action.

According to the UAE's First Long-Term Strategy Demonstrating Commitment to Net Zero by 2050 (issued in 2023), there are more than 30 existing strategies across all sectors. Additional policies, such as the

[12] Boroujerdi, N, Anderson, G., Muray, D., Evans, M. The Challenges and Opportunities in Europe's Oil & Gas, CCUS and Hydrogen Sectors. Retrieved from https://www.woodmac.com/news/opinion/challenges-and-opportunities-europes-oil-gas/.

[13] Mills, R. The National News, 2023. Retrieved from https://www.thenationalnews.com/business/comment/2023/09/11/how-the-uae-is-advancing-efforts-to-strengthen-its-carbon-capture-commitment/.

[14] ADNOC Allocates 15 Billion to Low-Carbon Solutions. Retrieved from https://www.adnoc.ae/en/news-and-media/press-releases/2023/adnoc-allocates-15-billion-to-low-carbon-solutions.

Clean Energy Strategic Target 2035 in Abu Dhabi and the UAE National Hydrogen Strategy, were updated during the development of the net zero pathway.[15]

Below are the details of a few key policies and strategies:

- The UAE Vision 2021, which sets the national agenda and the key performance indicators for the UAE's social, economic and environmental development, and aims to achieve a sustainable environment and infrastructure, a diversified and competitive knowledge economy, and a responsible and proactive role in addressing global challenges.
- The UAE Energy Strategy 2050, which sets the targets and the roadmap for the UAE's energy sector and aims to increase the UAE's contribution of clean energy to 50% of the total energy mix, reduce the carbon footprint of power generation by 70%, and improve energy efficiency by 40% by 2050.
- The UAE Green Agenda 2015–2030, which sets the framework and the action plan for the UAE's green development and aims to enhance the efficiency and competitiveness of the UAE's economy, conserve natural resources and biodiversity, and mitigate and adapt to climate change.
- The UAE Climate Change Plan 2017–2050, which sets the vision and the objectives for the UAE's climate action and aims to enhance the UAE's resilience to climate impacts, reduce greenhouse gas (GHG) emissions, and contribute to the global efforts to combat climate change.
- The UAE Nationally Determined Contribution (NDC) under the Paris Agreement, which outlines the UAE's voluntary commitment and contribution to the global goal of limiting the temperature rise to well below 2 °C and specifies the steps the UAE will take to reduce its GHG emissions by 23.5% by 2030 compared to the business-as-usual scenario.

[15] UAE Ministry of Climate Change and Environment, 2023, The United Arab Emirates' First Long-Term Strategy (LTS) Demonstrating Commitment to Net Zero by 2050. Retrieved from https://unfccc.int/sites/default/files/resource/UAE_LTLEDS.pdf.

These policies demonstrate the UAE's leadership and ambition in pursuing a low-carbon and sustainable development path supporting the adoption of CCUS as a key means to achieve these goals.

At present the UAE does not have a specific legislation or regulation governing CCUS, but rather relies on a combination of existing laws and regulations that cover various aspects of the CCUS value chain, such as environmental protection, energy, oil and gas, water, and land use. Some of the relevant laws and regulations include:

- The Federal Law No. 24 of 1999 for the Protection and Development of the Environment, which sets the general principles and standards for environmental protection and conservation and requires the issuance of environmental permits and licenses for any activity that may have an impact on the environment.
- The Federal Law No. 1 of 2019 on the Integrated Waste Management, which regulates the management of waste, including hazardous waste, and imposes obligations on waste generators, transporters, and treatment facilities.
- The Federal Law No. 2 of 2018 on the Regulation of the Petroleum Sector, which regulates the exploration, production, transportation, storage, and marketing of petroleum and petroleum products, and grants the Supreme Petroleum Council (SPC) the authority to issue policies and directives for the petroleum sector.
- The Federal Law No. 11 of 1981 on the Regulation of the Water Resources, which regulates the use, conservation, and protection of water resources, and requires the approval of the Ministry of Energy and Industry (MEI) for any activity that may affect the water resources.
- The Federal Law No. 5 of 1985 on the Civil Transactions, which regulates the ownership, transfer, and use of land and property, and defines the rights and obligations of the landowners and the users.

In addition to the federal laws and regulations, each of the seven emirates that constitute the UAE has its own local laws and regulations that may apply to CCUS activities within its jurisdiction. For example, Abu Dhabi, the largest and most populous emirate, has issued several laws and regulations that are relevant to CCUS, such as:

- The Law No. 19 of 2005 on the Regulation of the Water and Electricity Sector, which establishes the Regulation and Supervision Bureau (RSB) as the regulator of the water and electricity sector, and requires the issuance of licenses for the generation, transmission, distribution, and supply of water and electricity.
- The Law No. 4 of 2010 on the Regulation of the Abu Dhabi Water and Electricity Sector, which amends the Law No. 19 of 2005 and introduces the concept of renewable energy and clean energy and requires the RSB to set targets and incentives for the production and consumption of renewable and clean energy.
- The Law No. 11 of 2016 on the Establishment of the Department of Energy, which replaces the MEI as the competent authority for the energy sector in Abu Dhabi, and grants it the power to issue policies, strategies, plans, and regulations for the energy sector, including CCUS.
- The Law No. 13 of 2016 on the Regulation of the Petroleum Sector in the Emirate of Abu Dhabi, which regulates the exploration, production, transportation, storage, and marketing of petroleum and petroleum products in Abu Dhabi.
- The Law No. 21 of 2005 on the Establishment of the Environment Agency—Abu Dhabi (EAD), which establishes the EAD as the environmental regulator and protector in Abu Dhabi and grants it the authority to issue environmental permits and licenses, monitor and enforce environmental compliance, and conduct environmental impact assessments.
- Federal Decree-Law no. 11 of 2024, on the Reduction of Climate Change Effects, which sets out requirements for the competent authorities to determine annual targets for emission reduction for all sectors at the national level in accordance with the national pathway to climate neutrality.[16]

As illustrated by Al-Saleh et al. (2023), the following stakeholders may be involved in the development of the CCUS market: in Abu Dhabi

1. Government Agencies:

[16] Government of the UAE. Federal Decree-Law no. 11 of 2024, on the Reduction of Climate Change Effects. Available from: https://uaelegislation.gov.ae/en/legislations/2558.

a. Government of Abu Dhabi
b. Supreme Petroleum Council
c. Department of Economic Development
d. Environmental Agency-Abu Dhabi
e. The Executive Affairs Authority
f. Abu Dhabi Water and Electricity Authority
g. Mubadala
h. Masdar Carbon Management Unit

2. Research Institutes/Universities:

a. Masdar Institute of Science and Technology
b. Petroleum Institute
c. UAE University
d. University of Oslo
e. Norway Institute of Air Research
f. United States Department of Energy
g. Massachusetts Institute of Technology
h. Herriot Watt-University
i. Rice University
j. Stanford University
k. University of Texas at Austin
l. National Energy and Water Research Center

3. Private firms:

a. Siemens
b. Statoil
c. Praxair Industrial Gases LLC
d. Mustang Engineering
e. J C Penny

4. CO_2 emitters:

a. Emirates Steel Industries
b. Taweelah Asia Power Company
c. Emirates Aluminum
d. Hydrogen Power Abu Dhabi

Other emirates, such as Dubai, Sharjah, and Ras Al Khaimah, have also issued their own laws and regulations that may affect CCUS activities within their territories.

4 Challenges and Opportunities of CCUS in the UAE

To facilitate the development of a CCUS infrastructure network in the UAE, the regulation of transport and storage activities is currently being considered. In addition, as the GCC starts to develop its hydrogen strategies, the use of CCUS will be integral to the establishment of blue hydrogen production capabilities in the region. Given the production of blue hydrogen still generates CO_2 in the value chain (albeit, blue hydrogen is less carbon-intensive than gray hydrogen), the deployment of such technologies must be supported by investment into carbon–neutral energy sources. If not, criticisms of nations and oil and gas (particularly state-run) companies' failure to directly tackle the adverse impacts of climate change will remain. A balance between emitting less and removing more will be key to strike in the region.[17]

Despite the progress and potential of CCUS in the UAE, there are challenges that need to be considered for its wider adoption and deployment. These include:

- The high cost and complexity of CCUS technologies, which require significant capital investment, operational and maintenance expenses, and technical expertise.[18]
- The lack of a clear and consistent regulatory and policy framework for CCUS, which creates uncertainty and risk for project developers and investors.[19]
- The limited public awareness and acceptance of CCUS, which may affect the social license and stakeholder engagement of CCUS projects.[20]

[17] Russel, D., Ridge, S., Ahmad, I. COP28 Ahead: Carbon Capture, (Utilisation) and Storage. Retrieved from https://cms-lawnow.com/en/ealerts/2023/11/cop28-ahead-carbon-capture-utilisation-and-storage.

[18] IEA. (2019). Carbon Capture, Utilisation and Storage. Retrieved from https://www.iea.org/reports/carbon-capture-utilisation-and-storage.

[19] GCCSI. (2018). Global Status of CCS 2018: At a Critical Juncture. Retrieved from https://www.globalccsinstitute.com/resources/global-status-report/.

[20] Bui, M., Adjiman, C., Bardow, A., Anthony, E., Boston, A., Brown, S., … & Fennell, P. (2018). Carbon Capture and Storage (CCS): The Way Forward. Energy & Environmental Science, 11(5), 1062–1176.

- The competition with other low-carbon technologies, such as renewable energy, hydrogen and energy efficiency measures, which may offer more attractive or feasible options for decarbonizing the UAE's energy system.[21]
- The cost of adding CCUS must make commercial sense. The high capital cost associated with the implementation of CCUS may restrict the willingness of external investment without sufficient certainty of future returns. As part of this, the general lack of a legal regulatory framework for CCUS in the region will need to be addressed. The UAE's CCS policy package includes plans for the introduction of Carbon Contracts for Difference ("Carbon CfDs") which will offer long-term revenue stability (akin to similar contracts seen in the UK and globally for renewables).[22]
- The right incentives also have to exist (tax credit and CFD schemes, etc.) to encourage first and early movers to invest more and scale the industry. The industry would also welcome detailed proposals in respect of the gas shipping and transportation regulatory model. A constant theme is the need for collaboration and partnership at all levels of the industry. This is still a nascent technology, subject to ongoing refinement and there are considerable benefits in sharing experience, technological know-how and co-operating within clusters and hubs.[23]
- A key challenge for CCUS (as with the vast majority of new energy technology) is ensuring the development of an adequate regulatory framework. Existing petroleum and environmental legislation can be utilized to an extent but certain issues (for example, allocation of long-term liability for sequestrated CO_2) often require that bespoke regulations be put in place.[24]

[21] IRENA. (2019). Renewable Energy Outlook: UAE. Retrieved from https://www.irena.org/publications/2019/Nov/Renewable-Energy-Outlook-UAE.

[22] Russel, D., Ridge, S., Ahmad, I. COP28 Ahead: Carbon Capture, (Utilisation) and Storage. Retrieved from https://cms-lawnow.com/en/ealerts/2023/11/cop28-ahead-carbon-capture-utilisation-and-storage.

[23] Ashrurt. (2024). Powering Change Technologies Fuelling the Future, page 30, Retrieved from https://www.ashurst.com/en/powering-change-technologies-fuelling-the-future/.

[24] Kapinos, M. (2021). Carbon Capture, Utilization and Storage—What is the Big Deal? Retrieved from https://www.akingump.com/en/insights/blogs/speaking-energy/carbon-capture-utilization-and-storage-what-is-the-big-deal.

- As illustrated by Liu et al., 2022,[25] the following key regulatory issues and challenges are applicable to a number of member countries within the GCC:
  - Regulatory issues: (a) no explicit law or regulation on CCUS with some general laws loosely applicable to CCUS; (b) No formal permitting or authorization process for CCUS projects; (c) under-developed technical standards and environmental management criteria; (d) no clear provisions on liability, including transfer of liability, liability from site closure and monitoring, and liability for contingency;
  - Integration of CCUS in climate mitigation strategy: (a) varying degree of commitments to CCUS by member states of GCC as climate mitigation strategy; (b) early stage of CCUS development spanning from research to project demonstration; (c) no immediate plan to enforce policy instruments to reduce emissions; (d) incremental and pragmatic CCUS projects relying on public funding; (e) concentration of CCUS for EOR due to economic incentives; (e) significance of transboundary movement of CO_2 after sequestration;
  - Ensuring consistency with international law: (a) most member states of the GCC have not ratified the London Protocol (except Saudi Arabia); (b) lack of common regulatory grounds to regulate transboundary issues; (c) strong implications on regional cooperation due to lack of common regulatory denominator.
  - Similarly, Al-Saleh, 2023, reinforces that globally, CCUS regulatory frameworks are in the formative stages of development. Based on the interviews carried on his study, a favorable CCUS regulations and financial mechanisms should be developed to govern CCUS activities in Abu Dhabi. This can encourage increased participation of current and future stakeholders in CCUS developments in Abu Dhabi. At present, the Environmental Agency-Abu Dhabi has set guidelines for regulating the general effluent emissions from CO_2 capture facilities.

[25] 1.Li, Q., Chen, ZA., Zhang, J-T., Liu, L-C., Li, XC., Jia, L. (2016). Positioning and Revision of CCUS Technology Development in China. Int. J. Greenhouse Gas Control 46, 282–293.

Such guidelines should be adhered to when designing CCUS facilities in Abu Dhabi.
- Other impediments for the development of CCUS technologies in Abu Dhabi include: (a) there is no financial subsidy in place to further encourage investment in CCUS-based projects; (b) knowledge development (e.g., under-developed CCUS-based capability); (c) it is unclear when CCUS supporting legislation and regulation will be published; (d) market creation (e.g., the market has been slow to develop given the absence of CCUS-based regulations and other factors); (e) resources mobilization (e.g., lack of CCUS-skilled workforce).[26]

However, there are also several opportunities to accelerate the development and deployment of CCUS in the UAE. These include:

- The strategic vision and leadership of the UAE government, which has set ambitious targets and plans for reducing greenhouse gas emissions and enhancing energy security and diversification[27];
- The abundance of accessible natural and geological resources, such as oil and gas fields, saline aquifers, and carbonate rocks, which can provide ample sources and sinks for CO_2 capture and storage[28];
- The existing and emerging markets and applications for CO_2 utilization, such as EOR, low-carbon products, and synthetic fuels, which can create value and revenue streams for CCUS projects[29];
- The regional and international cooperation and collaboration on CCUS, such as the GCC Carbon Capture and Storage Network, the

[26] Al-Saleh, Y., Auktor, G., Varkey, V. (2023). Challenges and Opportunities for the Emerging Carbon Capture, Utilisation and Storage Innovation System in the United Arab Emirates. Retrieved from https://www.researchgate.net/publication/263497190_Challenges_and_opportunities_for_the_emerging_carbon_capture_utilisation_and_storage_innovation_system_in_the_United_Arab_Emirates.

[27] UAE. (2017). UAE Energy Strategy 2050. Retrieved from https://u.ae/en/about-the-uae/strategies-initiatives-and-awards/strategies-plans-and-visions/environment-and-energy/uae-energy-strategy-2050.

[28] Al Hosani, K., Al Kindi, S., Al Shamsi, K., Al Dhaheri, N., & Al Muhairi, A. (2018). CO_2 Storage Potential in the United Arab Emirates: A Review. International Journal of Greenhouse Gas Control, 78, 57–67.

[29] Mac Dowell, N., Fennell, P., Shah, N., & Maitland, G. (2017). The Role of CO_2 Capture and Utilization in Mitigating Climate Change. Nature Energy, 2(4), 17071.

Clean Energy Ministerial CCUS Initiative, and the Mission Innovation CCUS Challenge, which can provide platforms and mechanisms for knowledge sharing, capacity building, and technology transfer[30];

- Using CCUS recognizes the existing energy mix in the Middle East and offers a route to move toward net zero. It is certain that CCUS alongside renewable and nuclear energy generation will be vital for reaching net zero and decarbonization targets across the Middle East[31];
- Chemical absorption is one of the most widely used methods for capturing CO_2. Post-combustion capture techniques are being studied, which includes CCUS. Research on post-combustion CO_2 capture has just been underway in the last twenty years. Its principal goal is to enhance the effectiveness of the CO_2 absorption method while significantly lowering specific operational difficulties, including the deterioration of solvents and corrosion[32];
- While the development of vertically integrated structures can be difficult to establish in nascent industries, certain GCC countries have been successful in this approach, being led by state-owned companies which are able to deploy government influence and large capex toward projects. There is also significant potential for the establishment of CCUS hubs using shared transport and infrastructure given the region's existing assets such as natural gas fields and industrial clusters[33];
- The GCC region has some unique advantages in deploying CCUS, including the potential storage sites offered by the vast oil and

[30] GCCSI. (2019). GCC Carbon Capture and Storage Network. Retrieved from https://www.globalccsinstitute.com/initiatives/gcc-carbon-capture-and-storage-network/.

[31] Russel, D., Ridge, S., Ahmad, I. COP28 Ahead: Carbon Capture, (Utilisation) and Storage. Retrieved from https://cms-lawnow.com/en/ealerts/2023/11/cop28-ahead-carbon-capture-utilisation-and-storage.

[32] McLaughlin, H., Littlefield, A., Menefee, M., Kinzer, A., Hull, T., Sovacool, B., Bazilian, B., Carbon Capture Utilization and Storage in Review: Sociotechnical Implications for a Carbon Reliant World. Retrieved from https://www.sciencedirect.com/science/article/pii/S1364032123000710.

[33] Pole, S. (2023). Too Little, Too Late? Challenges to Deploying CCUS at Scale. Retrieved from https://www.lexology.com/library/detail.aspx?g=a63ae6e3-5953-4c5b-8a94-d09378cfde2b&utm_source=lexology+daily+newsfeed&utm_medium=html+email+-+body+-+general+section&utm_campaign=lexology+subscriber+daily+feed&utm_content=lexology+daily+newsfeed+2023-11-30&utm_term=.

gas reserves, and the win–win opportunities through the combined use of CCUS for EOR to promote the roll-out of CCUS in this region.[34]

- As illustrated by Sagoff, 2024, the use of generative artificial intelligence (AI) may provide R&D capabilities useful in manufacturing CCUS capture, transportation, and storage infrastructure.
- Moreover, Chen et al. (2020), highlights that AI technologies may offer a powerful tool to model the complex behavior of energy systems and optimize the operation of carbon capture facilities (such as those adjoining power plants), while balancing the input of high penetration renewable power in future low-carbon energy systems.

5 Lessons Learned

While the UAE has been a pioneer in the development of CCUS technologies to date, to ensure continued growth, a number of steps will need to be put in place, including: the development of CCUS regulation and attracting experienced CCUS professionals from developed CCUS markets overseas to ensure effective implementation, training, and overall management of the issues around such technologies.

According to the International Energy Forum (2024),[35] the main challenge related to the creation of a clear regulatory framework is the establishment of cohesive and complementary industry standards to allow CCUS to scale faster. Other ideas would be effective implementation of emission reduction targets through market incentives such as carbon price discovery mechanisms and government support measures such as contracts for difference or tax allowances.

The implementation of CCUS policies requires global cooperation and coordination to ensure that the technology is implemented effectively, efficiently, and equitably. This requires international agreements and cooperation between governments, industries, and other stakeholders.

[34] Li, Q., Chen, Z.A., Zhang, J.-T., Liu, L.-C., Li, X.C., Jia, L. 2016. Positioning and Revision of CCUS Technology Development in China. Int. J. Greenhouse Gas Control 46, 282–293.

[35] International Energy Forum. (2024). IEF Publishes Reports on MENA Circular Carbon Policies to Advance Collaboration and Climate Goals. Retrieved from https://www.ief.org/news/ief-publishes-reports-on-mena-circular-carbon-policies-to-advance-collaboration-and-climate-goals.

There are also challenges associated with the permitting process for CCUS projects. The current permitting process [in many GCC countries can be lengthy and complex, requiring approvals from multiple agencies and stakeholders. This can lead to delays and increased costs for CCUS projects.

Furthermore, the International Energy Forum, 2024, recommends immediate actions for the UAE in the following aspects for further development of CCUS technologies: (a) financial incentive for pursuing CCUS; (b) knowledge development of CCUS-based capability); (c) more transparent in policy-making process; and having further CCUS-skilled workforce.

Finally, research also shows that the use of AI can be an important differential for the enhancement of these technologies.

6 Conclusion

CCUS is a vital technology for reducing greenhouse gas emissions and mitigating climate change, especially in countries like the UAE that have a high dependence on fossil fuels and a high-carbon intensity. The UAE has demonstrated its commitment and capability to implement CCUS projects and has several existing and planned projects that can showcase its leadership and innovation in the field.

However, there are still certain challenges that need to be addressed for CCUS to reach its full potential and scale in the UAE. Therefore, it is essential to develop and implement a comprehensive legal framework, coherent strategy, and opportunities to implement AI automation that can address the technical, economic, regulatory, social, and environmental aspects of CCUS, and leverage the opportunities and drivers that can enhance its value and viability.

CHAPTER 21

CCUS in the United Kingdom

Nathaniel Babajide

1 Introduction

Carbon capture utilisation and storage (CCUS), sometimes also referred to as carbon capture storage (CCS), remains one of the essential instruments for mitigating climate change and reaching net-zero emissions goals globally (Dziejarski et al. 2023; International Energy Agency (IEA) 2023). Since the signing of the Paris Climate Agreement, efforts to limit the average global temperature from rising to 1.5 °C have continued to witness significant progress spurred by supportive policies and rising investment in CCUS development in different regions around the globe (Isabela and Hirdan 2021; eSilva and Costa 2021). As reiterated in Sarnoff (2021), the technology entails capturing carbon dioxide (CO_2) directly from the air of large point sources, such as power generation and industrial infrastructure, and compressing and transporting (via pipeline, rail ship, or truck) for an array of applications or injection into depleted geological formations/reservoirs.

N. Babajide (✉)
Aberdeen Business School, Robert Gordon University, Aberdeen AB10 7QE, UK
e-mail: babajidenathaniel@yahoo.com; n.babajide@rgu.ac.uk

E. G. Pereira et al. (eds.), *Carbon Capture Utilization and Storage*, Sustainable Development Goals Series,
https://doi.org/10.1007/978-3-031-81272-9_21

With more than 200 million tonnes of anthropogenic CO_2 successfully injected into the underground rock formation, CCUS has been identified to play strategic roles in global decarbonisation efforts in several ways, including deep emissions offset in hard-to-abate sectors, such as cement, steel or chemicals, diversification of energy supply sources, and energy security through effective use of fossil and other fuels such as biomass, waste, and low-carbon hydrogen production, and abating technically difficult CO_2 from the atmosphere to achieve balance emissions outlook (UKCSS Research Community 2022; Global CCS Institute 2023).

Amongst CO_2 capture technologies, post-combustion, oxyfuel, and precombustion remain the well-known processes and studies are replete on their applications for various purposes (Yusuf and Ibrahim 2023). Principally, with CO_2 being used in different industries, including food and beverage processing, there already exist large-scale processes for capturing, channelling, storing, and utilising this gas. Nonetheless, Kuckshinrichs and Hake (2015) noted that most of the carbon capture processes developed recently are targeted at applications in fossil-fired power plants and large-scale industrial processes in cement, petrochemical (oil refineries), iron and steel, and hydrogen-producing industries (Leeson et al. 2017).

Recent estimates by IEA 2023 indicated that about 1 billion tonnes worth of storage capacity is required globally by 2030 for a net-zero pathway to be consistent with achieving 1.5 °C goals. Whilst global carbon removal capacity receipt stood at $13 billion in 2023, Garden (2023) projected that global investment in carbon removal capacity to be in the range of $100–$400 billion by 2030 and reaching net zero by 2050 would require $6 trillion of investment, starting with a minimum 2030 investment of half a trillion. Nonetheless, the Global CCS Institute (2018) revealed that the available underground storage resource globally is far higher than what is needed to meet climate targets, with countries such as Canada, Australia, China, and the UK identified as having some of the largest CO_2 storage potentials globally.

Given its distinctive geological advantage, industrial infrastructure, expertise, and experience (in the oil, gas, and petrochemicals sector), the UK has the capacity to safely store up to 78 billion tonnes of CO_2 under its seabed and has continued to make significant efforts in developing CCU S technologies into a profoundly valuable national asset to expand its economy base to £5 billion per annum and achieve set net-zero goals by 2050. On an annual basis, the UK has the capacity to store up to 20–30 million tonnes of CO_2 by 2030, the equivalent of removing 4 to 6 million cars from UK roads every year (Department for Energy Security and Net Zero 2024). In addition, the UK government's commitment and a significant investment of up to £20 billion in the carbon capture industry positions it as one of the largest in Europe.

Further, CCUS deployment constitutes a significant area of focus given the UK's intention to harness socio-economic growth prospects, including generating exporting and job creation opportunities whilst transitioning to net-zero emissions. Also, the country has been identified to be comparatively advantageous in the aspect of production, innovation, and utilisation of CCUS products, services, and technologies. Amongst myriad of clean and emerging technologies, CCUS deployment has been reported to yield above-average returns. It possesses a huge growth potential in scaling up of investment required to reach its net-zero ambitions. Against this backdrop, this chapter examines the historic CCUS situation and challenges in the UK.

The chapter is thus organised as follows: This brief introduction is followed by an overview of the UK GHGs and CO2 emissions trends and drivers in Sect. 2. Several CCUS main projects, global market outlook and financing options, regulatory frameworks, and implementation strategies by the UK government are discussed in Sect. 3. Section 4 discusses some of the key opportunities and challenges associated with CCUS deployment in the UK and globally. Section 5 concludes the study.

2 Synopsis of UK GHGs and CO_2 Emissions: Trends and Drivers

As Allan et al. (2023) underlined drastic consumption of fossil fuels, i.e. coal, oil, and natural gas, as the leading source of GHGs, especially CO_2 anthropogenic emissions, the latest annual estimates in carbon brief analysis (2023) revealed that the country's total annual territorial GHGs emissions plummeted to reach a new record of 383 million tonnes of carbon dioxide equivalent (MtCO2e) in 2023, 53% below 1990 levels and the lowest since 1879. Figure 1 further confirms an intermittently declining trend in GHGs and CO_2 emissions since the 1990s, whereas the latter stood at 303 MtCO2e in 2023, denoting approximately 50% decline from the 1990 estimate of 603.6 MtCO2e (DESNZ 2023).

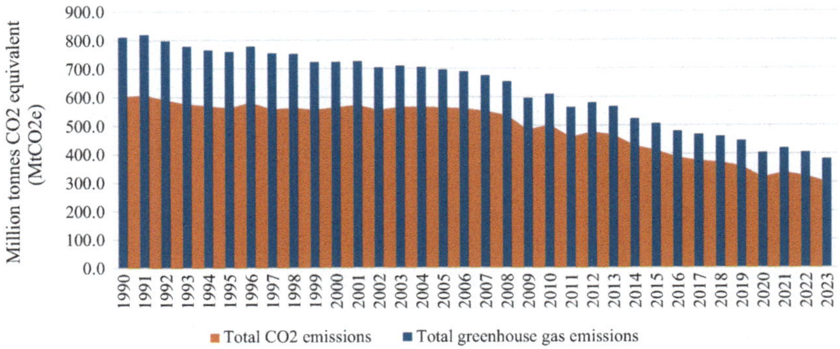

Fig. 1 UK historic GHGs and CO_2 emissions outlook, 1990–2023 (*Source* Generated by Author from UK government energy data [DESNZ 2023])

In terms of the sources, anthropogenic (human) activities, especially combustion of fossil fuels, principally oil, gas, and coal, remain the leading CO_2 emitting sources that accounted for 48% and 43%, respectively, in 2023, followed by coal (6.1%) and cement (1.6%). IEA50 (2023) further reported that the country's per-capita CO_2 emissions nosedived by 48% from 8.84 to 4.55 tCO_2/capita between 2000 and 2022. At sectoral levels, domestic transport remains the country's highest emissions-stimulating sector, responsible for over a quarter (29%) of the country's total emissions in 2023, followed by buildings and product uses (20%), industry (14%), electricity generation (12%), agriculture (12%), and fuel supply (8%) whilst waste generation and other activities including land use, land use change, and forestry accounted for the remainder (5%), as illustrated in Fig. 2.

The key drivers of this recent fall in UK emissions have been declining gas demand and coal utilisation in power generation, both of which fell by 11% and 23%, respectively, in 2023 (DESNZ 2023; Carbon Brief's Analysis 2024). Other drivers include a rise in energy and different prices that resulted in low demand for heating coupled with rising shifts in fuel use from coal to gas and renewables in power generation. Although the UK's fuel mix is changing, the dominance of natural gas still constitutes a major challenge, and the country needs to shift from fossil-based energy sources with proven emission reduction technologies, such as the intensive sequestration and storage of CO_2, alongside the invention of completely new valuable products (utilisation) to achieve its net-zero

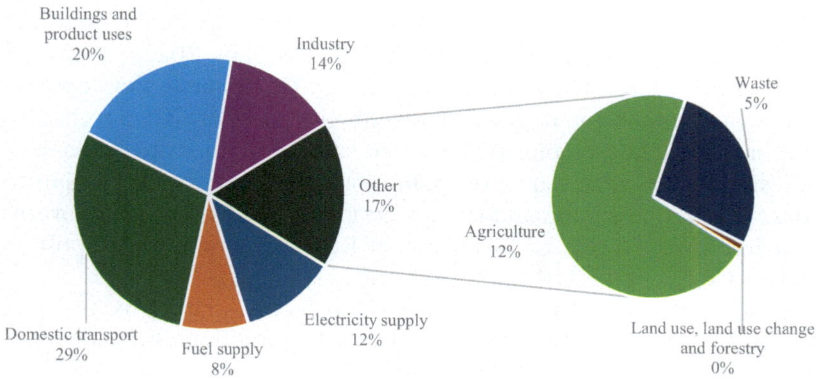

Fig. 2 Breakdown of UK GHGs emission by sector, 2023 (*Source* Generated by Author from UK government energy data [DESNZ 2023])

targets. Thus, the critical role of CCUS deployment has been acknowledged by the UK government with diverse policies, actions specific to CCUS, incentive mechanisms, and investment efforts increasingly gaining momentum in the last decade.

3 Snapshot of UK CCUS Initiatives and Main Projects

CCUS initiatives are essentially targeted at not only positioning CCUS as a veritable CO_2 mitigating option but also to proffer regulatory standards/frameworks, enabling policies, robust partnerships/collaboration, knowledge sharing, investment opportunities (both near and longer-term), and fostering synergy amongst stakeholders (including government, industry, financial organisations, and investors). As noted in the CCS Association (2023), the UK is growingly developing several strategies and initiatives to accelerate commercial deployment of CCUS across key regions and sectors of the economy.

Matschoss and Repo (2018), Yang et al. (2019), and Nielsen et al. (2022) noted that the UK's emphasis on CCUS as a viable route for decarbonising at critical levels and climate mitigation option is evident in how these technologies have been promoted through several initiatives devised including trading schemes, tax credits, environmental directives, patent tax relief, UK-EU Trade and Cooperation Agreement, research funding, amongst others (Yang et al. 2019; Shapps 2023). A critical component of UK CCUS initiatives is the CCUS research, development (R&D), and innovation targeted at lowering the costs of CCUS technologies through the development of cost-effective and efficient components alongside exploring new areas of applications cum support mechanisms to reduce the transportation and storage cost of CO2. As such, the government has invested over £130 million in R&D and innovation boost to develop CCUS in the UK since 2011.

In sum, the UK CCUS industry's long-term strategy is to store over 50Mt per annum by 2035, thereby supporting domestic industries' decarbonisation efforts and exploring associated export potentials. The explicit focus is on capturing 20–30 million tonnes of CO_2 every year (DESNZ 2023). This builds on the government's suites of industrial cluster sequencing processes, to build a robust capture network across the UK.[1] With the majority of carbon emissions emanating from UK industrial clusters, the country aims to establish CCUS in industrial clusters by 2030 by setting up "SuperPlaces" in key areas such as the North East, the Humber, North West, Scotland, and Wales as illustrated in Fig. 3.

Having legislated its 2050 net-zero target in 2019, the UK government confirmed £1 billion of government investment to support CCUS development in four industrial clusters. DECC (Department of Energy and Climate Change) (2012) and Cooper and Hammond (2018) noted that proposed CCS clusters hold huge prospects of offering integrated CO_2 pipeline networks, forming multiple branches connecting each individual source to a central hub and main pipeline, thereby sharing CCS infrastructure. The primary support for the North West and East Coast clusters, targeted to begin operation in the mid-2020s, was to be followed by the Northeast Scotland and the Humber clusters aimed to come on board by 2030. This profound support for the industrial cluster was supplemented by the 2023 Spring Budget announcement of a long-term funding package of about £20 billion in support of UK CCUS deployment and to assist the government in achieving its climate objectives (see Spring Budget 2023; Serin 2023).

[1] This cluster sequencing process represents a logical path towards CCUS deployment in the UK. Clusters sequenced for the mid-2020s are regarded as "Track-1" whilst slated to come online by 2030 are "Track-2".

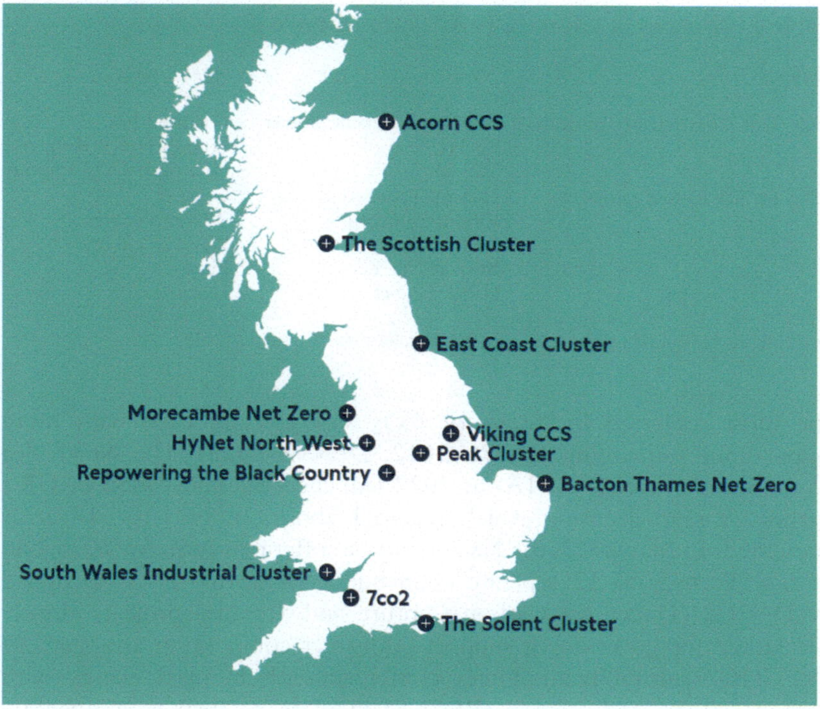

Fig. 3 Overview of the UK's proposed CCUS Projects across its major industrial hubs/regions (*Source* CCS Association [2023])

Furthermore, the government is committed to accelerating the development of CCUS technologies through the Track 1 process which aims to establish 2 new clusters with selected final eight projects considered for negotiations to support this ambition (through appropriate business models) which is presented in Table 1. It is thus evident that a variety of large-scale CCUS projects envisioned by the UK are aimed at capturing emissions from several sectors ranging from power generation, cement manufacturing, hydrogen production to waste to energy conversion, amongst others.

As Hawkes (2019) mentioned the existence of 574 possible storage sites across the UK, storage under both the North Sea and the North-east part of the Irish Sea, home of depleted oil and gas fields remains

Table 1 Synopsis of UK track 1 CCUS cluster projects

Cluster	*Name*	*Application*
The East Coast cluster projects	Net Zero Teesside Power	Gas fired power
	Bp H2 Teesside	Low carbon hydrogen
	Teesside	Hydrogen CO_2 Capture
The HyNet Cluster projects	Hanson Padeswood	Cement
	Viridor Runcorn	Waste-to-energy
	Protos Energy	Waste-to-energy
	Buxton Net Zero	Lime
	HPP1 HyNet	Hydrogen

Source Author's construct from BEIS (2019) and Global CCS Institute (2023)

the most preferred locations for CO_2 storage, and even with more opportunities emerging (DECC 2012; DESNZ 2023). The North Sea Transition Authority (NSTA) in 2022 launched the first round of UK's carbon storage licensing which received about 26 bids for 13 areas presented. In May 2023, the NSTA made an offer of awards for 20 carbon storage licenses to 12 separate companies. Added to this is a myriad of DACCS (Direct Air Carbon Capture and Storage) projects already earmarked for government support, a distinct competition arrangement for GHG removal technologies combined with biomass energy with carbon capture and storage (BECCS) in place to further compensate for other GHG emissions in the UK. All the aforementioned projects in historic Britain industrial clusters alongside new carbon storage areas with licenses recently issued by the NSTA would make a significant contribution towards the realisation of UK targets of storing 20–30 million tonnes of CO_2 annually by 2030.

3.1 CCUS Global Market Outlook and Financing Options

BloombergNEF's (2023) market outlook revealed that the CCUS market continues to witness notable expansion and profound spread into other markets. Whilst the global investments in CCUS projects and technologies hit $6.4 billion in 2022, the report observed a rapid diversification in the CCUS landscape from the historic dominance of natural gas processing plants to deployment in hard-to-abate sectors specifically cement, iron and steel, and power which are collectively projected to account for roughly 85% of global aggregate by 2050. Recent estimates in Mckinsey (2024) put the total number of CCUS projects operating globally at 68, with another 39 under construction and 533 at varying planning phases in 2023, indicating about threefold capacity increase from 2020, and with further potential to increase by 12 times, if all the announced projects/capacities in the pipeline are completed by 2030. Thus, the industry is envisaged to grow at a compounding annual rate of 18% thereby capable of capturing 420 million tonnes per annum by 2035, denoting 1.1% of present annual emissions emanating from fuel combustion and industrial processes globally.

North America and Europe jointly account for the biggest share (over 80% of announced capacity) principally driven by policy support and incentives. In the latter (Europe), a total of 119 CCUS projects are situated/proposed, with 109 (92%) at an early or advanced stage of development, 6 (5%) under construction, and the remaining 4 (3%) facilities already operational (Global CCS Institute 2023). Mckinsey (2024)'s focus on "hard-to-abate" sectors revealed that 35 CO2 transportation and storage networks are underway together with a catalogue of other specialised CCUS facilities, including 20 hydrogens, ammonia, and fertiliser facilities, 19 power generation and heat facilities, 17 cement facilities, and 15 biomass to power/heat infrastructure. In terms of market composition, the US remains the market leader with about 40% market share, followed by the UK (16%) and Canada (12%), whilst Australia, the Netherlands, and China are expected to hold 3–4% market share each by 2035.

As for capture costs, the value ranges from \$20-\$28 per tonne of CO_2 in industries (such as ethanol, ammonia, and natural gas processing) with a high concentration of CO_2 in off-gases, to between \$72 and \$80 in industrial sources, namely cement (\$80), hydrogen (\$79), and steel (\$72)—adding the transport and storage fees of \$20–\$50 per tonne of CO_2 would bring the total costs to \$92–\$130 per tonne of CO_2 whilst transport costs were further estimated to be two to four times higher for companies considering shipping CO_2 in liquid form. Given these high costs, the CCUS market, especially in the UK with huge prospects, requires worthwhile financing strategies cum support (discussed below) to accelerate growth and foster commitment to net-zero objectives.

3.2 CCUS Financing Strategies/Options

CCUS technologies undeniably come at a significant cost and IEA estimated that up to $13 trillion of private sector investment in CCUS will be required to achieve its "net-zero emissions by 2050 scenario" projections. Thus, finance constitutes an integral consideration in CCUS projects investment; this section enumerates some of the notable approaches/strategies often used in securing funding for these emerging technologies in the UK and other notable countries around the globe. Climatech (2023) stipulated the customary approaches to range from government funding (via subsidies, grants, and tax incentives) and public–private partnerships (PPPs), to project financing. IEA50 (2024) divulged that the government funding for the CCUS project through ongoing subsidy programmes in Europe and the United States reached over USD 20 billion in 2023. Similarly, the UK government set up Clean Growth Fund in 2020 to support low-carbon technologies, including CCUS which has continued to drive private sector investment and reduce the cost of capital.

In a similar vein, the PPPs help to bring private sector investors and governments together thereby sharing various risks and costs of CCUS projects. As lenders need a thorough understanding of the project's risks and returns, Climatech (2023) espoused other potential strategies that could be adopted (by government, institutions, companies, and other private entities) to finance CCUS projects including purchasing carbon credits and carbon bonds, project partnership (sharing costs, expertise and financial burden), and crowdfunding through online fundraising for CCUS technologies projects.

Having recounted that the majority of completed projects to date were financed by owners or equity investors, including private equity firms and off-takers such as chemical and utility companies, instead of banks, Beetz et al. (2023) opined that banks and others need to explore different (and even develop new) financing options that could be more apt for CCUS projects. Nonetheless, the study showed that some banks (early-mover banks) are deploying alternative financing options, notable amongst which are tax equity finance, equity investments, and mezzanine finance. Beetz et al. (2023) further noted that financial institutions are additionally offering finance indirectly through support for public–private investment initiatives for CCUS and hydrogen projects.

As for myriads of projects being currently developed in the North Sea (which are at different stages of maturity), ING (2022) divulged that a considerable proportion is being developed by strong partners relying on (own) balance sheet financing and that off-balance sheet financing (post construction) represents another attractive option, but the latter requires apt structuring from inception alongside balanced risk mitigation. The report thus highlighted the primary financing considerations for CCUS projects. Project agreements to be bankable from financing, commercial, and legal perspectives to deliver a successful process. Moreover, BloombergNEF (2023) and S&P Global (2024), Mergers and acquisitions represent another financing strategy observed amongst oil majors in recent years, and the trend is expected to continue with many more making a sharp move to acquire CCUS hubs and assets in advantaged geographical locations and industrial zones globally, especially in Europe and North America. Amongst such acquisitions in recent times were ExxonMobil's acquisition of Denbury, Occidental's acquisition of Carbon Engineering, just to mention but a few.

3.3 *Extract of UK Legal Frameworks, Legislations, and Regulations in Relation to CCUS*

As described by the Natural Resource Governance Institute (2015), a legal framework encompasses a set of rules, rights, and obligations governing the behaviour of actors, such as companies, governments, and citizens, as set forth in a set of legal documents that may include the constitution, legislation, regulations, and contracts. As presented in DECC (2012), the UK has made significant progress in establishing a broad legal framework for CCUS. Global CCS Institute (2018) report further revealed that the country already had over sixty-five (65) laws that are applicable across the CCUS project cycle.

Aside from having a history of participating in a plethora of international instruments that indicate how CO_2 storage, especially offshore storage, could be regulated, the country has a national act and specific regulations that focus on how to deal with CO_2 storage issues. Some of the earliest legislation includes the Convention on the Prevention of Marine Pollution by Dumping of Wastes and Other Matter (1972), the Convention for the Protection of the Marine Environment of the North-East Atlantic signed (1992), EC Directive on Geological Storage of Carbon Dioxide, National legislation is the Energy Act 2008, Planning Act 2008, just to mention but few (e Silva and Costa 2021).

For instance, the Energy Act 2008 details the national regulatory framework regarding licensing for offshore CCS in the UK—a vital legislation which e Silva and Costa (2021) espoused to have significantly transformed numerous energy infrastructure components and market regulation in the UK. Also, through its Climate Change Act, the country sets legally binding five-year caps on emissions (carbon budgets) and thereafter publishes relevant policies and measures to meet its targets. Relatedly, the UK's Planning Act (2008) offers a catalogue of rules on infrastructure that apply to CCUS pipelines and facilities. Further, the Committee on Climate Change (2020) emphasised that the UK committed to a net-zero emissions target by 2050 as it committed to innovation, productivity, and industry competitiveness alongside new institutional governance to achieve this lofty goal.

e Silva and Costa (2021) noted that the UK's commitment to energy transition and adoption of low-carbon measures has made it develop an integrated energy and climate policy framework. Recent updates on the government-proposed frameworks regarding transport and storage, power, and industrial CC business models have been provided in DESNZ (2023). Despite the significant efforts made by CCUS UK policies and legislative framework over the past decade, studies that examined CCUS legal frameworks and standards have underlined the need for further action (e Silva and Costa 2021). Speaking on creating a supportive regulatory environment and incentives to invest in CCUS, Grant Shapps, Secretary of State for the Department for Energy Security and Net Zero, echoed a variety of other strategies that position the UK as one of the world's most attractive businesses and investment environment to include "the Energy Bill" that creates the legislative framework for the economic regulation of CO_2 transport and Storage, "the powering Up Britain initiative" that supports country's ambitions of strengthening energy security, harnessing economic opportunities and delivering on net-zero responsibilities, and "the Green Finance Strategy" to solidify the country's position as a leader in the global green finance market and provide the financing needed, amongst others regulatory and supportive strategies (Shapps 2023).

4 X-RAYING AREAS OF UK CCUS OPPORTUNITIES

With CCUS being an integral component to meeting its net-zero targets and encapsulating several opportunities for micro- and macroeconomic expansion, the UK needs to benchmark its comparative advantages in the area of production and explore innovations surrounding CCUS products, services, and other technologies. As the technology amidst a portfolio of clean and emerging technology areas is rapidly becoming cost-competitive and yielding above-average returns in public investments, it portends huge millions to attract investors, as well as public and private sector funding and investment in CCUS projects. Likewise, as countries around the world are striving to achieve their carbon neutrality targets, CCUS projects represent one of the choicest technologies to deploy in that regard, and thus investment is expected to soar considerably over the next few decades. This is further confirmed by the IEA's projection that CCUS projects are expected to account for about 15% of global emissions reductions by 2050, which in turn offers colossal investment opportunities to the UK government, companies, and individuals to contribute towards environmental sustainability. Also, there exist diverse areas in the emerging UK CCUS supply chains, including decarbonisation in multiple domestic industries and developing an apt market structure for carbon capture and export markets.

As reiterated in DESNZ (2023), there exist massive economic and technical opportunities in the UK's North Sea resources and assets that could be harnessed to facilitate achieving net-zero emissions. Also, this essential technology offers a cream of potential employment opportunities and has significant implications for developing requisite skills and capabilities (including delivering higher technical education, green skills, and local industrial and technological expertise) across the value chain to position the country on the path of economic recovery and sustainable growth. Studies have also confirmed the capability of the sector to potentially preserve over 53,000 jobs in energy-intensive industries by 2030, and with significant investment, CCUS could create up to 50,000 jobs for the UK economy.

Crucially, CCUS represents a promising instrument for driving low-cost hydrogen production and combating emissions in heavy (hard-to-abate) sectors, such as cement, steel, or chemicals manufacturing sectors. Additionally, the CCUS clusters hold manifold prospects of not only providing integrated CO2 pipeline networks that could form multiple branches linking individual sources to a central hub and main pipeline, thereby becoming the main sharing CCUS infrastructure. Aside from huge prospects in supporting the decarbonisation of other parts of CO_2 transportation, such as trucks and ships, the economics of scale of integrated pipelines network could considerably lower the transport cost, especially from trivial CO_2 emitting sources across the country. In addition, Hammond (2013) underlined the opportunities in CCUS clusters to reduce supply disruption, transaction costs, and investment risks inherent in permitting and installing numerous point-to-point pipeline systems. Other opportunities have also been identified in the aspects of raising awareness on deleterious impacts of climate change on our planet which prompted the growing drive for environmental sustainability.

4.1 *Identified Challenges of CCUS Development in the UK*

Several studies have pointed out the core challenges of CCUS development globally to range from high operating and capital costs, low commercial efficiency of CCUS technological schemes, inadequate financial and public incentives mechanisms, poor regulatory framework to limited public awareness cum negative perception (Budinis et al. 2018; Ilinova and Cherepovitsyn 2021). Zhang (2021) asserted that some of these challenges have resulted in criticisms and growing scepticism on the possible role of the CCUS as a viable technology to mitigate climate change, given its slow pace of development and multifaceted risks. Other aspects of scepticism have been expressed in the literature, including storage risks, multiple uncertainties, and concerns regarding the long-term efficacy of the technology in curbing the menace of climate change (Arning et al. 2019). IEA (2020) further emphasised the public resistance to storage (chiefly onshore storage) as a critical challenge observed with CCUS project development, particularly in Europe. This is due to the perception of CCUS as a fossil fuel-based technology that is competing with renewables (for public and private investment), whereas in practical terms it has robust synergies with renewable energy.

Besides, there have also been extensive viewpoints involving the interconnected technical, social, economic, and political issues associated with CCUS projects (Arning et al. 2020). Whilst DECC (2012) and Griffin et al. (2016) presented transport and storage costs as a major challenge than that associated with capture costs, Hammond (2013) submitted commercial financing CO_2 pipelines as another significant hurdle to overcome for CCUS to become a mature, commercially viable, and widely adopted technology. Given the UK's prior failed major demonstration competitions, the Centre for Economic Performance (CEP) (2021) pointed out the inconsistent policy environment as another bottleneck militating against the UK's CCUS development to date. Similarly, IEA (2020) attributed the slower pace of advancement with most planned CCUS projects to inherent commercialisation concerns, policy inconsistencies, the inadequacy of incentives/emissions penalties, and the limited significance of CO_2 as an industrial input in some sectors.

Moreover, CCUS as a complex and capital-intensive technology exposes investors to a myriad of investment risks ranging from project, finance, geology, and technology to market risks. A critical part of the CCUS investment risks is the uncertainty about the carbon pricing market. The high volatility of carbon prices can make it a challenge for investors to forecast or aptly determine their future returns on CCUS projects. Added to this is the lack of a universally acceptable carbon pricing mechanism which accordingly makes it difficult for investors to suitably assess and compare the economic feasibility of diverse CCUS projects for sound investment decisions. Lastly, the economic downturn and lower oil prices constitute another set of threats capable of not only disrupting CCUS investment plans but also rendering CO_2 deployment for EOR (enhanced oil recovery) from the UK's matured oil fields unattractive.

5 Conclusion

As evident from the foregoing, CCUS technologies portend huge potential for reducing GHG emissions and decarbonising economies at national and regional levels. This chapter examines the CCUS situation in the UK, covering core areas including current trends, proposed projects, country-wide assets/plans, finance and support mechanisms, regulatory landscape, intrinsic challenges, and emerging opportunities. It additionally provides a glimpse into global CCUS market performance, market size and composition, major movers/drivers, cost dynamics, and evolving prospects. Given the UK's distinctive geological advantage, industrial infrastructure, expertise, and experience (in the oil, gas, and petrochemicals sector), the study divulged its capacity to safely store up to 78 billion tonnes of C O_2 under its seabed coupled with ongoing profound efforts in developing CCU S technologies into a valuable national asset, foster socio-economic growth, and achieve set net-zero goals by 2050. It further explores the national regulatory framework regarding licensing for offshore CCS in the UK matched with government commitments to accelerate the development of CCUS technologies and illustration of a variety of large-scale CCUS projects envisioned by the UK government to capture emissions from hard-to-abate sectors, principally from power generation, cement manufacturing, hydrogen production, and waste to energy conversion industries.

Aside from the itemised areas of opportunities, the study pointed out the core challenges of CCUS development to range from high operating and capital costs, inadequate financial, and public incentives mechanisms, and poor regulatory framework to limited public awareness cum negative perception. With finance representing one of the significant challenges faced by companies and investors, the study discusses some of the prominent approaches which range from government funding (via subsidies, grants, and tax incentives), PPPs to project financing. It offers useful insights into alternative and emerging financing strategies including carbon credits and carbon bonds, project partnership, crowdfunding, tax equity finance, equity investments, and mezzanine finance. The above glitches, therefore, underline the critical need for more proactive and strategic action to mitigate those risks and unlock the enormous growth opportunities associated with these technologies acknowledged as "a necessity, not an option" for the UK to meet its net-zero targets.

References

Allan, R. P., Arias, P. A., Berger, S., Canadell, J. G., Cassou, C., Chen, D., & Zickfeld, K. (2023). Intergovernmental Panel on Climate Change (IPCC). Summary for policymakers. In Climate change 2021: The physical science basis. Contribution of working group I to the sixth assessment report of the Intergovernmental Panel on Climate Change (pp. 3–32). Cambridge University Press

Arning, K., Offermann-van, H. J, Linzenich, A, Kaetelhoen, A., Sternberg, A., & Bardow, A. (2019). Same or different? Insights on public perception and acceptance of carbon capture and storage or utilization in Germany. Energy Policy, 125, 235–249

Arning, K., Offermann-van, H. J., Sternberg A, Bardow, A., & Ziefle, M. (2020). Risk-benefit perceptions and public acceptance of carbon capture and utilization. Environment Innovation Social Transition, 1 (35), 292–308

Beetz, E., Hegnsholt, E., Rakhou, E., Clayton, C., Jamal, Z., & Juman, T. (2023). Breaking the finance barrier for hydrogen and carbon capture. Energy Transition. Available at https://www.bcg.com/publications/2023/breaking-the-barriers-in-financing-hydrogen-and-carbon-capture (Accessed on 22 June 2024)

BloombergNEF. (2023). CCUS market outlook 2023: Announced capacity soars by 50%. Available at https://about.bnef.com/blog/ccus-market-outlook-2023-announced-capacity-soars-by-50/ (Accessed on 20 June 2024)

Budinis, S., Krevor, S., Mac Dowell, N., Brandon, N., & Hawkes, A. (2018). An assessment of CCS costs, barriers and potential. Energy Strategy Review, 22, 61–81

Carbon Brief's Analysis. (2023). Analysis: UK emissions in 2023 fell to lowest level since 1879. Available at https://www.carbonbrief.org/analysis-uk-emissions-in-2023-fell-to-lowest-level-since-1879/ (Accessed 20 May 2024)

Carbon Brief's Analysis. (2024). Analysis: State of the climate: 2024 now very likely to be warmest year on record. Available at https://www.carbonbrief.org/state-of-the-climate-2024-now-very-likely-to-be-warmest-year-on-record/(Accessed 20 October 2024)

Carbon Capture and Storage (CCS) Association. (2023). CCUS in the UK. Available at https://www.ccsassociation.org/all-news/ccsa-news/ccus-delivery-plan-update-2023/ (Accessed 16 October 2024)

Climatech. (2023). Carbon capture project financing: Strategies and challenges. Available at https://verde.ag/en/carbon-capture-project-financing/#:~:text=Strategies%20for%20financing%20CCUS%20projects,and%20cost%20between%20various%20stakeholders (Accessed 17 June 2024)

Committee on Climate Change. (2020). Reducing UK emissions: 2020 Progress report to parliament. Available at https://www.theccc.org.uk/publication/reducing-uk-emissions-2020-progress-report-to-parliament/ (Accessed 16 September 2024)

Centre for Economic Performance (CEP). (2021). Seizing sustainable growth opportunities from carbon capture, usage and storage in the UK. Available at https://www.lse.ac.uk/granthaminstitute/publication/seizing-sustainable-growth-opportunities-from-carbon-capture-usage-and%20including%20essential%20technology (Accessed 20 May 2024)

Cooper, S. J., & Hammond, G. P. (2018). 'Decarbonising' UK industry: towards a cleaner economy, Proceedings of the Institution of Civil Engineers, Energy, 171 (4), 147–157

DECC (Department of Energy and Climate Change). (2012). CCS roadmap: Supporting deployment of carbon capture and storage in the UK. DECC, London, UK

Department for Business, Energy and Industrial Strategy (BEIS). (2019). UK energy in brief. Available at www.gov.uk/government/statistics/uk-energy-in-brief-2019 (Accessed 2 May 2024)

DESNZ. (2023). Report on the proposed CCUS Transport and Storage Regulatory Investment Model Support Scheme by DESNZ. Available at https://www.gov.uk/government/publications/report-on-the-proposed-ccustransport-and-storage-regulatory-investment-model-support-scheme-by-desnz (Accessed 21 October 2024)

Department for Energy Security and Net Zero. (2024). Carbon capture, usage and storage industrial carbon capture business models update for track-1 expansion and track-2. Available at https://assets.publishing.service.gov.uk/media/661530cdc4c84d6602346a13/ccus-iccc-business-models-updateapril-2024.pdf (Accessed 10 September 2024)

Dziejarski, B., Krzyżyńska, R., & Andersson, K. (2023). Status of carbon capture, utilization, and storage technologies in the global economy: A survey of technical assessment. Fuel, 342, 1–22, https://doi.org/10.1016/j.fuel.2023.127776

e Silva, I.M.M., & Costa, H.K.M. (2021). The United Kingdom's experience in Carbon Capture and Storage projects: The current regulatory framework and related challenges. In Carbon capture and storage in International Energy Policy and Law (pp. 141–153). Elsevier

Garden, L. (2023). 2024 will be a huge year for the carbon capture, usage and removal sector. Available at https://www.greenbiz.com/article/2024-will-be-huge-year-carbon-capture-usage-and-removal-sector

Global Carbon Capture and Storage Institute Ltd. (2018). Fact sheet geological storage of CO2. Available at https://www.globalccsinstitute.com/wp-content/uploads/2018/12/Global-CCS-Institute-Fact-Sheet_Geological-Storage-of-CO2.pdf (Accessed on 20 May 2024)

Global CCS Institute. (2023). Global status of CCS 2023—Report & executive summary. Available at https://www.globalccsinstitute.com/resources/publications-reports-research/global-status-of-ccs-2023-executive-summary/ (Accessed 2 May 2024)

Griffin, P. W., Hammond, G. P., & Norman, J. B. (2016). Industrial energy use and carbon emissions reduction: A UK perspective. WIREs Energy and Environment 5 (6), 684–714.

Hammond, G. (2013). No access carbon dioxide capture and storage faces a challenging future. In Proceedings of the Institution of Civil Engineers—Civil Engineering, 166 (4), 147–147

Hawkes, A. (2019). CCS in the UK: 3rd time lucky? Lecture in energy transition research & innovation conference, 1–2 October, University of São Paulo, Brazil https://www.ccsassociation.org/discover-ccus/ccus-in-action/ (Accessed 16 May 2024)

IEA (2020). CCUS in Clean Energy Transitions. Available at https://www.iea.org/reports/ccus-in-clean-energytransitions(Accessed 26 September 2024)

IEA. (2023). Global energy review: CO_2 emissions in 2021. IEA, Paris. Available at https://www.iea.org/reports/global-energy-review-co2-emissions-in-2021-2 (Accessed 10 May 2024)

IEA50. (2023). United Kingdom Total CO_2 emissions from energy. Available at https://www.iea.org/countries/united-kingdom/emissions (Accessed 20 May 2024)

IEA50. (2024). Carbon capture, utilisation and storage. Available at https://www.iea.org/energy-system/carbon-capture-utilisation-and-storage (Accessed 1 May 2024)

Ilinova, A., Romasheva, N., & Cherepovitsyn, A. (2021). CCUS initiatives: Public effects and "combined value" performance. Resources, 10 (6), 61. https://doi.org/10.3390/resources10060061

ING. (2022). 2022 Climate Report. Available at https://www.ing.com/Newsroom/News/2022-Climate-Report-1.htm(Accessed 25 October 2024)

International Energy Agency (IEA). (2023). 'Net Zero Roadmap: A Global Pathway to Keep the 1.5 °C Goal in Reach. Available at iea.blob.core.windows.net/assets/13dab083-08c3-4dfd-a887-42a3ebe533bc/NetZeroRoadmap_AGlobalPathwaytoKeepthe1.5CGoalinReach-2023Update.pdf, p.132

Isabela, Morbach Machado e Silva, Hirdan Katarinade Medeiros Costa, (2021). Chapter 7—The United Kingdom's experience in Carbon Capture and Storage projects: the current regulatory framework and related challenges, Carbon Capture and Storage in International Energy Policy and Law, pp. 141–153

Kuckshinrichs, W., & Hake, J. (2015). Carbon capture, storage and use: Technical, economic, environmental and societal perspectives. Carbon capture, storage, and use. Available at https://www.semanticscholar.org/paper/Carbon-Capture%2C-Storage-and-Use%3A-Technical%2C-and-Kuckshinrichs-Hake/e88a7e6a175b0c1b281a90307cadf1b5e30ecafe (Accessed on 20 October 2024)

Leeson, D., Mac Dowell, N., Shah, N., Petit, C., & Fennell, P. S. (2017). A Techno-economic analysis and systematic review of carbon capture and storage (CCS) applied to the iron and steel, cement, oil refining and pulp and paper industries, as well as other high purity sources. International Journal of Greenhouse Gas Control, 61, 71–84.

Matschoss, K., & Repo, P. (2018). Governance experiments in climate action: empirical findings from the 28 European Union countries. Environmental Politics, 27 (4), 598–620.

Mckinsey. (2024). Global energy perspective 2023: CCUS outlook. Available at https://www.mckinsey.com/industries/oil-and-gas/our-insights/global-energy-perspective-2023-ccus-outlook (Accessed on 20 June 2024)

Natural Resource Governance Institute. (2015). Legal framework navigating the web of laws and contracts governing extractive industries. Available at https://resourcegovernance.org/sites/default/files/nrgi_Legal-Framework.pdf (Accessed 20 May 2024)

Nielsen, J.A., Stavrianakis, K., & Morrison, Z. (2022). Community acceptance and social impacts of carbon capture, utilization and storage projects: A systematic meta-narrative literature review. PLoS ONE, 17 (8), e0272409. https://doi.org/10.1371/journal.pone.0272409

S&P Global. (2024). Mergers and acquisitions fueling innovation and growth in the CCUS sector. Available at https://www.spglobal.com/commodityinsights/en/ci/research-analysis/mergers-and-acquisitions-fueling-innovation-and-growth.html (Accessed on 20 June 2024)

Sarnoff, J. D. (2021). Chapter 11—Negative-emission technologies and patent rights after COVID-19, Carbon Capture and Storage in International Energy Policy and Law, Climate Law Pages, 205–231. https://doi.org/10.1016/B978-0-323-85250-0.00020-7

Serin, E. (2023). What is the UK's policy approach to carbon capture, usage and storage (CCUS)? Climate change and the UK, The London School of Economics and Political Science. Available at https://www.lse.ac.uk/granthaminstitute/explainers/what-is-the-uks-policy-approach-to-carbon-capture-usage-and-storage-ccus/#:~:%20atmosphere (Accessed 20 May 2024)

Shapps, G. (2023). CCUS Net Zero Investment Roadmap Capturing carbon and a global opportunity April 2023. Available at https://assets.publishing.service.gov.uk/media/64a29b7d06179b00131ae94e/ccus-investment-roadmap.pdf (Accessed on 20 May 2024)

Spring Budget. (2023). UK Government Policy paper. Available at https://www.gov.uk/government/publications/spring-budget-2023/spring-budget-2023-html#:~:text=To%20increase%20resilience%20to%20future,the%20Climate%20Change%20Agreement%20scheme(Accessed 04 September 2024)

UKCSS Research Community. (2022). CCS for Energy Security and Net Zero—Allowing effective use of fossil fuels and biomass. Available at https://mailchi.mp/ukccsrc/february2022-8117607?e=2795e446a4

United Kingdom Planning Act. (2008). United Kingdom, Northern Ireland, Scotland, Wales. Available at https://www.legislation.gov.uk/ukpga/2008/29/contents (Accessed on 20 May 2024)

Yang, X., Heidug, W., & Cooke, D. (2019). An adaptive policy-based framework for China's Carbon Capture and Storage development. Frontiers of Engineering Management, 6, 78–86.

Yusuf, M., & Ibrahim, H. (2023). A comprehensive review on recent trends in carbon capture, utilization, and storage techniques. Journal of Environmental Chemical Engineering, 11 (6), 111393. https://doi.org/10.1016/j.jece.2023.111393

Zhang, H. (2021). Regulations for carbon capture, utilization and storage: Comparative analysis of development in Europe, China and the Middle East. Resource Conservation Recycle, 173, 105722

CHAPTER 22

CCUS in the United States of America

Owen L. Anderson

1 Introduction

Following passage of the Inflation Reduction Act in August 2022, with its CCUS tax credits, new CCUS projects were announced weekly through the early summer of 2023. Thereafter, enthusiasm waned due to high

* © Owen L. Anderson. Owen L. Anderson is a Distinguished Oil & Gas Scholar and Co-Academic Director of the Kay Bailey Hutchison Energy Center at The University of Texas at Austin and the Eugene Kuntz Chair in Oil, Gas & Natural Resources Emeritus and George Lynn Cross Research Professor Emeritus at The University of Oklahoma. Portions of this chapter are adapted from an article published in Summer 2023, Carbon Sequestration: A Fresh Look at an Essential Tool in the War on Climate Change, Oil, Gas & Energy Law Intelligence, OGEL 3 (2023)—Carbon Capture and Storage. OGEL Journal, https://www.ogel.org/journal-browse-issues-toc.asp?key=102 (2023).

O. L. Anderson (✉)
The University of Texas, Austin, TX, USA
e-mail: oanderson@law.utexas.edu

The University of Oklahoma, Norman, OK, USA

E. G. Pereira et al. (eds.), *Carbon Capture Utilization and Storage*, Sustainable Development Goals Series,
https://doi.org/10.1007/978-3-031-81272-9_22

costs, lack of public acceptance, and the failure of important states, like Texas, to pass needed legislation to encourage CCUS investment. As with oil, natural gas, and electricity, CCUS has been hampered by the bottleneck of transportation. Just as rural landowners have opposed new oil and gas pipelines and new electric transmission lines, they have opposed CO_2 pipelines needed to connect capture sites with sequestration sites. The Texas legislature, for example, has so far failed to enact regulatory legislation needed to spur sequestration beneath private lands—in particular regulatory compulsory unitization of storage reservoirs or acquisition of both transport and storage rights by eminent domain. In addition, CCUS costs have yet to decline. Thus, many of the announced projects are delayed or on hold, and a few projects have been at least temporarily abandoned.

As elsewhere in the world, capture costs are the highest component; however, storage costs in the United States are location-dependent. Offshore sequestration costs more than onshore storage due to increased costs of drilling, equipment, transport, and sequestration facilities, but offshore storage also has advantages of scale, less conflicting uses, and government ownership of pore space, rather than the subdivided and somewhat fractionalized private ownership of pore-space reservoirs encountered onshore.

A 2022 report by the U.S. Governmental Accountability Office[1] confirmed that the United States has abundant geologic storage capacity. Saline formations account for 95% of the potential storage sites.[2] This report cites a study by the National Energy Technology Laboratory, which concludes that CO_2 storage can be safe, secure, and permanent[3] and storage costs are estimated at $5/ton along the Gulf Coast, $7/ton in the Permian Basin, $11 per ton in the Williston Basin, $12/ton in California, Anadarko, Southern Oklahoma, and Illinois Basins, and $20/ton

[1] Decarbonization: Status, Challenges, and Policy Options for Carbon Capture, Utilization, and Storage. U.S. GAO, https://www.gao.gov/products/gao-22-105274.

[2] *Id.* at 36.

[3] *Id.* at 37, citing T. Warner et al., Overview of Potential Failure Modes and Effects Associated with CO_2 Injection and Storage Operations in Saline Formations, DOE/NETL-2020/2634 (Pittsburgh, Pa.: National Energy Technology Laboratory, Oct. 16, 2020), Technical Report Template and User Guide (energy.gov), https://www.energy.gov/index.php/lpo/articles/overview-potential-failure-modes-and-effects-associated-co2-injection-and-storage.

in the South Florida Basin and along the Atlantic Coastal Plain.[4] In each of these regions, costs are likely to decline as the sequestration industry matures; however, cost declines have yet to materialize—perhaps because most projects have yet to reach the FEED stage of development.

In the recent past, because the geology of depleted reservoirs is better understood than the geology of much larger saline reservoirs and because depleted reservoirs may have existing infrastructure that might be used for storage operations, the focus was on appraising and using depleted petroleum reservoirs for CO_2 storage.[5] More recently, however, owing to concerns that abandoned oil and gas wells could provide a pathway for CO_2 leakage from depleted petroleum reservoirs, the focus is shifting to much larger saline reservoirs,[6] which pose a reduced leakage threat, provide much greater storage capacity, and greater economies of scale.[7]

The United States Geological Survey has identified areas with "most [CO_2] storage potential."[8] These areas include coastal basins from Texas to Georgia but also significant areas in Alaska, the Rocky Mountain Region, and the Northern Great Plains.[9] Although large blocks of federal

[4] GAO-22-105274, Decarbonization: Status, Challenges, and Policy Options for Carbon Capture, Utilization, and Storage (Sept. 2022) at 38, https://www.gao.gov/assets/gao-22-105274.pdf.

[5] CCUS: 2023 will be a milestone year, Wood Mackenzie, p. 3, https://www.woodmac.com/news/opinion/ccus-2023-outlook/.

[6] Rystad Energy, PowerPoint Presentation (sf-asset-manager.s3.amazonaws.com), https://sf-asset-manager.s3.amazonaws.com/97637/392/5783.pdf.

[7] In 2009, a paper published by the Global CCUS Institute estimated 923 Gt CO_2 global capacity from existing "disused" petroleum reservoirs and estimated 10,000 Gt CO_2 global capacity in saline reservoirs, an estimate that is well short of current capacity estimates. CO2 storage in depleted gas fields. Global CCS Institute, https://www.globalccsinstitute.com/resources/publications-reports-research/co2-storage-in-depleted-gas-fields/. A more recent 2024 study suggests much larger gross capacities reduced somewhat for safety, trapping mechanisms, and reservoir performance limitations and reduced further due to regulation and economic factors. Jianquio Leng et al., A comprehensive review of efficient capacity estimation for large-scale CO_2 geological storage. ScienceDirect, https://www.sciencedirect.com/science/article/pii/S2949908924001353.

[8] Which area is the best for geologic carbon sequestration?, U.S. Geological Survey (usgs.gov), https://www.usgs.gov/faqs/which-area-best-geologic-carbon-sequestration.

[9] *Id.* For an assessment of onshore areas, including state coastal waters, see USGS Circular 1386: National Assessment of Geologic Carbon Dioxide Storage Resources—Results, https://pubs.usgs.gov/circ/1386/. For an assessment of optimal storage depths, see Stefan Iglauer, Optimal storage depths for structural CO_2 trapping, Int's J. of

land comprise much of Alaska, due to high transport costs and environmental concerns, little CO_2 sequestration is likely to occur in Alaska except perhaps direct-air capture (DAC) or CO_2 capture from Alaska industries. Federal land in the Rocky Mountain Region tends to be intermingled with privately owned parcels. Most of the lands in the Northern Great Plains and within the onshore Coastal Basins are privately owned and tend to be subdivided into smaller parcels, and many of these parcels are owned by two or more cotenants.

A major key to cost reduction will be the development of CCS hubs—"a cluster of emissions facilities that share the same CO_2 transportation and storage or utilization infrastructure."[10] In addition, CO_2 transportation hubs will serve to standardize CO_2 transport and make it more efficient. The U.S. Department of Energy is providing support for FEED studies for regional CO_2 transportation. A potential downside to standardization may be that the transported CO_2 will need to be of similar quality and concentration.

2 Some Key CCUS and CCS Projects

The "U" in CCUS stands for "utilization." In context, this means utilizing CO_2 in a manner that will prevent the utilized CO_2 from being emitted into the atmosphere.

Currently, the largest "use" of CO_2 is for enhanced oil recovery (EOR).[11] Once injected into an oil reservoir, the CO_2 increases the pressure in the oil reservoir, which pushes more oil toward lesser pressure

Greenhouse Gas Control 82 (Oct. 2018), Optimum storage depths for structural CO_2 trapping. ScienceDirect, https://www.sciencedirect.com/science/article/abs/pii/S1750583617306606.

[10] The world needs to capture, use, and store CO_2. McKinsey, https://www.mckinsey.com/industries/oil-and-gas/our-insights/the-world-needs-to-capture-use-and-store-gigatons-of-co2-where-and-how.

[11] For an explanation of this process, see Enhanced Oil Recovery, Department of Energy, https://www.energy.gov/fecm/enhanced-oil-recovery#:~:text=The%20four%20projects%20are%3A%201%20Improved%20Mobility%20Control,2%20Foam%20for%20CO%202%20-EOR%20Application%20. For more detail, see Carbon Dioxide Applications for Enhanced Oil Recovery Assisted by Nanoparticles: Recent Developments. ACS Omega, https://www.energy.gov/fecm/enhanced-oil-recovery#:~:text=The%20four%20projects%20are%3A%201%20Improved%20Mobility%20Control,2%20Foam%20for%20CO%202%20-EOR%20Application%20.

zones, i.e., cones of reduced pressure around well bores. The injected CO_2 also blends with the reservoir crude oil, improving the oil's ability to migrate through the reservoir into wellbores.[12] However, the bulk of CO_2 used for EOR has come from naturally occurring CO_2 produced from wells—many located in southern Colorado and New Mexico. This CO_2 is transported by pipeline for use in EOR—mostly in oil reservoirs located in southwest Texas. Initially, the primary purpose of utilizing CO_2 was to enhance oil recovery without a specific intent to also sequester the CO_2 to address climate change. Moreover, the sequestration of CO_2 as a side-benefit of increased oil production lowers the net benefits of this type of CO_2 utilization.

Nevertheless, CO_2 is being captured from various plants and used in EOR with the intent to sequester the CO_2. For example, CO_2 from a coal gasification plant in North Dakota is captured and piped 200 miles to Saskatchewan for EOR and sequestration in the Weyburn oilfield. Additional CO_2 from this plant is slated for sequestration without EOR. The University of North Dakota Energy and Environmental Research Center is exploring the possibility of using captured CO_2 for injection and sequestration into the Bakken shale oil formation in North Dakota to both increase oil recovery and sequester CO_2.[13]

A 2023 report identified 13 operational CO_2 capture projects in the United States, of which 12 were utilizing and sequestering the CO_2 for EOR.[14] However, eight under-construction projects in the United States will capture CO_2 for geologic sequestration.[15] Of an additional 70+ United States CO_2 capture projects described as being in advanced development, only three were destined for EOR.[16] Of nearly 60 United States CO_2 capture projects described as being in early development, only two

[12] Can CO_2-EOR Really Provide Carbon-Negative Oil?—Analysis. IEA, https://www.iea.org/commentaries/can-co2-eor-really-provide-carbon-negative-oil.

[13] Nidhal Badrouchi, Evaluation of CO_2 Enhanced Oil Recovery in Unconventional Reservoirs (2022), Evaluation Of CO_2 Enhanced Oil Recovery In Unconventional Reservoirs (und.edu), https://commons.und.edu/cgi/viewcontent.cgi?article=5321&context=theses; James A. Sorensen et al., Characterization and Evaluation of the Bakken Petroleum System for CO_2 Enhanced Oil Recovery, SPE-178659-MS/URTeC:2169871 (2015).

[14] Global-Status-of-CCS-Report-Update-23-Nov.pdf (cloudinary.com), https://res.cloudinary.com/dbtfcnfij/images/v1700717007/Global-Status-of-CCS-Report-Update-23-Nov/Global-Status-of-CCS-Report-Update-23-Nov.pdf?_i=AA.

[15] *Id.*

[16] *Id.*

were destined for EOR.[17] The Clean Air Task Force has a website that is supposed to list all announced CCUS projects.[18] The site lists projects by type and projected capture capacity, including those in development and those that are operational. Of those that are operational, most store the CO_2 in association with enhanced oil recovery.

To further reduce CO_2 emissions, other "uses" for CO_2 are under study. The key is to make CO_2 a valuable commodity, not just a waste product, so that capturing and using CO_2, rather than emitting it into the atmosphere becomes a sound business decision. Besides EOR, another direct use of CO_2 is for carbonating beverages but some of this CO_2 is emitted when the beverage is opened. Other uses, however, may include transforming CO_2 biologically or chemically for use in manufacturing synthetic fuels, polymers, and other products. The manufacturing of such products is likely to result in some further CO_2 emissions and some products may only recycle CO_2 for later emitting. So, the goal of developing these products is to reduce net CO_2 emissions. Other uses might count as storage, not necessarily permanent but at least relatively long term—such as incorporating CO_2 into cement or using it as a water substitute in the curing process resulting in some of this CO_2 mineralizing into calcium carbonate. In addition, studies suggest that CO_2 could be used to develop a substitute for cement, which is a hard-to-decarbonize commodity. Lehigh Hanson, Inc. and Fortera Inc. announced plans to capture CO_2 directly from a new-build cement plant and store it within the cement product itself. The objective is to reduce CO_2 emissions by 60%. CO_2 could be used to accelerate the growth of algae, which can absorb CO_2 faster than other sources of biomass. CO_2 could also be used

[17] *Id.*

[18] U.S. Carbon Capture Project Map. Clean Air Task Force (catf.us), https://www.catf.us/ccsmapus/. For an updated global list, see Global-Status-of-CCS-Report-Update-23-Nov.pdf (cloudinary.com), https://res.cloudinary.com/dbtfcnfij/images/v1700717007/Global-Status-of-CCS-Report-Update-23-Nov/Global-Status-of-CCS-Report-Update-23-Nov.pdf?_i=AA. For an updated summary of CCS Technologies, see State of the Art: CCS Technologies 2024. Global CCS Institute, https://www.globalccsinstitute.com/resources/publications-reports-research/technologycompendium2024/. Improved CCS technologies are rapidly evolving. For example, in July 2024, University of Texas researchers announced a new method for creating carbon dioxide hydrates—substances that can trap carbon dioxide like ice at high speeds and eliminating the need for harmful chemicals. University Of Texas Presents Fastest New Carbon Storage Technology (carbonherald.com), https://carbonherald.com/university-of-texas-presents-fastest-new-carbon-storage-technology/.

to make carbon composites, carbon fiber, and graphene—products that could substitute for certain carbon-intensive products.[19]

The petroleum industry is expected to play a major role in CCUS.[20] Among the most attractive sequestration sites are the Texas Gulf Coast and the Great Plains. The Texas General Land Office (GLO) in September 2021 announced a CO_2 sequestration "lease" to Talos Energy, Inc. and Carbonvert, Inc. covering 40,000 acres of state-owned submerged lands in the Gulf of Mexico.[21] Subsequently, Chevron acquired a 50% stake in this project, and TotalEnergies acquired Talos Low Carbon Solutions, a Talos' CCS subsidiary. Since then, the GLO has granted additional CO_2 sequestration leases.[22] In October 2021, the State of Louisiana issued two "operating agreements" covering five sequestration sites.

Summit Carbon Solutions, partially funded by Continental Resources, Inc., plans to capture CO_2 from over 30 ethanol plants across five states and ship the gas by pipeline to North Dakota for sequestration.[23] Due to the high concentration of nearly pure CO_2, capturing CO_2 from ethanol plants is especially attractive. A with other CCS projects, Summit has had difficulty getting pipeline easements causing the completion date to pushed back.

In August 2022, Occidental (OXY) announced that its affiliate, 1Point-Five, will begin detailed engineering and early site construction for its first

[19] This paragraph summarized Climate change: 6 uses for CO_2 that could cut emissions and make money. Vox, https://www.vox.com/energy-and-environment/2019/11/13/20839531/climate-change-industry-co2-carbon-capture-utilization-storage-ccu, which in turn summarizes various other reports.

[20] The IEA has a CCUS projects explorer database that is periodically updated to track CCUS projects worldwide, CCUS Projects Explorer—Data Tools. IEA, https://www.iea.org/data-and-statistics/data-tools/ccus-projects-explorer.

[21] Press release, cmr-george-p-bush-announces-new-coastal-partnership-for-carbon-sequestration1.pdf (texas.gov), https://glo.texas.gov/the-glo/news/press-releases/2021/september/cmr-george-p-bush-announces-new-coastal-partnership-for-carbon-sequestration1.pdf (Sept. 23, 2021).

[22] Press release, Commissioner Buckingham Secures $10 Billion for Texas Students with Historic State Land Carbon Capture and Storage Leases, https://www.glo.texas.gov/the-glo/news/press-releases/2023/august/commissioner-buckingham-secures-10-billion-for-texas-students-with-historic-state-land-carbon-capture-and-storage-leases.html (Aug. 22, 2023).

[23] Summit Carbon Solutions. Summit Carbon Solutions, https://summitcarbonsolutions.com/.

large-scale DAC (CO_2 removal[24]) plant in West Texas.[25] Expected to be operational in late 2024, the plant is part of a joint venture with BlackRock and is slated initially to capture 500,000 metric tons of CO_2 per year and scale up to 1 million tons per year.[26] In November 2022, OXY and 1PointFive announced a joint DAC and sequestration project that will be located on the fabled King Ranch along the Texas Gulf Coast. The DAC project will remove up to 30 million metric tons of CO_2 per year, and the sequestration project will store up to 3 billion metric tons of CO_2 from DAC as well as CO_2 captured and transported to the King Ranch from nearby industrial emitters.[27] OXY/1PointFive anticipates deploying 100 such plants worldwide by 2035, and possibly up to 135 plants with increased global policy support.[28] In March 2024 the U.S. Department of Energy announced the availability of up to US$24 million for research into technologies that can remove CO_2 directly from the air.

In North Dakota, Minnkota Power Cooperative, a wholesale supplier of electricity to rural electric distribution cooperatives, announced Project Tundra, a plan to capture CO_2 from a coal-fired power plant and sequester the CO_2 in saline reservoirs located in Central North Dakota.[29]

Several announced projects have encountered local opposition, e.g., the Summit Carbon solutions project, mentioned above,[30] the Wolf Carbon Solutions proposed a project to capture CO_2 from Iowa ethanol plants and pipe the CO_2 to Southern Illinois for sequestration,[31] and the Navigator CO_2 Ventures project to capture CO_2 from sources in

[24] "Removal" refers to removing CO_2 that is already in the atmosphere and then sequestering or utilizing it in a way that reduces the volume of atmospheric CO_2.

[25] Direct air capture must cut costs to succeed (pemedianetwork.com), https://pemedianetwork.com/carbon-economist/articles/ccus/2022/direct-air-capture-must-cut-costs-to-succeed/.

[26] *Id.*

[27] Oxy, King Ranch to develop CO_2 sequestration hub. Oil & Gas Journal, https://www.ogj.com/energy-transition/article/14285053/oxy-king-ranch-to-develop-co2-sequestration-hub.

[28] *Id.* 1PointFive will use DAC technology developed by Carbon Engineering Ltd. Carbon Engineering, Direct Air Capture of CO_2, https://carbonengineering.com/.

[29] Project Tundra (projecttundrand.com), https://www.projecttundrand.com/.

[30] Summit Carbon Solutions. Summit Carbon Solutions, https://summitcarbonsolutions.com/.

[31] About Wolf. Wolf Carbon Solutions, https://wolfcarbonsolutions.com/about-wolf/.

five states for CO_2 sequestration in Southern Illinois,[32] Each of these projects has suffered major setbacks due to opposition from landowners and environmentalists to CO_2 pipelines. The possible failure of these projects further suggests that the future of near-term CCUS projects may be largely confined to coastal areas where much of the CO_2 transport and sequestration will occur offshore and near emission sources. Elizabeth McGinley, a partner in Bracewell, who specializes in providing legal advice for CCUS projects, has observed:

> While multi-state pipeline and storage systems have hit significant development obstacles, CCUS projects with (i) storage in close proximity to the emitter (including projects where the storage site is at or very near the site of the emitter for use exclusively by the emitter) and/or (ii) the ability to utilize existing gas pipelines (even if the CO2 is shipped at a lower pressure)—are likely to achieve the most success in the short term.[33]

As of late November 2023, an IEA database anticipated that the U.S. will have planned CO_2 storage capacity of about 130 Mt by 2030; however, operational capacity will only be about 20 Mt—the same as 2023.[34] Currently, about 2/3 of operational and planned capture projects are associated with natural gas processing, but that will change by 2030, with about equal capture capacities associated with natural gas processing, biofuels, hydrogen/ammonia, and power/heat. The balance, about 20% of capture capacity, will be associated with other fuels, cement, heavy industry, and DAC.

Owing to the small number of announced CCUS projects that are in an advanced development stage, little can be definitively said about the appropriate CCUS business model for US projects. Deciding on the appropriate business model requires careful consideration of the eligibility requirements for the necessary permits and for CO_2 sequestration incentives. For example, under the U.S. Underground Injection Control regulations for CO_2 injection (Class VI) wells, the well owner or operator

[32] How Midwest Landowners Helped to Derail One of the Biggest CO_2 Pipelines Ever Proposed. Inside Climate News, https://insideclimatenews.org/news/05112023/landowners-fight-co2-pipeline-midwest-navigator/.

[33] Email in my files from Elizabeth McGinley, Dec. 27, 2023.

[34] CCUS Projects Explorer—Data Tools. IEA, https://www.iea.org/data-and-statistics/data-tools/ccus-projects-explorer.

must meet and maintain a bond or letter of credit, updated as necessary annually, until it obtains an approved site-closure certificate and an approved post-injection site care plan sufficient to cover the costs of any necessary corrective, remedial, or emergency work, the plugging of injection wells, post injection monitoring and site closure, and emergency response.[35]

Under the 45Q tax credit provisions, the taxpayer that owns the equipment that captures and sequesters CO_2 in secure geologic storage may claim the tax credit, but the taxpayer need not own the facility that would otherwise emit the captured carbon.[36] Parties engaged in CCUS may allocate tax credits among various qualified taxpayers through partnership agreements, and contracts for CCUS services are allowed, subject to certain regulatory requirements to qualify for the tax credits.[37] These tax credits may be sold for cash to a third party.[38]

From a sequestration operator's perspective, affiliates to an operator should separately hold components of the CCUS value chain. Over consolidation of CCUS into a single business entity, such as a petroleum exploration company, may make adapting to changing CCUS business practices more difficult. And many CCUS business models are likely to be built around CCUS as a service provided to two or more CO_2 emitters. Some companies may specialize in capture, pipeline companies may build and operate CO_2 pipelines, and affiliates of petroleum companies might develop and conduct the sequestration component. To achieve the necessary economies of scale, a sequestration operation is likely to take delivery of supercritical CO_2 from a variety of point sources, including DAC facilities, perhaps using two or more pipelines to transport CO_2 to a transportation hub, akin to the Henry Hub for natural gas. Even where a single entity (e.g., the operator of a power plant or refinery) conducts CCUS operations as part of a vertically integrated enterprise, the entity is likely to have affiliates operate distinct components along at least parts of the CCUS value chain.

[35] 40 CFR § 146.85.

[36] 26 U.S.C.A. § 45Q(f)(3)(ii).

[37] *Id.* at § 1.45Q-1(h)(2). Perhaps the most significant limit concerns provisions for liquidated damages *Id* at § 1.45Q-1(h)(2)(i).

[38] Text—H.R.5376—117th Congress (2021–2022): Inflation Reduction Act of 2022 (Congress.gov). Library of Congress, https://www.congress.gov/bill/117th-congress/house-bill/5376/text.

3 Legal Framework, Policy, and Norms

3.1 45Q Tax Credits

CCUS incentives are provided by "45Q tax credits," which were added to the tax code in 2008 and later expanded as part of the Bipartisan Budget Act of 2018.[39] These credits are available for "carbon oxides"[40] captured from an "industrial source"[41] or by "direct air capture," subject to certain annual capture thresholds,[42] and sequestered in "secure geological storage." If carbon oxides are captured and properly sequestered from an "industrial source," then that source becomes a "qualified facility"[43] for purposes of the 45Q tax credits.

CCS was further incentivized by the Inflation Reduction Act (IRA),[44] signed by President Biden on August 16, 2022. Before recent opposition to CO_2 pipelines, the Rhodium Group estimated that the IRA has the potential to reduce CO_2 and other greenhouse gas emissions 31% to 44% below 2005 levels by 2030 as well as promote the development of new green technologies.[45]

[39] Text—H.R.1892—115th Congress (2017–2018): Bipartisan Budget Act of 2018 (Congress.gov). Library of Congress, https://www.congress.gov/bill/115th-congress/house-bill/1892/text.

[40] The Bipartisan Budget Act, *id.*, expanded the 45Q tax credits to carbon oxide—a term that includes carbon dioxide, carbon monoxide, and carbon suboxide; however, this chapter uses the more familiar term, carbon dioxide, abbreviated as CO_2.

[41] An industrial source includes an "an electricity generating facility, that produces a carbon oxide stream from a fuel combustion source or fuel cell, a manufacturing process, or a fugitive carbon oxide emission source that, absent capture and disposal, injection, or utilization, would otherwise be released into the atmosphere as industrial emission of greenhouse gas or lead to such release" but excludes carbon dioxide production wells. 26 C.F.R. § 1.45Q-2(d).

[42] 26 U.S.C.A § 45Q(d)(2) and 26 C.F.R. § 1.45Q-2(g)(1).

[43] 26 U.S.C.A. § 45Q(c)-(d).

[44] Text—H.R.5376—117th Congress (2021–2022): Inflation Reduction Act of 2022 (Congress.gov). Library of Congress, https://www.congress.gov/bill/117th-congress/house-bill/5376/text.

[45] A Congressional Climate Breakthrough. Rhodium Group, https://rhg.com/research/inflation-reduction-act/. *See also* U.S. Department of Energy Office of Policy, 8.18, https://www.energy.gov/sites/default/files/2022-08/8.18%20InflationReductionAct_Factsheet_Final.pdf?stream=top. In its draft Carbon Removals Certification Framework (CRCF), the European Commission proposes an alternative to tax credits—the trading of captured carbon as credits in carbon offset markets. EU must clarify

Compared to the 2008 law, the IRA substantially decreased the capture thresholds,[46] enhanced the tax credits, and extended the required construction commencement date.[47] The tax credits may be claimed for 12 years from the date the qualified carbon-capture facility is placed into service provided construction begins before January 1, 2033.[48] The available tax credits are divided into base and bonus credits and will be inflation-adjusted starting in 2027 from base-year 2025. The bonus credit amounts, which are 5 times the base credits, are available to taxpayers who meet certain wage and apprenticeship guidelines.[49] The combined base and bonus credits, unadjusted for inflation, are: $85/metric ton for carbon oxides captured from industrial sources ($180/metric ton for DAC) and sequestered; and $60/metric ton for carbon oxides captured from industrial sources ($130/metric ton for DAC) and utilized for enhanced oil recovery (EOR) or for producing low- or zero-carbon fuels, chemicals, building materials, or other products.[50]

Taxpayers may elect to receive direct payments in lieu of tax credits for the first five years after carbon capture equipment is placed into service, but tax-exempt entities, such as states, municipalities, Indian Tribes, and cooperatives may receive direct payments for 12 years.[51] A taxpayer may sell its tax credits for cash, and the cash received is not taxable.[52]

CO_2 removal rules, https://www.pemedianetwork.com/transition-economist/articles/decarbonisation/2022/eu-must-clarify-co2-removal-rules/?id=57726876.

[46] For example, the capture threshold for a qualified power plant decreases from 500,000 metric tons of emitted CO2/annum to 18,750 metric tons; however, a power plant capture project must be designed to capture at least 75% of its emitted CO2. *Id.* Section 13104.

[47] *Id.*

[48] *Id.*

[49] *Id.* Federal Register: Prevailing Wage and Apprenticeship Initial Guidance Under Section 45(b)(6)(B)(ii) and Other Substantially Similar Provisions, 2022-26108.pdf (govinfo.gov), https://www.govinfo.gov/content/pkg/FR-2022-11-30/pdf/2022-26108.pdf.

[50] *Id.*

[51] Text - H.R.5376 - 117th Congress (2021–2022): Inflation Reduction Act of 2022 (Congress.gov). Library of Congress, https://www.congress.gov/bill/117th-congress/house-bill/5376/text.

[52] *Id.* at § 6418. In addition, the IRA makes $5.8 billion available for technological improvements that reduce greenhouse gas emissions from certain "energy-intensive industrial processes." § 50169. For a study of why the enhanced tax credits

In addition to federal tax credits, the State of California provides market-based incentives for lowering the carbon intensity of fuels through its Low Carbon Fuel Standard (LCFS) program.[53] By lowering carbon intensity, participants can earn credits that can be used or sold to offset higher carbon intensity fuels. Qualified CCUS projects, including qualified projects located outside California, may earn credits. However, among other requirements, these California incentives are not available to an operator unless the operator must prove the existence of a binding agreement among relevant parties that prohibits drilling and extraction that penetrates the confining layer above the sequestration zone within the AOR [area of review] to ensure public safety and permanent CO_2 storage.[54] Thus, CCUS projects within active oil and gas exploration and development areas will have difficulty qualifying for the California program.

3.2 *Underlying CCUS Law*

3.2.1 Federal Law

Federal regulations govern CO_2 sequestration wells, called Class VI wells[55] under the Underground Injection Control provisions of the Safe

were needed, see Capturing Investment: Policy Design to Finance CCUS Projects in the U.S. Power Sector—Center on Global Energy Policy at Columbia University SIPA. CGEP, https://www.energypolicy.columbia.edu/publications/capturing-investment-policy-design-finance-ccus-projects-us-power-sector.

[53] For an explanation, see: Understanding California's Low Carbon Fuel Standards Regulation—RMI and California's Cap-and-Trade Program and Low Carbon Fuel Standard Go Together Like Peanut Butter and Jelly. Union of Concerned Scientists (ucsusa.org), https://rmi.org/understanding-californias-low-carbon-fuel-standards-regulation/.

[54] State of California Carbon Capture and Sequestration Protocol under the Low Carbon Fuel Standard, § 9(c). Geologists and engineers have informed me that wells can be safely drilled through a sequestration reservoir at an additional cost. Thus, the California tax credits seem aimed more at stymying further petroleum development rather than protecting the integrity of sequestration reservoirs.

[55] Class VI wells are defined in 40 C.F.R. § 144.6. For a summary of the science of underground storage, see Manal Al-Safi et al., A review on underground gas storage systems: Natural gas, hydrogen and carbon sequestration, 9 Energy Reports 6251 (Elsevier Dec. 2023), A review on underground gas storage systems: Natural gas, hydrogen and carbon sequestration. ScienceDirect (https://www.sciencedirect.com/science/article/pii/S2352484723009794) and Sunil Kumar et al., A comprehensive review of value-added CO_2 sequestration in subsurface saline aquifers, 81J. of Nat. Gas Science and Engineering

Drinking Water Act—the principal United States law protecting underground sources of drinking water. Under the Class VI well regulations, effective in 2010, geologic sequestration "means the long-term containment of a gaseous, liquid, or supercritical carbon dioxide stream in subsurface geologic formations."[56] In general, these regulations provide extensive site study and characterization requirements, robust well construction requirements, well operational requirements, monitoring requirements to ensure the integrity of the injection well and reservoir, financial responsibility requirements, and ongoing reporting and recordkeeping requirements.[57] A closure and monitoring period is projected to extend 50 years, subject to an approved alternative period based on demonstrating that safe drinking water is not endangered.[58] To secure an approved closure, the operator must provide a site-closure report certifying the nature, composition, and volumes of the supercritical CO_2 stream, that injection wells have been properly plugged, and that regulatory agencies having authority to regulate future drilling of wells in the area have been properly notified.[59] Additionally, the operator must record in real property records a notice that the affected lands have been used to sequester CO_2, including a notice of where a survey of the area is available for inspection at a state, local, or tribal agency and applicable regional office of the U.S. Environmental Protection Agency (EPA). The survey must disclose the volumes injected, a description of the injection reservoir, and the period of injection operations.[60]

States may seek primacy to regulate Class VI wells. To secure primacy, the state must have regulations at least as stringent as the federal regulations.[61] States anticipating CO_2 sequestration projects, especially oil

103437 (Elsevier Sept. 2020), A comprehensive review of value-added CO2 sequestration in subsurface saline aquifers. ScienceDirect, https://www.sciencedirect.com/science/article/abs/pii/S1875510020302912.

56 40 C.F.R. § 146.81(d).

57 40 C.F.R. § 144 Subparts A-E and 40 C.F.R. § 146 Subpart H.

58 40 CFR § 146.93.

59 *Id.* § 146.93(f).

60 *Id.* § 146.93(g).

61 *See* eCFR: 40 CFR Part 146—Underground Injection Control Program: Criteria and Standards, https://www.ecfr.gov/current/title-40/chapter-I/subchapter-D/part-146, Subpart H. As of August 2024, the Council on Environmental Quality had not issued final guidance on CCUS, apart from a proposed draft. See Federal Register: Carbon Capture,

and gas producing states, are likely to seek primacy. As of August 2024, Louisiana, North Dakota, and Wyoming had achieved primacy for Class VI wells, designating their respective petroleum conservation agencies as the lead agency. Other states, including Alaska, Arizona, Oklahoma, Texas, and West Virginia are in the process of seeking primacy over Class VI wells. In oil and gas producing states, the state oil and gas regulatory agency is likely to be the principal state regulator of CO_2 sequestration and Class VI wells[62]; however, the state regulatory agency with authority over CO_2 capture and transport will vary from state to state.

The federal regulations do not comprehensively address all issues identified by the IEA. Except for federal lands, state law will determine issues like pore-space ownership, ownership of injected CO_2, and long-term liability or the release of liability. The federal government could preempt these liability laws but has not done so to date.

3.2.2 Select State Laws

Texas. In Texas, the principal statutes dealing with CO_2 sequestration unrelated to enhanced oil recovery operations or certified clean energy or clean coal projects are codified in the Texas Natural Resources[63] and Water[64] Codes. Subchapter K of the Texas Health and Safety Code governs carbon sequestration beneath the state's territorial waters, which are managed by the General Land Office (GLO) and is the source of the GLO's authority to execute CO_2 sequestration leases or to engage in projects on its own.[65]

Utilization, and Sequestration Guidance, https://www.ecfr.gov/current/title-40/chapter-I/subchapter-D/part-146.

[62] The Interstate Oil and Gas Conservation Commission, a governmental organization consisting of oil and gas producing states, has prepared a regulatory model for CO2 sequestration. *See*, CO2 Storage: A Legal and Regulatory Guide for States, CO2 Phase II Summary 12-17-2007 (oklahoma.gov), https://oklahoma.gov/content/dam/ok/en/iogcc/documents/issues/carbon/co2_storage_summary-road_to_a_greener_energy_future-2007.pdf.

[63] Tex. Nat. Res. Code §§ 121.001 to 121.004 (storage and extraction of anthropogenic carbon dioxide).

[64] Tex. Water Code §§ 27.040 to 27.050 (storage and associated injection of anthropogenic carbon dioxide).

[65] Tex. Health and Safety Code §§ 382.501 to 382.510.

Under the principal statutes, the Texas Railroad Commission is to seek primacy to administer and enforce federal law[66] and is given jurisdiction and rulemaking power over both onshore and offshore injection and sequestration of carbon dioxide.[67] The Commission must issue necessary permits before a party may drill and operate an anthropogenic carbon dioxide injection well[68] or operate a geologic storage facility.[69] No permit may be issued until the Commission provides an applicant with a "letter of determination" that the injection well or storage facility will not injure freshwater strata and that the storage formation is not a freshwater sand. This letter may be issued following a review of the area of review and the plans for corrective action, monitoring during injection and post injection, and post-injection site-care.[70] In addition, the Commission must issue a letter of determination that the injection well or storage facility will not interfere with a previous or existing Class I injection well or associated waste plume or with any well authorized by the Commission.[71] A permittee must annually provide the Commission with satisfactory evidence of financial responsibility.[72]

[66] Tex. Water Code § 27.048 (providing that "the railroad commission shall seek primacy…."). Pending primacy approval, parties seeking to sequester CO2 unrelated to enhanced oil recovery must file applications with both the Railroad Commission and the EPA. On May 3, 2022, the Railroad Commission approved proposed regulations for public comment, approved submission of a pre-application for Class VI wells, and requested Governor Abbott to seek formal primacy approval. Approvals have not come quickly, but the pace is accelerating. It took nearly five years for EPA to approve North Dakota's application; but Wyoming received its approval in about ten months.

[67] *Id.* at §§ 27.041 and 27.047. State rules must be consistent "with applicable rules or regulations adopted by the United States Environmental Protection Agency or another federal agency governing the injection and geologic storage of anthropogenic carbon dioxide." *Id.* at § 27.048. *See also* Tex. Health and Safety Code § 382.502 (authorizing the Commission to develop rules for the "location, construction, maintenance, monitoring, and operation of a carbon dioxide repository" in compliance with federal rules) and Water Code ch. 27, subch. F).

[68] The conversion of an existing well to a CO_2 storage well is governed by Texas Injection Well Act, Tex. Water Code ch. 27, as referenced in Tex. Nat. Res. Code § 91.802.

[69] Tex. Water Code § 27.043. For permit requirements, *see id.*, § 27.051.

[70] *Id.* at § 27.046.

[71] *Id.* at § 27.04611.

[72] *Id.* at § 27.050.

The Commission may impose fees to cover the costs of permitting, monitoring, or inspecting CO_2 injection wells and storage facilities and fees to enforce and implement its regulatory authority.[73] Fees, including any civil penalties,[74] are to be deposited in an anthropogenic carbon dioxide storage trust fund.[75] Together with additional funds received by the Commission, the purpose of the trust fund[76] is to assure financial responsibility "for the proper management of an anthropogenic carbon dioxide injection well or geologic storage facility."[77] The Commission may use the fund for permitting, inspecting, short- and long-term monitoring, investigating, recording, and reporting on geologic storage facilities and associated injection wells; for remediating related mechanical problems; for repairing mechanical leaks at geologic storage facilities; for plugging abandoned anthropogenic carbon dioxide injection wells; for related training and technology transfer; and for related compliance and enforcement activities.[78]

Ownership of stored CO_2 is deemed the property of the storage operator and its heirs, successors, or assigns but is subject to contracts, property interests, and other laws to the contrary.[79] The Commission may adopt rules for the removal of anthropogenic carbon dioxide from a storage facility for commercial or industrial use.[80]

Several bills were introduced in the Texas legislature in 2023 that would have substantially amended existing law; however, no major legislation was enacted. As introduced, one unenacted bill[81] contained provisions clarifying ownership of pore space and of CO_2; providing

[73] *Id.* at § 27.045.

[74] For penalties and other enforcement mechanisms, see Tex. Water Code Chapter 27, Subchapter F.

[75] *Id.* at § 27.045. Tex. Nat. Res. Code § 121.003 established the trust fund.

[76] Tex. Water Code § 121.003.

[77] *Id.* at § 27.073.

[78] Tex. Nat. Res. Code § 121.003.

[79] *Id.* at § 121.002.

[80] *Id.* at § 121.004.

[81] Texas Legislature Online—88(R) Text for SB 2107, https://capitol.texas.gov/billlookup/Text.aspx?LegSess=88R&Bill=SB2107. *See also*, Texas Legislature Online—88(R) Text for HB 4484, https://capitol.texas.gov/billlookup/Text.aspx?LegSess=88R&Bill=HB4484.

for more robust regulation of CO_2 sequestration, including compulsory "integration" (akin to regulatory pooling/unitization) of geologic storage reservoirs; and addressing long-term liability for sequestered CO_2. Another unenacted bill[82] would have limited a CO_2 injector's liability for trespass, nuisance, or conversion,[83] but the bill did not address long-term liability relating to sequestered CO_2. Current Texas statutes address long-term liability for CO_2 storage only respecting state-owned submerged lands.

The most important existing law in Texas addresses CO_2 sequestration beneath the state's territorial waters.[84] Subject to securing regulatory primacy from the U.S. EPA, the Railroad Commission is given regulatory and rulemaking power over CO_2 sequestration beneath state waters.[85] The Bureau of Economic Geology at the University of Texas at Austin is charged with assisting the Land Commissioner in identifying potential locations for a CO_2 "repository."[86] The Land Commissioner must then recommend suitable locations to the School Land Board (SLB) for its final determination.[87] The SLB is empowered to issue "requests for proposals" to "lease" suitable locations for the transport and storage of CO_2[88] and to establish a CO_2 storage fee.[89] The Railroad Commission is authorized to establish "standards for the measurement, monitoring, and verification of the permanent storage status."[90] The Bureau of Economic Geology acts as the scientific advisor and must review and provide data to the Board regarding "any measurement, monitoring, and verification

[82] Texas Legislature Online—88(R) Text for HB 4557, https://capitol.texas.gov/billlookup/Text.aspx?LegSess=88R&Bill=HB4557.

[83] The draft legislation provided that stored CO_2 is not a pollutant or a nuisance and shields a party capturing or storing CO_2 from a nuisance action. Regarding a trespass or conversion action for the lateral migration of CO_2 or for environmental damage resulting from its release into air or surface water, a claimant may not recover noneconomic damages without proving actual damages and other particulars.

[84] Tex. Water Code ch. 27, subchapter C-1, and Texas Health and Safety Code ch. 282, subchapter K.

[85] Tex. Health and Safety Code § 382.502.

[86] *Id.* at § 382.503.

[87] *Id.*

[88] *Id.* at § 382.504.

[89] *Id.* at § 382.505.

[90] *Id.* at § 382,506.

of the permanent storage status....”[91] When permanent storage of CO_2 has been verified and when the storage location has met all applicable state and federal requirements for closure, the Board acquires title to the stored CO_2 on behalf of the permanent school fund[92] and relieves the CO_2 producer from “liability for any act or omission regarding the carbon dioxide in the carbon dioxide repository.”[93]

Louisiana. In Louisiana, the Louisiana Geologic Sequestration of Carbon Dioxide Act governs CO_2 sequestration.[94] The Act provides for comprehensive regulation by the Commissioner of Conservation,[95] but the State Mineral Board is responsible for negotiating a CO_2 sequestration operating agreement for state-owned lands.[96] The Commissioner is empowered to “regulate the development and operation of storage facilities and pipelines transmitting carbon dioxide to storage facilities, including...the issuance of certificates of public convenience and necessity for storage facilities.”[97] Among the Commissioner’s express powers is to determine whether a given reservoir is “suitable and feasible” for CO_2 storage.[98] The Commissioner is expressly authorized “to perform any and all acts necessary to carry out the purposes and requirements of the federal Safe Drinking Water Act, as amended, relating to this state’s participation in the underground injection control program established under that act with respect to the storage and sequestration of carbon dioxide.”[99]

Significantly, upon the issuance of a certificate of public convenience and necessity, which certificate does not cause the operator to become a common carrier or public utility,[100] the Louisiana Act allows the use of eminent domain to acquire storage and related rights.[101] In general, the

[91] *Id.*

[92] *Id.* at § 382.507.

[93] *Id.* at § 382.508.

[94] La. Stat. Ann. tit. 30, ch. 11.

[95] *Id.* at § 1104.

[96] *Id.* at § 209.

[97] *Id.* at § 1104 A.(1).

[98] *Id.* at § 11.04 C.(1).

[99] *Id.* at § 1106 A.

[100] *Id.* at § 1107.

[101] *Id.* at § 1108.

use of eminent domain may not prevent drilling through the storage reservoir by persons having a right to drill[102]; however, in Caldwell Parish (the Louisiana version of a county), when exercising eminent domain, an operator may acquire rights to exclude drilling through the storage reservoir if no drilling or operating of petroleum wells within the area boundaries of the storage facility to depths below the base of the storage reservoir has occurred for five years and if all formerly productive reservoirs below the storage reservoir are no longer capable of producing in paying quantities.[103] This targeted amendment was apparently passed to make CO_2 sequestration projects in that parish eligible for the previously mentioned California incentives for CCUS projects.

Other States. For other examples of state laws governing carbon sequestration, see, e.g., Ky. Rev. Stat. Ann. §§ 353.800 to 353.812; Miss. Code Ann. ch. 53.11; N.D.C.C. ch. 38-22; W. Va. Code ch. 22-11a and 22-11b.; and Wyo. Stat. Ann. §§ 35-11-313 to 35-11-318. Unlike Louisiana, most other states rely on compulsory unitization of storage space, not eminent domain.

4 Concerns and Challenges

4.1 Ownership/Control of Pore Space Offshore

Owing to the costs of transporting CO_2 and of operating in a deepwater environment, most carbon sequestration is likely to occur well within the U.S. territorial sea and perhaps mostly beneath coastal state territorial waters. Obviously, identifying the party having initial ownership or control of the pore-space rights will be critical. For an offshore project, this will generally be the coastal state government adjacent to the offshore storage area or the federal government for storage areas beyond a state's territorial sea.

Under the United Nations Convention on the Law of the Sea (UNCLOS), a nation has sovereignty over its "territorial sea," including "its bed and subsoil"[104] measured seaward from established baselines,[105]

[102] *Id.* at § 1108. B.

[103] *Id.*

[104] UNLCOS Art. 2(2).

[105] *Id.* at Art. 5 and Art. 7.

and to any "internal waters" landward of such baselines.[106] Although the United States has not ratified UNCLOS, it has been honoring the treaty's provisions regarding the use of the seabed and its subsoil. Consistent with UNCLOS, the United States has claimed a 12-nautical-mile territorial sea.[107] Under UNCLOS, a state enjoys nearly absolute sovereignty over its 12-nautical-mile territorial sea; however, in the United States each coastal state controls the first three nautical miles seaward from the baselines, except Texas and Florida which, for historical reasons, control a three-marine-league (nine-nautical mile) territorial sea along their respective Gulf of Mexico coastlines.

Beyond its territorial sea, but also measured from the same baselines, the United States claims control of an exclusive economic zone (EEZ) of up to 200 nautical miles.[108] In addition, where existent, the United States claims an extended continental shelf beyond its EEZ in accord with UNCLOS.[109]

Regarding federal waters, the U.S. Bureau of Safety and Environmental Enforcement governs health, safety, and environmental practices regarding CO_2 sequestration, and the U.S. Bureau of Ocean Management governs the acquisition of CO_2 sequestration rights; however, these agencies have yet to publish transport and sequestration regulations.[110] When this chapter was written, the United States Bureau of Ocean Energy Management (BOEM) and the Bureau of Safety and Environmental Enforcement (BSEE) were still working on carbon sequestration regulations for depleted hydrocarbon reservoirs and saline aquifers located on the Outer Continental Shelf (OCS) beyond state territorial waters.[111]

[106] *Id.* at Art. 8.

[107] Presidential Proclamation 5928 (Dec. 27, 1988).

[108] UNCLOS Arts. 56 and 57.

[109] UCLOS Art. 76.

[110] For updated information, see Carbon Sequestration | Bureau of Safety and Environmental Enforcement (bsee.gov), https://www.bsee.gov/carbon-sequestration and Carbon Sequestration | Bureau of Ocean Energy Management (boem.gov), https://www.boem.gov/about-boem/regulations-guidance/carbon-sequestration.

[111] The Infrastructure Investment and Jobs Act, § 40307, Public Law No: 117–58 (Nov. 15, 2021), PUBL058.PS (congress.gov), https://www.congress.gov/117/plaws/publ58/PLAW-117publ58.pdf, gives the Secretary of the Interior the authority to grant a lease, easement, or right-of-way on the Outer Continental Shelf (OCS) for long-term sequestration of carbon dioxide. *See also* Executive Order 14008 of January 27, 2021, Federal Register/Vol. 86, No. 19/Monday, February 1, 2021/Presidential Documents,

A May 2022 presentation, by Michael Celata, Regional Director of the Gulf of Mexico Regional Office of BOEM, estimated OCS CO_2 storage capacity beneath federal waters in the Gulf to be 89 Gt in the Lower Miocene formation and 199 Gt in the Upper Miocene formations.[112]

4.2 Ownership/Control of Pore Space Onshore

Regarding onshore storage, a GOA report notes that "[f]ewer than half [the states] have defined pore-space rights."[113] Some commentators have argued that deep pore space should be declared a public resource in the United States,[114] but such a declaration is unlikely and would be of dubious constitutionality. Numerous states have recognized private ownership of pore space by legislation[115] or by court decision.[116] So far,

eo-14008-tackling-climate-crisis-home-abroad.pdf (energy.gov), https://www.energy.gov/sites/default/files/2021/02/f83/eo-14008-tackling-climate-crisis-home-abroad.pdf.

[112] U.S. Dept of the Int., Bureau of Ocean Energy Management, Offshore Carbon Storage (May 19, 2022), https://www.boem.gov/sites/default/files/documents/about-boem/regulations-guidance/BOEM-5th-International-Conference.pdf.

[113] GAO-22-105274, Decarbonization: Status, Challenges, and Policy Options for Carbon Capture, Utilization, and Storage (September 2022) at 40, GAO-22-105274, Decarbonization: Status, Challenges, and Policy Options for Carbon Capture, Utilization, and Storage, https://www.boem.gov/sites/default/files/documents/about-boem/regulations-guidance/BOEM-5th-International-Conference.pdf.

[114] *See generally*, Robert Zadick, The Public Pore Space: Enabling Carbon Capture and Sequestration by Reconceptualising Subsurface Property Rights. 36 Wm & Mary Envtl L& Policy Rev. 257 (2011); AB Klass and EJ Wilson, Climate Change, Carbon Sequestration, and Property Rights, 2010 U. of Ill. L. Rev. 363 (2010); and James and Thomas R. DeCesar, An Evaluation of Eminent Domain and a National Carbon Capture and Geologic Sequestration Program: Redefining the Space Below, 45 Wake Forest L. Rev.261 (2010); Will Reisinger, N. Moser, T. Dougherty, and James D. Madeiros, Reconciling King Coal and Climate Change: A Regulatory Framework for Carbon Capture and Storage, 11 Verm. J. of Envtl. L. 1 (2009); and John Sprankling, Owning the Center of the Earth, 55 U.C.L.A. L. Rev. 979 (2008).

[115] *See, e.g.*, Indiana, Ind. Code § 14-39-2-3; Kentucky, Ky. Rev. Stat. § 353.800; Montana, Mont. Code § 82-11-180; Nebraska, Neb. Rev. Stat. § 57-1604; North Dakota, N.D. Cent Code § 47-31-03, Oklahoma, Okla. Stat. tit. 60 § 6; Utah, Ut. Code., § 40-6-20.5; West Virginia, W. Va. Code § 22-11B-18; and Wyoming, Wyo. Stat. § 34-1-152. *See generally* CCUS Legal and Regulatory Database—Data Tools. IEA, https://www.iea.org/data-and-statistics/data-tools/ccus-legal-and-regulatory-database.

[116] *See, e.g.*, Cassinos v. Union Oil Co., 18 Cal Rptr.2d 574 (Cal. App. 1993); U.S. v. 43.42 Acres of Land, 520 F. Supp. 1042 (W.D. La. 1981); Dep't of Transportation v. Goike, 560 N.W.2d 365 (Mich. App. 1996); Burlington Res. Oil & Gas Co. LP v.

state legislation and courts have generally declared that the surface owner owns pore space, subject to the mineral owner's right of reasonable and necessary use for petroleum and mineral development. Professor Barry Barton has debunked the notion of a so-called English rule supporting mineral-owner title to pore space, which should help bolster the prevailing view that surface owners own pore space.[117] Thus, where mineral rights have been severed from surface ownership, courts will likely determine that pore space remains with the surface owner except perhaps when an express right to use pore space unrelated to mineral development is expressly severed. Nevertheless, a few states, by legislation, have limited the ability to sever the pore space from the surface[118] except for "leasing" the right to engage in sequestration operations.[119]

Lang and Sons Inc., 259 P.3d 766 (Mont. 2011); Mosser v. Danbury Res., Inc. 898 N.W.2d 406 (N.D. 2017) Ellis v. Arkansas Louisiana Gas Co., 450 F. Supp. 412 (E.D. Okla. 1978); Brown v. Continental Res., Inc., 2021 WL 6755489 (D.S.D. 2021); and Lightning Oil Co. v. Anadarko E&P Onshore, LLC, 520 S.W.3d 39 (Tex. 2017).

For commentary, see Owen L. Anderson, Geologic CO2 Sequestration: Who Owns the Pore Space, 9 WYO. L. REV. 97 (2009), reprinted in LANDMAN (January/February 2010); Stefanie L. Burt, Who Owns the Right to Store Gas: A Survey of Pore Space Ownership in U.S. Jurisdictions, http://www.duqlawblogs.org/joule/wp-content/uploads/2016/07/Who-Owns-the-Right-to-Store-Gas-A-Survey-of-Pore-Space-Ownership-in-U.S.-Jurisdictions-.pdf; and Pore space ownership issues for CO2 sequestration in the U.S., https://doi.org/10.1016/j.egypro.2009.02.258. *See also* Ruth Ivory-Moore, Pore Space Rights. U.S. Overview (Global CCUS Inst. May 2022), Brief-Pore-Space-Rights-5.25.pdf, https://www.globalccsinstitute.com/wp-content/uploads/2022/05/Brief-Pore-Space-Rights-5.25.pdf (containing a bibliography of other papers addressing pore-space ownership in the United States) and Tara K. Righetti, Correlative Rights and Limited Common Property in the Pore Space: A Response to the Challenge of Trespass in Carbon Capture and Sequestration, 47 ENV'T L. REP. NEWS & ANALYSIS. 10420, 10428 (2017). For criticism of the Kentucky Smallwood case, above, and see generally Mark A. Imbrogno, Pipedream to Pipeline: Ownership of Kentucky's Subterranean Pore Space for Use in Carbon Capture and Sequestration, 49 Louisville L. Rev. 291 (2010).

117 *See* DONALD ZILLMAN et al., The Law of Energy Underground: Understanding New Developments in Subsurface Production, Transmission, and Storage, ch. 2, *The Common Law of Subsurface Activity: General Principle and Current Problems* by Barry Barton (Oxford 2014) (concluding that English common law supports surface-owner title to pore space. A few American courts have mentioned, but none have expressly adopted, a so-called English rule that favored pore-space ownership in mineral owners, see, e.g., Central Kentucky Natural Gas Co. v. Smallwood, 252 S.W.2d 866, 868 (Ky. 1952).

118 *See, e.g.*, N.D. Cent Code § 47-31-05.

119 *See, e.g.*, N.D. Cent Code § 47-31-06.

Some commentators[120] have suggested that an exception to surface-owner pore-space ownership might be lands patented under the U.S. Stock-Raising Homestead Act of 1916 (SRHA),[121] which applies to over 70 million acres of patented land in the Western United States. Patents issued under this Act are subject to a reservation in the United States of "all the coal and other minerals."[122] In my opinion, because the reservation is limited to minerals, the Act should not be construed as reserving pore space.[123] Eventually, this question may have to be resolved by the U.S. Supreme Court. The answer may turn on whether the Court decides that the text of the Act is sufficiently clear or whether resort to the Act's legislative history is necessary.[124]

[120] *See, e.g.*, Kevin L. Doran and Angela M. Cifor, Does the federal government own the pore space under private lands in the West? Implications of the Stock-Raising Homestead Act of 1916 for geologic storage of carbon dioxide, 42 Environmental L. 527 (2012).

[121] Pub.L. No. 64–290, 39 Stat. 862, codified at 43 U.S.C. § 299.

[122] 43 U.S.C. § 299.

[123] Owen L. Anderson, Geologic CO_2 Sequestration: Who Owns the Pore Space?, 9 Wyo. L. Rev. 97, 138 (2009). Although I could find no federal case law directly on point, two decisions of the Interior Board of Land Appeals provide limited guidance. See Phillips Petroleum Co. 105 I.B.L.A. 345, 349–350 (1988) (confirming that federal ownership of minerals does not include ownership of pore space, but the case does not indicate whether the land at issue was patented under the SRHA) and Mallon Oil Co., 104 I.B.L.A. 145, 150 (1988) (regarding acquired federal land and holding that pore space belongs to the surface owner, applying presumed Montana law). By analogy, City of Kenai v. Cook Inlet Natural Gas Storage Alaska, LLC, 373 P.3d 473 (Alas. 2016) supports by analogy federal ownership of pore space under SRHA patents. Using rules of statutory construction, and not rules of deed construction, the Alaska Supreme Court held that a state statute required the state to reserve pore space. The court applied the rules of statutory construction that construes ambiguities in public land grants strictly against the grantee. Thus, the SRHA question may turn on whether the U.S. Supreme Court regards the federal statute requiring the reservation of minerals as ambiguous.

[124] The legislative history does contain some language that might imply that a patentee was to receive only rights in the surface and not the subsurface—e.g., "The purpose of section 11 is to limit the operation of this bill strictly to the surface of the lands described and to reserve to the United States the ownership and right to dispose of all minerals underlying the surface thereof." In holding in a five to four decision that gravel was a mineral under the Act and thus reserved to the United States, Justice Thurgood Marshall cited this legislative history when stating: "In the present case this principle applies with particular force, because the legislative history of the SRHA reveals Congress' understanding that the mineral reservation would "limit the operation of this bill *strictly to the surface of the lands*." Watt v. Western Nuclear, Inc., 462 U.S. 36, 60 (1983), (emphasis in original), citing, H.R.Rep. No. 35, 64th Cong., 1st Sess. 4, 18 (1916).

On a related point, in *Entek GRB, LLC v. Stull Ranches, LLC*,[125] the federal Court of Appeals for the Tenth Circuit, in an opinion written by the then future Supreme Court Justice Neil Gorsuch, held that a mineral lessee could use the surface of land patented under the SRHA to access minerals on nearby SRHA lands through a federally approved unitization plan. In essence, he ruled that each reservation of minerals under an SRHA patent was part of a unified reservation of minerals in all SRHA lands. If the Supreme Court eventually rules that the United States controls the pore space to lands patented under the SRHA, then implementing an onshore carbon-sequestration project regarding adjoining parcels of SRHA patented lands would be somewhat simplified through a federally unitized carbon-sequestration project that would permit unified access rights to the SRHA parcels.

In contrast to SRHA lands, however, the vesting of pore-space ownership in private surface owners, rather than in private severed mineral owners, will make most CCS projects easier to implement because severed minerals have tended to become increasingly fractionalized into smaller and smaller undivided fractions through generational transfers. For a large sequestration project encompassing a large geographic area, locating and securing sequestration rights from what may be hundreds, perhaps thousands, of owners of small fractional mineral interests would be challenging. Moreover, mineral ownership of pore space would further hinder public acceptance of CCS projects in rural areas.

Although the nature of carbon storage rights has yet to be classified, it seems most closely akin to a servitude (easement), which would generally require the unanimous consent of all pore-space owners.[126]

Note that Marshall's comment on this legislative history ignores the second half of the sentence, which describes the Act's purpose as reserving minerals, not pore space. To reason that a mineral reservation should be broadly construed to reserve gravel as a "mineral," hardly justifies reasoning that pore space is a mineral.

125 763 F.3d 1252 (10th Cir. 2014). For a thorough analysis of *Entek*, see Tara K. Righetti, *Surface Access to Severed Federal Minerals*, 61 Rocky Mt. Min. L. Inst. 8-1 (2015).

126 *See, e.g.*, Tex. Mortg. Co. v. Phillips Petrol. Co., 470 F.2d 497 (5th Cir. 1972) (applying Texas law). On the other hand, pore space rights could be classified akin to an oil and gas lease—either a fee simple determinable or profit-à-prendre. In most states, one fractional co-tenant owner, acting unilaterally, may develop oil, gas, and minerals without violating the rights of other fractional co-tenant owners provided that the developing co-tenant does not prevent the other fractional co-tenant owners from exercising their

Further complicating mineral-owner title to pore space is possible separate ownership of specified minerals in many tracts—for example, in many circumstances the ownership of coals and coal-seam gas is divided.[127] Consider that 45Q tax credits are available for storing carbon in non-minable coal seams, many of which might contain coal-seam methane. If the mineral owner owns the pore space, then which mineral owner—the coal owner or the gas owner—owns the right to use the coal seam for CO_2 storage?[128]

4.3 *The Right to Use and Exclude the Use of Pore Space*

4.3.1 *Rights Among Pore-Space Owners*

Perhaps the most essential component of a CCS project is "secure geologic storage," which for tax-credit purposes "includes, but is not limited to, storage in deep saline formations, petroleum reservoirs, and unminable coal seams."[129] Secure storage means that the injected CO_2 must not escape the geologic formation into which it is injected for storage. If it does escape, then the federal government may recapture the tax credits.[130] For onshore projects especially, securing the geographical footprint for large sequestration reservoirs may be challenging—depending, in large part, upon how the courts treat subsurface intrusions across subsurface property lines.

development rights. *See*, Prairie Oil & Gas Co. v. Allen, 2 F.2d 566 (5th Cir. 1924 (applying Oklahoma law). The obvious problem with applying this latter view to pore-space use is that the other fractional co-tenant owners could not effectively use the same pore space. On the other hand, if classified as a servitude, unanimous consent of all fractional pore-space owners could prove difficult; however, regulatory unitization or authorizing eminent domain, discussed in subsection 3, below, could resolve this co-tenancy and related problems.

[127] *See, e.g.*, Amoco Production Co. v. Southern Ute Indian Tribe, 526 U.S. 865 (1999).

[128] *Cf.*, NCNB Texas Nat'l Bank v. West, 631 So.2d 212 (Ala. 1993) (as between coal owner and oil and gas owner, the right to produce coal-seam methane depends on the place of capture—whether captured directly from the coal seam or after the gas has migrated from the seam); United States Steel Corp. v. Hoge, 468 A.2d 1380 (Pa. 1983) (coal owner has the right to exploit coal-seam methane); and Central Natural Resources, Inc. v. Davis Operating Co., 201 P.3d 680 (Kan. 2009) (oil and gas owner has the right to exploit coal-seam methane).

[129] 26 C.F.R. § 1.45Q-3(a).

[130] *Id.* at § 1.45Q-5(f).

The key question is whether a CCS operator must acquire storage rights from all pore-space owners that suffer intrusion by a CO_2 plume, intrusion by displaced brine, or even a change in pressure gradient. From a purely efficiency and economic standpoint, the necessary acquisition of storage rights should be limited—e.g., to the owner of the land where an injection well is located. From a sequestration-integrity standpoint, however, the acquisition of storage rights needs to encompass lands beyond a small-tract injection site. But the question is: how much land?

The contract area in one Texas GLO (General Land Office) CO_2 sequestration lease comprises a state-owned offshore area encompassing almost 41,000 acres (about 16,600 hectares). Regarding the Louisiana CO_2 sequestration operating agreements, one encompasses three tracts comprising a total of 122,000 acres (about 49,400 hectares), and the other agreement encompasses two tracts comprising a total of 44,500 acres (about 18,000 hectares). Each tract in these operating agreements ranges from about 12,000 to 57,000 acres (about 4,900 to 23,000 hectares), with three of the five areas each comprising about 32,000 acres (about 13,000 hectares) each. The size of each tract suggests that the sequestration operator is acquiring rights to an entire, or at least a substantial portion of, a confined storage reservoir. In these Texas and Louisiana examples, acquiring the sequestration right was straightforward because the state government owned and controlled the pore space beneath the entire acreage.

The goal of a CCS project is secure sequestration—ideally, in perpetuity. Under United States policy, to assure secure geologic storage, a CCS project developer must investigate and model an "area of review" (AOR)—the area surrounding the proposed project that contains sources of safe drinking water that might be endangered by the project.[131] A project developer must provide scientific proof that injected CO_2 will not escape into safe drinking water[132] or into the atmosphere. In other words, the storage reservoir must have an impermeable seal that prevents CO_2 from escaping. Thus, the operator must investigate whether anything, including existing well bores, whether abandoned or active, or faults may

[131] 40 CFR § 146.84(a).

[132] An injection well must be cased and properly cemented from the surface through the base of the deepest safe drinking water source, *id.* at § 146.86(b)(2) and (b)(5).

compromise the seal. In some cases, a developer may have to remediate old, abandoned wells to prevent them from becoming an escape pathway for sequestered CO_2. Areas where existing saltwater disposal wells have induced earthquakes are unlikely candidates for CO_2 storage because of the presence of faults.[133] Future drilling, unrelated to the CCS project, poses further risks to the integrity of a storage reservoir. For large storage reservoirs, more than one sequestration operation might be conducted—perhaps by different operators, which will further complicate permitting, determining optimal, yet safe, injection rates and total cumulative injections, and monitoring the storage reservoir for leaks.[134] A CO_2 sequestration operator and its successors may need access over the entire AOR—an area that may have to be changed over time[135]—to perform any necessary remediation of existing wells, active or abandoned or other leakage pathways. This access right, which should include associated testing and monitoring rights, will need to be long term.

To assure the integrity of a sequestration reservoir, more robust property rights should extend throughout the Maximum Monitoring Area (MMA)[136]—the smaller area within an AOR that consists of the area that will be occupied by the free phase CO_2 plume until stabilized,[137] plus a ½-mile buffer zone surrounding the Active Monitoring Area (AMA). The AMA:

> is the area that will be monitored over a specific time interval from the first year of the period (n) to the last year in the period (t). The boundary of the active monitoring area is established by superimposing two areas:

[133] In Texas, the Railroad Commission has authority to regulate the underground disposal of saltwater and other petroleum wastes and may exercise this authority to address induced seismic activity in designated areas. *See* 16 Tex. Admin. Code § 3.9 (6)(A)(vi) and § 3.46 (d)(1)(f)]. These rules are likely to be extended to Class VI disposal wells.

[134] Class VI injection facilities and transport pipelines must be monitored and tested for leaks through an EPA-approved Monitoring, Reporting, and Verification Plan (MRV). 40 CFR § 98.448.

[135] The AOR must be reevaluated every 5 years during the active life of the project. 40 CFR § 146.84(b)(2)(i).

[136] 40 C.F.R. § 98.449.

[137] Over time, some of the injected CO_2 stabilizes by either dissolving into the in situ brine or becoming mineralized via a chemical reaction with the reservoir at 10, Technical Report Template and User Guide (energy.gov), https://www.energy.gov/sites/default/files/2021/01/f82/DOE-LPO_Carbon_Storage_Report_Final_December_2020.pdf.

(1) The area projected to contain the free phase CO_2 plume at the end of year t, plus an all-around buffer zone of one-half mile or greater if known leakage pathways extend laterally more than one-half mile.
(2) The area projected to contain the free phase CO_2 plume at the end of year t + 5.[138]

Over the active life of a sequestration project, if the CO_2 plume migrates beyond the projected AMA, then the AMA will need to be expanded.

The EPA has illustrated the AOR, the MMA, and the AMA with the following graphic[139]:

[138] 40 C.F.R. § 98.449.

[139] Technical Support Document for Injection and Geologic Sequestration of Carbon Dioxide: Subparts RR and UU, Geologic Sequestration and Injection of Carbon Dioxide: Subparts RR and UU Technical Support Document (epa.gov), https://www.epa.gov/sites/default/files/2015-07/documents/subpart-rr-uu_tsd.pdf, U.S. EPA, Office of Air and Radiation, GHG Reporting Program, Fig. 5-1 (November 2010). At appropriate reservoir temperature and pressure, the density of supercritical CO_2 is typically .5 to .7 times the density of water. Thus, due to buoyancy injected supercritical CO_2 will migrate upward to the seal and then gradually spread out laterally. *See generally* CO2 plume migration and dissolution in layered reservoirs, Gege Wen et al., CO_2 plume migration and dissolution in layered reservoirs, 87 Int'l. J. of Greenhouse Gas Control 66 (2019), CO_2 plume migration and dissolution in layered reservoirs. ScienceDirect, https://www.sciencedirect.com/science/article/abs/pii/S1750583619300246, and Keyla Gonzalez et al., Monitoring the CO_2 Plume Migration during Geological Carbon Storage using Spatiotemporal Clustering (essopenarchive.org), https://essopenarchive.org/users/536933/articles/599243-monitoring-the-co2-plume-migration-during-geological-carbon-storage-using-spatiotemporal-clustering.

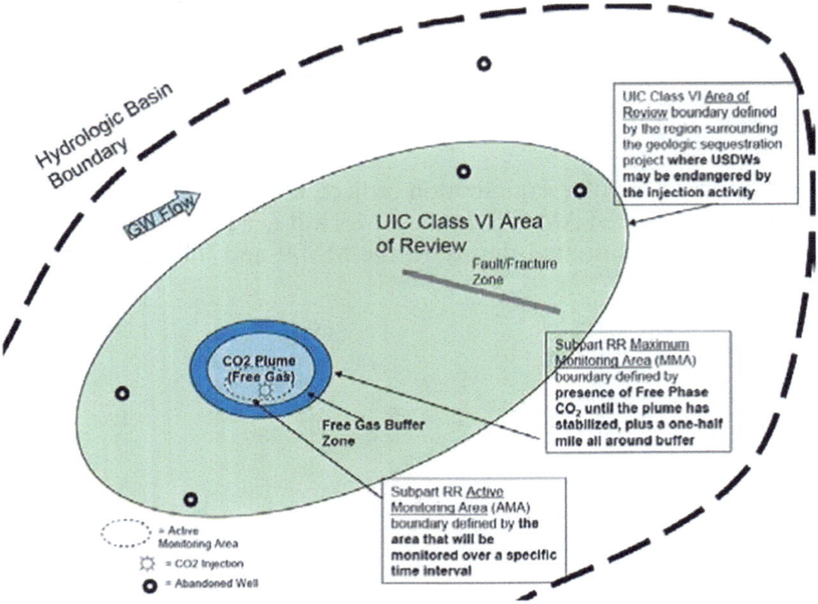

Despite the graphic, the EPA has not specified the property interest and rights a sequestration operator will need to secure in AORs, MMAs, or AMAs. Rather, the nature and extent of the necessary property rights are left to state law.

The major investment risk is the risk associated with leaks of CO_2 or of displaced brine into underground sources of drinking water. Leakages into abandoned, possibly unknown, legacy well bores pose the biggest risk.[140] Sequestration laws and regulations must address predicting the likelihood, frequency, and impact of leaks; leakage mitigation, remediation steps, and the allocation of associated costs to the various stakeholders; and the accumulation of data to support possible insurance to cover at least some of the costs.[141]

[140] 12-Sahar_Bakhshian_Combining-Quantitative-Leakage-Risk-Assessment.pdf (utexas.edu), https://sites.utexas.edu/energy-week/files/2023/04/12-Sahar_Bakhshian_Combining-Quantitative-Leakage-Risk-Assessment.pdf.

[141] *Id.* Costs include monitoring, mediation, and remediation, and the risks of interruption of sequestration operations, business interruption of upstream capture sources, third-party liability, and associated legal costs. *Id.*

At appropriate reservoir temperature and pressure, the density of supercritical CO_2 is typically 0.5 to 0.7 times the density of water. Due to the buoyancy of supercritical CO_2, it will migrate upward to the seal and then gradually spread out laterally. Imagine the CO_2 plume taking the shape of an inverted bell whose sound bow gradually widens both laterally and vertically as additional CO_2 is injected into the reservoir. The rate of migration will depend on the porosity, permeability, and relative homogeneity of the reservoir and on injection rates and pressure.[142]

Accordingly, a project developer needs property rights and regulatory protections that limit any activities that might compromise long-term CO_2 sequestration and that allow monitoring of the storage area for leakage within the entire MMA, which as previously indicated, may expand over time. Securing onshore property rights will be challenging due to private ownership, subdivisions of tracts, the proliferation of co-tenancies, and significant uncertainties about the rights of private landowners to protect against subsurface intrusions.

Ideally, to save on CO_2 transportation costs, a storage reservoir should be relatively close to the point sources of CO_2. Most point sources will be close to urban areas, where lands are commonly subdivided into small parcels. Few parcels will be owned in unencumbered fee simple absolute and mineral rights are highly likely to have been severed from the surface in many of these areas. Even though a surface owner may own the pore space beneath her surface, mineral owners retain a right of reasonable use of the subsurface to facilitate petroleum and mineral operations. Valuable information about the geology of an AOR will be available from prior petroleum development within an AOR area, but prior and especially present and future petroleum operations may compromise secure CO_2 sequestration. Possible competing injection projects, whether involving CO_2 or other fluids including toxic wastes, present further risks of compromising storage integrity.

[142] *See generally* Gege Wen et al., CO_2 plume migration and dissolution in layered reservoirs, 87 Int'l J. of Greenhouse Gas Control 66 (August 2019) CO_2 plume migration and dissolution in layered reservoirs. ScienceDirect, https://www.sciencedirect.com/science/article/abs/pii/S1750583619300246 and Keyla Gonzalez et al., Monitoring the CO_2 Plume Migration during Geological Carbon Storage using Spatiotemporal Clustering (essopenarchive.org), https://essopenarchive.org/users/536933/articles/599243-monitoring-the-co2-plume-migration-during-geological-carbon-storage-using-spatiotemporal-clustering.

Increasing subsurface pressure will occur when CO_2 displaces brine, which will migrate ahead of the CO_2 plume. Thus, a given injection project could interfere, or at least affect, another nearby injection project or a nearby extraction project—especially any within the same reservoir. Ideally, one sequestration project should encompass an entire closed reservoir, but due to the size of saline reservoirs that will encompass numerous tracts, and in some cases cross state boundaries, multiple CO_2 sequestration projects may inject CO_2 into a common reservoir. Thus, robust regulation will be necessary to prevent harmful interference among distinct projects.

Although regulations can help prevent activities that might compromise storage integrity, they do not address property rights, which are left to state common law. Unfortunately, the scope of private property rights in the subsurface, particularly trespass and nuisance law, is unsettled in many states, especially regarding the lateral migration of injected substances across private subsurface boundaries.[143] The American Law Institute recently addressed subsurface trespass law in the forthcoming RESTATEMENT (FOURTH) OF PROPERTY, but, in my opinion, the current draft does not strike the right balance between a property owner's right to exclude and a property owner's correlative right to use the deep subsurface.[144]

[143] *See generally* Joseph A. Schremmer, Subsurface Trespass in the Restatement (Fourth) of Property (July 31, 2024 draft), https://papers.ssrn.com/sol3/papers.cfm?abstract_id=4912423; Joseph A. Schremmer, *Subsurface Trespass: Private Remedies and Public Regulation*, 101 Neb. L. Rev. 1005 (2023); Joseph A. Schremmer, A Unifying Doctrine of Subsurface Property Rights, 46 Harv. Envtl. L. Rev. 525 (2022); Joseph A. Schremmer, Pore Space Property, 2021 Utah L. Rev. 1 (2021); Joseph A. Schremmer, Getting Past Possession: Subsurface Property Disputes as Nuisances, 95 Wash. L. Rev. 315 (2020); Owen L. Anderson, Legal and Commercial Models for Pore-Space Access & Use for Geologic CO_2 Sequestration, 72 U. PITT. L. REV. 701 (2011) (co-author with Gresham); Owen L. Anderson, Subsurface "Trespass": A Man's Subsurface is Not His Castle, 49 WASHBURN L. J. 247 (2010); Owen L. Anderson, Lord Coke, the Restatement, and Modern Subsurface Trespass Law, 57th Ann. Inst. on Min. L. 22 (L.S.U. 2010), revised and reprinted 6 Tex. J. Oil Gas & Energy L. 203 (2011); Owen L. Anderson, Geologic CO_2 Sequestration: Who Owns the Pore Space, 9 WYO. L. REV. 97 (2009), reprinted in LANDMAN (January/February 2010).

[144] For a thorough and thoughtful critique of the RESTATEMENT draft, see Joseph A. Schremmer, Subsurface Trespass in the Restatement (Fourth) of Property (July 31, 2024 draft), https://papers.ssrn.com/sol3/papers.cfm?abstract_id=4912423. For an article that pre-dates the current draft RESTATEMENT by more than 10 years, see Owen L. Anderson, Lord Coke, the Restatement, and Modern Subsurface Trespass Law, 57th Ann. Inst. on

The most favorable case law for the underground injection industry is *Chance v. BP Chemicals, Inc*,[145] an Ohio Supreme Court opinion, which also represents the best public policy for balancing the correlative rights of neighboring owners. As matter of first impression, after finding that a regulatory permit did not insulate the permittee from all liability, the Ohio Supreme Court held that BP Chemicals could not be held liable for the infliction of emotional distress or for punitive damages. The court also rejected the landowners' claims for nuisance, fraud, and ultra-hazardous activity, noting that BP Chemicals was not negligent. The court further held that the landowners could not recover for subsurface trespass because their assertions of harm were too speculative.[146]

In sharp contrast to the Ohio Court's reasoning in *Chance*, in August 2022, the North Dakota Supreme court, in *Northwest Landowners Ass'n v.*

Min. L. 22 (L.S.U. 2010), revised and reprinted 6 Tex. J. Oil Gas & Energy L. 203 (2011) (containing a much simpler suggestion for addressing subsurface trespass in a future RESTATEMENT).

[145] 670 N.E.2d 985 (Ohio 1996) (addressing trespass). In Ramirez v. Akzo Coatings, Inc., 791 N.E.2d 1031 (Ohio App. 2003), the court extended *Chance* to apply to a claim of private nuisance. *Chance* has been favorable cited in the context of underground gas storage, Baatz v. Columbia Gas Transmission, LLC, 929 F.3d 767 (6th Cir. 2019) and in a groundwater storage dispute by the Colorado Supreme Court, Bd. Of Cty. Com'rs of Cty. of Park v. Park Cty. Sportsmen's Ranch, LLP, 45 P.3d 693 (Col. 2002). Several courts have ruled that property owners may not recover for an alleged decline in property values resulting from alleged underground contamination. *See, e.g.*, *Smith v. Kan. Gas Serv. Co.,* 169 P.3d 1052 (Kan. 2007); *Adkins v. Thomas Solvent Co.*, 487 N.W.2d 715 (Mich. 1992); *Walker Drug Co., Inc. v. La Sal Oil Co.,* 972 P.2d 1238 (Utah 1998). *See also*, Boudreaux v. Jefferson Island Storage & Hub, 255 F.3d 271, 275 (5th Cir. 2001) (holding that injected salt water that migrated beneath plaintiff's property was not actionable in trespass absent actual measurable damages); and Raymond v. Union Texas Petroleum Corp., 697 F. Supp. 270 (E.D. La. 1988) (also regarding migrating salt water). In Boehringer v. Montalto, 142 Misc. 560, 254 N.Y.S. 276 (S.Ct. 1931), the court held that a sewer installed 150 feet beneath the surface was not an encumbrance because the depth of the sewer was "beyond the point to which the [land]owner could conceivably make use of the property" but if it were an encumbrance, the landowner would be entitled to only nominal damages. Indiana has a statute that is consistent with *Chance. See* Ind. Code § 14-39-2-12.

[146] *See also* West Edmond Salt Water Disposal Ass'n v. Rosecrans, 226 P.2d 965, 970 (Okla. 1950) (refusing to find that migrating salt water constituted a trespass).

State,[147] found several state statutes[148] enacted in 2019 unconstitutional and in so doing made unnecessarily overly broad statements about subsurface trespass law. The statutes that were held unconstitutional, inter alia, allowed petroleum operators having a regulatory permit to inject fluids into pore space without permission from, and without paying compensation to, surface owners and further barred tort actions for the injection and migration of fluids into pore space. Although the case result is not surprising, the court dicta about subsurface trespass law is troubling. The dicta observed that "North Dakota law has long established that surface owners have a property interest in pore space.... Allowing [free temporary or permanent use of pore space][149] ...takes away one of the most treasured property rights because it takes away landowners' right to exclude oil and gas operators from trespassing and disposing waste into their pore space"[150]; that "[s]urface owners have a right to compensation for the use of their pore space for disposal and storage operations,"[151] and although recognizing that mineral owners have a right of reasonable use of the surface estate, including the subsurface, that right did not extend beyond a pooled drilling unit or unitized field.[152] These broad statements conflict with the Ohio court's holding and reasoning in *Chance*. Nevertheless, given that the North Dakota court's focus was on the unconstitutionality of state statutes, *Nw. Landowners* may not be as broad as the court's opinion suggests in a case that involves an actual claim of subsurface trespass.

147 Nw. Landowners Ass'n v. State, 978 N.W.2d 679 (N.D. 2022). *See also* Mosser v. Denbury Resources, Inc., 898 N.W.2d 406 (N.D. 2017) (answering certified questions regarding saltwater disposal into pore space).

148 N.D.C.C. §§ 38-08-25; 38-11.1-03; and 47-31-09.

149 *Id*. at 691.

150 *Id*. at 692.

151 *Id*. at 692–93.

152 *Id*.

Although many courts have rejected private nuisance claims,[153] the viability of subsurface trespass claims remains uncertain in many states. For example, the Texas Supreme Court continues to duck the issue of

[153] *See, e.g.*, Arkansas, Miller v. Janiski, 705 S.W.2d 442 (Ark. 1986) (depreciation in property value is insufficient basis to sustain abatement of, or award damages for, nuisance); California, Koll-Irvine Center Property Owners Assn. v. County of Orange, 29 Cal.Rptr.2d 664 (Cal. App. 1994) (decline in property value without actual interference with property rights does not support a claim of private nuisance); Kansas, Smith v. Kan. Gas. Serv. Co., 169 P.3d 1052 (Kan. 2007); ("to maintain an action for nuisance, a landowner must establish an interference with the owner's use and enjoyment of the property which is separate and distinct from the claim that the property's value has diminished because of marketplace fear or stigma"); City of Louisville v. Munro, 475 S.W.2d 479, 482–83 (Ky. 1971) (diminished property value will not support private nuisance because a nuisance must disturb physical comfort or be offensive to physical senses); Michigan, Adkins v. Thomas Solvent Co., 440 Mich. 293, 311, 487 N.W.2d 715 (1992) (property depreciation alone does not set forth a cognizable claim in nuisance); Mississippi, Leaf River Forest Products v. Ferguson, 662 So. 2d 648, 663 (Miss. 1995) (no recovery for a decrease in property value caused by public perception without accompanying physical harm to property); North Carolina, Twitty v. State, 354 S.E.2d 296 (N.C. App. 1987) (private nuisance is more than threat of contamination; threat must amount to substantial interference with the use of the property, mere diminution in value is not enough); Pennsylvania, Golen v. Union Corp., 718 A.2d 298, 300 (Pa. Super. 1998) (alienability and diminution in value are relevant to calculate damages but are not cognizable injuries of themselves); Walker Drug Co., Inc. v. La Sal Oil Co, 972 P.2d 1238, 1244 (Utah 1998) (inability to use property as collateral to secure a mortgage loan because of fears arising from contamination of adjacent properties is not sufficient to support a nuisance claim); Virginia, Adams v. Star Enterprise, 51 F.3d 417, 422–23 (4th Cir. 1995) (applying Virginia law—no recovery for a private nuisance which is not visible or otherwise capable of physical detection from plaintiff's property); and Edgcomb v. Lower Valley Power & Light, 922 P.2d 850, 860 (Wyo. 1996) (diminution of value of property is not interference with the use of property so as to constitute a nuisance).

whether subsurface migrating plumes are actionable in trespass or another tort.[154]

The case law addressing the lateral migration of natural gas injected for storage should answer the question of ownership and control of injected CO_2. The better-reasoned cases have held that gas stored underground, including gas that migrates beyond the subsurface boundary of the gas-storage operator's property, continues to belong to the gas-storage operator.[155]

4.3.2 *Rights of Sequestration Operators vis-à-vis Petroleum Operators*

Another issue is whether severed mineral owners may impede effective and efficient CO_2 sequestration. Given the optimal depth of CO_2 sequestration (about 1,300 meters[156]), a major concern is a petroleum drilling operation that drills into and through a reservoir suitable for CO_2 storage, whether such drilling occurs before, during, or after sequestration operations. Geologists and engineers advise that carbon sequestration and petroleum operations can coexist if petroleum operators properly isolate their well bores within the sequestration reservoir—even though the risk of leakage does increase.[157] Alternatively, petroleum operators could

[154] *See, e.g.*, Environmental Processing Systems, L.C., v. FPL Farming Ltd., 457 S.W.3d 414 (Tex. 2015); and FPL Farming Ltd v. Environmental Processing Systems, L.C., 351 S.W.3d 306 (Tex. 2011) (addressing a migrating plume of hazardous waste). In Regency Field Services, LLC v. Swift Energy Operating, LLC, 622 S.W.3d 807 (Tex. 2021), the court seemed sympathetic to providing relief where an injection caused actual damages to petroleum wells, but the court remanded the matter on limitations ground. *Cf*. Coastal Oil & Gas Corp. v. Garza Energy Trust, 268 S.W.3d 1 (Tex. 2008) (holding that petroleum operator was not liable to neighboring royalty owners for drainage caused by hydraulic fracturing because the rule of capture shielded the operator from liability). In Railroad Comm'n v. Manziel, 361 S.W.2d 560 (Tex. 1962), the court addressed trespass in dicta, holding that if the Commission properly authorizes secondary recovery operations to increase oil production, then no trespass occurs where injected water migrates beneath neighboring lands even when the migrating water displaces oil.

[155] See, e.g., Lone Star Gas Co. v. Murchison, 353 S.W.2d 870 Tex. App.—Dallas 1962, writ ref'd.

[156] Stefan Iglauer, Optimal storage depths for structural CO_2 trapping, Int's J. of Greenhouse Gas Control 82 (Oct. 2018), Optimum storage depths for structural CO_2 trapping. ScienceDirect, https://www.sciencedirect.com/science/article/abs/pii/S1750583617306606.

[157] To secure California tax credits, which are available for carbon-sequestration projects within and outside of California, drilling through the sequestration reservoir is not

bypass a CO_2 storage reservoir by way of horizontal drilling. Of course, either approach increases the cost of petroleum operations. Since a significant portion of CO_2 emissions result from petroleum use, climate activists might argue that these costs are appropriate additional costs of engaging in the oil and gas extraction business.

Logically, the law should provide that a sequestration operator and a petroleum operator should reasonably accommodate each other. However, some states have enacted statutes that subordinate carbon sequestration to petroleum development,[158] which might prevent a reasonable accommodation to achieve a balanced result. Owing to financing and other requirements, the sequestration operator may need to resolve such conflicts during the planning stages of a sequestration project. Due to strategic bargaining by the multitude of mineral owners, many of whom may own small fractional co-tenancy interests in the minerals, reaching an accommodation bargain will be challenging—especially for geographically large sequestration sites and in states where CO_2 sequestration is expressly subordinate to petroleum drilling. In some circumstances, a CO_2 sequestration operator (or the petroleum operator) may have to spend capital on remediating old oil and gas wells or on reworking existing ones. Because of these challenges, especially that oil and gas wells may provide a pathway for CO_2 leakage, major CO_2 sequestration projects are focusing on using more secure and larger saline reservoirs, ideally located away from present or past oil and gas plays, rather than on using depleted oil and gas reservoirs or from engaging in EOR projects associated with CO_2 sequestration. In the end, a regulatory approach to resolving sequestration-operator/mineral-developer conflicts may be the best approach.

4.3.3 Regulatory Unitization of Pore Space or Acquisition by Eminent Domain

Some states are legislating a regulatory approach analogous to compulsory unitization of oil and gas reservoirs.[159] Under this legislation, the

permitted. State of California Carbon Capture and Sequestration Protocol under the Low Carbon Fuel Standard, § 9(c).

[158] *See, e.g.*, N.D. Cent Code § 47-31-08 and 09(1).

[159] See, e.g., Baldwin's Kty. Rev. Stat. § 353.806 and .808; Ind. Code § 14-39-2-4 (authorizing "integration"); N.D.C.C. ch. 38-22, in particular, § 38-22-10 (authorizing "amalgamation); and W. Va. Code § 22-11B-19 (allowing a "collective storage order").

regulatory agency having jurisdiction over CO_2 sequestration may force unwilling pore-space owners into a unitization arrangement to allow a CO_2 sequestration project to move forward. The regulatory unitization order would dedicate the needed pore space to the sequestration project, and the affected pore-space owners would be compensated for the use of the pore space by the sequestration operator as determined in the order. The current statutes offer no guidance to the regulatory agency on how to determine the compensation, but only provide that each pore-space owner be fairly compensated.

In contrast to a unitization approach, Louisiana has authorized the use of eminent domain to acquire pore-space rights.[160] If properly used, either approach should avoid most trespass concerns. However, absent "quick take" processes, eminent domain could be time-consuming and expensive.

5 Lessons Learned

In the United States, the major lesson has been that the public has yet to appreciate and embrace the need for CCS. Leading the opposition to CCS are rural Americans,[161] where the bulk of onshore CO_2 transport and sequestration will occur, and environmental interest groups, which regard CO_2 sequestration as a means of prolonging the use of fossil fuels. Their immediate focus is to delay and discourage large-scale CCS project development. In general, their preferred approach is to oppose CO_2 pipelines.

Another lesson is that the expected decline in capture costs has yet to materialize. Delays in implementing planned projects, whether due to opposition or other factors, always increase costs. And costs are unlikely to come down until at least a few large-scale projects are constructed and become operational, so that the nascent CCS industry can gain the information needed to develop innovative efficiencies.

These descriptive terms are no doubt used as substitute terms for pooling and unitization because oil and gas pooling and unitization practices have become controversial in several states.

[160] La. Stat. Ann. tit. 30, § 1108.

[161] Many rural Americans also oppose new transmission lines, which are vital to electrification using renewable energy.

A final lesson is that investors have been reluctant to commit resources to CCS ahead of the 2024 presidential election. Given the election outcome, continued reluctance may be anticipated. Skepticism over climate change has been one of the hallmarks of right-wing Republicans. And many Democrats are anti-fossil fuel. Although much of the petroleum industry and large CO_2 emitters now recognize the linkage between CO_2 emissions and climate change, they are unlikely to invest in CCS or other climate-change mitigation operations without reliable long-term government backing and regulatory, tax, and policy certainty.

6 Conclusion

Absent a major "green" energy breakthroughs and their rapid deployment, the world will continue to rely on large supplies of oil, natural gas, and natural gas liquids for at least the next several decades. Thus, large-scale CCS is necessary for the world to avoid potentially catastrophic climate change. Both federal and state governments must enact laws to facilitate the transport and sequestration of CO_2, and courts should clarify subsurface trespass law in a manner that will facilitate beneficial uses of the deep subsurface.

Meanwhile, the global-warming clock will keep ticking. On September 14, 2023, the National Aeronautics and Space Administration (NASA) announced that the Northern Hemisphere's meteorological summer was the hottest on record since global temperature records began in 1880. The months of June, July, and August 2023 were on average 0.23 °C warmer than any other meteorological summer since 1880 and 1.2 °C warmer than the average summer between 1951 and 1980.[162] 2024 was even warmer.[163]

[162] NASA Announces Summer 2023 Hottest on Record—Climate Change: Vital Signs of the Planet, https://climate.nasa.gov/news/3282/nasa-announces-summer-2023-hottest-on-record/.

[163] See NASA Analysis Confirms a Year of Monthly Temperature Records. NASA, https://www.nasa.gov/earth/nasa-analysis-confirms-a-year-of-monthly-temperature-records/.

CHAPTER 23

Conclusion

Eduardo G. Pereira, Alberto J. Fossa, and Thomas L. Muinzer

This book has striven to be a comprehensive and timely resource on the development of Carbon Capture, Utilization, and Storage (CCUS) technologies. As the world's global average temperature continues to rise as a consequence of climate change,[1] CCUS technology has emerged as an instrumental tool in the transition toward sustainable development and the race against climate change. This book not only focuses on

[1] Leon Wang, Leigh Wang, Yang Li, John Wang, *A century-long analysis of global warming and earth temperature using a random walk with drift approach*, Decision Analytics Journal, Volume 7, 2023, 100237, ISSN 2772-6622, https://doi.org/10.1016/j.dajour.2023.100237, page 8.

E. G. Pereira (✉) · A. J. Fossa
Institute of Energy and Environment, University of São Paulo (IEE-USP), São Paulo, Brazil
e-mail: egp.portugal@gmail.com

A. J. Fossa
e-mail: afossa@newencreative.com.br

T. L. Muinzer
University of Aberdeen, Aberdeen, UK
e-mail: thomas.muinzer@abdn.ac.uk

E. G. Pereira et al. (eds.), *Carbon Capture Utilization and Storage*, Sustainable Development Goals Series,
https://doi.org/10.1007/978-3-031-81272-9_23

filling a critical gap in literature by providing an in-depth examination of CCUS technology, but also acts as a stepping stone for future research and practical application. The book's most significant contribution is in addressing the gap in regulatory knowledge of CCUS developments, and in providing an introduction to CCUS and fundamental issues concerning sustainable development, which engages the need to meet the demands of the SDG 2030 goals and CCUS' role in achieving the Paris Agreement's net-zero emissions goal will ultimately be determined by technical improvements, CCUS incorporation into national climate programs,[2] and global cooperation within the Paris Agreement framework.[3]

The book acknowledges the critical role of financing in propelling CCUS forward. A key takeaway is that financing for CCUS projects heavily relies on government support, primarily through tax incentives or carbon market mechanisms. Emerging economies face significant hurdles in achieving net-zero emissions and transitioning to renewable energy. While substantial investments, estimated at $2.4 trillion annually by 2030,[4] are required to overhaul the energy sector, the upfront costs of renewable projects and the complexities of grid integration and energy storage pose formidable challenges. These obstacles, coupled with the need for immediate energy solutions, often lead developing nations to rely on fossil fuels.

To stimulate CCUS industry development, governments have employed a dual approach, offering incentives and benefits to encourage investment and operation, while simultaneously imposing strict regulations and penalties to ensure compliance with environmental and performance standards (Fig. 1).

With only 25 countries having CCUS facilities, 40.5% of the CCUS projects in development are based in the US, followed by 35.7% in

[2] (*Unfccc.int*), https://unfccc.int/process-and-meetings/the-paris-agreement/nationally-determined-contributions-ndcs#:~:text=NDCs%20embody%20efforts%20by%20each,that%20it%20intends%20to%20achieve, accessed 16 October 2024.

[3] 'Making Mission Possible: Delivering a Net-Zero Economy—Etc (*Energy Transitions Commission*, 21 September 2022), https://www.energy-transitions.org/publications/making-mission-possible/#:~:text=September%202020&text=In%20its%20new%20report%20Making,fuels%20combined%20with%20carbon%20capture, accessed 16 October 2024.

[4] Songwe V., N. Stern and A. Bhattacharya, 2022, *Finance for Climate Action: Scaling Up Investment for Climate and Development.* London: Grantham Research Institute on Climate Change and the Environment, London School of Economics and Political Science.

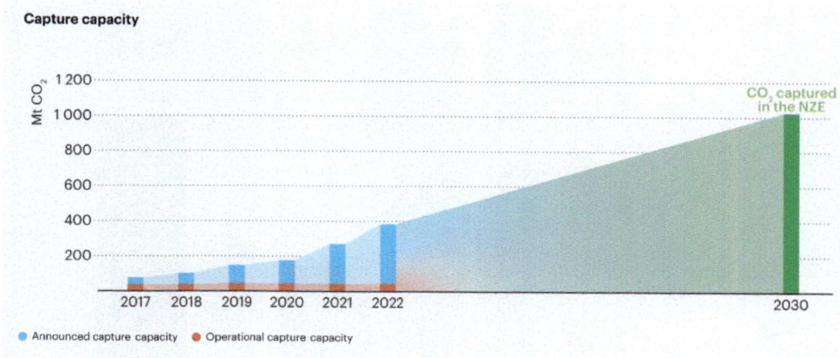

Fig. 1 Carbon Capture Capacity Projection; Global CO_2 capture capacity could reach net-zero emission (NZE) levels by 2030 if current growth trends continue and all announced projects are realized. Accelerating project timelines, especially for CO_2 storage development, is crucial to achieving this goal (*Source* IEA (2023), CCUS, IEA, Paris https://www.iea.org/reports/ccus, Licence: CC BY 4.0)

Europe, 6% in China, 6.5% in Canada, 7.7% in the Asia Pacific, with the remaining 3.6% being based elsewhere around the world.[5] Due to the high upfront capital cost, CCUS technologies have provided an option that has tended to be exclusive to the Global North to date, with developed countries exhibiting a stronger economic ability to incentivize the establishment of CCUS industries in their territories. Countries like the US and Canada[6] have initiated tax incentives to attract investors, encouraging the development of CCUS industries within their borders. More narrowly, the US, in recognizing CCUS as a crucial tool for climate and economic development, has used 45Q tax-incentivized credits, focused on public funding and R&D, and addressed environmental justice concerns via the Justice 40 program (Fig. 2).

[5] IEA, Global CCUS projects in development by region or country, 2021, IEA, Paris https://www.iea.org/data-and-statistics/charts/global-ccus-projects-in-development-by-region-or-country-2021, IEA. Licence: CC BY 4.0.

[6] https://distribution-a617274656661637473.pbo-dpb.ca/8e95e1ac78923bcec809e769bbe39a85e5258ad4582499199a27ab26687f8627.

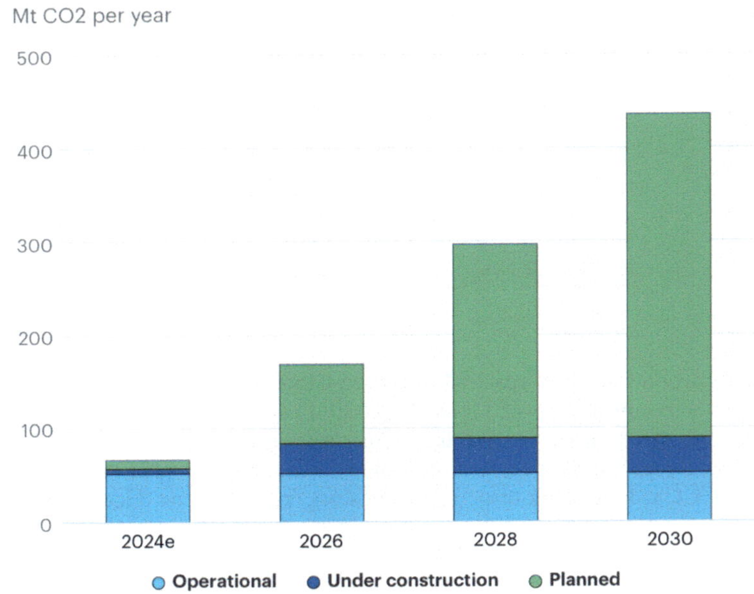

Fig. 2 Operational and planned capture capacity worldwide. Includes all operational, under construction, and planned CO_2 capture facilities with an announced capacity of more than 100,000 t per year (or 1000 t per year for direct air capture facilities) and an announced timeline. Calculated by combining the estimated capacity of capture, full chain, and CCU projects in the IEA CCUS Projects Database. 2024e includes all announced capacity which could come online by the end of 2024. IEA (2024), CCUS Projects Explorer, IEA, Paris https://www.iea.org/data-and-statistics/data-tools/ccus-projects-explorer

Similarly, Canada has also introduced tax credits for several CCUS schemes and has enforced stringent carbon-pricing regulations on carbon-emitting industries, and offsetting credits within Canada's Greenhouse Gas Offset Credit System.[7] Such approaches may further incentivize CCUS, with the potential inclusion of direct air capture and sequestration and the introduction of provincial-level policies like Alberta's framework

[7] https://www.canada.ca/en/environment-climate-change/services/climate-change/pricing-pollution-how-it-will-work/output-based-pricing-system/federal-greenhouse-gas-offset-system.html.

support CCUS through regulatory amendments and funding for CCS projects, the most recent being the passing of a $2 billion fund to encourage and expedite the design, construction, and operation of CCS projects.[8]

There is a gap in regulating and developing the utilization part of CCUS, apart from the enhancement of the recovery element. Carbon taxation offers a relatively straightforward economic tool for reducing greenhouse gas emissions. By building on existing fuel tax structures, countries can more easily implement this policy. Carbon taxes hold the potential to significantly contribute to achieving the Paris Agreement goals,[9] particularly when integrated into the Agreement's five-year Nationally Determined Contributions (NDCs) scheme. Additionally, government subsidy schemes also play a hand in attracting investors, as the government can compensate them in the case of lower than anticipated returns on a CCUS project; this scheme has been implemented by the Dutch government's sustainable energy transition subsidy scheme (SDE++), targeting technologies which lower CO2 emissions.[10]

Being the largest GHG emitter worldwide,[11] China, an economic giant rivaling the United States has emphasized much importance of climate reduction mechanisms given its position as being one of the largest contributors to climate change and carbon emissions and its rising energy demands and consumption. Unlike any other country being recently exposed to CCS and CCU technologies, China has made remarkable progress in advancing legal frameworks, procedural and technical considerations, and establishing commitment to the CCUS movement.

As of 2024 March, China has introduced 100 policy documents at both state and local levels.[12] These legal frameworks offer financial incentives to encourage investment, define rights and responsibilities of stakeholders, establish environmental protection measures, and ensure the

[8] Carbon Capture and Storage Funding Act, Statutes of Alberta, 2009 Chapter C-2.5, Current as of December 15, 2022.

[9] Paris Agreement, Article 4, paragraph 2.

[10] SDE++2023, Stimulation of Sustainable Energy Production and Climate Transition, Commissioned by the Dutch Ministry of Economic Affairs and Climate Policy, https://english.rvo.nl/sites/default/files/2023-09/BrochureSDE2023English.pdf.

[11] Climate Watch, available at https://www.climatewatchdata.org/countries/CHN?end_year=2020&start_year=1990

[12] https://zhuanlan.zhihu.com/p/606785412

safe and responsible operation of CCUS facilities. By addressing these aspects, CCUS laws and regulations contribute to mitigating climate change and promoting sustainable development. Despite rapid development, Chinese CCUS faces several challenges. Key technologies for large-scale full-chain systems need to be established, costs remain high, existing projects are primarily small-scale, and policies are insufficiently supportive. Public investment, while driving current projects, suffers from low competitiveness and economic challenges. Market incentive policies like financial guarantees and tax breaks are lacking, hindering the commercialization of CCUS technologies in China.

Similarly, placed in Southeast Asia, Indonesia is the largest GHG emitter in the region. With a vast potential for carbon dioxide storage, Indonesia is well-positioned to become a regional hub for CCS/CCUS technology. The government's recent enactment of MEMR Regulation 2/2023, which focuses on implementing CCS/CCUS in the upstream oil and gas sector, demonstrates a strong commitment to this initiative. Furthermore, the drafting of a presidential regulation for wider sector implementation signals Indonesia's intention to leverage CCS/CCUS as a key strategy to achieve its carbon emission reduction goals. While widespread implementation may not be imminent, Indonesia's favorable conditions, coupled with strategic regulatory support and incentives, position it for future success. By addressing regulatory gaps, providing financial incentives, and fostering collaboration among stakeholders, Indonesia can create a conducive environment for the growth and adoption of CCS/CCUS technologies. This will not only contribute to its domestic climate goals but also establish Indonesia as a leader in the global transition to a low-carbon economy.

In the EU, an umbrella approach has been approached by introducing several regulatory frameworks such as the "European Green Deal"[13] and "Green Deal Industrial Plan," which aim to create a supportive environment for scaling up net-zero technologies. It has put forward a

[13] Communication from the Commission to the European Parliament, The European Council, The Council, The European Economic and Social Committee and the Committee of the Regions, The European Green Deal, https://eur-lex.europa.eu/legal-content/EN/TXT/?uri=CELEX:52019DC0640.

Net Zero Industry Act[14] to incentivize CCUS technologies and established an Innovation Fund, which is supported by the EU's carbon pricing system, and Horizon Europe to promote the development and deployment of innovative low-carbon technologies including CCUS. The TEN-E (Trans-European Networks for Energy) policy[15] further encourages the development of cross-border carbon networks and infrastructure projects that benefit from financial support under the Connecting Europe Facility. Government support is crucial for the development of CCUS projects, especially the capital-intensive transportation and storage infrastructure. However, this reliance on government backing introduces significant investment risk. Changes in government policies or priorities could jeopardize projects, particularly those without independent revenue streams.

In addition, Norway and Britain's Frigg Agreement, a unique instrument in resource sharing between states, has been instrumental in the development of the CCUS industry. It established a framework for cooperation on carbon capture and storage (CCS) projects, including the sharing of expertise, resources, and infrastructure. This agreement has been crucial in advancing the technology and demonstrating its potential to reduce greenhouse gas emissions. Additionally, the Frigg Agreement has helped to attract investment and support for CCS projects, which is essential for their commercialization and widespread deployment.

In regard to the Middle East, the UAE, while lacking a specific CCUS regulation, has indirectly supported its deployment through various policies and strategies. These initiatives align with the country's broader commitment to sustainable development, energy diversification, and climate action. The UAE Energy Strategy 2050[16] has set the goal of

[14] REGULATION (EU) 2024/1735 OF THE EUROPEAN PARLIAMENT AND OF THE COUNCIL of 13 June 2024 on establishing a framework of measures for strengthening Europe's net-zero technology manufacturing ecosystem and amending Regulation (EU) 2018/1724, https://eur-lex.europa.eu/legal-content/EN/TXT/PDF/?uri=OJ:L_202401735.

[15] Regulation (EU) 2022/869 of the European Parliament and of the Council of 30 May 2022 on guidelines for trans-European energy infrastructure, amending Regulations (EC) No 715/2009, (EU) 2019/942 and (EU) 2019/943 and Directives 2009/73/EC and (EU) 2019/944, and repealing Regulation (EU) No 347/2013, https://eur-lex.europa.eu/legal-content/EN/TXT/PDF/?uri=CELEX:32022R0869.

[16] https://u.ae/en/about-the-uae/strategies-initiatives-and-awards/strategies-plans-and-visions/environment-and-energy/uae-energy-strategy-2050.

reaching 50% clean energy in the total energy mix, a 70% reduction of the carbon footprint, and a 40% improvement in energy efficiency by 2050. Additionally, in its 2023 First Long-term Strategy Demonstrating Commitment to Net Zero by 2050,[17] the UAE outlined over 30 strategies across different sectors. To further strengthen its net zero pathway, the country also updated existing policies like the Clean Energy Strategic Target 2035 in Abu Dhabi and the UAE National Hydrogen Strategy. While the UAE lacks specific legislation governing CCUS, it relies on a combination of existing laws to regulate various aspects of the CCUS value chain. These laws cover environmental protection, energy, oil and gas, water, and land use. Key laws include Federal Law No. 24 of 1999 for environmental protection, No. 1 of 2019 for waste management, No. 2 of 2018 for the petroleum sector, No. 11 of 1981 for water resources, and No. 5 of 1985 for land ownership and use.

Similarly, Qatar, heavily reliant on oil and natural gas for its energy needs, has the world's third-largest natural gas reserves and the thirteenth-largest oil reserves. The energy sector accounts for a substantial 96.5% of its GHG emissions, with electricity and heat production contributing over half of these emissions. Qatar is unique however in the sense that, one of the main challenges faced by states in terms of the high demand of upfront capital is not an obstacle for it. Qatar's strong financial position and commitment to sustainability provide an opportunity for the country to invest in CCUS research and development, potentially becoming a leader in this field. However, a relevant issue remaining is the need and requirement of a proper legal framework with detailed specifics to avoid any gaps in the practices of carbon capture.

In Nigeria, following the trend of pledging carbon neutrality, the country has set its goal to reach zero emissions by 2060.[18] Although, like many other countries, due to the novelty of CCUS, Nigeria too, does not have a particular legal framework on CCUS technology. CCUS practices are however supported by several legislations indirectly. The Climate Change Act provides several measures on carbon-reducing practices, such

[17] UAE Ministry of Climate Change and Environment, 2023, The United Arab Emirates' First Long-Term Strategy (LTS)Demonstrating Commitment to Net Zero by 2050. Retrieved from https://unfccc.int/sites/default/files/resource/UAE_LTLEDS.pdf.

[18] https://statehouse.gov.ng/news/at-cop26-president-buhari-pledges-net-zero-emissions-by-2060-says-nigeria-will-maintain-gas-based-energy-transition/.

as the application of a carbon tax and emissions trading system,[19] stipulation of fines and charges as disincentives for flouting climate change mitigation and adaptation obligations,[20] and carbon budgeting whereby periodic carbon budgets are set in line with Nigeria's NDCs.[21] The Petroleum Industry Act (PIA) indirectly supports the development and deployment of CCS in Nigeria. By requiring petroleum concessionaires to prepare environmental management plans,[22] the PIA encourages the adoption of carbon reduction technologies like CCS. Additionally, the PIA's prohibition on unlawful gas flaring and the NUPRC's right to take associated gas promotes the capture and utilization of natural gas rather than flaring.[23] The NUPRC also actively supports the development of CCS in Nigeria, proposing regulations that would require petroleum mining lessees to provide CCS services using dry wells or abandoned reservoirs on their lease areas.[24] Without clear frameworks for rights, responsibilities, and liabilities, there is no motivation for industries to invest in capital-intensive carbon capture and storage (CCS) technologies. Additionally, high costs associated with infrastructure development and a lack of policy incentives pose significant challenges. These factors inhibit the industrial scaling of CCS in Nigeria.

In the Caribbean, Trinidad and Tobago, an island country boasting a long history in the oil & gas industry, has been a silent leader in the unitization of the CO2 sector. Trinidad and Tobago's "Trinidad method" has been known as an early method of compression of CO2 for utilization, with Ghana seeking their assistance on it. Similarly, Trinidad and Tobago's enhanced oil recovery projects have been in the industry since the 1970s, showing their much earlier acceptance of CCUs practices.[25] The country

[19] Climate Change Act of Nigeria, section 15(1)(e).

[20] Climate Change Act of Nigeria, section 15(1)(e).

[21] Climate Change Act of Nigeria, section 19(1).

[22] Petroleum Industry Act, op cit., section 79(2)(h) & 102.

[23] Petroleum Industry Act, op cit., section 104.

[24] Draft Acreage Management (Drilling and Production) Regulations 2022, regulation 36(3), https://www.nuprc.gov.ng/wp-content/uploads/2022/08/Nigeria-Acreage-Management-Drilling-and-Productions-Regulations.pdf

[25] University of Trinidad and Tobago (2018) "Postgraduate Research Seminar: Carbon Dioxide Enhanced Oil Recovery in Trinidad and Tobago" https://utt.edu.tt/index.php?articles=1&article_key=6943&wk=11&show_archive=1#:~:text=CO2%2DEOR%20projects%20in,pipelines%20were%20compromised%20by%20corrosion.

was one of the first pre-climate awareness states, having focused on carbon reduction discussions and practices since the 2010s, namely the Carbon Reduction Strategy Task Force (CRSTF) which presented the CCUS concept for the creation of economic opportunities.[26] In the following years the country's policies and actions have created a hospitable environment for CCU/CCS industries to establish and flourish, there however is a common obstacle for the need for specialized policies to address gaps in the regulatory frameworks.

Similarly, Australia has embraced its Indigenous roots and has ensured unique measures for the protection of indigenous groups/lands and their identity. Aboriginal freehold land is a distinctive landholding system in the Northern Territory, Australia. Established in 1976 through the Aboriginal Land Rights Act 1977, it converts former reserves into permanent, inalienable land held by Aboriginal land trusts. These trusts represent Traditional Owners (TOs) and negotiate agreements for mining and petroleum activities on the land. Such activities require consent from both the TOs and the federal minister, with a formal agreement necessary between all parties involved. While there's no specific regulatory framework for GHG/CO2 sequestration or storage in underground formations in the Northern Territory as of January 2024, petroleum activities are governed by the Petroleum Act 1984, Petroleum Regulations 2020, and related codes of practice. The process for obtaining exploration licenses or permits involves multiple steps, including consent from Traditional Owners and approval from the federal minister. The Petroleum Act applies to petroleum titles but not CO2 storage, leading to a complex regulatory landscape that combines both codes of practice and schedules. Aboriginal freehold land is a distinctive landholding system in the Northern Territory, Australia. Established in 1976 through the Aboriginal Land Rights Act 1977, it converts former reserves into permanent, inalienable land held by Aboriginal land trusts. These trusts represent Traditional Owners (TOs) and negotiate agreements for mining and petroleum activities on the land. Such activities require consent from both the TOs and the federal minister, with a formal agreement necessary between all parties involved.

[26] Ministry of Energy and Energy Industries, "Carbon Capture Utilization and Storage (CCUS) https://www.energy.gov.tt/our-business/carbon-capture-utilization-and-storage-ccus/.

Brazil's massive offshore oil and gas deposits, as well as its developing industrial sector, present major opportunities for CCUS.[27] The country's geological characteristics, notably the pre-salt layer of the Santos Basin, present intriguing potential for CO_2 storage.[28] Petrobras, Brazil's state-owned oil corporation, has been at the forefront of CCUS initiatives (especially for CO_2 reinjection into its offshore operations to improve oil recovery).[29] The Brazilian government is interested in CCS/CCUS cutting greenhouse gas emissions and meeting its climate targets under the Paris Agreement.[30] However, the regulatory environment for CCS/CCUS in Brazil is still emerging,[31] with debates concerning special laws to control these operations underway even though new legislation was recently approved about "fuels for the future," which also covers specific issues about CCS. Despite the promise, there are still hurdles, such as high implementation costs, infrastructure development, and public acceptability.[32] As Brazil expands its CCS/CCUS capabilities, it has the potential to become a pioneer in this technology in Latin America, contributing to both energy security and climate mitigation initiatives.

[27] 'Carbon Capture, Storage, and Utilization Technologies (CCUS) and their Importance to the Energy Transition in Brazil' (*Tecnologias Ccus no Brasil (Versão Português)*), https://www.ibp.org.br/publicacoes/tecnologias-ccus-no-brasil-versao-portugues/, accessed 17 October 2024.

[28] João Paulo Pereira Nunes, Gabriel S. Seabra, Luis Carlos de Sousa, A review of CO2-injection projects in the Brazilian Pre-Salt—Storage capacity and geomechanical constraints, International Journal of Greenhouse Gas Control, Volume 137, 2024, 104232, ISSN 1750-5836, https://doi.org/10.1016/j.ijggc.2024.104232 (https://www.sciencedirect.com/science/article/pii/S1750583624001750).

[29] Contributor, 'Brazil's Petrobras Expands Carbon Capture at Offshore Fields' (*Gas Outlook*, 1 February 2024), https://gasoutlook.com/analysis/brazils-petrobras-expands-carbon-capture-at-offshore-fields/, accessed 18 October 2024.

[30] (*Global status of CCS 2023—scaling up through 2030*), https://www.globalccsinstitute.com/wp-content/uploads/2024/01/Global-Status-of-CCS-Report-1.pdf, accessed 17 October 2024.

[31] George Augusto Batista Câmara, José Célio Silveira Andrade, Luiz Eraldo Araújo Ferreira, Paulo Sérgio Rocha, Regulatory framework for geological storage of CO_2 in Brazil—Analyses and proposal, International Journal of Greenhouse Gas Control, Volume 5, Issue 4, 2011, Pages 966-974, ISSN 1750-5836, https://doi.org/10.1016/j.ijggc.2010.12.001 (https://www.sciencedirect.com/science/article/pii/S1750583610001817).

[32] https://www.gov.br/planalto/en/latest-news/2024/10/lula-enacts-fuel-of-the-future-law-201cbrazil-will-drive-the-worlds-largest-energy-revolution201d accessed on the 18th of October 2024.

Furthermore, financing for a CCUS project depends on several factors, including its revenue model, feasibility, government support, technical challenges, and risks. Funding can come from various sources: (1) Equity: Investors contribute their funds for project development. Their motivations range from seeking returns to decarbonizing their portfolios; (2) Debt Finance: Commercial banks may offer loans to developers with existing revenue streams or utilize project finance for off-balance sheet development; (3) Capital Markets: Developers can access large capital pools through public share offerings or issuing bonds. This is typically used for established developers with diversified revenue streams; (4) Government Support: Governments can provide grants, loan guarantees, or direct investment in projects. They can also create policies that indirectly support CCUS, such as implementing carbon pricing schemes or undertaking the development of transportation and storage infrastructure. This support can be crucial, especially in the early stages, but it can introduce risk if government commitment weakens; (5) Public–Private Partnerships: These partnerships share risk between the government and private developers, making projects more attractive to investors.

The specific mix of financing entirely depends on the individual project and its circumstances according to the country it is based in and relevant state policies, which may create a progressive and hospitable regulatory environment or may hinder development. Excluding political risks, CCUS projects present unique challenges for financiers due to their novelty and complexity. The evaluation of CCUS projects involves a comprehensive assessment of various risks. These include technological risks related to the project's novelty, scalability, and reliability, geotechnical risks associated with potential ground instability and seismic activity, supply risks stemming from uncertainties in supply agreements and resource availability, and ESG and reputational risks concerning the project's environmental and social impacts. In all, risk management of CCUS technologies typically requires the undertaking of extensive due diligence practices and the production of tailored risk mitigation plans.

This book has also provided a comprehensive examination of regulation and development in the utilization of CCUS technologies within the evolving landscape of energy transition and climate change mitigation strategies. The case studies presented in Part Two have offered valuable insight into best practice scenarios and challenges encountered in implementing CCUS initiatives across diverse global jurisdictions. These cases highlighted the need for a standardized approach while acknowledging the unique circumstances of each region.

However, a critical gap has emerged—the underdevelopment of regulations and frameworks for CCUS utilization beyond enhanced oil recovery (EOR). While EOR has demonstrably benefited from captured carbon,[33] the vast potential of CCUS applications in other sectors remains largely untapped. Without developing regulations for safe and permanent storage of utilized carbon, standardization for CCUS products, policy frameworks, and life cycle assessments, one cannot fully harness the potential of CCUS for a sustainable future; thus, concerted efforts are required to bridge this regulatory gap.

CCU technologies have a peculiar role in mitigating climate change as they transform CO_2 into valuable products, creating a circular carbon economy. Two major technological routes, physical and chemical, offer promising applications, from fuel production to construction materials. The contribution of each product to climate mitigation depends on the duration of carbon storage. Technology readiness levels for CCU technologies range from early research and development to demonstration and deployment. Achieving adoption stage and scaling up requires significant investment in research, development, and infrastructure.

CCUS technical complexity introduced by inconsistent abbreviations, terminology, and concepts impairs communication and policy formulation. Standardization is a relevant tool to overcome these barriers and foster the successful deployment of CCUS technologies as it facilitates transparency, comparability, and regulatory compliance, enabling scaling up projects and integrating them into climate strategies. Carbon capture and utilization (CCU) presents unique standardization challenges, differing from carbon capture and storage (CCS). International cooperation through standardization forums like ISO/TC 265 and CEN/TC 474 are fundamental to develop and harmonize CCUS standards.

[33] Dai, Z., Middleton, R. S., Viswanathan, H. S., Fessenden-Rahn, J., Bauman, J. H., Pawar, R. J., Lee, S., McPherson, B. J. An Integrated Framework for Optimizing CO2 Sequestration and Enhanced Oil Recovery. *Environ. Sci. Technol.* 2014, *1*, 49–54.

The widespread adoption of CCUS is hindered by economic and societal challenges. High costs,[34] lack of infrastructure,[35] and public perception[36] issues have impeded progress. Policy interventions, such as financial incentives, carbon pricing, and regulatory frameworks, are essential to address these barriers. While countries like China and the EU have made strides in implementing CCUS policies,[37] geopolitical tensions and uncertainties about the future of CCUS systems complicate matters.

A more holistic approach is necessary to overcome these challenges. This book reveals the importance of understanding CCUS as a complex socio-technical system. This perspective emphasizes the interconnectedness of technological, economic, social, and environmental factors. By examining CCUS through this lens, one can identify critical research areas going forward, such as public acceptance, innovation systems, and policy design.

To fully realize the potential of CCUS, a comprehensive research agenda is required. This should include a deeper understanding of the sociotechnical dimensions of CCUS,[38] the development of sophisticated models for assessing CCUS portfolios, and exploration of the geopolitical implications of large-scale CCUS deployment.

International agreements like the Paris Agreement have catalyzed a global push to address climate change through coordinated policies and laws. This underscores the fact that achieving net-zero emissions requires a collective, international approach. While CCUS is not a silver bullet for climate change, it can play a vital role in decarbonizing hard-to-abate sectors. By addressing technical, economic, and social challenges,

[34] IEA. Energy technology perspectives 2015: Mobilising innovation to accelerate climate action. *International Energy Agency*, 2015.

[35] Bowen, F. *Carbon capture and storage as a corporate technology strategy challenge. Energy Policy* 2011, *39*, 2256–2264; Babin, A., Vaneeckhaute, C., Iliuta, M. C. *Potential and challenges of bioenergy with carbon capture and storage as a carbon- negative energy source*: A review. *Biomass Bioenergy* 2021, *146*, No. 105968.

[36] IEA. *Energy Technology Perspectives 2020*; International Energy Agency, 2020.

[37] https://www.globalccsinstitute.com/wp-content/uploads/2022/09/CCUS-for-China-Net-Zero-Future-0922.pdf.

[38] McLaughlin, Hope, Littlefield, Anna, Menefee, Maia, Kinzer, Austin, Hull, Tobias, Sovacool, Benjamin, Bazilian, Morgan, Kim, Jinsoo & Griffiths, Steve. (2023). *Carbon capture utilization and storage in review: Sociotechnical implications for a carbon reliant world*. Renewable and Sustainable Energy Reviews. 177. 113215. 10.1016/j.rser.2023.113215.

and fostering international cooperation, the deployment of CCUS can be accelerated in a way that contributes to a sustainable future. Ultimately, the success of CCUS depends on a multidisciplinary effort involving scientists, policymakers, industry leaders, and the public. By working together and adopting a systems-based approach, the full potential of this critical technology can be unlocked.

The large-scale deployment of CCUS technologies could be an important tool for achieving the ambitious climate targets outlined in the Paris Agreement. However, significant hurdles impede its widespread adoption. The absence of robust legal and regulatory frameworks,[39] lack of economic incentives, and insufficient public awareness are key barriers.

To overcome these challenges, a multi-pronged approach is recommended. Firstly, increased investment in research and development (R&D) is crucial to improve capture efficiency, reduce costs, and explore novel applications for captured CO_2. This fosters understanding and mitigates concerns surrounding CCUS implementation.

Secondly, the establishment of robust and adequate legal and regulatory frameworks at both national and international levels is paramount. These frameworks should promote transparency, establish clear liability frameworks, and govern monitoring and long-term storage. Implementing technical oversight committees throughout the CCUS lifecycle, from capture to utilization, is also recommended. These committees can ensure qualified personnel oversee each stage and implement best practices in technology, environmental considerations, safety, and verification. Furthermore, fostering public awareness through educational campaigns and media engagement is essential. By promoting an understanding of CCUS benefits, public support for its implementation and sustainable practices can be increased. This broadens advocacy efforts and facilitates the inclusion of CCUS in government policies and funding initiatives.

Finally, developing nations are encouraged to explore grant opportunities offered by international organizations like the Global CCS Institute, the International Energy Agency, and the Secretariat of the

[39] Zhang, H. *Regulations for carbon capture, utilization, and storage: Comparative analysis of development in Europe, China and the Middle East. Resour. Conserv. Recycl.* 2021, 173, No. 105722; Gährs, S., Knoefel, J. *Stakeholder demands and regulatory framework for community energy storage with a focus on Germany. Energy Policy* 2020, *144*, 111678.

UN Framework on Climate Change. Collaboration with research institutions, industry partners, and NGOs can strengthen grant applications and increase funding success.

In conclusion, this book serves as a springboard for further exploration and development of CCUS technologies. By bridging the knowledge gap, highlighting the financial challenges, and advocating for a more balanced approach to utilization, this book empowers policymakers, researchers, and industry leaders to take concrete steps toward a sustainable future with CCUS. The road ahead necessitates continued investment in research and development, innovative financing mechanisms, and robust regulatory frameworks encompassing the entire CCUS value chain. By working collaboratively, we can harness the power of CCUS to mitigate climate change and pave the way for a cleaner and more sustainable future. In terms of the global disparity between the north and south, it is through access to grants for developing countries, categorization of facilities for UN-listed developing countries, and proper and accountable governing by developing states that widespread CCUS adoption can be achieved. Robust legal and regulatory frameworks, economic policies, and public support through education and community engagement also have a key role to play. Such measures assist in leveling the international playing field, which has been subject to disproportionate imbalance since the Industrial Revolution. Here, the West has benefited historically by taking advantage of its opportunity to be a heavy emitter as part of its efforts to lay the foundations of Western democracy and economic growth.

The manufacturer's authorised representative in the EU is Springer Nature Customer Service Centre GmbH, Europaplatz 3, 69115 Heidelberg, Germany. If you have any concerns regarding our products, please contact ProductSafety@springernature.com

Printed and bound by CPI Group (UK) Ltd, Croydon, CR0 4YY
30/09/2026
02225567-0006